W9-BST-113

OFFICIAL
DISCARD
SONOMA COUNTY LIBRARY

OFFICIAL
DISCARD
SONOMA COUNTY LIBRARY

ALSO BY MARIANA GOSNELL

Zero Three Bravo:
Solo Across America in a Small Plane

Ice

THE NATURE, THE HISTORY, AND THE USES OF AN ASTONISHING SUBSTANCE

MARIANA GOSNELL

ALFRED A. KNOPF, NEW YORK 2006

THIS IS A BORZOI BOOK
PUBLISHED BY ALFRED A. KNOPF

www.aaknopf.com

Owing to space limitations, permissions credits can be found following the index.

Library of Congress Cataloging-in-Publication Data
Gosnell, Mariana.
Ice : the nature, the history, and the uses of this astonishing substance / Mariana Gosnell.—1st ed.
p. cm.
Includes index.
ISBN 0-679-42608-6
1. Ice—Popular works. I. Title.

QC926.32.G67 2005
551.3'1–dc22 2005045126

Manufactured in the United States of America
Published November 20, 2005
Second Printing, January 2006

For Doug, Steve, Kathy, and Scott, my nephews and niece,
whose earliest exposure to ice outside a glass of lemonade
was the frozen surface of the Olentangy and
Scioto rivers in Columbus, Ohio

Let the Brazos
Freeze solid! And the Wabash turn to a leaden
Cinder of ice!

—John Ashbery,
"Into the Dusk-Charged Air"

As if everything in the world is the history of ice.

—Michael Ondaatje,
Coming Through Slaughter

Contents

ICE

INTRODUCTION

Keep your eye clear
as the bleb of the icicle

—Seamus Heaney,
"North"

IT STARTED WITH an ice cube. I happened to be looking at an ice cube floating in a glass of iced tea one day when it came to me: hard water—what a concept! I stared at the cube, and in the center of it I saw a little round whitish fuzzy-edged ball, like a wad of spider silk—what *was* that? Looking more closely, I noticed the ball had some wavy, petal-like layers in it and a smoky trail leading out toward one edge. Most of it was gloriously clear, though, with only tiny sprays of elongated bubbles, and light shone off the bubbles and through the clear part and played on the outside surface. Such a strange and beautiful object!

Not long afterward, I went on a story assignment to an "ice island" in the Canadian High Arctic, where ice experts from oil companies were doing large-scale tests on ice. Before they set up drilling rigs in waters off the coasts of Newfoundland and Alaska—commonly called "ice-infested" waters, as if the ice were rats or insects—they needed to know more about what would happen if ice hit a rig or if a tanker hit ice. The island was a 3,000-year-old, five-mile-long slab of mostly freshwater ice which had broken from an ice shelf off Ellesmere Island eight years earlier and had been drifting in the company of sea ice ever since. By the time the Canadians set up a research camp on it, it was fewer than 600 miles from the North Pole. Except for a few low peaks sticking up off Ellef Ringnes, a rocky island in the distance named for a Norwegian beer manufacturer who supported early Arctic expeditions, the landscape surrounding the camp—the ice island, the sea ice attached to it—was more or less flat in all directions. And except for those low peaks and a few dark nuggets lying on top of the snow cover, clues that Arctic foxes had been coming around, the landscape was all white under 24-hour-a-day sunlight.

The tests were meant to mimic a sudden collision between ice and a metal rig or tanker, so that researchers would have an idea what damage the ice could do to the metal and the metal to the ice. After sawing a ten-foot-deep, 300-foot-long trench in the ice, engineers lowered into its glowing blue-green interior what was essentially a giant hammer (an indenter) with a metal plate about the thickness of a ship's hull on its face and wires running to computers from sensors on the plate. Somebody flipped a switch, and with a shudder and a restrained crunching noise, the indenter hit one wall of the trench with a million pounds per square inch of force, equal to the weight of a thousand small cars sitting on top of each other, one engineer said. When the indenter was pulled away from the wall minutes later, the metal plate had a dent in it the size of a saucer and the wall had a circle of crushed white ice on it the size of the plate. Running through the white, however, was a large swirl of the same blue-green as the surrounding wall, indication that the ice there had *not* been crushed; in outline it looked rather like a map of Newfoundland.

Before the next test, a three-day blizzard set in, and nobody could go anywhere without risk of being lost in a whiteout. The ice experts, from universities and government as well as oil companies, sat around in their plywood huts with socks drying on the bunk beds and enormous cooking pots of snow heating on the stoves (for wash water), telling me things about ice when I asked, some things sounding like tall tales: ridges of sea ice sticking 60 feet up off the surface of the ocean; icebergs the size of pianos moving as fast as men running; a wire being pulled through a bar of solid ice without cutting it in two; a beam of ice hanging in a cold room with a weight on one end and stretching, like taffy in slow motion, until it was more than twice as long as before; ice gripping boulders on river bottoms and carrying them up to the surface . . . Ice is fascinating, they said. In some ways unique. Changeable, idiosyncratic, complex.

It is more brittle than glass. It can flow like molasses. It can support the weight of a C-5A transport plane. A child hopping on one leg can break through it. It can last 20,000 years. It can vanish in seconds. It can carve granite. It can trace the line of a windowpane scratch. It can kill peach buds. It can preserve mammoths for centuries, peas for months, human hearts for hours.

One ice expert, a physicist from India whose name in Sanskrit means "crystal clear," told me something that sounded particularly far-fetched: that ice is not, as a person would reasonably assume, a low-temperature material, the ultimate in cold things, but a high-temperature one. Even the coldest piece of glacial ice is close to its melting point, within 50°F of it. Steel, when we encounter it, unless we work in a foundry, is 2,700°F from its melting point; metamorphic rocks are over 3,000°F from theirs. Some of the paradoxical aspects of ice derive from this happenstance. Ice, in its own terms, is hot.

Also, it floats, or "swims," in its own melt. This rare behavior comes from the way water molecules are linked in ice crystals, so that they form wide-open hexagons. Since a crystal's outward shape usually reflects its molecular structure, we get the six-sided snow crystal. "These glorious spangles," Henry David Thoreau wrote of fresh snow crystals lying upon a meadow, "the sweepings of heaven's floor. And they all sing, melting as they sing, of the mysteries of the number six; six, six, six." The word "crystal" itself comes from the Greek for "clear ice," *krystallos,* first given to the mineral quartz because the ancients thought it actually *was* ice, water frozen in extreme cold.

Once I became hooked on the subject, I learned that several great men in the history of science and thought have given attention to ice, along with everything else they were doing. Michael Faraday, pioneer in electricity and magnetism, studied the surface properties of ice after pondering why it is that snow crystals glom together into a snowball. Astronomer Johannes Kepler concluded (and was probably the first in Europe to do so) that snow crystals are hexagonal and symmetrical and not by chance, but although he tried hard to figure out why, he couldn't. Mathematician René Descartes also tried (and was wrong), yet he made what were perhaps the first scientifically accurate drawings of snow crystals. Physicist Galileo Galilei demonstrated during a public debate on the nature of floating bodies that *regardless of its shape* ice floats, something his audience must not have appreciated before because when he shoved a large slab of ice under water and released it and it bobbed to the surface, they applauded and cried out. Philosopher Francis Bacon, wanting to test the preservative powers of cold, bought a hen from a poor woman and had her take out the innards, so the story goes, then stuffed the gutted chicken with snow, got a chill from handling the snow, fell ill, and died within days. Bacon had once written, "Whosoever will be an enquirer into Nature let him resort to a conservatory of Snow or Ice."

Enquirers into ice nowadays include not only physicists and engineers like those on the ice island but biologists (who want to know, for instance, why fish in Antarctica don't freeze and frogs in Wisconsin do, and survive); astronomers (who study the ice in comets and Saturn's rings); chemists (who can tell by analyzing long cylinders of glacier ice what air temperatures were like when Erik the Red sailed to Greenland and settlers there grew cabbages); meteorologists (who are interested in ice's role in producing lightning and ozone holes); geologists (for whom ice is a type of rock, more accessible for study than other rocks since it's right on the earth's surface, in a relatively shallow layer); climatologists (ice affects climate by reflecting sunlight and changing sea levels); and paleoclimatologists. We are still in an ice age, just a warm period within it.

One paleoclimatologist points out that it is a "remarkable coincidence, if you are not a big believer in Gaia [the idea that life itself has sculpted Earth to be hospitable to life], that our planet is perched on the triple point of water," the temperature at which all three phases of water, solid, liquid, and vapor, can coexist and interact, a situation he calls "crucial to our climate." Our climate, an oceanographer adds, "which we can consider something of a cosmic accident," provides a range of air temperatures that lies mostly between "−40°C and +40°C [−40°F and +104°F]," and it is in the center of that range that ice forms. At this time in Earth's history, a tenth of the land area is covered with ice, most of it in the Antarctic and Greenland ice sheets. A quarter of the land has ice buried inside it, in summer as well as winter. Every continent except Australia has glaciers on it. And 7 percent of the ocean carries a layer of ice on its surface.

As I was enquiring of the enquirers for this book, I learned about the many places outside the regions of the two poles (and what has been called the Third Pole, the peak of Mount Everest) where ice can be found, some places I had never dreamed of: in hibernating painted turtles, inside caves in France, in tropical clouds, in living rose bushes, in mosquito eggs, in halos around the sun.

Global warming, it seems, is making inroads in much of the world's ice. Arctic sea ice is thinning, permafrost in the ground is thawing, glaciers are retreating. Twenty years from now, Mount Kilimanjaro won't have a white top on it. The North Pole may be ice-free much of the time, just as ancient mariners thought it was all of the time. (One 19th-century American, already missing several toes due to frostbite, set out to force his ship through the ring of ice he fervently believed surrounded an "open polar sea" and to sail with ease from there right on up to the North Pole, convinced, as many "learned physicists" were, that "the sea about the North Pole cannot be frozen"; his ship got as far as the west coast of Greenland before it was stopped cold by ice.)

During tests after the blizzard on the ice island stopped, ice in the trench failed by crushing, flaking, pressure melting, and sintering (welding without melting, what happened to the blue-green swirl). Within two years, the island itself, which the Canadians had expected to last for up to 40 years, carrying researchers around and around the Arctic Ocean on circular currents, was pushed by wind into a southerly path and, unobserved during a long winter night, split into three unusable pieces.

It was the kind of surprise that awaits those of us who would seek to understand this strange, beautiful substance in our midst, this solid phase of life's essential liquid, with its wide range of behaviors, changeable nature, and multiple guises.

CHAPTER ONE

LAKES

That the glass would melt in heat,
That the water would freeze in cold,
Shows that this object is merely a state,
One of many, between two poles.

—Wallace Stevens,
"The Glass of Water"

TO SOMEONE WHO LIVES in the temperate north, the event seems simple enough. When the air turns cold, the water in ponds and brooks, lakes, rivers, ditches, and water glasses freezes on top. If the air stays cold, the ice on top gets thicker. It may get thick enough, in lakes and ponds at least, that as the Christian knight told the disbelieving Saracen in Sir Walter Scott's novel *The Talisman,* a horse could carry you "over as wide a lake as thou seest yonder spread out behind us, yet not wet one hair above his hoof." To the Saracen, this preposterous claim was evidence that the knight belonged to a nation that loves to laugh. "Neither the Dead Sea," he scoffed, "nor any of the seven oceans which environ the earth will endure on the surface the pressure of a horse's foot." He was misinformed; several of the Earth's seas and many thousands of its lakes and ponds can take the pressure of a horse's foot in winter, and more besides. Yet there are things that happen when a body of water turns to ice that seem almost as improbable, when you find out about them, as the knight's claim, made under a knot of palm trees by a plashing fountain, seemed to the Saracen.

The most improbable thing is that the ice is *on top.* Almost all other substances are denser and heavier as solids than as liquids, typically 10 to 20 percent heavier, but ice is about 9 percent lighter. Drop other solids in tubs filled with their liquid form, and they will sink; drop ice in (or shove it under, as Galileo did), and it will float (or bob up). As temperatures get colder, molecules generally move around less vigorously and pack more closely together than at warm temperatures, but when water gets cold enough to freeze, the molecules move apart. Lucky for us.

As the water freezes, molecules take up precise positions in a lattice, repeated in all directions and over many layers, to form an ice crystal. The pattern is hexagonal, with an oxygen atom and its two attached hydrogen atoms occupying each of the six corners.* A single layer of identical hexagons would look like tiles on a bathroom floor, or maybe a chicken-wire fence. Stacked layers produce rows of identical hexagonal cages. Because of the angle at which bonds hold the molecules apart, each cage has a large void in it. It is this openness, the hole at the heart of each submicroscopic hexagon, that accounts for the fact that ice floats. If it did not, we probably wouldn't be here to notice. Lakes and seas would freeze from the bottom up, and once seas froze they would have a very hard time melting. Aquatic life over much of the world wouldn't survive or wouldn't have gotten started in the first place. Human life might not have evolved.

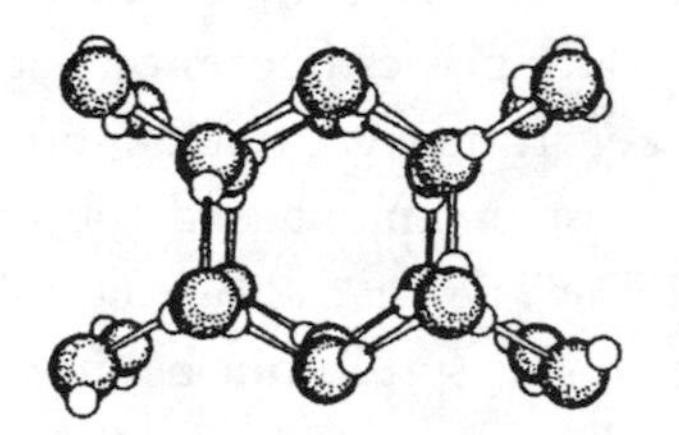

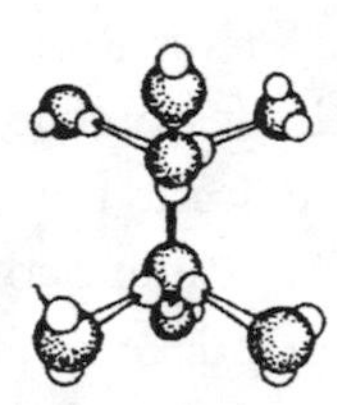

Ice lattice showing arrangement of oxygen (large balls) and hydrogen atoms: views along hexagonal axis (left) and perpendicular to it.

Another improbability related to water's freezing is that we get to see it happen. If water were like most other substances, it would freeze at much lower temperatures than it does (by one estimate, at about −150°F instead of 32°F) and boil at much lower temperatures (by one estimate, at −112°F instead of 212°F). Ordinarily, molecules in a liquid move about randomly, bumping into one another and bouncing away, repelled on contact. But water molecules often stick to others because of the unusual way they are charged, positively at one end and negatively at the

* In addition to two other oxygen atoms in the same hexagon, each oxygen atom is joined to an oxygen atom in another hexagon in the same layer and another in a layer above or below it—in effect, to its four nearest neighbors.

other.* Since opposites attract, water molecules link up readily in the liquid, an oxygen atom in one molecule bonding to a hydrogen atom in another.

These hydrogen-to-oxygen bonds between molecules are called "hydrogen bonds." And although they are much weaker than the hydrogen-to-oxygen bonds *within* molecules, they are strong enough to account for most of water's strange properties, including its high heat capacity (think how slowly an ocean warms in summer and cools in autumn; think how far north the Gulf Stream goes and how much heat it still has to give to Europe when it gets there); its high surface tension (think of the globular perfection of a cloud droplet); its absorption of infrared radiation (think of water vapor's large contribution to the greenhouse effect); and its high freezing and boiling points. The reason water has a high heat capacity is that it takes a lot of energy to break the hydrogen bonds between molecules and free the molecules to move faster. The reason water has a relatively high boiling point is that an enormous amount of heat is needed to rupture enough hydrogen bonds to turn the liquid into vapor.

And the reason that water has a high *freezing* point is that, as heat is withdrawn, the hydrogen bonds grow so much more robust that they lock molecules into positions in ice crystals at a higher temperature than would be expected if water were an ordinary substance. It is the strength of the hydrogen bonds, then, that enables water to convert to ice at the relatively benign (to warm-blooded creatures) temperature of 32°F. It is their assertiveness that allows us to walk on frozen water and chop up pieces of it to drop into our drinks. Hydrogen bonds put ice in our world.

Then we were on the roof of the lake.
The ice looked like a photograph of water.

—David Berman, *"Snow"*

One winter, I rented a cabin in Elkins, New Hampshire, so I could watch a lake freeze. Although I grew up in the temperate north (Ohio)

* The asymmetrical charge comes about this way. The oxygen atom, being larger than the hydrogen atoms, has a greater pull on the electrons they share and in addition has two sets of unshared electrons on the side of the molecule away from the hydrogen atoms. This gives the molecule's electron cloud roughly the shape of an X (or a jack, or, in one scientist's view, a "four-horned snail"). The two ends of the X with hydrogen atoms are positively charged; the two ends of the X without hydrogen atoms are negatively charged.

and had seen some puddles and brooks ice over, I had never seen a larger body of water freeze up. The lake was Pleasant Lake, probably named in summer by someone with a hamper full of beer; in winter, temperatures around it run as low as −30°F. The lake is a mile and a half long by three-quarters of a mile wide (narrow enough that at 10 o'clock each morning for several days straight I saw a deer appear on my side of the lake and swim across to the other side, which oddly looked exactly like my side) and over 100 feet deep at its deepest point. The long line of the lake lies on a northwest-to-southeast heading, the way the North American ice sheet was going when it scooped out the basin for it 20,000 years ago, and the way the prevailing winds blow now, toward the post office and town beach.

The lake is also only a short drive away from one of the world's premier centers for the study of ice, snow, and frozen ground, the U.S. Army's Cold Regions Research and Engineering Laboratory (CRREL), whose declared mission is to solve technical problems in cold regions, most of which "arise because water changes into ice." The scientists and engineers at CRREL work on such problems as how to build a bridge over weak ice; what's the thickest ice that submarine conning towers can break through; how to keep helicopter blades from icing up; and what composite makes the best gliding surface for skis. Along the way they also do basic work, on, for instance, how glaciers move over their beds and salt drains out of sea ice. I hoped they could answer some of my noodling questions about what was going on at Pleasant Lake.

> Ice is a fit subject for contemplation.
>
> —Henry David Thoreau, *Walden*

It could be like watching paint dry, I am thinking, or maybe an orange grow old, the object of my attention just gradually getting stiff. I am warned that any major changes will probably take place in the middle of the night, when temperatures usually run lowest and I won't be able to see much. I am prepared for only modest thrills. I read up.

I learn that before a lake—any small, freshwater, temperate lake—freezes, the water in it usually "turns over." In summer, the top layer, heated by the sun and therefore lighter than the water below, stays on top, except as winds disturb it. But as the days grow shorter and the air cools, the top layer of water cools too, growing denser and heavier until it sinks and is replaced by another layer, which cools in contact with the air, sinks, and is followed by another layer, which upon chilling descends with the rest. Eventually the entire water column, or the entire upper part of it, reaches the same temperature, the one at which water is at its densest and heaviest, 39°F—not, as would be expected if water were an ordinary substance, at 32°F.

This improbability was recognized at least 300 years ago, although not the reason for it. In 1870, William Roentgen, the discoverer of X-rays, seeking a reason, proposed that liquid water contains two kinds of molecules, simple molecules of water and molecules of ice. As water temperatures fall, the proportion of ice molecules will rise, and since ice molecules take up more space than simple molecules, the water will expand in the cold. He was not the first to advance the idea that there's ice in water. A dozen years earlier, H. A. Rowland, after noting the "remarkable" fact of water's contraction at 4°C (39°F), observed that "the water hardly seems to have recovered from freezing."

Some scientists still hold the view that water is a partially broken-down ice structure. In his book *Meditations at 10,000 Feet,* physicist James S. Trefil writes that he likes to express the view by saying that "*water never quite forgets that it was once ice.*" "The positions that water molecules take in ice crystals," he explains, "are duplicated in the liquid." Although the molecules hold these positions only briefly, with any individual molecule joining an icelike structure then "flitting" to another, there will be more of these structures to flit to as the water temperature gets closer to freezing.

Nobody can yet prove Trefil and Roentgen right or wrong. Even at this late date, scientists don't know what the structure of water is. (The structure of ice, by contrast, is well known, all molecules in their assigned places.) "There's a lot of dissension," Stuart A. Rice, professor of chemistry at the University of Chicago, points out. "The volume of the literature on water is like an ocean in itself." One reason for the uncertainty is that the current methods of studying water, X-ray and neutron diffraction, don't give a precise picture. "It's like looking at a statue where the arms are moving," Sidney W. Benson, Distinguished Professor of Chemistry, Emeritus, of the University of Southern California in Los Angeles, states. "You can measure the distance between the shoulder and the elbow and between the fingers and the toes, but you get a blurry pattern. The molecules don't keep their orientation in space. They break apart in nanoseconds."

Pronouncing the idea of water as a "disordered" ice structure "dead as a doornail," Benson worked out his own theoretical model. In it, water consists of cubes and rings "at all temperatures up to and beyond the boiling point." Some of the cubes join together, forcing each other into loose assemblies; "two cubes might be joined at a corner or an edge." The assemblies have big holes in them, so they take up more room than cubes. At 39°F, "the magic number for water," he says, the open structures "start to dominate."

Other models of water include one where it contains a mix of five different species of molecules—having anywhere from zero to six hydrogen bonds each—and one in which simple water molecules inhabit the cavities of hydrogen-bonded clusters of molecules. Rice and Mark G. Sceats, now at the University of Sydney in Australia, have developed a "random network model," in which the hydrogen bonds are continuous (they don't break) but distorted (they bend and twist). Bonds that are strong and taut and linear in ice are relaxed and floppy in water, Rice explains. The twisting of the bonds creates strain, and the strain causes energies and frequencies in the network to shift, so that many different configurations form, with many different angles between them. Yet the molecules are connected, Rice notes, "like a bedspring."

Whatever the structure of water turns out to be—a Hungarian scientist is reportedly working on an electron diffraction system that might sharpen the picture—there is general agreement that the hydrogen-bonded association of molecules in the liquid is responsible for its being densest and heaviest at 39°F. Below 39°F, water expands because the greater clustering effect from increased hydrogen bonding between slowed-down molecules overrides the more-efficient-packing effect of the slowed-down molecules.* *Above* 39°F, the water expands because molecules are moving faster and taking up more room, and hydrogen bonds are being broken so there are fewer clusters. The water keeps on expanding as it is heated, all the way to the boiling point.

thou the waters warp

—William Shakespeare, *As You Like It*

Once the temperature of the water column in a lake reaches 39°F, any water on top that's chilled by the air will be lighter than the water

* As the ice gets colder than 32°F, it starts shrinking and getting heavier again—slightly. At 14°F, for instance, ice is 0.1 percent more compact and dense than it was at 32°F.

underneath and won't sink but will stay on top, in excellent position to be further chilled by the air. It can take months for a lake to turn over, but once it has, the surface layer, in constant contact with the wintry air, can give up a lot of heat in a hurry. In a small lake or pond, all it may take for the top layer—thinner than a birch leaf floating on water, a few ten-thousandths of an inch deep—to change into ice is a single calm, cold, clear night.

> *The brief sun flames the ice, on pond and ditches,*
> *In windless cold*
>
> —T. S. Eliot, *"Little Gidding"*

The first time I go out to check on the lake is Thanksgiving Day, when every sensible person is in. Air temperature is 14°F and the winds are gusting to 35 miles an hour, but the water temperature is 44°F; the lake hasn't turned over yet. Still, I find some ice, not on the lake but next to it. The winds have blown water off the lake surface onto rocks and low-hanging branches along the shore. Cold air, cold rocks: the water probably froze in a flash. On the lake, with the large storage tank of warmer water beneath it, the top won't freeze so easily. The ice on the rocks is clear and shiny and looks like floor sealant. If the wind spraying goes on and on, I read, and the splash-ice builds in layers, it's called "candle dipping."

Water on the branches has frozen into icicles, which are as regularly spaced as rake tines or fringe on parlor lamp shades. When talking about ice, you can block that metaphor but not those similes; the material takes so many evocative forms that the images keep on coming, a few even making it into the scientific lexicon: pancake, grease, candle, bullet, plate, slob, honeycomb. Some of the icicles are two feet long and swinging in the gusts, like wind chimes, but without sound. I spot a twig with a blob of ice at the end, built around a bud. I break it off and suck on the ice like a lollipop, wondering: how can ice be so refreshing on a day so bitter?

The following day, air temperature is up to 37°F, and the innocent prettiness of the splash-ice is gone. The slim tines, fringes, and chimes have fattened and coarsened into root vegetables, teats, and mittens. The sealant on rocks is cloudy now, like . . . like skim milk, or a blind man's eye. The heat of the sun's rays has opened up countless, minuscule cracks in the ice, which scatter later rays and turn the ice opaque.

A different kind of ice shows up on shore a few days later, after strong winds have blown through. Slightly creamy in color, spittly, airy, crisp on the outside but with an odd, dry softness underneath, like a dustball below somebody's bed, it's so insubstantial that a handful of it collapses on my palm. The winds must have whipped the water into foam which the waves flung onto the beach, where it froze in a long, trembling snake of a line, another harbinger of what will happen on the lake itself before long.

ice across its eye as if
The ice-age had begun its heave.

—Ted Hughes,
"October Dawn"

Although light winds can hasten the moment of consummation on a lake by dragging cold air over the surface and pulling heat out of the water, heavy winds can delay the moment by folding the chilled top layer under and bringing warmer water up. An ideal setup for freezing is a night during a cold snap with no wind and no clouds. On a cloud-free night, water can give up an enormous amount of long-wave radiation to the black body of the sky. "The sky just *wants* to take the heat out of the water," CRREL geologist Anthony Gow explains. "It's all a question of getting rid of energy quickly." The water must get rid of not only enough energy to lower its temperature to the freezing point but a lot more besides. Ordinarily, it takes the loss of only one calorie for the temperature of one gram (0.035 ounce) of water to be lowered one degree C (1.8°F), but for a gram of water to change into ice once its temperature has been lowered to the freezing point, it must give up nearly *80 calories* more.

The forfeiture of those 80 calories doesn't lower the water's temperature one whit, since the loss doesn't go toward reducing the speed of the molecules' movement, which is what temperature measures. It goes into increasing the number of hydrogen bonds between molecules. In water at 32°F, about 15 percent of the molecules will be bonded, while in ice at the same temperature 100 percent of the molecules will be. The transition from liquid to solid is accomplished by the removal of those extra calories (and once the removal is complete, the transition is abrupt). In order for a gram of ice to melt once its temperature has risen to 32°F, it must *gain* 80 calories, the heat going to bending or breaking hydrogen bonds in the ice lattice. That's why a glacier whose temperature is 32°F can still be in decent shape; a lot more heat is needed to push it over the edge into melting.*

The energy involved in changing phases is known as "latent heat," so named by a 17th-century Scottish chemistry professor who found that the heat couldn't be measured by a thermometer, as ordinary "sensible heat" can. Latent heat, he wrote, "appears to be . . . concealed within the water." (It's also "latent" because the water will give back to the ice when the ice melts any heat removed when it froze.) Concealed or not, latent heat is real, and the main business of freezing is getting rid of it. There needs to be something in the environment—air, rocks, adjacent water, rooftops, skin—that is colder than the water that is about to freeze and that can absorb the 80 calories; otherwise, the water will stay water.

The next time I check on the lake, it looks ready to boil. Which is odd since it is supposedly ready to freeze. Steam is rising off the surface in long white plumes, some of them swirling around like dust devils. Another way of getting rid of energy in a lake quickly, besides relinquishing it to a dark, clear sky, is to have a polar air mass move in over the top of it, slowly, not stirring up the water, which is what I presume is happening. The water is releasing heat and moisture into the frigid air, where the moisture is condensing into "steam fog." Over the ocean, it's called "sea smoke," in Norwegian fjords "frost smoke," in polar regions "Arctic frost smoke."

Maybe the lake is giving up the ghost before my eyes, I think. Maybe these plumes are its final, humid exhalations. . . . The most popular simile for a lake is "like a living creature," and it's easy to fall into thinking along those lines. "A lake reacts much like a warm-blooded animal," John W. Miller writes in an essay, "Knowing Ice," based on his childhood memories. "In summer it perspires to stay cool; in winter it grows a whalelike hide of ice against the outdoors. The colder the weather, the thicker its hide, and in extreme cold it keeps a fleece of dry snow for extra warmth."

In only one small, secluded spot does Pleasant Lake breathe its last. When I round a corner of the shoreline I see, in a bay where the beach slopes gently into the lake and the water is only a few inches deep, what looks like a windowpane. I am excited out of all proportion to the event. *First lake ice!* It is as thin as a windowpane, about the size of a win-

* Water also requires large amounts of energy to change between its other phases, from solid to vapor, vapor to solid, liquid to vapor, etc. For example, for liquid water to turn to vapor once its temperature has reached the boiling point of 212°F, it needs to gain an additional 500 calories.

dowpane, and as clear as a windowpane. Through it I can see four pebbles and a mangled oak leaf. I believe I can see them better than I would be able to in summer, when even on quiet days water exhibits a faint tremor. In an essay based on *his* reminiscences, "The Pond at the Center of the Universe," Gene Logsdon compared fresh pond ice to a "giant television screen, tuned to nature's own PBS station." One day he went to the pond to watch his nieces and nephews play ice hockey and found all of them lying face down on the ice instead. They were gazing through the ice at "a trio of snapping turtles, their carapaces as big as meat platters, very clearly visible scarcely two feet below, lolling on the pond bottom as if it were June. If we all lay there without movement [he joined them], fish would congregate under our bodies, obeying an instinct to hide under logs." Other people have compared being on clear lake ice to standing on air or riding in a glass-bottomed boat.

Terry, a friend in Elkins, calls this first ice "cat ice," meaning a cat could walk on it but nothing else could. According to the *Concise Oxford English Dictionary,* cat ice is "thin ice unsupported by water," but Terry insists that in southeast England, where he grew up, there's plenty of cat ice and nobody requires there to be air under it.

The fact that the ice is clear shows that the water froze slowly. As ice crystals form, they reject impurities—air, salts, minerals, gases, bacteria—banishing them to the boundaries. Capacious as the crystal lattice is, there's not enough room for most particles to fit inside it. Therefore, no matter how filthy the water from which ice crystals are formed, they will be pure H_2O. ("As chaste as ice," said Hamlet.) When water freezes rapidly, air can be trapped at the boundaries as bubbles. But if it freezes slowly, the air has time to diffuse into the surrounding water. CRREL geologist Tony Gow has found a simple, linear relationship between the bubbliness of a piece of ice and the speed at which it froze.

There is no second ice at Pleasant Lake. A dismal week of sleet, snow, rain, and—twice—air temperatures above 50°F follows, and there is no ice anywhere around, no pane, cat, splash, or even foam ice.

I've known the wind by water banks to shake
The late leaves down, which frozen where they fell
And held in ice as dancers in a spell
. . . .
They seemed their own most perfect monument.

—Richard Wilbur, *"Year's End"*

Although impurities are rejected by ice crystals, they do play a role in producing them. Like all liquids, water needs something to get the freezing process started, and that something is usually a foreign particle: mote of dust, grain of salt, scrap of vegetation, speck of bacteria. "Water *wants* to be ice when the air is very cold," Charles A. Knight, senior scientist at the National Center for Atmospheric Research (NCAR), says. "It just has to learn how." It learns how by having a nucleus to organize its atoms around when they take up their positions in the crystal lattice, a pattern for them to follow. The best ice nucleators are those whose structure is most like ice. But in protected places, if the top of the water is cold enough, David M. Cole, research civil engineer at CRREL, says, "crud floating on the water will usually do."

Knight defines nucleation as "the spontaneous appearance of a little ordered domain in the midst of randomness." Ice is the ordered form of water. "Intuitively, it is quite clear," he explains in his book *The Freezing of Supercooled Liquids*, "that the first appearance of such a domain might be a chance process, water molecules falling into the right pattern on their own. It might be something like the children's game that consists of a little tray with shallow indentations in the bottom, into which a number of metal ball bearings must be rolled." The child tilts the tray to get the balls to slide into the little holes, but the balls already in place slip out—drat! "If we think of the child trying to solve this puzzle blindfolded . . . we may have an acceptable analogy to the chance aspect of nucleation."

In other words, a very long shot. The lowest temperature at which such "homogeneous" nucleation (it manages without ice nuclei) takes place is –40°F, a temperature at which bonding is so enhanced that some transient icelike clusters reach a critical size (50 to 100 molecules, by one estimate) and qualify as "ice embryos," capable of being nucleators themselves. But homogeneous nucleation takes place only when the water is very pure (lacking foreign particles) or is in the form of cloud droplets so tiny they're not likely to contain or encounter a foreign particle capable of nucleating them before their temperature gets down to –40°F. In lakes, water never gets anywhere near that cold because it is always less than pure and is nucleated "heterogeneously" (with the help of ice nuclei) first.

Even so, lake freezing doesn't take place at 32°F, the nominal freezing point of water, but below it, if only by a fraction of a degree. The liquid and solid are at equilibrium at 32°F, and there's no incentive for the water to make a phase change. So the water "supercools." (Some scientists prefer "undercools" or "subcools," considering those words more accurate, but "supercools" has won out.) Supercooled water is primed to freeze as soon as favorable nuclei present themselves. Some nuclei are already present in the lake, but others fall onto the surface as snow, dirt, even cold rain. The best ice nuclei—the best teachers—are ice crystals themselves. Once ice crystals start forming on a lake, they beget more ice crystals. "All you need is a start," Gow says. "Off goes the ice!"

It's a calm morning in early December, air temperature 27°F. I walk past a dock on Pleasant Lake and happen to glance at a dark patch of water beside it. Something tips me off. I step down for a closer look. What I see takes my breath away. The patch, about nine feet square, has a skin of ice on top of it so thin the dark water shows through, yet it's not smooth and blank and plain like that windowpane but highly textured, with lines running all over it, helter-skelter, a frenzy of very fine lines, some raised, some indented, some bent, some dagger straight, some branching like feathers (the barbs on one side only), a couple of them trailing a fan of fine threads, some outlining irregular four- and five-sided polygons, the edges beveled as if gouged out of metal, and rounded things like poker chips, angled slightly to the surface, their upper parts sharply defined and casting a tiny shadow but their lower parts dissolved as if in mud, and something that looks like a knot of wood, something that looks like a clam shell, like a boomerang, like the lobes of a broken heart.

I am dumbfounded. What I'm looking at seems too wild, too rich, too weird to be natural. Why *this* shape, I keep wondering, and not *that*? Why *here,* and not *there*? "Perhaps the most surprising fact in connection with the formation and growth of ice crystals is this," Wilson A. "Snowflake" Bentley wrote in 1907, "that so many diverse types form and grow, each perhaps in a different manner, at the same time within a given body of water, and apparently under the same identical conditions of temperature, air pressure, environment, etc." Yes! Seen up close there are tiny plots with tiny terraces on them, what looks like cross-bracing for a bridge, pyramids with the peaks lopped off, overlapping cloud banks. Intersections, striations, and truncations—it's cubism in nature! "Ice crystals of whatever type," Bentley added, "freely merge and freeze together, one to the other."

If I had been at the lake at the witching hour, whenever that was, 3, 4, 5 a.m., and had had a flashlight, what I might have seen was a very thin needle of ice suddenly shoot across the water from a starting point on the dock. "Once it goes, it really goes," Gow says. "Unimpeded, it can have a long run." (He has seen a needle run for several feet.) The needle might have zipped across the water for a foot or two, halted as latent heat released into the water at the freezing front warmed the water ahead, then, as the air chilled the water again, resumed its run until it *was* impeded by another needle which had been zipping along on another speedy course and blocked its path. After that, it might have branched, like the veins of leaves, extending into unfrozen water, or grown ribs itself or broadened, by thickening on one side, even as other needles were doing the same, until in a matter of minutes a network of crisscrossing ice needles lay on the lake surface, framing open patches of water, after which the open patches froze, flat and smooth as plates, sealing the cover, in that one spot.

Quite properly called a skim, the ice is probably less than a 20th of an inch thick. "When first formed, and for some time thereafter," Bentley wrote, "ice crystals are exceedingly thin . . . hardly thicker, in fact, than thin paper." (The ice even looks like paper in places, flocked wallpaper, the low relief highlighted by the early sun.) "Some crystals might shoot down into the water a little bit," Cole explains, "but it's mostly a 2-D phenomenon." The skim will thicken by the crystals' growing downward, which they will do more slowly than they grew sideways. Cole calls the downward growth of lake ice "arduous." It takes longer for the latent heat of freezing* to make its way through even a thin layer of ice by conduction than it does for the heat to pass directly to the air and dissipate. The water turns to ice in layers disk by disk, "very much like stacked Chinese coins," as one early researcher put it, producing a smooth underside. "Ask divers," Gow says. "In a lake, it's always smooth between ice and water."

This skim doesn't get a chance to thicken. Winds blow over the lake that night and snow falls on it the next morning, and when I go out there's nothing left of the ice I admired. That bizarre manifestation of an ordinary natural process has vanished. All that complicated molecular structure-building went for naught! Freezing is the quintessential ephemeral act. Gow explains that it often takes two or three episodes of freezeover before a lake achieves freezeup, distinguishing between the two. Freeze*over* takes place when all or a large part of a lake is covered by ice, freeze*up* when the ice cover is in place permanently (or until spring). In small ponds during a long cold spell the two events may be one, but in most lakes the ice comes and goes several times before it comes to stay.

* Also called the latent heat of solidification, crystallization, fusion, and congelation, all ways of saying "liquid turning into ice."

On this windy day, I settle for looking at splash-ice, which is first rate. Waves tossed the water so high that every post and boulder on the beach end is wearing an ice hat, ice garland, or ice crown studded with ice jewels. Icicles on branch tips are swollen to the size of pears, bobbing in the wind. Thomas and Marina Langlois, a husband-and-wife team who studied ice on Lake Erie from 1936 to 1964, wrote about the time they saw splash-ice form on the lake's "cliffed shoreline": "As the icing continues, individual blades of grass act as centers of ice growth and become nodules, making a smooth lawn as uneven as a rocky beach. This splash-ice is solid as it builds up from the ground, so it does not break off the blades of grass, nor the stems of such delicate herbs as harebells and rock asters."

A couple of days later, air temperature is 5°F and the lake is steaming again. Half the lake seems to be involved this time. In the rising wisps I see a minuscule but brilliant flash of light. I see another bright dot, another, and another, all sharp, mica-like glints. The very air is winking. Yet there's not a cloud in the sky. "In the winter when you turned your flashlight upward," scientist Paul A. Siple wrote in his book *90° South: The Story of the American South Pole Conquest,* "you saw ice crystals falling continually, like dust beams in the attic when the sun pours through small peepholes." He was referring to "diamond dust," tiny ice crystals that form near the ground directly from water vapor. The crystals have no arms, as snow crystals do, since they don't make the same long falls through the air that snow crystals do, attracting vapor on the way. Ice crystals are common in polar regions, where they float in the air like glitter over a dance floor, but they show up occasionally in nonpolar regions, too, when the air near the ground is very cold and stable, as it must have been that morning at Pleasant Lake.

> Above me is a cloudless blue sky, beneath is the sky blue, i.e., sky-reflecting ice, with patches of snow scattered over it like mackerel clouds.
>
> —Henry David Thoreau, Journal (*Winter*)

Parts of the lake that were steaming soon quit steaming, and they also quit mirroring the birches and pines on the opposite hillside. I conclude that those parts are now covered with a film of ice and further conclude that it was the diamond dust that seeded them. Not long afterward I see things growing on the new-grown ice. They are "frost flowers," each flower a bunch of ice crystals which are sticking up an inch or two off the surface and projecting petal-like in several directions. Frost of any kind forms when water vapor freezes directly onto a cold surface, like your lawn (after a damp, chilly night) or fresh ice on top of a lake. There are asters called frost flowers, but these look more like Christmas lilies. Frank Hurley, the photographer on Ernest Shackleton's 1914–16 expedition to the Antarctic, described the sea ice in which the ship *Endurance* was held fast and on which many frost flowers were arrayed and lit by the morning sun as "a field of pink carnations."

I discover I can tell the comparative ages of the swaths of ice the flowers are on (the water froze in patches, or swaths) by how large the flowers are and how close together. The youngest ice, which froze most recently, has a spangled look. The flowers there are mere dots, sprinkled sparsely and evenly over the dark ice ("evenly" because each flower takes moisture from a zone of air around it). Ice that formed slightly earlier has slightly larger flowers which are slightly closer together; by their presence they give the ice a dark gray look. The ice that formed earliest is almost white with blossoms. Thus the lake, where it is covered, consists of regions of ice shaded from black to nearly white by the time-delayed blooming of frost flowers.

Some of the ice is still around a few days later after a light snowfall, although the frost flowers are not. The white of the snow is marked instead with a large number of black circles and lines. It doesn't take much snow to weigh down an ice sheet so that it rides below the waterline and forces water up through holes and cracks. (After all, less than a tenth of an ice cover floats above the waterline anyway.) The water soaks the snow and turns it to slush; the slush freezes into "snow-ice"; and the snow-ice becomes part of the ice cover. Snow-ice is granular, white, and opaque from all the tiny air bubbles trapped inside. "It looks like frozen yeast somewhat," Thoreau said.

Snow is itself a form of ice, but because ice and snow have different properties and behave differently from each other, scientists distinguish between them. The big difference is air. On the ground, a layer of snow has channels of air running through it, but when the channels become compressed by the weight of overlying snow, as in a glacier, or are filled by water that freezes, as on a lake, the air channels turn into bubbles, unconnected to one another. When that happens, the snow is no longer considered snow. It is ice. Since this book is about ice, it will be about snow only when snow relates to ice, as sometimes happens on a frozen lake.

Over the course of a winter, half the thickness of a lake's ice cover may come from snow-ice ("white ice," as it's also called), the rest from lake ice ("black ice"), which forms from the direct freezing of lake water. Black ice is black by association, dark water showing through the transparent skin. (It's also called "congelation ice," because the water congeals rather than evolves from snow.) White ice and black ice grow in opposite directions, white ice upward, black ice downward. Occasionally, white ice forms on a lake without black ice as a base. One time after a heavy snowfall, the Langloises reported seeing a "snow-blanket, with folds made by ripples in the water," floating on the lake surface, "before the entire blanket, folds and all, became a solid sheet of ice."

On the white surface of Pleasant Lake, where the snow is still snow and not yet snow-ice, I count over a hundred of the dark circles. These are "slush holes," openings blackened around their edges as if ink were being pulled up from below and staining the snow instead of water. It's not known why slush holes develop where they do—*here and not there*—unless there's an obvious source of heat, like a dark pebble or stick lying on the ice, which absorbs sunlight disproportionately and starts melting the snow around it. (Once in the ice of a bay I saw leaves, cigarette butts, twigs, cigars, wood chips, dog droppings, ticket stubs, and strings lying in depressions two or three inches deep which had been melted out by the sun in the exact shape of the objects; even a blade of grass lay at the bottom of its own narrow, custom-melted groove.)

NCAR's Charles Knight has given some thought, however, to *how* slush holes form. When snow weighs down an ice cover, water is forced up not only through large openings but also through exceedingly small ones, such as holes along ice-crystal boundaries—often called veins—or in bubble tubes—sometimes called worms. At first, many of the small openings serve as pipes, carrying water to the surface, where it melts snow and ice, but some of the openings will naturally be a bit larger than others and will therefore carry more water and themselves be subject to more melting, with the result that they become still larger and capable of carrying more water, and so on. Through feedback, some holes in the ice that start out the size of pinpricks grow to the size of hula hoops.

Some of the slush holes on Pleasant Lake have wavy lines radiating from them, like the legs of dancing bugs, or (most unlovely image) the shreds of a blood clot. Knight calls them "stars." Gow calls them "spiders." Knight found the rays "puzzling" at first. "If one thinks of them as streams," he says, "the flow seems in the wrong direction, out of the holes instead of into them." After examining many stars, he concluded that they tend to form when snow depresses an ice cover, which causes water to rise through an opening in the ice and to flow outward from it, melting wavy rays out of the snow and exposing the black ice underneath.

I find a crooked line of these impish things running parallel to the shoreline of Pleasant Lake, some of them no more than a few inches across. "The writer finds himself surprised at the range of phenomena that occur when snow on lakes is transformed into slush," Knight wrote. In photos he took of slush holes on frozen mountain lakes, many of them looked like large poppies, almost too well proportioned to be natural. A few had concentric rings surrounding them, the ring closest to the hole frozen solid and blanched by frost. Several had inch-thick patches of dry snow immediately next to them but slush everywhere else, and some were buried entirely within white ice and registered only as shadows.

According to a Russian who visited a certain Japanese village some years ago, rice farmers there used "ice patterns" in a swamp to predict the quality of the coming harvest. "Cherry blossoms" and "palm trees" augured a good harvest, "centipedes" or "rice stalks" a bad one.

After a snowfall the next day, the spiders of Pleasant Lake have a vague, blurry look to them. Snow turned the land into a winter wonderland but the lake into a grayish sludge. At the post office, a man says he saw an ice-fishing shanty out on a nearby lake already. "Must have dropped it in by helicopter," he mutters. A woman says *she* saw a man walking far out on another lake holding a *child* by the hand! Everybody shudders, shakes heads. We've all read about the ski mobiler who fell through the ice on a lake farther north.

The only live creatures I've seen on Pleasant Lake so far are crows. What's in it for them? I wonder. In a chapter on tool-using in his book *The Minds of Birds,* Alexander F. Skutch recounts how, in parts of Norway and Sweden, hooded crows pull ice fishermen's lines out of holes in the ice to get at what's on the end. "Grasping the cord in its bill," Skutch wrote, "the crow walks slowly backward with the line as far as possible. Then it returns to the hole by walking forward with its feet on the line, thereby preventing it from slipping back into the water. If necessary, the bird repeats this sequence until the fish or bait comes within reach."

By the week before Christmas, only a few areas of open water are left on the lake. "This is it," Terry says. "I think this is it." While I'm away for Christmas and New Year's, winds strip every bit of ice off the lake and dump the slabs and chunks on the beach. "It looked like the ocean here," the postmaster informs me when I get back. "The waves were so full of broken ice they were almost white. Water froze as it shot off the crests of the waves." "It was the wildest thing," Terry agrees. But by the first week in January, the lake has been pounded with sleet and freezing rain and passed through several cold, cloudless nights, and again it is frozen. The question is, is it frozen up? I see a single, dark gray slot of water out in the middle, like an open zipper. Lakes, I learn, tend to freeze from the outside in, since they are shallowest around the edges, and since the edges provide nucleating points, and since cold air pours off the land, chilling water at the margins but not necessarily making it out to the middle.

Overnight, the water in the slot freezes. The lake is zipped up. Shut down, closed for the winter, out of commission, asleep under its tough hide. For the next few months, the ice, gray-white and scaly on top, from wind-hardened snow and refrozen slush, should keep all of the wind, and most of the sunlight, from reaching the water below.

> *Like frozen-over lakes whose ice is thin*
> *And still allows some stirring down within.*
>
> —Richard Wilbur, *"Year's End"*

Pleasant Lake is not *really* shut down. It is not stagnant. It is not unchanging. It is not still. It is not even *that cold*.

For instance:

• The temperature of the water under the ice not only doesn't go down, it may even go up. The ice cover partly insulates the water from the effect of cold air; sunlight trickles in through patches of clear ice; and organisms in the water and vegetation on the lake bed give off heat when they decompose. Water in contact with the underside of the ice sheet (and next in line to freeze) will be 32°F, but the bulk of the water will be warmer, about 39°F.

• The water under the ice lid is not immobile. One morning when Thoreau was chopping ice from Walden Pond, as he did every morning in winter so he could melt it for drinking water, he dropped his ax through the opening he'd made in the ice. Peering down into the water, he saw the ax lying on its head on the bottom, its upright handle "gently swaying to and fro with the pulse of the pond." Although wind doesn't move the water under an ice cover, the water does move. Streams flow in and out and springs bubble up, producing currents whose locations some ice fishermen try to remember, since currents weaken the ice.

• The ice keeps on getting thicker, although more and more slowly as winter goes on, until at some point it stops. According to one table, when air temperature is 23°F, it should take only a little over an hour for a third of an inch of ice to form on water that's 32°F. When the air temperature is −22°F, that should happen in a little over ten minutes. If ice kept up either pace, it would reach the bottom of the lake in a matter of weeks. But it doesn't keep up the pace. Ice is a poor conductor of heat, and as the ice thickens, the latent heat of freezing has farther and farther to travel to get to the cold air above it, and that takes longer and longer. According to the same table, when air temperature is 14°F, it should take fewer than two days for water to produce a layer of ice four inches thick, a week to produce ice 7½ inches thick, four weeks to produce ice 15 inches thick, and eight weeks to produce ice 25 inches thick. If there is snow on the ice—and snow is an even lousier conductor of heat than ice—the growth rate will slow further.

Yet snow giveth even as it taketh away. A layer of snow that slows the growth of black ice can, if it becomes wet and refrozen, speed the growth of white ice. These mirror effects roughly balance out, according to two Canadian researchers who studied lakes in Labrador: one kind of ice grows thicker when and where the other does not.

• The ice surface is not dormant. It is exceedingly active . . . at a molecular level. "When we consider a solid material," Victor F. Petrenko, professor of engineering at Dartmouth College, says, "we think that every atom or molecule can move just a little inside its stable position, but it's not the same with ice. Every molecule in ice rotates with high speed, like a propeller. That's not surprising for liquids, but it is for solid materials. In ice the speed can reach one million cycles per second." The vibration of the spinning molecule is often so great that the molecule breaks free of the hydrogen bonds holding it in the ice lattice and departs.

Even when ice is very cold, molecules are leaving and returning to the ice surface at furious rates. "I used to think the ice surface was something static and stuck around for a long time," Steven George, professor of chemistry at the University of Colorado at Boulder, says. "But it doesn't just sit there. It's extremely dynamic. The ice surface is constantly being eroded and redeposited, like topsoil. Some water molecules are evaporating even as others are raining down on the ice." At the extremely low temperatures characteristic of the stratosphere, George has measured the residence time of a molecule on the surface of a piece of ice at mere milliseconds. "Even at −100°C [−148°F]," he says, "you can lose one layer of molecules per second."

On Pleasant Lake in midwinter, the molecules that depart the ice surface will generally be counterbalanced by those that return. Toward the end of winter, however, or on sunny days during it, the ice will probably lose more molecules off the top than it gains, and the cover will thin from above. This process, whereby a solid turns directly to vapor without passing though a liquid stage, is called "sublimation."

• Lake-ice crystals grow larger . . . and even compete for territory. All snow-ice crystals, whether they're in the top, middle, or bottom of the snow-ice layer, are about the same size and the same lumpen shape. They're so close together they cannot grow much, except on warm days when they expand a little and attract vapor given up by other crystals. A snow-ice layer thickens by adding *numbers* of crystals. Lake-ice crystals, however, do become bigger, many of them a whole lot bigger. Gow has found single crystals that were between one and two feet long and extended right through the ice sheet. He has also found crystals that were nearly three feet *wide* at the bottom. Lake-ice crystals are in fact among the largest crystals in nature. (In theory, there is no limit to the size of an ice crystal, David Cole points out, "other than self-weight." On the Mendenhall glacier in Alaska, he notes, "beautiful single crystals" have grown to the size of human heads.)

As for competing for territory . . .

Crystals in general are known for their symmetry, and one way scientists have of describing the kinds of symmetry different crystals have is by drawing imaginary straight lines through them from opposite faces or corners so the lines meet in the middle. If a person were to rotate the

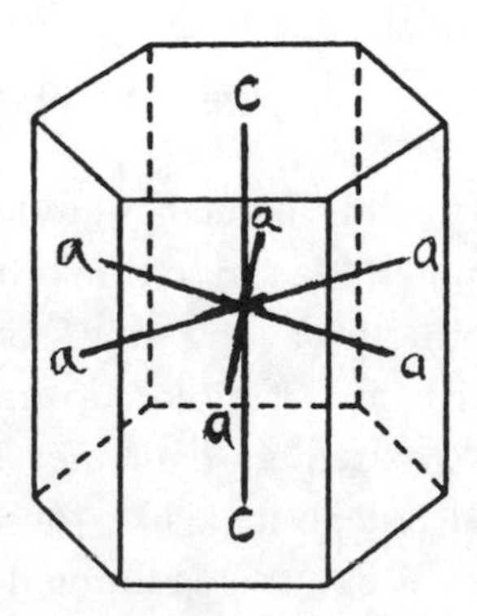

The *a*- and *c*-axes of an ice crystal

crystal around one of these lines, or axes, the crystal would look the same to him at various points in the rotation as it did at the start. A crystal of ice has two sets of these "axes of symmetry," one running perpendicular to the hexagon layers and three running parallel to the layers and radiating from the single axis like spokes around a wheel axle. The spokes are the *a*-axes, the axle the *c*-axis. (See diagram on previous page.) This fact is of more than academic interest; the way the axes in ice crystals are lined up in relation to the water surface affects the structure of the ice cover, and thus probably its strength. Those crystals whose *c*-axes lie parallel or horizontal to the water surface become needles, while those whose *c*-axes lie perpendicular or vertical to it become plates. (There can be slight deviations.)

On a lake the size of Pleasant, the ice that forms on a calm, cold night is typically a mix, with one-fifth of the surface covered by needles and four-fifths by plates. What determines which way the *c*-axes will lie and thus whether needles or plates will form, Gow discovered through extensive research, is the way the water is nucleated, from the air or the water. After spending several winters studying lakes and ponds in New Hampshire and Vermont as well as growing ice in tanks at CRREL, he found that crystals with their *c*-axes horizontal—the C_h's, as he calls them—form when the lake is seeded from the air, by multiple particles of snow, ice fog, ice bits blown off the land, etc. Crystals with their *c*-axes vertical—the C_v's—form in the absence of seeding, presumably nucleated by impurities already present in the water. Water temperature makes no difference, apparently. "Meteorological events immediately prior to freezing," Gow stated, determine the orientation of ice crystals "invariably."

The C_h's and the C_v's grow downward from the initial skim in markedly different ways. C_h crystals grow to be long, slender, and columnar, with well-defined boundaries. They're only a little wider at the bottom (an inch or so across) than they are at the top (a fifth of an inch or less across). By contrast, the C_v crystals grow to be coarse-grained, "fibrous," and irregularly shaped, with poorly defined boundaries; a group of them function almost as one "supercrystal." They are also much, much wider—up to 100 times wider—at the bottom than they are at the top.

Reading accounts of the growth of these crystals, you might think of war. Gow uses words like "invasion," "domination," and "elimination" to describe what goes on. Others refer to an "elbowing out," a "cutting off," and a "pinching out." To reach their impressive sizes, the C_v's often encroach on the growth path of the C_h's, which are less aggressive. On one New England pond, Gow discovered that the C_v's had "wedged out" the C_h's a mere three inches into the black-ice layer and forced them to stop growing altogether. Even in those parts of a lake where the skim consists overwhelmingly of C_h-oriented needles (which is often the case along the shore since ice nuclei are blown off the land onto the water), the downward growth of the abundant C_h's is "frequently interrupted by the sudden appearance of *c*-axis vertical crystals of highly irregular shape," which continue to grow rapidly, Gow notes, "at the expense of the *c*-axis horizontal crystals until the latter [are] entirely eliminated."

None of these hostile takeovers can be seen with the naked eye. Cut a chunk of ice out of a lake cover and it will look like just ice. But slice off a thin vertical section of the ice and view it through crossed polarizing filters under a white light, and you will be able to see individual crystals and judge how they grew or didn't grow. Most of the crystals will be brightly colored, red, green, blue, violet, orange, etc., according to the way their optical (*c*) axes are oriented. (Ice has two refractive indices, which cause light waves to travel through at different speeds, a phenomenon known as birefringence.) A vertical slice of lake ice under cross-polarizing filters will typically consist of, starting at the top, confetti-like dots of color—small, randomly oriented grains of snow-ice—then below it several long, slender rods of bright color (crystals of C_h ice) or else some dark C_v crystals which are separated from each other by only faint boundaries, "incredibly boring" under cross-polarization, Cole says. Below that you'd often see the C_h's being squeezed out by the muscular widening of the C_v's.

Why one kind of lake ice should dominate the other isn't clear, Gow says. Some investigators have found lakes that crystallize in one orientation one year and in the opposite orientation the next year. There may be years when the C_h's prevail. And the C_h's do a certain amount of wedging each other out, although their efforts are modest compared with those of the bullying C_v's.

• While the crystals are playing out their little war games, air bubbles aren't sitting idly by. All white ice has bubbles in it, and most black ice does, but the bubbles in black ice are fewer and allow some light through. The transparency of a piece of ice is determined by the area that, from the viewer's perspective, is covered by bubbles. If, for instance, a given amount of air is contained in a single bubble, the ice that it's in will be clear, while if the same amount of air in the same piece of ice is divided among several thousand bubbles per cubic inch, all of them reflecting light, the ice will appear white. Even snow-ice high on the "bubbliness index" and indisputably opaque, however, is not as air filled as you might think. According to Gow, only about 5 percent of even very white ice consists of air bubbles.

In black ice, bubbles tend to be tubes, not spheres. They grow longer during freezing, and in the direction of freezing. The more prolonged the cold the longer the tubes will be, and the greater the cold the bigger around the tubes will be, and the more abundant. If temperatures should rise while the ice is growing, a tube may narrow or become pinched off. If temperatures fall, it may widen or reappear deeper in the ice sheet. "You are looking at a kind of weather gauge," Gow explains. If you sawed pieces of ice out of the cover at regular intervals and examined the bubbles in it, you could read the history of weather over a winter at that spot. The reason you should remove ice at intervals is that bubbles change even after they have formed. Air "migrates" through the ice, Gow points out. The tubes "pulse" with changes in temperature. Long tubes separate into trains of shorter ones; shorter tubes turn into spheres; spheres become squatter.

Some bubbles don't come from air that's dissolved in the water but are added directly to it, by burbling springs, decaying weeds, and swimming animals. Muskrats, minks, beavers, and otters stir up bubbles as they paddle around, or leave breath bubbles behind them. You can recognize these bubbles because they are big. As they float upward through the water, they expand, since water pressure diminishes toward the surface, then "flop" up against the underside of the ice and flatten out, Gow notes. "They sit there like amoebae under pressure from below" and are frozen in like that. Once after several nights of subfreezing temperatures, I saw within the otherwise clear ice of Central Park's Turtle Pond a crooked tier made up of what looked like white dimes, one situated more or less above the other, each dime representing a single cold night, I presumed, and all, I presumed, arising from a particularly gassy set of rotting weeds.

• Fish keep on living and moving around in a lake even after ice has sealed the top shut. Oxygen is brought in by streams, or furnished by underwater plants that manage a little photosynthesis despite the low light under a cloudy ice cover, or provided by air through breaks in the ice. All species of North American game fish can survive the cold water temperatures of a frozen lake, Jim Capossela writes in his book *Ice Fishing,* although a few do become sluggish. Some head for the bottom once a lake freezes shut because water usually becomes very clear under an ice cover in the absence of waves that stir up particles already in the water or runoffs from land that dump more particles into it. And "with greater water clarity," Capossela points out, "comes greater vulnerability." One day Thoreau spotted an ice fisherman standing on an ice cover with a large yellow pickerel in his hands, and that got him to thinking about how "anxious" small fishes must be about the prospect of being swallowed by larger ones. "What tragedies," Thoreau mused, "are enacted under this dumb, icy platform in the fields!"

• An ice cover not only thickens over a winter but may also broaden. When the air gets chillier, the ice cover contracts—which can end up making it bigger. The contraction causes cracks to open up, which, if they are deep enough, may fill with water from below, after which the water freezes solid. With many such cracks, an ice cover may become too wide for the basin it's in and push up against the shore, depositing ice rubble on land.

What's that? Frost? Thunder?

—Anna Akhmatova,
"We're all drunkards here . . ."

When I gaze out at the ice on Pleasant Lake, I don't see any of this going on, not the battle of the crystals, the comings and goings of surface molecules, the migration of air bubbles, the growth of the cover up and down and sideways. The ice seems to be just lying there, inert as the granite hills. But when I listen, I know something is going on with the ice. One day I hear a wooden boat hitting a dock over and over, *chik, chuk, chik, chuk,* but there is no boat. Another day I hear a giant, phantom switch being whipped through the air, *whhhuuuh!* The ice is swelling and contracting in response to changes in temperature, opening cracks and closing them to relieve the stress.

Sound is where ice-as-living-creature similes are most convincing. Ice groans, bangs, squeaks, booms, whoops, grunts, thuds, gulps, and clunks. Over the winter I hear the noises of seals crying, trains rumbling, and wound-up springs unwinding (*boinggg!*). I listen to the murmur of a dinner-table conversation on the other side of a wall. Other people have heard creaking ship timbers, explosions, and drumbeats. (The drumbeats probably come from water bumping against the underside of ice that's arched upward.) Thoreau mentioned "a sort of belching, . . . somewhat frog-like."

Since contraction tends to take place at night when the air is coldest, the loudest sounds issue from the ice when shore-dwellers are trying to sleep. There's thunder and the *ka-chung!* of fireworks leaving launch pads. The first time I heard the booms in the middle of the night from my cabin near the ice edge, I thought the house was suffering structural failure.

In John J. Rowlands's book *Cache Lake Country: Life in the North Woods,* in which he described in detail the year he spent living alone in a cabin beside an unnamed lake, probably in Canada, he reported that the lake's ice cover there often gets to be four to five feet thick. When it does and the ice "begins to crack you know about it," he wrote, "for the noise is like many shotguns going off at the same time." He offered his own ice-as-living-creature image. "Sometimes you hear a grinding sound like a monster gritting its teeth. The Chief [an elderly Cree neighbor] says the lake is just stretching in its sleep."

CHAPTER TWO

RIVERS

and all springs by frost
Have taken cold, and their sweet murmure lost

—John Donne, *"Ecclogue"*

THE BIG DIFFERENCE, ice-wise, between rivers and lakes is a crystal that in an early incarnation looks like an eyeglass spectacle, a transparent, circular platelet that may be as much as 100 times larger in diameter than in thickness but the whole of it smaller than a snowflake; even after it has added truncated little dendrites to its edges, like the crimping around a soda bottle cap, rarely does it get bigger than a few hundredths of an inch across, the size of a black fly's egg.

Yet in concert with others of its kind, this crystal can reduce the water supply of a city to a trickle or shut off power to thousands of homes and businesses. "Nothing is more dramatic," an early student of ice observed, "than to witness a 30,000 kw hydroplant removed from the operating electrical network in less than an hour by the accumulation of these tiny frazil particles on the intake racks." "Frazil" (a French-Canadian word from the French for "coal cinders"; in Danish the particles are "swimming crystals," in German "free-swimming crystals," in Icelandic "swift ice") is born in turbulence, and all rivers are turbulent. "A river may *look* quiescent," George Ashton, former chief of the research and engineering directorate at CRREL, says, "but it is not. If you drop ink in it, the ink will swirl as it goes downstream."

Lakes, too, develop frazil, particularly large lakes with enough fetch to allow winds to roil the water; winds mimic the mixing effect of a river's flow. One day I spotted at the leeward end of Pleasant Lake, still mostly unfrozen at the time, several dozen small, slushy clumps, or "flocs," of frazil. They bobbed around for a while, looking variously like water lilies or ice cream balls. They drifted together to form larger flocs before becoming frozen into a sheet of lake ice, where they looked like paving stones. On the lake, the flocs of frazil were endearing little oddities. But in rivers, they can be downright dangerous (a word even sober scientists aren't afraid to use), at least to the engineering works serving the disproportionate number of people who have chosen to live along rivers and who rely on them for transportation, drinking water, and electrical power.

"Frazil is far more important than we first appreciated," Ashton admits. (Some people pronounce "frazil" to rhyme with "Brazil" but most to rhyme with "Basil," as in "Rathbone." By definition frazil crystals have to be "fine" and "in suspension.") "We used to think frazil was something that goes on only at the beginning of winter," Ashton continues, "then wasn't around much. Now we know that much of the ice on rivers is frazil, and it really piles up." The capacity for piling up is the main characteristic that makes it dangerous. Concentrations can reach nearly a million crystals per cubic yard of water. On videotape, frazil forming in a laboratory flume looks like a blizzard in progress. Fast-moving, open stretches of river become "ice-making factories," and accumulations of crystals downstream from them can be 60 feet thick and hundreds of feet wide. A tumbling mountain stream can produce frazil in quantities that exceed the depth of the water in summer, and the ice piles up higher than the banks.

> Snow'd all day, the Ice ran thick and air Cold.
>
> —William Clark, *Journal,* November 13, 1804

The reason most frazil crystals start off as round as an eyeglass spectacle is that the water they're tumbling around in is a uniform temperature, which draws off heat from all sides evenly. (Needles do appear in rivers in places where turbulence is low.) As the crystals grow, they assume new forms. "Frazil ice is always changing," Bernard Michel, professor of civil engineering at Laval University in Quebec, noted in his 1971 CRREL monograph, *Winter Regime of Rivers and Lakes.* The tiny disks evolve into irregular platelets that are only slightly less tiny, 0.16 to 0.4 of an inch wide, at most 0.6 of an inch wide. The disks and platelets collect into flocs that are often so loosely bound and porous they are half water. Some of the flocs bump into others and merge into pancakes, so named because onshore observers get the idea that the things are flat on the bottom as well as on the top, which is a mistaken impression since they actually have slushy globs of frazil hanging from their undersides and are more like hemispheres. Pancakes sometimes coalesce into floes; in parts of the Missouri River, floes of frazil get to be several hundred feet across.

Rivers generally produce a great deal of frazil while small lakes generally do not because the turbulent riverine flow keeps stirring upper, air-chilled layers of water into deeper ones, and the entire water column frequently supercools, and not just the paper-thin top layer, as on quiet lakes. The amount of supercooling involved is minuscule, only a few hundredths or thousandths of a degree below the freezing point. (Ashton is impressed, he writes, by the "small temperature differences [that have] seemingly disproportionate effects on river ice, as it forms, evolves, and decays.") Once some frazil crystals form, many usually do. As the crystals near the surface are carried deeper into the flow, they bump against other crystals or rocks or the shore and break into shards which act as seeds, causing more supercooled water to freeze and to produce more frazil crystals ("secondary nucleation"). Multiplication is "explosive," Ashton says. The ice-making factory is operating full tilt—as long as the water stays supercooled, that is. When the water warms those few hundredths of a degree back up to 32°F, production of frazil crystals stops, although the product sticks around, sometimes in river-clogging amounts. As soon as the water warms *above* 32°F, even by a fraction of a degree, the crystals melt away. It often happens that in the cool of one night ice crystals form, then in the heat of the next day they melt, and that night they form again.

While a frazil crystal is growing, or "active," it is extremely adhesive. Someone called it "tenacious." "There is no solid surface it won't stick to," Ashton insists, "unless that surface is warm." As water turns to ice, he points out, "it naturally grabs and forms a bond with everything it touches. The ice itself is the glue, at the point of contact." For centuries, fishermen have been lifting their baskets, nets, and traps out of northern rivers and finding them covered with a snowlike mush. A person can determine whether or not there is frazil forming in a river by lowering a frayed rope or steel chain with a weight tied to the bottom into the water and after several minutes pulling it back up. If it's covered with glistening little flakes, the answer is decidedly yes. "Frazil looks like the fine silver confetti you throw on something," Ashton declares. "That sparkle stuff."

Being confetti-sized, frazil crystals are easily carried along in the flow and distributed to almost any depth of the river, where, if they are active, they stick to any cold thing in their path, not only fishing lines but dock pilings, gate locks, anchors, weeds, rocks, and trash racks. Power companies install trash racks over the intake pipes of their plants to keep logs, stones, and oil cans from being drawn in along with the water, and frazil sticks to those trash racks like chewing gum. Other crystals come along and stick to those crystals and other crystals to those—the ice builds forward into the flow and outward to the sides and bridges the gaps between bars, creating a barrier between the water and the pipe, cutting off the plant's water supply and stopping its production of power.

The good news for people who count on having electricity in winter is that trash racks are only briefly vulnerable to frazil. The period of vulnerability may be as short as a few minutes at a stretch of river, or no more than one night each winter at a particular spot. As soon as the water is warmed those few hundredths or thousandths of a degree, by sunlight or the latent heat of freezing or the intercession of the power company (which may heat the racks), the crystals stop growing. Once they do, they stop clogging trash racks. "Passive" frazil crystals don't stick to anything. They are as inert as grains of sand.

> *"When I blow my breath about me,*
> *When I breathe upon the landscape,*
> *Motionless are all the rivers,*
> *Hard as stone becomes the water!"*
>
> —Henry Wadsworth Longfellow,
> *"The Song of Hiawatha"*

Contrary to what you might think, imagining clogged pipes, an ice cover doesn't usually get started at narrow spots in a river but at wide ones. Currents speed up where a river channel is narrow, and a current that's moving at two feet per second can break up an incipient ice cover. Where currents are particularly fast, frazil crystals may not gather into flocs or floes but instead form a general surface slush. Or they may remain suspended in the water column, leading some onshore observers to assume (another misconception) that because there isn't any ice visible on the river, there isn't any ice there.

While moving downriver, frazil crystals behave like fine, buoyant sediment, Ashton notes, only an "upside-down counterpart" of fine sediment. In places where the flow is slow, sediment will sink, frazil will rise. In places where the current causes fine sediment to settle on the inside of river bends, frazil will collect on the outside of the same bends. Near the inlets of freshwater reservoirs, frazil builds underneath the ice cover into fanlike deposits reminiscent of river deltas, only in reverse.

Once begun, a river cover progresses upstream by accumulating flocs and slush and pancakes that are heading downstream, until it reaches a set of rapids or other fast-flowing section of the river. There, any newly produced frazil is swept beneath the cover and carried along in the water until it reaches a calm spot, usually a deep pool, where it attaches itself to the underside of the cover, as a "hanging dam." Hanging dams can be gigantic, blocking most of a river's cross section and causing severe flooding on the land around it. The Saint Louis River regularly develops hanging dams which are 15 to 30 feet thick, the Saint Lawrence River dams 40 to 60 feet thick.

Frazil is not the whole story on a river. (On smaller rivers and streams and brooks, it may not even be the main story.) While frazil is doing its fast-forming, tumbling act in the middle, ice of a different kind is making a quieter debut along the shore. Near the edges of a river and in flat stretches, there are usually few eddies, and only the top layer of water chills enough to supercool, and ice develops there, as it does on lakes, as a surface sheet of black, or congelation, ice. While this black ice is growing downward, it is also growing outward, toward the more turbulent center, and if the river isn't particularly wide it may meet ice advancing from the opposite shore to complete the cover. I saw this happen over several cold mornings on a brook that runs into Pleasant Lake. First, bands of ice formed along the banks on either side, like two lacy white window curtains framing a blurred scene (water rushing downstream). Then, each day for several days afterward, the ice reached a little farther out into the water, until one day the curtains drew together and shut off the view.

FROST.
I make the unstable stable, binding fast
The world of waters prone to ripple past

—Christina Rossetti,
" *'All Thy Works Praise Thee, O Lord.'*
A Processional of Creation."

A river's ice cover usually ends up being (and looking like) a jumble—of black ice attached to the shoreline; of frazil flocs and pancakes; of aprons of black ice around boulders in midriver; of frazil clumps containing dirt and stones they've lifted off the bed; of black ice that grew downward from the underside of a thin layer of surface frazil; of old ice slabs broken up and reassembled and refrozen; of frozen slush; of snow-ice; and of no ice at all. Where there's no ice at all is where the river produces frazil. The frazil doesn't form in water underneath ice (although it's often carried there); to supercool, the water has to be exposed to air. A stretch of river that will, when it is covered, produce a layer of solid ice 18 inches thick will, when it is uncovered, produce a layer of frazil ice 18 *feet* thick.

The ice cover of a river thickens the way the ice cover of a lake does, by adding black ice to the bottom and white ice to the top, except that the river also continues to churn out frazil in those uncovered stretches, and the frazil adheres to the underside of the cover downstream and thickens it. In early winter, the water under the cover will be 32°F since it has been flowing past ice and is chilled to the temperature of the ice. But as winter goes on, some warmer water does make its way into the river—from rain or melted snow or feeder streams or the hot broth of industrial and municipal wastes. It can create an ice feature that few people have seen and until a couple of decades ago almost none had even suspected the existence of and scientists still have difficulty accounting for: "waves" on the underside of the ice, as regular as surf at the seashore.

It's another case of small temperature changes in a river having disproportionate effects: water only a few hundredths of a degree above the freezing point can transform the normally smooth underside of an ice cover into an undulating one, with the waves running crosswise to the flow. Ashton named the waves "ripples" since they remind him of ripples in sand. He wrote his doctoral thesis on ones he studied on a river in Iowa. He cut large blocks out of the ice cover with a saw, turned them over, found ripples, and measured the wavelengths of the ripples. He discovered that the wavelengths are related inversely to the velocity of the current. In sections of fast flow, the wavelengths are short, with as little as an inch between crests; in sections of slow flow, they're long, with as much as several feet between crests.

But once a wavelength is established, its size doesn't change, and once the ripples themselves are established, they don't go away—that is, the underside of the ice never again becomes smooth. The ripples do, however, move. As turbulent, above-freezing water flows by them, their troughs deepen and their crests sharpen . . . which is the opposite of what present understanding of turbulent heat transfer suggests *should* be happening. "It's counterintuitive," Ashton concedes. "The logic of our knowledge suggests the ripples can't exist." He discovered that the crests grow steeper on one side—the lee—than the other side, and this asymmetrical effect causes the waves to migrate downstream, very very slowly, at the rate of about an inch a day. The ice itself stays put; it's the wavy profile that moves.

Occasionally, the troughs melt all the way through the ice, and that produces a corrugated pattern on top, of dark stripes running from bank to bank, which is supposedly what revealed their presence to German scientist F. W. P. Lehmann, who was probably the first person to describe them, in the 1920s. In the absence of surface stripes, however, the ice cover gives no clue that on its undersurface there are waves moving along, ever so slowly. "Who looks under the cover of river ice?" Ashton asks, knowing the answer is "nobody else."

Harry Houdini probably didn't know about ripples, but he may have had something like them in mind when he conferred mythical status on one of his escape feats by claiming that the jump he made off a bridge into the open Detroit River on a chilly November day in 1906 while handcuffed was actually a plunge through a hole in ice. After extricating himself from two pairs of police handcuffs under water and failing (so he said) to find again the entry hole in the ice so that he could exit through it, as Milbourne Christopher noted in his book *Houdini: The Untold Story,* Houdini "turned his head sideways to inhale air trapped between the ice and the water," then "swam in ever-widening circles until he found the opening." In truth, an ice cover on a river usually lies right on top of the water and is flexible enough to ride down and up with changes in water level and thus is nearly always supported, except for close to shore where the ice may be attached to the banks and air gaps develop under it. Houdini was given to creative reminiscences; in later years, Christopher wrote, Houdini claimed his "under-ice escape" had taken place in Pittsburgh.

to reside
In thrilling region of thick-ribbed ice

—William Shakespeare,
Measure for Measure

Water flows not just under river ice but above it. When such water freezes, Germans call it "*aufeis,*" the term most scientists use. Hudson Stuck, an Episcopal bishop who traveled around Alaska as a missionary between 1904 and 1920 and wrote about his experiences in the book *Ten Thousand Miles with a Dog Sled,* called it "overflow ice." "In the lesser rivers, where deep pools alternate with swift shallows, the stream freezes solid top to bottom upon the shoals and riffles," he reported.

"Since the subterranean fountains that supply the river do not cease to discharge their waters in the winter, however cold it may be, there comes presently an increasing pressure under the ice above such a barrier. The pent-up water is strong enough to heave the ice into mounds and at last to break forth, spreading itself far along the frozen surface of the river. . . . Sometimes for many miles at a stretch the whole river will be covered with a succession of such overflows, from two or three inches deep to eight or ten, or even twelve; some just bursting forth, some partially frozen, some resolved into solid 'glare' ice. Thus the surface of the river is continually renewed the whole winter through."

What makes the phenomenon hard for people to understand, according to Stuck, is that "it would seem that after a week or ten days of fifty-below-zero weather . . . all water everywhere would be frozen into quiescence for the rest of the winter. Throw a bucket of water into the air, and it is frozen solid as soon as it reaches the ground. There would be no more trouble, one would think, with water. Yet some of the worst trouble the traveller has with overflow water is during very cold weather, and it is then, of course, that there is the greatest danger of frost-bite in getting one's feet wet."

And one's dogs' feet wet. "When the dog comes out of the water into snow again," the bishop remarked, "the snow collects and freezes between the toes, and if not removed will soon cause a sore and lameness. . . . So, whenever his dogs have been through water, the careful musher will stop and go all down the line, cleaning out the ice and snow from their feet with his fingers. Four interdigital spaces per foot make sixteen per dog, and with a team of six dogs that means ninety-six several operations with the bare hand . . . every time the team gets into an overflow. The dogs will do it for themselves if they are given time, tearing out the lumps of ice with their teeth; but inasmuch as they usually feel conscientiously obliged to eat each lump as they pull it out, it takes much longer."

Sometimes an ice layer from the most recent overflow of a series of overflows would support the sled dogs but not the sled, "so that the dogs were travelling on one level and the sled on another, and a man had to walk along in the water between the dogs and the sled for several hundred yards at a time, breaking down the overflow ice with his feet. At other times the thin sheets of overflow ice would sway and bend as the sled passed quickly over them in a way that gives to ice in such condition its Alaskan name of 'rubber-ice.' "

For Stuck, who kept on traveling around Alaska all through the winter, tending to his isolated parish flocks, "waterways are the natural highways. The more frequented routes," he noted, "gradually cut out the serpentine bends of the rivers by land trails, but in the wilder parts of the country, travel sticks to the ice." For Thoreau, it was the same, only not so wild. "I have three great highways raying out from one centre which is near my door," he wrote in his journal in February 1859. "I may walk down the main river, or up either of its two branches. Could any avenues be contrived more convenient? With the river I am not compelled to walk in the tracks of horses." One day, to the accompaniment of a jay's "steel-cold scream," he skated to Pantry Brook on ice that was thin and "all tessellated, so to speak, on which you see the forms of the crystals as they shoot."

The stern Uruguay chafes its banks,
A mass of ice. The Hooghly is solid
Ice. The Adour is silent, motionless.

—John Ashbery,
"Into the Dusk-Charged Air"

For those who would travel on a river not by skate or dogsled but by the usual conveyance, a boat, an ice-covered river is no highway. On February 3, 1805, when the Lewis and Clark expedition was on the upper Missouri, Meriwether Lewis wrote:

"The situation of our boat and perogues is now allarming, they are firmly inclosed in the Ice . . . the instruments we have hitherto used has been the ax only, with which, we have made several attempts that proved unsuccessfull. we then determined to attempt freeing them from the ice by means of boiling water which we purposed heating in the vessels by means of hot stones, but this expedient proved also fruitless, as every species of stone which we could procure . . . burst into small particles on being exposed to the heat of the fire. we now determined as the dernier resort to prepare a parsel of Iron spikes and attach them to the end of small poles of convenient length and endeavour by means of them to free the vessels from the ice." Lewis wrote no more of the alarming situation, so the iron spikes must have done the trick.

Almost two centuries later, ice still hinders boat passage in many waterways of the north—in the United States, Canada, Scandinavia, and Russia—or stops it altogether. The Volga, Ob, and Lena, great transportation routes to northern Russia in summer, are closed to traffic in winter. In the United States, the Mississippi River from Minneapolis–St.

Paul south to Galesburg, Illinois, is closed to boat traffic for three months of the year because of ice. So is the Missouri from Sioux City to where it joins the Mississippi. South of Galesburg the Mississippi is open all winter, but in the northern part of that southern stretch boats often have to push their way through ice to get ahead. The same is true on other navigable U.S. rivers that form ice yet stay open, including the Illinois, the Ohio-Allegheny-Monongahela system, the Columbia, and the Hudson. In the regions where those rivers run, according to Ashton, "the temperatures oscillate around the freezing point throughout the winter. It's like having several winters all in one." If the ice buildup is so great that the boats can't push their way through it, the boats just wait, and after several days, the ice usually clears.

However, there are penalties for continuing to operate on ice-prone rivers, according to a CRREL report (which points out that "ice-prone rivers" serve 19 American states containing 45 percent of the country's population). One of the penalties is that tow boats can't push as many barges through ice covered water as they can through open water; tow trains that are 30 or 40 barges long in summer may be only ten or even five barges long in midwinter. Also, tow boats have to move more slowly where there's ice, and burn more fuel. At locks, they may have to wait for hours while ice is "locked through." Ice bangs up many boats. All in all, ice's appearance on U.S. rivers adds a few hundred million dollars each year to the cost of shipping on them.

Shortly after midnight on January 2, 1998, a freighter from Lisbon carrying Portuguese, Spanish, and Italian wine, cheese, perfume, auto parts, building tiles, chemicals, and tomato paste entered the Port of Montreal, to great acclaim. The captain was presented with a gold-headed cane and had his picture taken for the newspaper. For over a century, the port had been awarding a gold-headed cane (or, in the early years, a top hat) to the captain of the first ship of the year to make it across the Atlantic and along 1,000 miles of the Saint Lawrence River to Montreal without a stopover. The first ship would usually arrive in Montreal about the middle of April, when ice in the Saint Lawrence loosened up naturally, but beginning in the early 1960s the delivery of container cargo to the North American heartland became such a lucrative business that Canadians decided to keep the river open to traffic all winter long, as far west as Montreal.

When the ship from Lisbon arrived three and a half months earlier than ships used to, press releases praised not only the "experience, training and sound judgment of the officers and crew" of the winning ship but the "imagination, ingenuity and determination" of the Coast Guard crews who subdued the ice and made "winter navigation a reality" on the Saint Lawrence. The crews used a combination of icebreakers, which broke up the ice in a central channel, and ice booms, which kept ice accumulating *outside* the channel, the rationale being that stable ice on either side of a channel means less ice drifting into it. Also, water in a narrow central channel flows faster than in a wide one, helping sweep broken ice along.

on a night
so cold that in the bustling city which had sprung from
his dreams, blocks of ice were seen floating on the Nile.

—Nicholas Christopher,
"31" (from *5° and Other Poems*)

The upper Niagara River, traveling fast on its way to the famous falls, stays open all winter (although not many boats take advantage of the situation; even people whose vessel of choice is a barrel prefer summer). That part of the river produces great quantities of frazil, an estimated three million cubic yards of it per day when air temperatures fall below 32°F. Power-plant operators don't worry much about the frazil (they heat their trash racks), but they do worry about ice: *lake* ice. Lake Erie drains into the upper Niagara, and in a typical winter the lake has 10,000 square miles of ice floating on it. The ice cover gets broken up in storms, and pieces of the ice are swept into the river, where they crash into docks and produce flooding along the banks and block water-intake pipes before plunging over the falls and piling up at the bottom as an "ice bridge."

In 1964, to keep ice from Lake Erie in Lake Erie, Canadian and American engineers started stringing an ice boom (logs chained to floating barrels) across the entrance to the Niagara. Now, as soon as the water at the city of Buffalo's intake reaches 39°F, they install the boom, on the assumption that ice will start forming in the lake soon afterward. Some ice does get through the boom during storms as winds shove the logs under, and ice bridges still form at the bottom of the falls, not only from lake ice but also from "river ice, frazil, slush and snow, a well-wetted mixture," A. H. Tiplin noted in *Geology of Our Romantic Niagara,* "which freezes into an agglomerate mass when it is tossed out of the water below the falls." Although the ice bridges are impressive these days, with water cascading behind them, in the years before the ice booms went up, the ice bridges at Niagara Falls were *really* something to behold.

"There were certain years when the ice stretched from shore to shore in a wild, rumpled mass," Pierre Berton wrote in his book *Niagara.* One of those years was 1899. "By January 20, seasoned veterans of earlier ice bridges declared this one safe," he related, "and small groups of thrill seekers headed out over the treacherous surface. The river below the Falls was fifteen hundred feet across and almost two hundred feet deep. Just behind was the full force of the cataract. Yet so strong was this frozen bridge that hundreds were able to cross from shore to shore. . . . The first shanty appeared on the ice on January 20, and others soon followed. If the ice held there would be curio shops, Indian tepees, photographers' shacks, makeshift saloons, and even buildings identified as 'hotels.' Since this was an international no-man's-land, liquor could be dispensed freely, if not cheaply.

"Old-timers who remembered previous ice bridges looked forward to the informal winter carnival—the crowds on the ice, singing and laughing, paying top prices for coffee and sandwiches, the cliffs echoing with their shouts. Men planted flags on hillocks to record that they'd been the first to clamber to the top. Others explored crevasses to estimate the thickness of the ice. Some of these were thirty to forty feet deep, suggesting that the ice itself, most of which was submerged, was more than one hundred feet thick."

Into this jolly setting stepped a young couple, not honeymooners but good friends, he a traveling salesman and she a student in a school of shorthand. He wanted to cross the ice bridge; she did not. He convinced her; he should not have. They picked their way "gingerly over the ice," Berton wrote. "Two hundred yards out . . . they found a boulder of ice and sat down to enjoy the scenery. Half an hour passed." He noticed other people had gone back to shore but still felt safe. She heard "a singing noise under her feet, he told her it was only her imagination. At last, feeling that they had seen all that could be seen, they started back toward the American shore. They had not gone far when, by gestures, a crowd on the bank indicated that they could not reach the boat dock; the ice had broken away from the shore, leaving a stretch of water too wide to cross. . . .

"They were now hurrying as fast as possible. Ahead lay a large fissure in the ice, three feet across." He threw chunks of ice into the water to make a bridge that she could cross on. "He could see black water a hundred feet below and knew that one unsure step would mean death for both. He prepared himself to jump across the gap when he heard a loud report like that of a cannon, followed by grinding and crashing. The great ice bridge had torn loose from its foundations and was starting to move downstream.

"Their only hope of escape lay on the Canadian side. . . . Almost immediately he felt the ice part beneath his feet." She fell between two large ice boulders. If he had "not been holding her with a sure grip, she would have gone to the bottom or been crushed between two grinding chunks of ice." They tried to grab hold of girders on the new Honeymoon Bridge but were swept past it. "They set off for the Canadian side, stumbling, often vanishing from sight in one of the gullies, then reappearing to the cheers of the crowd. Often they were forced to leap blindly into ravines five or ten feet deep." A couple of times he left her to scout a way through the ice; spectators, thinking he was abandoning her to a hideous death, shouted "Coward!" and swore to shoot him if he made it. He did make it; she made it too. "Willing hands," he related, "stood waiting to receive us and to congratulate us on our almost miraculous escape from certain death." Nobody shot him.

Within hours, "the waters of the river were again jammed solid," Berton reported. "The new ice bridge was larger and stronger than any that season. It remained in place for a record seventy-eight days."

Mighty mounds from tiny crystals grow.

CHAPTER THREE

GREAT LAKES

Roofed with ice the Big-Sea-Water

—Henry Wadsworth Longfellow,
"The Song of Hiawatha"

THE EFFECT OF SIZE on the ice of the North American Great Lakes can probably best be appreciated by considering the seiche. "Seiche" is a French-Swiss word for a tidelike movement in Lake Geneva, although even in Lake Geneva a seiche is not a tide. The Langloises, the husband-wife team who watched the snow blanket, folds and all, turn into an ice blanket, called it a "wind set-up." It is a sloshing back and forth of water in a lake basin, like the sloshing in a bathtub, except that the initiating force in the case of the lake is atmospheric rather than a child lunging after a rubber ducky. High winds or big changes in barometric pressure or both push water at one end of a lake up into a sort of bulge, which then swings back against the wind toward the other end, where it piles up while the water level at the first end drops, after which the bulge swings back toward the first end, and so on, back and forth, like a pendulum. The oscillations keep on, even after the storm has passed and the winds have died. Midway between the ends, water levels don't change much, since the water rocks around a nodal point, like a seesaw on a crossbar. In January 1942, after what must have been a particularly vicious squall, water at the Buffalo, New York, end of Lake Erie was 13½ feet higher than it was at Toledo, Ohio, 250 miles west (a record high slosh), while in between them, around Sandusky, Ohio, the water was most likely calm, or nearly so.

The Great Lakes experience seiches even in winter when ice covers a large part of the surface, and that does strange things to the ice. Wayne Nelson, who lives on Madeline, one of the Apostle Islands in Lake Superior, has seen ice in a bay between the southernmost of the islands and the Wisconsin mainland "ripple like a river" during a seiche. "Millions of gallons of water a second are going through there," he says. "The ice goes up and down and up and down until it rips at the shoreline." As an instructive wintertime amusement, C. Allen Wortley of the College of Engineering at the University of Wisconsin–Madison suggests drilling a hole in the ice of any bay on the Great Lakes during a seiche and watching the water in the hole rise and fall "like a yo yo." Shorter oscillations are often superimposed on the major one, like the harmonics of a plucked string, he says, and in shallow areas there could be as many as eight nodal points (a pretty complicated slosh). Also, wave pressure is sometimes so high that the water, instead of going up and down in the hole like a yo-yo, shoots up "like a geyser" and floods the surrounding ice. One very cold morning, Wortley was out on Lake Michigan taking measurements of the ice while dressed in street clothes, since he planned to watch the Super Bowl with friends after the outing, and water shot up so energetically from the hole he'd just drilled that it "quick-froze" his shoes to the ice. He had to rip the leather uppers off his shoes, leave the crepe soles stuck in the ice, and walk back to shore in his socks.

In the Langloises' posthumously published book, *The Ice of Lake Erie Around South Bass Island, 1936–1964,** they describe many of the ice features they discovered around their island, some features not generally found on smaller lakes like Pleasant. Although Erie is the second smallest of the Great Lakes, it is still large—larger than the state of Massachusetts. Lake Superior, larger than the state of South Carolina, is the largest freshwater lake in the world. Because the Great Lakes are landlocked, they qualify as lakes, but because they are large they exhibit some characteristics of rivers; strong winds and strong currents and even the Coriolis effect produce riverlike turbulence in them. In addition, they are connected to each other by several rivers, a canal, a strait, and a small lake and are themselves connected to the Saint Lawrence River, which empties into the Atlantic, so that they are in essence parts of one large river system, "many pools," the Langloises wrote, "in a long, oceanbound stream."

Among other things, the Langloises observed that water levels around their island fluctuated by as much as four *feet* over a period of

* Thomas Langlois, according to the paper's introduction, was a "gentle and courteous scholar," also an apparently curious and unprejudiced one, having turned his scientist's eye to, besides ice features, the migration of monarch butterflies, the structure of octagonal buildings, and the mating behavior of slugs.

days due to "seiche-induced" currents. They noted too that strong winds and waves and currents often interfered with the formation of an ice cover, or broke up or dislodged one when it did form. "A strong west wind lowers the level of the western end of the lake," they explained, "resting the sheet-ice on shoals and thus increasing the attachment of the ice to the land, . . . but a strong east wind raises the level and breaks these lee-shore attachments." One January day they recorded that a sheet of ice eight inches thick with 130 fishing shanties sitting on it "went out," apparently because strong northwest winds not only raised the water level, thus breaking the attachment of ice to land, but opened an area of the lake farther west that the dislodged ice could move into. (Forty of the shanties were gone for good.) On another January day, powerful winds broke up sheet ice along the south side of a point of their island, and "fragments of this hard ice were rolled around in the water . . . making a roar which we heard from over a mile away."

> Great Lakes ships have always had to battle ice. . . .
> Wind-driven ice is an implacable and unpredictable foe.
>
> —Dwight Boyer, *Great Stories of the Great Lakes*

Sometimes offshore ice ends up on shore. Wind drives slabs of ice across long, open stretches of the Great Lakes, where they gather momentum before slamming into land, splitting apart and buckling and sliding over and under one another, producing piles of rubble. "The severity of the ice piling and accumulation is often not appreciated," one scientist noted wryly, "because it occurs when observers prefer to be in more protected places." The Langloises, who apparently felt little need to be in protected places, noticed one wild February day "an extremely large panel of ice, perhaps 100 yards wide, shoved up against the rocky cliff at Thompson's Beach. . . . The panel slid up the face of the cliff, about thirty feet high, debarking big trees and shoving aside little ones." A fisherman's shanty, meant to shelter its owner against harsh weather, was still attached to the ice panel but "crumpled beyond repair."

Such "ice pushes" or "ice shoves" can take place almost without warning, according to the Langloises, and last as short a time as ten minutes. On several occasions they watched wind-driven waves toss "slush-balls" onto banks where, as the wind changed direction and the water level dropped, the balls were left stranded and froze solid. Wortley has seen "baseball-size" chunks of ice catapulted by waves 20 feet in the air and land in trees. On the southern shore of Lake Erie near Dunkirk, New York, as well as on the southern shore of Lake Ontario between Rochester and Oswego, New York, four geologists at the State University College in Fredonia, New York, discovered conical mounds of ice that were formed by ice "eruptions." They called the mounds "ice volcanoes" because, they explained, "their origin is in so many ways analogous to that of true volcanoes."

The conditions needed for ice volcanoes to form, they figured out, are air temperatures below freezing; lake water that's mostly unfrozen, "with only slush and a few floating ice blocks" in it; a wind blowing at least 35 feet a second; and shorefast ice already in place which wind-driven waves can break against. These elements combine in the following way: during winter storms, high winds drive waves against the edge of shorefast ice, with the force of the hits concentrated at slight indentations in the ice. As the waves keep pounding in, they erode the slight indentations into V-shaped channels. Then more waves hit the channel walls, sending spray, slush, and ice flying as high as 35 feet in the air. The erupted material freezes before it reaches the ground, or soon afterward, as it goes "flowing, sliding, and rolling down the flanks of the [developing] cone." A cone can form overnight, or in as little as a few hours, but it can't get to be any higher than water can be sent flying.

At Dunkirk, rows of ice volcanoes were lined up more or less parallel to the shore, the geologists found, with the ones in the rows farthest from shore the tallest, up to 15 feet high, and the ones closest to shore the lowest, under three feet high. Where several rows stuck around for a good while, they were apparently able to protect the lakeshore from erosion during the violent winter storms that produced them in the first place.

Although sacrificing the soles of his good shoes to the ice of Lake Michigan seems to have amused Wortley, at least in recollection, he is not at all amused by another consequence of wind-induced water fluctuations under a Great Lakes ice cover. When water levels rise in a bay during a seiche or other event, he explains, the ice cover often responds to the pressure from below not by breaking away from the dock pilings to which it is attached but by rising with the water and taking the pilings with it. The situation turns permanent when sediment slides into the holes in the bed that opened up when the pilings were hoisted, so that when water levels drop later on and the ice does break away from the pilings (leaving circular "ice bustles" around them), there aren't holes for the pilings to drop back into. This can happen over and over until by the end of winter the pilings are standing five, ten, even 20 feet higher than they were at the end of summer—or they are yanked out entirely. "If your harbor is a Great Lakes harbor," Wortley concludes, "even if it is in the most sheltered of coves, you are not safe from uplift forces."

One phenomenon the Langloises observed takes place only during very long cold winters. Objects—"sand, gravel, sticks, poles, or planks," they wrote—pass clear through the ice cover, from the bottom to the top. At the start of winter, lake levels are low and the young cover rests on elevated areas of the lake bottom, reefs, shoals, etc. Later, when water levels rise, the ice lifts away from the lake bottom and takes with it whatever objects it has the buoyancy to pull upward, and those objects become incorporated into the floating ice sheet. Over the winter, ice is added to the bottom and top of the ice sheet, black on the bottom and white on the top, as elsewhere, but in Lake Erie there's usually not enough white ice to make up for what's lost off the surface through melting and evaporation, with the result that the objects move upward through the ice sheet, without actually moving. Ice thickens under them and thins above them, causing them to inch their way toward the surface. Because they are darker than the ice, they also absorb heat from sunlight passing through the ice, which contributes to further melting of the ice on top until in late winter, presto! The sand, gravel, sticks, poles, and planks off the bottom appear on top of the lake, "having passed through the entire thickness of the ice sheet."

Year to year, lake to lake, and section to section of the same lake, Great Lakes ice covers vary greatly, in both thickness and extent. For example, the maximum extent of ice coverage of the five lakes during the last two weeks in February, which is the period of greatest ice buildup, ranged in the decades between 1960 and 1979 from 2 percent to 100 percent.

Latitude matters, not surprisingly. Going north, ice thicknesses and extent increase. In the middle of April in 1957, Dwight Boyer pointed out in his book *Great Stories of the Great Lakes,* there was still so much ice left on Lake Superior, the northernmost of the Great Lakes, that 100 freighters were trapped in it. The log of a Coast Guard icebreaker assigned to pump out and patch up several of the ice-damaged freighters included the entry: "Ice cakes found in flooded upper compartments." Ironically, Boyer pointed out, these salvage operations took place "on days when in cities a few hundred miles to the south forsythia was in full flower, daffodils were in bloom and hopeful gardeners were planting early peas."

Longitude also matters, perhaps surprisingly. The western halves of the Great Lakes tend to freeze before the eastern halves, and to develop thicker ice, probably because prevailing westerly winds blow cold air off the land. Depth certainly matters; Lake Erie's mean depth is only 62 feet, and during the last two weeks of February in the 1960s and 1970s the lake was on average 90 percent covered by ice while Lake Ontario, with a mean depth of 282 feet, was only 24 percent covered.

In the winter of 1948–49, ice was so extensive on Lake Superior that wolves were able to hotfoot it over to Isle Royale, 30 miles from the Ontario mainland. "Bold wolves—probably only a mated pair—" Les Line wrote in an article in *National Wildlife* magazine, "made a run over a solid ice bridge between Canada and the island. It was an extraordinary event. 'The weather has to be cold and calm enough for the lake to freeze over,' explains David Mech, a U.S. Geological Survey biologist. . . . 'Then you need wolves predisposed to make a long trek over the ice. . . . The chances of this happening are pretty remote, or Isle Royale would have had wolves before the 1940s.'" Isle Royale now has many wolves on it, all descendants of the first bold pair.

An almost 100 percent coverage by ice in the winter of 1993–94 inspired Wayne Nelson to attempt the first-ever crossing of Lake Superior by ultralight airplane. He claimed he hated to see all that emergency-landing space go to waste. Wearing a military ice-survival suit, he took off one morning in his open-cockpit homebuilt from Bayfield, Wisconsin, carrying five gallons of fuel; that was the maximum legal fuel limit but not enough to get him all the way across. His plan was to land on ice partway over and refuel by having a friend fly overhead in an airplane, lower a gas can on a rope, and "skid" the can across the ice to him.

Somewhere in midlake, he came across a "mini mountain range" of ice four and a half miles long and ten to 30 feet high. "It was the most gorgeous aqua-blue, probably the prettiest thing I ever saw," he says. One chunk of ice stuck up 35 feet in the air. "Not even a tank could climb that thing." He concluded that the wind had pulled sheets of ice on either side of a crack apart, then shifted "and brought those suckers back! It was very humbling to know how much energy that took."

After seven hours and 140 miles, Nelson landed at Thunder Bay, Ontario. He admitted that for one 20-mile stretch he had been scared because there was no landing surface beneath him; open water appeared where ice had been a day earlier. All winter long, during windy and stormy weather, the ice of all the Great Lakes keeps shifting around, making new openings, new ridges, new cracks. While airborne, Nelson discovered that he could tell just by looking down at the cracks how thick the ice was. According to the Langloises, some cracks on Lake Erie appear in the same places year after year. "Islanders know these cracks," they noted, "and when driving on the ice . . . carry planks to bridge [them] when they are open, and axes to cut through them when they are jammed shut into a ridge." According to Nelson, Madeline Islanders can draw maps of the recurring cracks on their side of Lake Superior. But the residents of Siberia's Lake Baikal go the residents of Erie and Superior one better: they give some of their regular cracks names.

(Anything bigger than . . . Michigan or Baikal,
Though potable, is an "estranging sea")

—W. H. Auden, *"Lakes"*

Lake Baikal is another great lake. " 'The sacred sea,' 'the sacred lake,' 'the sacred water,' " Valentin Rasputin wrote in his book *Siberia on Fire*, "—that is what native inhabitants have called Baikal from the beginning of time." (North America's Great Lakes have also been called seas; "*mers douces,*" Samuel de Champlain said of Superior when he came upon it.) Although Lake Baikal is smaller than Lake Erie in surface area, it is many, many times deeper, over a mile deep, the deepest lake in the world. It contains more water than all the Great Lakes combined, one-fifth of all the fresh water in the world. While the Great Lakes are shallow basins scooped out by an ice sheet a few thousand years ago, Baikal's basin is a rift that opened in the earth more than 25 million years ago. Today it is still widening and deepening, giving the lake a good shaking now and then.

The lake also gets a good shaking from strong winds that blow across Baikal much of the year. "The western *sarma* springs up out of nowhere," Colin Thubron wrote in *his* book *In Siberia.* "With hurricane force it pitches sheep and houses over the cliffs, and ices fishing-boats in freezing rain before sinking them. Only winter brings a kind of peace. Then the lake freezes so solid that it becomes a lorry road. But without warning, during sharp temperature changes, a six-foot crack may open underfoot and streak for up to eighteen miles across the ice, pulling down trucks and bulldozers to join the tea-caravans of Bactrian camels engulfed a century back."

One of Baikal's astonishments is the clarity of its water. "As the transparent and slightly alkaline water deepens," Thubron observes, "other colours are filtered from its light spectrum, until only blue, the least absorbent, remains." When he first saw the lake, from a bluff high above it (and it "became an ocean" to him), the blue was a "kingfisher-blue . . . All colour, from here, had refined to this drenching blue." Baikal's ice is likewise clear and—where winds have blown off the snow cover—blue. It is also thick. "In winter," Rasputin wrote, "when the transparent ice, swept clean by the winds, seems so thin that the water beneath it is alive and stirring, like under a magnifying glass; you're afraid to step on it, yet all the while it might be over one meter thick."

Late one March day, Rasputin walked far out onto the lake's frozen surface. "The pure light-blue sky grew flushed and spread out above me; a full moon hung to my right," he wrote in *Siberia, Siberia*. "But on the windswept clearings of ice the moon glimmered with a compressed light underneath me, too, and star sparks smoldered." He was alone.

"Long shafts of thunder came running at me under the ice," he went on, "exploding and rumbling directly beneath my feet, but I quickly got used to them and stopped being afraid. I crossed [an ice] road leading from one shore to another, which was staked out with fir trees standing gloomily and awkwardly in formation under the bright sky like bundled-up figures. Baikal spread out wider and wider before me. . . . I walked and walked. . . . It was as if I had accidentally entered some kind of enchanted kingdom of forces different from those we know, of different sounds and times that make up a different life. A solid mirror of bare ice stretched ahead and behind; it sloped like the sky and glowed just like the sky with all its lights, but they were bent and shaped like icicles. It was radiant above and radiant below, and a light blue radiance lay on the ice. . . . Baikal grumbled with a muffled sweetness, somewhere little ice bells tinkled in a strumming trickle, somewhere something flowed and subsided with a sigh."

In November, an ice cover starts forming in the northern part of Lake Baikal; it works its way south and by January or February has enveloped the southern part. In April it starts thawing, and strong winds help rip it apart. Rasputin noted, "In spring, Baikal, quivering from time to time, breaks up this ice with wide, bottomless cracks that can't be crossed on foot or by boat and then, pushing them back together, erects magnificent masses of light blue ice walls." Probably not unlike the mini–mountain range that Nelson spotted from his ultralight airplane, half a world away in Lake Superior: miles long, many feet high, and aqua-blue—the prettiest thing he ever saw.

CHAPTER FOUR

LOADING

Praise not . . . the ice until it has been crossed

—Havamal, Viking saga

IT'S WINTER, and on great lakes and small ones, on rivers and brooks and ponds, the ice has formed. What before was mobile and penetrable and uncrossable except by a swimmer or a boat is now hard and resistant and capable of holding up a skater or a truck. For as long as humans have been inhabiting cold regions, those who live near bodies of water that freeze in winter have been taking advantage of this God-given extension of their land, this new-made real estate, and using it for roadways, bridges, racecourses, rinks, battlefields, housing foundations, and platforms to hunt and fish off of.

But how, you might ask, can you know whether a body of ice will support you and your skates or your truck? How strong *is* ice? How strong is *this* ice you are about to step out upon? Anyone who has watched the scene in the 1938 Sergei Eisenstein film *Alexander Nevsky* in which Teutonic knights in full armor with lances and shields are fleeing from Russian peasants across the ice of an Estonian lake in the winter of 1242 and the ice (real ice; no Styrofoam in 1938) gives way under the knights and their steeds, breaking into jagged chunks which the men try to use as stepping-stones but are quickly dumped from as the chunks tilt, then when some soldiers try to crawl back up onto the ice and can't make it they slip down into the water where, after thrashing around, they sink for good, the tops of their crested helmets slowly disappearing into the water between floes. . . . Well, anyone who's seen that has a good, gut feeling for how loads can exceed the bearing capacity of ice.

Trust not one night's ice.

—George Herbert,
Jacula Prudentum

Some workers in the field use the "Gold Formula" to estimate how much weight a body of ice can support. The Gold Formula has nothing to do with the gold standard or with the good-as-gold quality users hope the ice will have but with Lorne W. Gold, researcher emeritus at the National Research Council of Canada, who in the early 1960s developed a formula for predicting how thick freshwater ice should be to support certain weights. While acknowledging that no formula can be foolproof since ice, "being formed in an outdoor, uncontrolled environment" and "at all times close to its melting point," is naturally variable, Gold recognized the need for a scientifically based guideline for practical use.

To get one, he enlisted the help of the Canadian pulp and paper industry, which cut logs in wintertime and transported and stored them on ice. A couple of the logging companies agreed to record for Gold every time during their operations that ice failed under a vehicle or horse plus the circumstances of that failure: weight of the vehicle (or horse), speed of its travel at the time, thickness of the ice, temperature of the air, etc. They also agreed to take note of some occasions when the ice did not fail.

Gold's formula for general use, which he worked out by plotting both breakthroughs and nonbreakthroughs, is $P = 0.025\ h^2$, with P the weight of the allowable load in tons, h the thickness of the ice in inches, and 0.025 the constant. So if you are approaching a lake with, say, nine inches of clear black ice on top of it, then by using the formula you would conclude that you shouldn't drive anything heavier than a two-ton automobile across it (your Lincoln Continental, for example). A companion formula, $P = 0.025\ (h + w/2)^2$, in which w stands for the thickness of white (or snow) ice, takes into account the widely held but unverified assumption that white ice is only half as strong as black ice; each inch of white ice would count as only half an inch of "effective" ice.

Other people have developed other guidelines to ice's loading capacity. One guideline simply squares the ice thickness and multiplies by 100 to get allowable load: one inch of ice can support 100 pounds; three inches, 900 pounds; 12 inches, 14,401 pounds (what's going on with that extra pound?), and so on. It illustrates well that ice strength increases exponentially with thickness. For users of ice who want something even simpler, here's a handy, popular, generic, one-stop table to suggest what ice of different thicknesses can support:

2 inches	a person on foot
3 inches	a group in single file
5 inches	a snowmobile
7½ inches	a passenger car (2 tons)
8 inches	a light truck (2½ tons)
10 inches	a medium truck (3½ tons)
12 inches	a heavy truck (8 tons)
15 inches	a heavier truck (10 tons)
30 inches	70 tons
36 inches	110 tons

Footnotes to such tables are not to be ignored: "Strength value of river ice is 15 percent less." "Distance between vehicles [should be] about 100 × thickness." And so on.

The ice isn't really holding up your truck (or horse), however. The water is. "The ice won't hold if there isn't any water against it," explains Brian Sigfusson, a builder of ice roads in northern Canada. "It's like a window pane. If you stood on ten window panes, you'd go through them all, but if you put them on water they might carry you. It's the upthrust against the glass that does it." What an ice sheet does is distribute a load over a broader area of water than the one defined by the object producing the load; the ice sags or bows under the object until the weight of the water it displaces is equal to the weight of the load. The sagging creates a boat-type hollow around the object; the thicker the ice, the wider the hollow. Also the thicker the ice, the shallower the hollow, which puts less stress on the ice, so it's less likely to break under a load.

Thickness, then, is the main factor where ice is concerned, but it's not the only one. For instance, when air temperatures have been above 32°F for more than six hours of the last 24, ice thicknesses need to be a third higher to support a given load. Also, if there's heavy snow on top of the ice, it can push the ice sheet down below the water line and force water up through cracks and holes, producing an effect like, as George Ashton puts it, a "leak in a boat." And like water in a boat, water on an ice cover will counter the upward pressure of water under the cover, and, like a boat, the ice won't be able to float as high or carry as much load as it did before it sprang the leak.

It ain't a super highway, but it's all that we got.
It's there when it's cold and it ain't when it's hot.

—Michael Faubion, *"Ice Road" (song)*

Few people are better equipped to talk about how much weight ice can bear than those, like Sigfusson, who build winter roads. Every year in the northern part of the northern hemisphere, in Canada, Russia, Alaska, and Scandinavia, men head out onto frozen water and frozen ground and build supply roads to settlements that are often too isolated, or on land that is too swampy, to merit year-round ones. It's perilous work. They drive onto lake covers in their heavy tractors and plow trucks, testing and preparing ice routes for the truckers who will follow. Their survival, as well as the survival of their equipment, depends on their assessment of the condition of the ice at the time. In case their assessment is wrong, they often drive with one hand on a door handle, sometimes with the door already open and their feet outside, resting on a fender.

"My own last-ditch survival secret," admitted Sigfusson's uncle Svein, who built ice roads before Brian did and wrote a book about it called *Sigfusson's Roads,* "was to be afraid at all times, developing a hair-triggered coward's reflex, with which I jumped off four sinking tractors in my 25 years." In the book he detailed what can happen to a person who is insufficiently afraid of ice. One time he and his men were opening an ice road to Reindeer Lake, Saskatchewan, where they intended to pick up a load of frozen fish, but the ice on the way there turned out to be full of cracks. Taking time to bridge the cracks had put them behind schedule and allowed a "swing" (tractor train) from another hauling company to catch up. The swing boss, "pompous" *and* "incompetent," wanted to pass Sigfusson's team and drive ahead onto ice the team hadn't yet prepared.

"That crack is unfit," Sigfusson told the swing boss, but the man sneered and, "as if on cue," one of his "skinners" (drivers) drove a five-ton tractor over the crack "disdainfully," making a show of manipulating the controls. When the ice held, the skinner gloated, then spun his tractor around and a short time later approached the crack again. "The last rays of sunset vanished and the night turned dark," Sigfusson recounted. As the lights of the skinner's tractor became "diffused through the vapour that hung over the crack," Sigfusson and his men felt a sense of "imminent doom."

"We watched helplessly," he wrote, as "for an instant the blurred light beams seemed to deflect upward. Then they slowly faded out. We ran to the spot and found only a fearful black hole in the ice. Far below, we could see a faint glimmer where the Cat's* headlights still shone through the murk. The cat was now nearly buried in 'loon shit'—the bush term for the soft muck at the bottom of the lake. The skinner's body was down there too; he had had an instant to jump but was not quick enough. He was caught in the suction of the sinking cat and was dragged straight to the bottom, 27 fathoms down. He probably perished within moments in the paralyzingly cold water. There was no chance of retrieving either the machine or the body."

Two more men drowned on ice roads that winter, and seven more tractors ended up buried in loon shit. Sigfusson himself "nearly went through the ice at least a hundred times." Being afraid at all times apparently wasn't enough to protect a man from bad ice; bad weather, bad luck, and bad timing played their parts. For 25 years, Sigfusson and his crews built 3,000 miles of road each winter through the low, wet, muskeggy, windswept bush country of northern Canada, across ice covered lakes with names like Big Trout, Jackfish, Bearskin, Pickle Crow, North Spirit, Whiskey Jack, and Red Sucker, as well as over the pieces of land that separated them. In the process, Sigfusson became in his own view a "serious student of ice. All the lakes in the Bush were my laboratory," he wrote. Among the things he learned about ice from making roads on it:

• Even ice that is four feet thick will "dry-crack" under a loaded tractor train. (Dry cracks are ones that don't go all the way through the ice, typically only two-thirds of the way, so that water doesn't rise in them.) Dry-cracking is "a perverse sort of sign that the ice is strong," Sigfusson claimed. The boom it makes "is like the voice of doom to greenhorns, but it reassures experienced men." If you can hear it over the noise of your tractor, that tells you the ice hasn't broken through, just as "the sound of thunder means lightning has struck elsewhere."

• "For reasons of its own," a dry crack "can turn unobtrusively into a wet crack."

• "Hollow ice," or ice unsupported by water, often develops along shorelines after a lake's water level drops. "The trick was to bring the trains onto the ice where the lake bottom sloped gradually and the mass of snow weighed the ice down to the earth."

• "Ice was generally stronger when it had more impurities (to the

* "Cat" for Caterpillar, even when the tractor's make isn't Caterpillar.

point where three inches of frozen mud would support a nine-ton tractor). But once a thaw set in, impure ice deteriorated . . . faster than clear ice."

With his two sons, Brian Sigfusson now builds the ice roads that his uncle Svein used to, with many of the routes going to Ojibway and Cree reserves which don't have summer roads. "Kingfisher to Wunninum Junction," he names some routes, sounding like a train dispatcher, "Wunninum to Kasabonika, Kasabonika to Long Dog." He too has become a student of ice:

- "You should cross a frozen river at the wide part," he says. "There'll be minimum current to erode the ice from the bottom. In narrow parts, the ice could be two feet today and 18 inches tomorrow and nothing to tell you the six inches are gone."
- Trafficked ice roads are better than untrafficked ones because the weak spots have already cracked and flooded and refrozen.
- "If you can't get the snow off fast enough, you won't get the width you need, and the road will dish downward. The pressure will bow the whole road down and it will flood right over, and you'll have a permanent scab, and water will be under there forever, and you'll get a bad road, and you'll have to throw that one away and try to make a good road somewhere else."

A "good" ice road is far wider than the vehicles that use it. The weight of the plowed snow that's pushed into piles on either side produces cracks five to ten feet inboard of the sides, and if the road is narrow the cracks will thus end up in the middle, causing it to dip. If the road is wide, the cracks will appear well away from the middle, and the middle will bow upward, or "crown." A good ice road has a hump in it. "The profile should be like a pyramid," Brian Sigfusson says. Or, as others suggest, like a loaf of bread or a submarine.

John Denison, the subject of Edith Iglauer's book *Denison's Ice Road,* built winter roads even farther north than the Sigfussons, opening up trails to gold, silver, and uranium mines in the Northwest Territories. Nine out of every ten miles of his roads passed over frozen lakes, including one that ran for hundreds of miles from Great Slave Lake to Great Bear Lake and on the way passed through 19 other lakes, including Beaverlodge, Rabbit, Fishtrap, and Gunbarrel. Denison told Iglauer that it is far more dangerous to build roads over lakes than over land but also far easier. At the latitudes where he worked, building ice roads consisted mostly of plowing snow off and keeping snow off, "so the frost [cold] can go down." The better the ice is kept plowed, he insisted, "and the more traffic there is, the thicker the ice gets." Ice on the roads could end up being twice as thick as ice elsewhere on the lakes, a circumstance that sometimes had an odd effect after the season was over.

One summer Denison was flying over a lake across which he'd made an ice road the previous winter, Iglauer wrote, and from the air he could see light strips on the lake bed under water, "a kind of light green with everything dark brown around it." The thick ice of the road had taken so much longer to melt than the rest of the ice that the growth of the vegetation had been "retarded."

The fool slides o'er the ice that you should break

—William Shakespeare, *Troilus and Cressida*

"As a rule," Iglauer quoted Denison, "it's the second guy who goes through and the first man over who cracks the ice." Often, that first guy is driving too fast. "It's smart to go slow when you're not sure," he advised, "about six miles an hour, so as not to bend the ice when you are coming near the shore. Because if you make a wave, it has no place to go, and it will snap back and break the ice." Vehicles make waves in water under an ice cover just as boats do in water with no ice cover. As a vehicle moves along, the boatlike hollow it creates in the ice sheet moves along with it (this can supposedly be seen from a helicopter), and the water the hollow displaces moves radially outward from it as a wave (which can be seen from a helicopter, too). How fast the wave is going depends on how deep the water is at any one spot and how thick and flexible the ice cover is, and if the wave's speed happens to coincide with the vehicle's speed, something like resonance occurs. "It's a little like going through the sound barrier," Don Haynes, chemical engineer at CRREL, explained. The load's deflection of the ice is amplified by the wave's deflection of the ice, and the bearing capacity of the ice is reduced to less than half of what it was before the vehicle came by. Two waves from two moving vehicles may also cross and set up their own ice-water oscillations, stressing the ice.

Every year, the province of Manitoba builds an extensive network of winter roads, more than 2,000 roads running 1,000 miles over ice and land. The roads are about 300 feet wide. They are numbered. They have signs for turnoffs. They have speed limits (for a large truck on a frozen lake or river, it's ten miles an hour). Hundreds of vehicles pass over the roads. What, you might wonder, would the Saracen warrior, who wouldn't believe a lake could bear the weight of a single horse's hoof, think of *that?*

Or of any of the following true tales?

• For four years, from 1880 to 1884, a railroad ran on the ice of the Saint Lawrence River, and it was the real thing, too, with iron rails, steam engines, passenger cars, cabooses, fur-covered seats, schedules, round-trip tickets, and freight (coal, timber, hay). The Grand Trunk Railway of Canada owned a railroad bridge over the river, but it refused to let its competitors use it, so one competitor with lines running north of the Saint Lawrence and another with lines running south of the river decided to link up for full service, sending trains across the watery divide on side-wheel ferries in summer and on ice in winter.

Despite some ridicule and entreaties to desist ("In the interests of humanity, . . . let us not forget that life is involved," one citizen wrote to a newspaper, signing himself "Safety"), workers started laying track across the frozen river in early January of 1880. First, they bored holes to determine the thickness of the ice in various places, then laid out a path that curved slightly to take advantage of the thicker spots. On that path they laid, crosswise, timbers varying in length from 12 to 24 feet so the bearing surface would not only be large but have an irregular boundary, to spread the load. At right angles to the timbers they set others, in two rows running parallel to the direction of the route, then they put down crossties, pounded in spikes, and laid light iron rails on top. Finally, they pumped water between the bottom timbers, which would, when it became ice, have the same strengthening effect as sand and gravel do on railbeds on land.

Inaugural day was January 31, when a small steam engine with a stuffed beaver mounted above the engineer's head chugged away from Montreal on the north shore pulling two cars behind it adorned with evergreens and flowers and filled with official guests, including ladies in large hats, toward Longeuil on the south shore. Thousands of onlookers stood on the ice and banks, some of them "joking sceptics and sombre men of warning," according to a later account. The weight of the train, 60 tons, caused water to spurt up through cracks and drill-holes in the ice, alarming a passenger who at the end of the ride accepted celebratory refreshments with the remark, "This is better than groping after our dead bodies under the ice."

During the following winter, which was mild, an engine pulling 17 cars jumped the track and—to the accompaniment of a fireman's shout,

"Jump for your lives!"—broke through the ice and sank. Yet there were no dead bodies under the ice on that occasion nor on any others during the entire operation of Canada's first "trans-glacial" rail service. At the end of four seasons, Grand Trunk gave in and allowed one competitor to use its bridge, and the Canadian government bought out the other.

• In the winter of 1657–58, during the First Northern War, King Charles X Gustav of Sweden and his troops crossed ice that their Danish enemies believed was impossible to cross and triumphed by surprising them. "The sea ice was thin," George K. Swinzow wrote in a CRREL paper called "On Winter Warfare," "and the Danish forces felt safe, separated from Charles." Charles, however, had a plan. He carefully tested the ice at a strait called Little Belt, reinforced it by laying down planks and straw and flooding the surface, and waited until the floodwater froze. His strategy was to spread his men, horses, and baggage over the ice so their weight wouldn't be concentrated during the crossing, but one of his cavalry commanders apparently had another idea. He thought that if his unit hurried over the ice, the ice could support a heavier load. The commander forthwith dispatched "a closed column at full speed" across the ice, and he, his horses, and his men all fell through and perished, while the ice-savvy king made it safely across and stunned the Danes, sitting in false security behind the thin-ice barrier. "What is thin ice for one," Swinzow concluded, "may be just thick enough for another."*

• In her book *Glacier Pilot,* Beth Day explains what happened when pioneer bush pilot Bob Reeve became lost in ice fog while flying a young family to its new home in Nome, Alaska. He brought his Fairchild down "between towering spruce trees" and made an emergency landing on a frozen river (the Kateel, as it turned out). After securing the Fairchild (it was Reeve's practice to chop holes in the ice and freeze ropes into the holes), he set up camp on the ice for his passengers, one of them a four-month-old baby. The air that night was so cold, Day wrote, that "hot tea three feet away from the fire froze within minutes." The next day the ice fog lifted slightly, and Reeve chopped another hole in the ice, this time to find out which direction the water in the river was running, reasoning that even though he didn't know

* Although Napoleon's retreat from Russia during the winter of 1812–13 is often credited to, or blamed on, the "harsh Russian winter," Swinzow argues that had that winter been harsher, or at least as harsh as "normal" (it was "rather mild and late"), there would have been ice on the River Berezina which could have served as a bridge, and losses to the French army would have been "significantly less." As it was, the French had to cross the Berezina "over rather poor [non-ice] bridges, with great losses of men, horses, and cannon."

which river he was on, somebody was bound to be living downstream along it. He scraped frost off his wings, heated his engine, took off, and followed the river all the way to an Eskimo village, where he landed, this time on a frozen creek.

> When the pond is frozen I do not suspect the wealth under my feet. How many pickerel are poised on easy fin fathoms below the loaded wain.
>
> —Henry David Thoreau, Journal (*Winter*)

There's a lake in Minnesota with a French name—Mille Lacs—but an un-French pronunciation—Mill Ax—which in the middle of winter has 65,000 fishing shanties on top of it. That's reportedly more houses than 95 percent of the towns in Minnesota have. "Nowhere in the world has this many ice houses," Steve Johnson, owner of the Mille Lacs Bait and Tackle Shop, asserts. The typical fishing shanty measures 10 feet by 16 feet, but many on Mille Lacs are larger, *much* larger, with upper stories, insulated walls, stoves, bunk beds, couches, TVs, bathrooms, and bars.

"These are not huts," Johnson insists, "not shanties, not shacks. These are *houses.*" The reason why so many fishermen, who really just need protection from the elements while they tend their tip-ups, should go so far beyond the basics is that the ice offers them a building lot they do not have to buy or lease. "A person of modest means who could never afford a dream home with a lake view can, in winter, enjoy a luxurious ice shelter on the best lake in the area," Jim Capossela points out in his book *Ice Fishing.* Although many fishermen move their houses to different ice locations every week, others don't move theirs until spring when they are required to take them off the lake. Some fishermen spend entire weekends on the ice, arriving after work on Friday and not leaving until Sunday night, and a few spend the whole winter there. "A frozen lake is a great place for social experimentation," Garrison Keillor noted in one of his *Prairie Home Companion* radio monologues. "There are no property lines. There's a deadline. Everything is temporary."

How does the ice take it all? A fishhouse can weigh four tons. On a fine weekend each house can have as many as four vehicles parked around it. When there's an event, cars and trucks converge; one year, 1,000 were lined up beside a snowmobile race track on the ice. "It looked like the Kmart parking lot," a fisherman says. What about all those fishing holes perforating the ice, an average of six holes per homeowner? "People will go into an area and make it look like a cribbage board until they catch a fish," says John Nelson, an employee of the Bait Shop gas station. "They'll literally exhaust an area."

So how *does* the ice take it? For one thing, central Minnesota is a seriously cold place in winter, and the ice on Mille Lacs can be a foot thick by Christmas and thicker after that. Also, the lake is large, 132,516 acres. In addition, as Johnson points out, "Everything gives. The ice just sags." When a load is short term, ice responds elastically; it changes shape under the stress yet regains its previous shape once the stress is gone. But under long-term loads, the ice doesn't fully regain its old shape once it has deformed. A parked van makes a "dimple" in the ice, as one fisherman describes it, and "several vehicles standing together make a bigger dimple." A fishhouse too can make a big dimple. As long as the house stays on one spot, the ice continues to deform under it. For that reason, each house is kept at least 50 feet from any other house. The ice isn't damaged by the long-term load, unless its bending limit is exceeded, but it is altered. "All the molecules try to adjust their orientation to decrease the influence of the load," Dartmouth's Victor Petrenko explains. The molecules realign themselves and reconnect to form a new structure.

> People have a false sense of security. They think the ice is a parking lot. They forget that they are on *water.*
>
> —Stephen Estabrooks, former chief of public facilities and public events support, National Capital Commission, Ottawa

Every year several vehicles go through the ice of Mille Lacs, and Jim Staricha makes much of his living in winter pulling them out. The winter of 1994–95 was "a pretty good one" (for him): "Fourteen vehicles, four snowmobiles, and a tractor." Most vehicles fall through because of "pie shapes," he concludes. Pie shapes are pieces of ice shaped like pizza slices. "If a piece is round or square," Staricha says, "it won't sink when you hit it, but if you get your vehicle over the point of a pie shape, it'll drop right down." Pie shapes form in steps, under short-term "point" loads, thus:

Step 1: A car drives across a frozen lake. The ice under it bows downward, which causes the bottom of the ice to bend so far that a crack opens up in it, with the orientation of the crack determined by chance. Step 2: A second car passes over the spot that has a crack underneath and produces another crack, at right angles to the first. Step 3: Other cars roll over the same spot and produce cracks that bisect the wedges formed by the first two cracks, until at some point a car passing over the spot has only a bunch of wedges under it. The car's driver doesn't know this, however. The cracks have opened only on the underside of the ice. Step 4: The wedges are so narrow that the area of greatest stress moves from the bottom of the ice to the top and away from the central point so that the next car that passes over the spot causes a circular crack to open up which runs through all the radial cracks, like the rim around a pizza pie.

The ice might still support a car, if it is thick, since friction between uneven spots on the crack's opposing faces could keep the wedges from separating. "Nevertheless," one scientist notes drily, "it would be foolhardy to place much faith in such a situation." If the ice isn't thick, the pizza slices will be more or less free floating, and any weight could tilt them and allow a car wheel to drop through. One recent winter, according to Staricha, a 34-year-old man drove out onto Mille Lacs in the middle of the night with a friend, allegedly to steal a fishhouse. The owner of the fishhouse owed him $500 and wouldn't pay up. As the men were hauling the fishhouse away, their truck hit what Staricha surmises was a piece of pie ice. The truck fell in the water, after which the piece of ice must have righted itself and snapped shut "like a trap door." The friend made it out but not the one who was owed. Staricha, who had the job of recovering the body as well as the truck, concluded that the driver had tried to save his vehicle. "It isn't worth saving *anything,*" he says. The man still had his car keys in his hand. "He must have thought he had all this *time.*" But when the man got around to kicking at the front windshield, the windshield was above his head and he couldn't get purchase.

Besides, Staricha noticed, the window on the driver's side was rolled up. It's considered smart when driving on ice to keep your windows open. It's also considered smart to unfasten your seat belt and unlock your doors. On truly dicey ice, Allen Wortley keeps his door not only unlocked but ajar. He also takes off his parka. "You need as much exit time and freedom of movement as possible," he says. Some people take the doors—or even the tops—clear off the cars they drive on ice.

And I'll stand—God help me!—on this ice,
However light and brittle it is

—Anna Akhmatova, *"So many requests . . ."*

There are things you can do to increase your chances of not falling through ice yourself. Say you're standing at the edge of a frozen lake or river and want to walk out upon it but aren't sure what it will support. You could start by drilling a hole near shore, sticking a ruler down the hole and measuring the ice thickness, then using a chart or formula to decide if that thickness is great enough, for you, your snowmobile, or your car. Drill more than one hole; lakes freeze unevenly. In some regions, the practice is to drill a hole every 100 feet on lakes, every 50 feet on rivers.

Once you're on the ice, test what's ahead by hitting it sharply with a "spud" (long-handled chisel). If the spud doesn't go through, you probably won't either. In Alaska, Bishop Stuck used an ax to test the ice ahead of his dogsleds. The rule, he said, was "ice was not safe unless it took three blows of the ax to bring water." Some scientists have their own not very scientific methods for judging the strength of the ice they want to work on. Don Haynes uses a rock. "It's a very corny test," he says sheepishly. "We take a rock that weighs maybe 10 or 15 pounds, throw it at the ice, and if it bounces, the ice is okay. If it goes through or imbeds, it isn't."

When Wayne Nelson flew an ultralight over Lake Superior, he used a brick. Half a brick, actually. His father, a commercial pilot, taught him the trick. Whenever his father wanted to land a small airplane on the ice of Lake Superior to drop off supplies for a fisherman, he'd toss a brick out the window. He calculated that a brick dropped from 100 feet would break through ice five and one half inches thick but not ice six inches thick. He needed six inches to land on. Since Nelson's ultralight weighed less than an airplane and needed only three inches of ice to land on, he concluded that half a brick dropped from 50 feet would do.

O'er the ice the rapid skater flies,
With sport above and death below

—Pierre Charles Roy,
under a print of skaters,
translated from French
by Samuel Johnson

A few winters ago, a man we'll call Joe Berg went skating on Pleasant Lake on his way home from work. He was by himself. Air temperature was 5°F and the ice was a foot thick. "It was smooth, the most beautiful ice you can imagine," Berg recalls. "I was blasting along on it and the next thing I knew I put a foot down, and it was, 'Oh shoot [not his real expletive], I'm going to die.' It was quick. It went from perfectly solid ice to instantly no ice. I panicked. I thrashed, like anybody who goes into water of that temperature. You are going to die trying. I tried to kick with my skates, but it was like I had no feet. Feet are like paddles but skates have a rounded surface; they don't give a kick.

"I knew where the good ice was; I had skated there. So I followed my tracks, beating the ice with my arms the whole way, until I got to a place where I drove down and my arms didn't go through. It was okay as long as I was swimming, but when I tried to get up on the ice, I slid back and immediately took a dunk. It created a big wave, and all the ice got wet, like greased butter. There was no grip.

"Every time I slid back, my head went under water and I sucked in water. The cold shock makes you breathe in when your face hits water. On the tenth try . . . I don't know how many tries . . . I realized it wasn't going to work and I was going to die. After that I got calm. I decided to take off my skates and use the blades as an edge to drag myself up on the ice. First, I threw both gloves off and lurched up at the ice and gripped it with my fingernails. I slid back in the water, then my fingernails caught on a crack. All I needed was one little crack. I spread my arms so they took the weight evenly. The ice was super-duper thin. It was sagging. I knew I wouldn't have another try. I knew I couldn't let the ice break.

"I got my chest up on the ice and was trying to get my stomach up but I couldn't move any more because I had a hockey puck in my pocket and it caught on the edge of the ice. There was no way to lurch back to free the puck—I'd slip off. I decided to use my right hand to free the puck but I couldn't leave all my weight on just the left hand—that would break the ice—so I laid my face down on the ice, took the weight with my cheek, and reached my hand around and got hold of the puck and freed up the jacket. Once I'd done that, I inched up onto the ice. I did not stand up. I crawled all the way to shore."

One of Berg's mistakes, he later realized, was to skate where a brook enters the lake with a flow so strong that in springtime you can trace its course through the water by the movement of pollen on top. "Springs can weaken the ice," Capossela pointed out in his ice-fishing book. "Currents can weaken the ice. . . . Carp wallows . . . can weaken the ice. . . . If open water is nearby, wave or wind action can weaken the ice." Dams can weaken the ice, and so can pipes under causeways, bends in rivers (one safety manual advises taking extra care "where the Elbow enters the Bow"), and large, dark objects. Rafts, docks, and boats frozen into the ice can absorb enough sunlight that the heat renders ice weak for several feet around them. "Take note, sometime, of where a goose or a swan has just relieved itself," Capossela advises. "Come back six hours later and those droppings may already be two inches down into the ice. Come back three days later and they may be almost through the ice!" Although swan droppings probably won't put you in danger, "a small patch of leaves, a few feet square" just might. One time Capossela was walking on a lake, tapping the ice in front of him with a spud, when he spotted ice that "just didn't look right." Sure enough, "the spud went right through!" Leaves blown onto the lake's surface, he deduced, had melted halfway through the ice and made it "treacherously thin."

O'er the treacherous ice he followed,
Wild with all the fierce commotion
And the rapture of the hunting.
But beneath, the Evil Spirits
Lay in ambush, waiting for him,
Broke the treacherous ice beneath him
Dragged him downward to the bottom,
Buried in the sand his body.

—Henry Wadsworth Longfellow,
"The Song of Hiawatha"

If you realize that the ice you're walking on is very thin or weak, keep your movements slow and smooth and your center of gravity low. You might try the polar-bear shuffle, sliding your feet across the ice with your weight on both at once, if you can manage it, so one heavy foot won't punch through. "As silly as it sounds in winter," Wayne Morris, chief of the Calgary (Alberta) Fire Department, says, you might also consider wearing a life preserver or personal flotation device while on the ice. Whenever Morris goes ice fishing with friends on mountain lakes, they not only wear life preservers but rope themselves together, like climbers in crevasse country, "something it would be almost impossible to get the average guy to do."

Take along a pair of "ice picks," either wood sticks with nails protruding from the ends or else just very big nails. Hang them on a string around your neck or stuff them in a chest pocket. Then if you fall through a hole, you can claw your way onto the ice without having to depend on your fingernails. Keep your clothes on. Experts used to advise undressing because of the weight of wet clothes (Joe Berg says his weighed 50 pounds), but now they advise staying dressed because of the warmth. Don't thrash, if you can help it. "Most clothing traps a fair bit of air," Morris explains, "and the more you thrash, the more you force the air out and you lose flotation." What you should do is face the last ice that held you and lay your arms on top of it or else grab the edge, and if that ice breaks away, reach for the next piece and hold on to it. Keep reaching and holding on until ice does not break, then start clawing. Also kicking, as if you were doing the crawl stroke; the momentum can carry your body partway up onto the ice. Once you're partway up, slowly bring a leg or the side of your body up, too, then roll away from the hole.

Kicking may also help if you're in moving water, as in a river, because it could elevate your body slightly so your center of gravity is high enough as you hit an ice edge that the current won't pull you under the ice. (Ordinarily, only about a tenth of your body floats above the surface of fresh water, as if you were yourself a piece of ice.) Even if you kick, however, your prospects are not great in a river, Capossela warns. "Forget about surviving a fall through ice into a moving river," he states flatly. "It can take no more than six feet of water and an old bluegill pond [bluegill ponds can weaken the ice] to put you on the other side of the Ouija board." For if you fall in, chances are excellent that you will be carried downstream from the hole you went through to a place on the river where there are no holes to come back up through: the stuff of very bad dreams.

If someone else has fallen though and you want to help him, find a tree branch or something else that the person in the water can grab and hold on to while you pull. A hockey stick, belt, board, coat, ladder, rope, sled, or even small boat will do. Stand on shore or lie on good ice and push the branch, rope, whatever, to the victim, or else throw it to him, "underarm, like in softball," a volunteer fireman recommends. The object can even be an inflated fire hose, which the victim can work his way along since the thing floats. "A last resort," Morris says, "is a human chain."

And I—could I stand by
And see You—freeze—
Without my Right of Frost—
Death's privilege?

—Emily Dickinson,
"I Cannot Live Without You"

When trying to save someone else, don't fall through the ice yourself. Every winter you hear about would-be rescuers' having to be rescued, or failing to be rescued. A six-year-old boy fell through slushy ice on a pond in Central Park late one afternoon, about 50 feet from shore. Four joggers, responding to his sister's loud screams, crawled out onto the ice, pushing a ladder before them. All four fell through the ice. One by one they struggled back to shore. "I'm cold, I'm cold, I can't feel my legs," the boy yelled. His head sank in the water, came back up. "Oh Jesus, please please Jesus," his sister cried. Next, a rollerblader crawled out onto the ice, on the side away from where the joggers had gone through. He too fell through the ice, telling himself, he later admitted, "This is going to end up looking really stupid." He lost the feeling in his hands. The boy shouted, "I can't swim, I'm going to drown."

A transit policeman tiptoed across the ice. *He* fell in. Then the rollerblader felt something under his feet: the pond bottom. He worked his way over to the boy and got him halfway onto the ladder the joggers had pushed out. Somebody on shore threw a piece of orange plastic fencing to him, and he caught the end of it. The policeman, still in the water, grabbed the other end of the fencing and pulled. The boy, the ladder, and the rollerblader were about halfway to shore when the city rescue squad arrived, tossed out a life ring, and pulled all of them the rest of the way in. The policeman made it back on his own.

The story has the tone of a farce, but a couple of years earlier a fireman died trying to save a dog that fell through ice in a Central Park pond. Animals aren't a cinch to rescue; for one thing, most of them can't grasp. For another, like many humans, they tend to panic and not respond well to instructions ("kick!"). In Hugh Gray's 1809 *Letters from Canada,* he described the method carriage drivers used to save horses when they fell through ice on the Saint Lawrence River. (On ice, pulling a carriage was usually so easy, Gray commented, that the horses had "little else to do than to get out of the way of the carriage.") The rescue involved strangling the horses, briefly. Before setting out, the driver put a noose around the horse's neck, and if the horse broke through the ice and began thrashing, the driver as well as the passengers got out of the carriage, took hold of the end of the rope with the noose on the end, and tugged "with all their force." Within seconds, the horse, deprived of oxygen, was floating on its side in the water, motionless, and in that docile state could be hauled out onto the ice. There the noose would be loosened, the horse would resume breathing, the driver and passengers would climb back in the carriage, which the horse would once again pull with ease over the ice, "as much alive as ever."

When their vehicles fell through the ice, Sigfusson and Denison went to a lot of trouble to retrieve them; the profits of an entire season of ice-road building could be lost whenever one or two vehicles were. Denison's crew once spent several summer months trying to locate a tractor that had gone through the ice of Great Slave Lake the previous winter, then spent several weeks the next winter pulling the tractor up. Using explosives, they opened a hole in the new ice, Edith Iglauer explained, then lifted the tractor out using gin poles and winches. When the tractor was finally free of the hole and parked on the ice with its motor running and the men were standing in a caboose having coffee after their labors, "they heard a crackin' noise," Denison related, "and they ran outside. There was the Cat, goin' chug, chug, chug, and then clunk! It was gone again. Disappearin' right through that ice and they couldn't do ennathing about it! All that work for nothing! I think what happened, when they blasted the ice they must have fractured it, and the vibration of the motor running shook down through."

Iglauer called Denison "Sisyphus" because of the monumental struggle he went through to build ice roads over and over again, winter after winter, knowing that every single one of them would be gone in a few months, knowing that the heavy weight he had pushed up the hill would, inevitably, roll down it. This seemed more of the same, onto the ice, down through the ice, chug, chug, chug, *clunk!*

CHAPTER FIVE

BREAKUP

The bluebird comes and with his warble drills the ice
and sets free the rivers and ponds and frozen ground.

—Henry David Thoreau, Journal (*Winter*)

AT 7:00 ON A WEDNESDAY morning in March 1992 in Montpelier, Vermont, Lee Duberman of the About Thyme Cafe on State Street was setting out muffins for the breakfast rush of statehouse workers. A few doors away at the First in Fitness health club, Carole and Bill Shouldice were lifting weights before opening their Country Store on Main Street. At that moment, a few blocks south, several chunks of ice slipped quietly from the positions they had been occupying in the Winooski River beside the Pioneer Street bridge and floated away. They passed under the Bailey Avenue bridge, drifted toward a bend in the river, and stopped. According to *Ice and Water,* a brochure put out by the city chronicling what happened that day, "the ice had jammed, shutting the door on the river as decisively as the closing of a bank vault." The waters of the Winooski, kept from pursuing their usual route, took up a new one, up and over the northeast bank of the river, out onto low ground between Main and State streets, then into State and along its side streets. Meanwhile, in a pincer movement, water from the Winooski's North Branch was finding its own new route, running down Elm Street, carrying mud, garbage, ice, and a Dumpster with it.

By 8:00, a mere hour after the first ice chunks slipped their moorings, "the Winooski River had claimed the city," *Ice and Water* reported. Downtown Montpelier was a lake, State Street was a "roiling gray river," and the side streets were "canals." In the water, as if to gloat, "ice floes three feet thick sashayed through town." In some spots, the water stood six feet high. "It rose so *fast,* that's what blew everyone away," Duberman told the *Barre-Montpelier Times Argus.* Employees at the Vermont Historical Society raced to build a wall out of boxes of 1989 tax forms to hold back the water long enough that they could move rare books, quilts, and antiques to high ground. Volunteers at the Savoy Theater, standing in thigh-deep water, had to form a "video brigade" to carry thousands of film tapes out of the flooded basement. Schools were closed, 120 businesses shut down, 200 people evacuated from their homes and businesses. A man and woman, using a shovel and a board for paddles, canoed down Main Street to rescue a friend stranded in his office. On the streets around them were rowboats, inflatable rafts, motorboats, kayaks, and a Jet Ski.

To get rid of the ice blocking the river, officials considered blasting it with explosives, spreading coal dust on it from an airplane to darken its surface so it would melt faster, and pounding it to bits with a heavy instrument. They went with the pounding. While several backhoes clawed at the ice blocks, an 82-ton crane dropped a 26-foot-long I-beam on them, over and over. This dislodged some ice and allowed some water to flow through, but before long a new ice dam formed downstream from the old one and sent water levels in Montpelier to the highest levels of the day. Around 5:15 p.m., the new ice dam gave way and released ice slabs 20 to 30 feet long, which slammed into a railroad trestle bridge, twisting half of it so it faced up and down the river rather than across it. By 6:00 p.m., "all the water had drained out of Montpelier," *Ice and Water* stated, "like the contents of a bathtub with the plug pulled."

Everyone knows about flooding on rivers when they're wide open and wet and running, but few people know about flooding on rivers when they're covered with ice. "*Nobody* pays attention to ice jams," Kathleen White, research hydraulic engineer at CRREL, states firmly. One reason, as CRREL's George Ashton explains, is that ice floods tend to be "very local"—at only one stretch of the river, say, while two miles in either direction nothing is happening—as well as "very brief"; once the ice plug goes, the water goes with it. Yet ice floods can produce water levels far higher than open-water floods do. Each year in the United States alone, ice floods cause many millions of dollars in damage to waterside property, plus unreckoned losses from disruptions to navigation and power production.

There are two kinds of ice jams: floating and grounded. With a floating jam, some water backs up behind the ice blocks but some gets through. With a grounded jam, ice stops the water flow over the whole width and depth of the channel, and the only water that gets through seeps between chunks, as in a rock dam. Ice jams don't tend to form where the river narrows, as you might expect, but where it widens and the bed flattens. That's because in wide, flat stretches the flow relaxes its hydraulic hold on the ice and allows it to settle against obstacles, including bridge piers, sandbars, gravel bars, islands, the outside of sharp bends—and other bodies of ice. Ice works as well as any other obstacle as a collection point for ice, particularly if the ice has been frozen firmly to the banks or bed or underside of the ice sheet, as frazil.

Most ice jams take place in the spring, but they can form any time after freezeup, even during freezeup, when clumps of frazil clog river channels "like giant Sno Cones," as CRREL's Don Haynes put it. In Baltic, Connecticut, in January 1994, the heart of one of the coldest winters in recent memory, the temperature of the air over the Shetucket River shot up 58 degrees in a single day, from −5°F to 53°F. The river responded as if touched with a hot wand. The ice " 'went out' . . . with a snap and a boom and surge," the *Norwich Sunday Bulletin* reported. "Within 10 minutes, residents say, there was four feet of water in the dugouts at Babe Blanchette Memorial Field. A powerful mix of water and ice chunks raced through T. J.'s Cafe." Room-sized ice chunks snapped "full-grown trees 'like toothpicks.' " Children skating on an iced-over ballfield near the river barely managed to skate ahead of an advancing wall of water and ice. "I was the first one out," said a six-year-old. "I'm never going to see a river ever, ever again."

The deterioration of a river's ice cover in spring goes by the violent-sounding name of "breakup" even when the ice simply melts or slips quietly away. Breakup is a very different process from freezeup, ice engineer Bernard Michel pointed out, as well as a more complicated one. For instance:

Freezeups usually work their way upstream, breakups downstream.

Freezeups have a "gradual and easy" character, breakups an "abrupt, rapid," and "steeple-chasing" one.

Freezeups can cause "local inundations of limited extent," breakups "destructive icy floods and waves."

The timing of freezeups is determined by water temperature in the river stream, the timing of breakups by "runoff awakening on the watershed."

Runoff awakening comes about mainly by warmer air temperatures in the watershed area starting to melt the snow cover there. The snowmelt, as well as runoff from the rain that is more likely to fall in spring than snow is, drains into the streams that feed the river, and drains particularly fast if the ground is still frozen. Water level in the river rises as a result, putting pressure on the ice cover from below, causing it to arch and crack. This gradual, small-scale fracturing and weakening of the river's ice cover is called "conditioning" or "pre-breakup."

Another process is going on at the same time as conditioning but at a different rate: the ice sheet floats downstream fairly quickly and splits into floes. Where water flow is slow, cracks in the ice appear near shore, depriving the central part of the cover of its main anchor points and allowing it to float free, while leaving some of the shore ice intact. Where water flow is fast, the ice is often attached to boulders on the riverbed, and the heightened water level causes that ice to crack in a checkerboard pattern. Water rises through the cracks and floods the ice surface, melting the snow on top. Some pieces of the checkerboard also break loose and float downstream and pile up against the upstream edge of a still-sturdy ice sheet in a slower-moving section of the river. Michel called each pile of ice chunks with its supporting ice sheet an "ice-reach." By the end of the pre-breakup period, the river consists of a series of ice-reaches, each holding back a little more water than the next one downstream. In a diagram, the river resembles a staircase, with shallow steps.

Wrapped each in ice the bodies go
Bump-bumping downriver in spring
From the high, cracking hills
Where the encasing's done. For all their numbers, though,
The current-clogging multitudes, there's not a trace
Of any body once the melting's
Taken place.

—Brad Leithauser, *"Sub-arctic"*

The stage is set for the second phase of breakup, often called "the drive," presumably because herds of large, white, moaning objects moving along reminded someone of a cattle drive. There is no drive if there hasn't been much snow or rain to raise water levels, in which case the ice will just deteriorate until it is "gently" pushed along by the water flow, according to Michel. Yet most rivers in cold regions do stage an annual ice drive, and it's usually a tumultuous affair. Unlike a cattle drive, an ice drive tends to take place very fast. The way it happens is that the pressure of the added water, as well as melting of the ice sheet from above (by the sun) and from below (by warm water), weakens the ice in one of the ice-reaches to the point where the ice pile at its upstream end breaks through the ice sheet supporting it, and pieces of the sheet move downstream to the next intact ice-reach, where they are stopped and form a jam. In this way, ice jams form at several ice-reaches, until one of the larger jams gives way, and "its impact carries all others along its course," Michel wrote, "freeing the river of ice in a matter of hours."

The surge of ice-bearing water that opens up the river in one go is sometimes called a "liberating wave." The ice does not move out quietly. "Most river breakups occur when there is still a strong ice cover," Michel observed, and once the cover is converted into very large floes, the floes "are rammed on the banks with a tremendous force to form heaps of ice. Some of these floes are projected into the air, then fall back and break. . . . This general ice movement makes a grinding noise that can be heard well in advance of its coming."

In *Cache Lake Country,* John Rowlands referred to the "thunder of ice going out," which he found "unmistakable," and probably sweet as well, since the sound heralds an event which "next to the geese coming north means more to us than anything else that happens in early spring." "Us" included the Cree chief who once stood with him and watched a violent river drive. "We saw a jam on the rocks," Rowlands wrote, "and the water flooded over the banks and rushed down through the woods in a frothing torrent. The jam broke and great blocks of ice, some as big as a wagon bed, came thundering down the main stream. You wouldn't know there were any rapids then, just tossing gray ice riding water black and cold and powerful."

Ice blocks torn loose during breakup can be even larger than wagon beds, as wide as the river itself. "All breakups start with big plates," Michael Ferrick, research hydrologist at CRREL, points out. But as the plates careen downstream, they collide, spin around, flip over, and slide under and on top of each other, getting broken up in the process. "Ice attacking ice," Ferrick says. "Most rivers behave like a grinder." He compares the momentum of river ice in breakup to that of a moving freight train.

In her book *Uncle Tom's Cabin,* Harriet Beecher Stowe caught some of the drama of a river in dynamic breakup as little Eliza, clutching her young son, made her escape from bounty slave hunters by running across the Ohio River on bobbing ice blocks:

"In early spring, river swollen and turbulent, great cakes of floating ice were swinging heavily to and fro in the turbid waters," Stowe wrote. "On Kentucky side of Ohio river . . . the ice had been lodged and detained in great quantities, and the narrow channel which swept round the bend was full of ice, piled one cake over another, thus forming a temporary barrier to the descending ice, which lodged, and formed a great, undulating raft, filling up the whole river. . . .

"Nerved with strength such as God gives only to the desperate, with one wild cry and flying leap, [Eliza] vaulted sheer over the turbid current by the shore, on to the raft of ice beyond. . . . The huge green fragment of ice on which she alighted pitched and creaked as her weight came on it, but she stayed there not a moment. With wild cries and desperate energy she leaped to another and still another cake;—stumbling—leaping—slipping—springing upwards again! Her shoes are gone—her stockings cut from her feet—while blood marked every step; but she saw nothing, felt nothing, till dimly, as in a dream, she saw the Ohio side."

> We found a number of carcases of the Buffaloe lying along shore, which had been drowned by falling through the ice in winter and lodged on shore by the high water when the river broke up about the first of this month.
>
> —Meriwether Lewis, *Journal,* April 13, 1805

The final stage of breakup is the "wash." Once the drive has swept ice from the river's main channel, there's often ice still connected to shore, with sheared-off edges, "vertical walls of ice of smooth, masonry-like form," in Michel's description. The wash consists of moving water and ice knocking the walls of ice to pieces and bearing the pieces away, cleaning up the mess.

Breakups don't always happen in this way one-two-three. Ashton says, "Break-up is a progression only in a very general sense. Like spring, it arrives at different times in different places." If the mouth of the river is in the south where temperatures are warmer than at the source, breakup may move upstream instead of down. Or the midstretch of a river might be the first to lose its ice if that's where a major tributary enters. Or widely separated parts of the same river could go out all at the same time.

I have never seen a river stage a drive (never heard the thunder), but I did see a river shortly after a drive when there was lots of ice debris along the edges for a wash to tidy up. The river was the Tanana in Nenana, Alaska, where the famous Nenana Ice Classic is held every year. To participate in the ice classic, you lay a $2 bet on the time that the ice on the river will go out, naming not only the day and hour of breakup but the minute as determined by the exact instant that a four-legged wooden tripod, set on the ice in midriver with lines running from it to a clock, a siren, and a butcher's cleaver on shore, moves 100 feet downstream because the ice under it does. When that happens, the clock stops and the siren wails and the butcher's cleaver drops Rube Goldberg–like onto a rope holding the tripod to a support on land.*

The year I visited Nenana, 243,000 people, including some from ice-poor states like Hawaii and Florida, placed almost half a million dollars in bets on when the ice would go out on the river. Freezeup had been late that year and the winter very mild, by Alaskan standards. "We didn't even hit 30 below!" Jan Hnilicka, manager of the ice classic, exclaimed. Yet the ice on the river was 58 inches thick, the thickest she could find in records of the event, which went back to 1917, the year when trappers and fishermen in outlying camps first rode into town to work the pool. She reasoned that the ice was thick because the winter had been windy, and whenever snow fell on the ice it got blown off, leaving the ice exposed to air, "mild" as the air was.

At 11:01:44 p.m. Alaska Standard Time on April 29, a time nobody picked, the tripod toppled. Four people or groups of people who bet on 11:00 and five who bet on 11:02 (including the "2 p.m. Coffee Club of Nome") split $260,000 (the rest went to charity). One of the winners, Robert Sibold, an electrician in Fairbanks, confided that in making his choice he had consulted the almanac, tide charts, and measurements of rainfall in the mountains and applied his own theory about the contribution of adjacent water tables to the river's flow. He also placed 300 to 400 other bets on other times.

In Hnilicka's judgment, what caused the breakup—the *coup de glace,* as it were—was anchor ice popping up off the river bottom. In that stretch of the Tanana, she explained, water near shore tends to freeze all the way down to the bed, "which is what ties the ice down during breakup." The day before, she had seen a piece of dirty ice suddenly appear in what had been a break in the ice. "It had sticks and roots and silt on it," she says. "It must have popped up and turned turtle. That was probably what allowed the whole mass to go."

Five hundred people were standing on the river's edge when the main piece of ice went, and other chunks began following it. "The chunks looked like whales," Hnilicka says, "big animals floating down.

* There are other popular ways of observing ice-outs on northern lakes and rivers, including "plunge contests" in which an old refrigerator or clunker of a car is set out on the ice and bets are taken on which day it will fall in the water.

You could see them crash into each other. People yelled and cheered for these chunks when they crashed. It was very exciting." When I arrived half a day later, I found only a slow, steady stream of ground-up ice floating down the river's central channel. But not far away downstream, there was a large cove with countless ice chunks of all sizes crammed into it, like rubble in a bombed-out city. Violence had produced them, yet they lay dead still: a floating jam. From shore I could see out beyond them in the distance the central channel of mashed-up ice still streaming by, and along with the smaller pieces a large, flat, high-floating white floe, looking less like a whale, I thought, than Huck Finn's raft, drifting free and easy, heading toward who-knows-what adventures farther down the river.

> *breaks up in obelisks on the river,*
> *as I stand beside your grave.*
>
> —Ai, *"Ice"*

Most of the ice blocks along shore were large ones, three or four feet across, yet some were coming apart in the light breeze. Crystals were dropping away from them, falling and shattering, with a delicate, high-pitched, tinkling sound. The blocks were casualties of "candling," a process associated with spring thaws in which sunlight dissects ice into individual crystals by melting the ice between them. Ice in the minuscule spaces between crystals melts before the crystals themselves do because they contain impurities, deposited there by the developing crystals, and impurities lower the melting point of water. The melting-point difference between pure and impure ice can be as little as a thousandth of a degree, yet in this, as in so many other aspects of the story of freshwater ice, a little means a lot.*

The process is called candling because the long vertical crystals of lake or congelation ice that tend to form on the margins of rivers instead of frazil remind some people of bunched candles. They remind Wayne Nelson of "double-wide" pencils. "You can grab a piece of this ice, give it a tap, and it'll break into millions of uniform fragments," he says. "*Billions* of fragments. A single one will float like a pencil will. Think how

* Once the sun starts to melt and enlarge the boundaries, you can see them—and thus the shape of the crystals they enclosed—by laying a piece of paper over bare lake ice and rubbing it with a crayon, pressing the crayon tip into the boundary indentations. The result—a welter of irregular crystal shapes and sizes—might remind you of a map of 19th-century middle Europe.

many of them it would take each standing on edge to form an ice sheet five miles across!"

Candled ice, he points out, is "probably the most dangerous, least forgiving, and fastest-to-change of all ice." After a cold night, during which the crystal boundaries refreeze and the ice sheet regains some of its strength, "it takes very little warming if the sun comes out, as little as 20 minutes, before the ice goes back to being deadly." Even after several nights *and* days of subfreezing temperatures, Nelson claims, "three chops of a chisel held at a 45-degree angle will get you through a foot and a half of candled ice. The crystals will pop right out of their holes, like icicles being squeezed out of your hand."

Contemplating a sheet of candled ice while standing on shore, you could be misled into thinking it would support you. The ice sheet keeps most of its original volume since the amount of ice between crystals makes up only a tiny portion of the total. But looking down at the ice while standing on it, you'd know you're in trouble. "When you drill a test core," declares Ferrick, who has drilled many, "it looks like a bunch of spiderwebs." To some people it looks like a honeycomb, another name for candled ice.

All honey-combed, the river ice
was rotting down below;
The river chafed beneath its rind
with many a mighty throe.

—Robert Service,
"The Ballad of the Black Fox Skin"

When there's still ice on a lake but it's too weak to support a car or even a snowmobile yet not so weak that a boat can get through it, some people get around in "icejammers," also called "windjammers" and "windsleds." Nearly always homemade, an icejammer usually consists of a boatlike shell with wood or Teflon runners underneath, an airplane engine and propeller mounted on a post behind the driver (sometimes with a metal guard to keep the driver's head out of the prop's arc), and a steering wheel connected by cables to an air rudder at the back. If the ice breaks under it, it floats in the water long enough for its engine to carry it up onto better ice. Wortley described the ride he got in a windsled that Wayne Nelson drives when the ice road from Madeline Island to the Wisconsin mainland becomes unreliable: "You're sitting right next to the engine and being choked with gasoline fumes," he explained.

"There aren't any springs, and you feel every little crack, bump, and discontinuity in the ice. When the thing gets up to 70 or 80 miles an hour, it's very [he searches for the word] *exciting.*"

"Then one day," Nelson says, "boom! You've been looking at ice, and the next morning, water. On that day, we once again pass the torch, from the windsled to the ferry."

March 8. I'm still at Pleasant Lake. The betting sheet for which day the ice will go out is posted on the wall of Marshall's Autobody Shop, across the street from the post office. Dave Marshall defines ice-out as the day when in his judgment his boat can get from one end of the lake to the other without touching a single piece of ice. The winter has been mild, by New Hampshire standards, and bettors figure ice-out will be early. On this March day, most of the ice is covered with water, from melted snow, so that it looks unsafe but isn't necessarily. In spring, even good ice can have water on top, sometimes a lot of water. "You're driving in three-inch waves with water splashing all around and you can't even see the ice," Nelson complains.

Although many people assume otherwise, the ice is sometimes safest where the meltwater stands deepest. "On a beautiful day when the snow starts to melt," according to Nelson, "sport fishermen with their four-wheelers and snowmobiles look around for the ice with the least amount of water on it and head for it. But that's probably where the water drained out because the ice is candled all the way through to the bottom!"

March 9. Overnight, the meltwater on top froze, and much of the ice has a shellacked look. With all the temperature changes, the ice cover is making thudding noises, like a big hollow drum.

March 13. In front of my cabin, a band of water a few feet wide has opened up between the land's edge and the ice's edge. Ice on a lake usually melts at the shoreline first, in part because the ice tends to be thinner there than farther out, and in part because the land warms up faster than the lake and gives off heat.

March 19. Before placing my bet, I decide to see how thick the ice is. Using Marshall's auger, I drill a hole about 100 feet from shore, drop a rope with a stone tied to the end through it, pull up on the rope until the stone catches on the underside of the ice. Thirteen inches. Average date of ice-out on Pleasant Lake is April 16. I put my money on April 7.

March 21. I can see holes near the edge of the ice cover so neatly rounded they could have been augered. Meltwater drains through these holes, which probably formed over an old crack. On Lake Superior, if the sun is out, according to Nelson, "by 10 or 11 a.m. you start to see a film of water running into little indentations in the ice." Ice-road builder Svein Sigfusson called such porous ice "wormwood" and wasted no time getting across it, "hauling with all deliberate speed while we still had a road to haul upon."

March 23. The crows are back. They just stand, portentous, black on white, on ice halfway across. Near shore, I notice the ice has long, delicate, white filaments of cracks in it, preserved as if in jelly.

March 27. The ice cover probably resembles a sandwich by now. The top has crusty-looking patches on it, left behind when the meltwater drained off, a few of the patches light blue, reflecting the sky. Under the top layer is probably another one, consisting of meltwater, or else ice made soggy with meltwater. Under that layer, the rest of the way down, is probably good, solid, intact ice. But I'm guessing.

March 30. Like many things in decline, ice is not always lovely. On Pleasant Lake, it is now dirty white, dirty-gray, and dirtier-gray, pocked and dull under a cloudy sky.

April 5. After a night spent rattling the cabin, the wind is gusting to 40 miles an hour. On lakes, strong winds play the role that runoff water in feeder streams does on rivers. They produce currents which put pressure on the underside of the ice, cracking it. A woman at the post office talks about the year that ice on the lake went out in less than an hour. Winds pushed the ice from one end of the lake to the other as if they were sweeping it, she says, then tossed the broken bits on shore, to the accompaniment of clinking sounds, like breaking glass. "Quite musical, really."

Even with the gusty winds, the ice cover on Pleasant Lake seems to be holding fast. I'm beginning to feel proud of its staying power. It is looking increasingly dark, though, as if stained by smoke.

April 6. At 9:30 a.m., I hear a hum. I look out a window of the cabin, thinking the hum might be coming from an idling truck, but I don't see a truck. The hum continues and I keep looking around for something that could be making it. Slowly it dawns on me that the hum—like an ascending growl, growing louder all the time—is coming from the lake. *The ice is humming!* I stand transfixed, as if a tree had talked. Over the winter I had heard single noises, chirps, thunderclaps, glubs; this is a concert. Sounds issuing from all over the ice and mingling into one loud, strange song. This, I think, is the "music of the spheres."

The hum's tone shifts slightly, to a higher muffled one, like a distant squadron of bombers returning to home field. There's nothing ominous about it, no quality to remind a person standing on shore in a state of rapture that the ice cover is a dying entity. Just the steady, reverberating hum. It lasts for two hours. By 11:30 a.m., the lake is quiet again.

April 7. This is the day I bet on. The edge is 30 to 35 feet from shore, but the ice is definitely there. I notice the edge is as sharp as if it had been cut with lawn clippers. I lose.

April 8. At 6:30 a.m., air temperature is 21°F. The area of open water in front of my cabin is half-covered with a skim of ice, one of those paper-thin skims of early winter, with the same sort of ornate patterning. "Although most important during the cooling-off phase," the Langloises wrote, "the quick formation of a skim of ice on quiet water during very cold, still nights . . . may be seen *most often* during the warming-up phase" (italics theirs). By 8:30 a.m., the skim has melted.

April 10. The ice is humming again. I hear wind funneled down a canyon, a jet passing high above, the occasional rat-tat-tat of small-arms fire. The wind must be putting pressure on the whole ice sheet. *Something* will happen. Wind not only causes ice to crack but speeds melting by stirring currents underneath the ice, bringing warm water up. Sometimes the melting is so rapid that people believe the impossible, that the ice sank. A woman I meet at the supermarket tells me that some years the ice on nearby Lake Sunapee "just turns over and goes to the bottom."

April 11. The ice cover is still holding on, but it's pulled so far away from the banks that it resembles an area rug. Like a rug, it is frayed at the edges and lies low, barely sticking up above the surface of the water. By now, it's probably half water. It makes a high, blowing sound. That afternoon, I fly over it in a small airplane and take note of its large, loose shape, its mixed shades of white and gray, its long cracks. It still covers almost two-thirds of the surface, or more than I had thought.

April 12. The ice is gone. When I wake up, the ice is utterly gone. There is no ice on Pleasant Lake. I walk all the way around the shore and don't find a single nugget of ice, not a scrap, not a splinter, not a clue that yesterday afternoon, ice covered two-thirds of the surface. I check the beach and don't see the slightest remnant washed up onto it. There was no wind during the night; "the ice just sank," they'd say. It stole away in the dark. I feel a pang. From one shore to the other the water is the same dull brown shade under a cloudy sky, the wavelets all the same insignificant height. Boring, I think, knowing that this is not a popular sentiment.

I miss it already. I miss the ice.

CHAPTER SIX

ALPS

Supple glass. Fluid stone.

—Thomas Wharton,
Icefields

"LET US APPROACH THE ALPS," John Tyndall beckons his readers in *The Forms of Water,* a book based on notes from a lecture he gave to a young audience ("juvenile auditory") in 1871 on the subject of mountain glaciers. The book's stand-in for the audience was a single pupil of his imagination, "the abstract idea of a boy," for whom the British physicist developed, he was surprised to realize as their joint excursions on the printed pages ended, "an affection consciously warm and real." I picture the little fellow wearing lederhosen and a cap knitted by his mother and smiling shyly when his teacher announces on one of their climbing jaunts, "We reach a place called the Chapeau, where, if we wish, we can have refreshment in a little mountain hut."

The purpose of the jaunts, onto several of the great Alpine glaciers—the Mer de Glace, the Unteraar, the Talèfre—was to introduce reading and listening audiences to the sights, sounds, and even touch (underfoot) of glaciers and to explain something of their makeup and behavior as scientists had come to understand them. For by 1871, the new science of glaciology* was in full swing, and so many investigators of the ice were crawling over glaciers, trying to pry loose their secrets, particularly secrets from eras when glaciers in the Alps were thought by many to have been far larger, that there was talk of "glacier-madness." Laymen too were crawling over glaciers; when Mark Twain visited the Alps a few years after Tyndall's lecture, he found "gangs of excursionizing tourists" ascending Mont Blanc by way of the Glacier des Bossons and its giant-icicle-fringed "buttresses of ice."

* Glaciology is, strictly speaking, the study of ice of all kinds, but it is generally used to mean the study of glaciers.

One day Tyndall and his small companion set out for the Mer de Glace, carrying alpenstocks no doubt. From the village of Chamonix, in Savoy, it is but "a short hour's walk" to the source of the Arveiron River—which turns out to have been recently *of* ice, and to issue from ice. Meltwater (it is summer) exits through an arch at the end of the Glacier des Bois, the lower portion of the Mer de Glace, so much meltwater that it feeds the river. The ice arch must have beckoned because Tyndall warns, "Do not trust the arch in summer. Its roof falls at intervals with a startling crash, and would infallibly crush any person on whom it might fall." The hikers skirt "curious heaps and ridges of debris" in front of a high ice wall at the glacier's end, then climb the slope beside it and survey the scene ahead.

"The glacier descends a steep gorge," Tyndall reports, "and in doing so is riven and broken in the most extraordinary manner. Here are towers, and pinnacles, and fantastic shapes wrought out by the action of the weather, which put one in mind of rude sculpture . . . From deep chasms in the glacier issues a delicate shimmer of blue light.* At times we hear a sound like thunder, which arises either from the falling of a tower of ice, or from the tumble of a huge stone into a chasm." So chaotic is the glacier surface that even "the best iceman would find himself defeated in any attempt to get along it"; therefore, the two hew to the edge of the glacier for a while, climbing partly on rock next to it and partly on rock on top of it—"a ridge of singularly artificial aspect."

When they reach the Mer de Glace (a not quite appropriate name, Tyndall gently chides the name givers, since "the glacier here is much more like a *river* of ice than the sea"), they spot, perched on a crag on the opposite side, the Montanvert inn, "well known to all visitors to this portion of the Alps," and head for it. The ice on the way there is less chaotic, and at first it's dirty, "but the dirt soon disappears, and you come upon the clean crisp surface of the glacier. . . . From a distance it resembles snow rather than ice," a resemblance caused by "minute fis-

* The reason that clear ice appears blue is that it absorbs most of the red and orange light that enters it. "White minus orange-red is blue," says Charles Braun, professor of chemistry at Dartmouth College (who with a colleague showed that absorption is caused not by hydrogen bonding, as had been thought, but by the vibrations of the water molecules themselves). In Antarctica, there's a type of ice known officially as "blue ice"; it is old, cold, hard, and generally bare, from winds that constantly sweep its surface clear of snow. "The color of a windy sea and calm milk, equally mixed," Michael Parfit wrote of the blue-ice blue. Areas of Antarctic blue ice can be as big as 400 square miles or as small as a turkey platter.

Ice that appears white usually has a lot of bubbles in it, which reflect the incoming light.

suring" of the surface under the heat of the sun. "*Within* the glacier," he says reassuringly, "the ice is transparent."

From the inn (where perhaps they pause for a little more refreshment?), the prospect is grand. "Looking straight up the glacier the view is bounded by the great crests called La Grande Jorasse, nearly 14,000 feet high." (Twain, looking straight up the same stretch of glacier from the same famous inn, is more cordial to the idea of glacier as *mer:* "At this point it is like a sea whose deep swales and long, rolling swells have been caught in mid-movement and frozen solid.") Facing east, the two climbers trek up the tributary Glacier du Géant, on ice that is less compact than the ice farther down, "but you would not think of refusing to call it ice." Ahead looms an icefall, the ice counterpart of a waterfall. As the slope steepens, "the ice becomes more and more fissured and confused. We wind through tortuous ravines, climb huge ice-mounds, and creep cautiously along crumbling crests, with crevasses right and left."

Using an ax, Tyndall carves steps in the steep ice walls (in those days some people made their living carving steps for gentlemen climbers), and by swerving from one side of the glacier to the other—the center was impassable—they reach the top of the cascade. There they find still more cracks in the ice, which "yawn terribly." They cross them on bridges of snow, with care, and rope, and thoughts of mortality. "For not only are the snow bridges often frail, but whole crevasses are sometimes covered, the unhappy traveller being first made aware of their existence by the snow breaking under his feet. Many lives have thus been lost, and some quite recently."

Continuing upward, they find themselves on snow, not ice. The snowfield is a relatively smooth, gently inclining plateau. After a long ascent of it, they stand at last on the Col du Géant and "now look over Italy."

It is the moment for celebrating science. The foundation of all we know today about glaciers, living and long dead, was laid in the nineteenth century in these Alps, by robust climber-scholars like Tyndall, and by Tyndall himself. "They were no idle scamperers on the mountains that made these wild recesses first known," he lectures his by-now-probably-tuckered-out companion. "It was not the desire for health which now brings some, or the desire for grandeur and beauty which brings others, or the wish to be able to say that they have climbed a mountain or crossed a col, which I fear brings a good many more; it was a desire for *knowledge* that brought the first explorers here."

Although locals knew otherwise, since their forebears had seen glaciers impinge on their farms and settlements, earlier in the 19th century most people, including scientists, considered glaciers to be little more than picturesque oddities. In a lecture to a New York audience a few years before Tyndall's, Louis Agassiz, a Swiss biologist, recalled that glaciers were thought of as a mere "accident in the study of physical geography . . . curious accumulations of snow formed on the summits of certain mountains" which added to the beauty of the scenery. It was one local, a chamois hunter, who got the idea that glaciers must once have filled the entire valley where he hunted instead of only part of it and who shared his idea with a bridge engineer, who in turn presented the idea at a scientific meeting, where it was met with indifference (the Great Flood was thought responsible for scoring rocks and moving them about). The engineer did win one convert, however, a Swiss naturalist, who eventually converted the skeptical Agassiz, who took up the cause with a vengeance and converted other scientists.

In *A Tramp Abroad* Twain expressed the point of view of the "scoffers": "You might as well expect leagues of solid rock to crawl along the ground as expect solid leagues of ice to do it." Scoffing would come easy to anyone who hadn't spent time in glacier country; unlike rivers of water, rivers of ice flow so slowly they're rarely caught doing it. Trying to see a glacier actually in motion has been likened to trying to see the hour hand on a clock moving. Yet in the Alps, evidence of glacier motion is abundant, as explorer-scientists discovered, and as Tyndall points out, in a strategy he called "Questioning the Glaciers":

• First line of evidence for movement: the gently sloping snowfield that Tyndall and his friend climbed below the Col de Géant was not piled as high as Mont Blanc (at 15,771 feet, the highest mountain in the Alps). All glaciers form from snowflakes, which are transformed over time, by compression and recrystallization, into ice.* Some of the snow that falls in the upper, coldest reaches of mountains during winter melts the following summer, but more remains, and the next winter and sum-

* Snow changes slowly into ice, the colder the setting, the slower the change. First, new-fallen snow crystals settle within a fluffy snowpack, moving closer to each other and becoming rounded as molecules evaporate off their sharp points and condense in their hollows (where there are more molecules for them to join up with). These small, round crystals, which, once they have lasted through a summer melt season, are called *névé* in French and *firn* in German, bond together, or sinter, under the weight of more snow above them as well as from the continued migration of vapor molecules, until eventually all air channels are blocked. When there remain in the pack only individual air bubbles, unconnected to each other, the permeable snow has become impermeable ice.

mer this happens again, and so on, year after year. Tyndall calculated that if the snow were to keep piling up at, say, two feet a year, in 5,000 years the snowfield below the Col de Géant would stick up higher than Mont Blanc. But it does not.

Meanwhile, on the lower, warmer parts of the glacier, not only does more of the snow that falls there during winter (and in scanter quantities than at higher up) melt but so does some of the ice that underlies the snow, to form "running brooks which cut channels in the ice, and expand here and there into small blue-green lakes," as well as perhaps a river that pours through an ice arch. Since in the lower reaches the amount of snow that falls is not enough to make up for the amount of snow and ice that melts there, that portion of the glacier should waste away entirely before long. But it does not.

The reason neither of these things happens is that ice moves from the area of accumulation (high up) to the area of wastage (lower down), or as today's glaciologists, who are fond of using terms from economics, put it, from the area of income to the area of expenditure. The ice flows downhill, under the force of gravity. If you were to follow a single crystal from the instant it fell near the mountain peak to the instant it was changed into a water droplet that fed the Aveiron River at the bottom, you would see the crystal slowly being buried in the upper part of the glacier as snow fell on top of it; changing with others into ice; traveling the length of the glacier; and being exposed at the bottom when ice above it melted or evaporated away.* In the Mer de Glace, the ice crystals that form the inviting but untrustworthy arch at the end of its companion glacier might have begun life as snowflakes high up the mountain hundreds of years earlier. As the crystals aged and descended, they grew until at the end some might have been as big as basketballs.

• Second line of evidence for glacial movement: those "curious heaps and ridges of debris" that Tyndall and his little partner skirted at the end of the Glacier des Bois, plus the "ridge of singularly artificial aspect" they trod on beside it. Both ridges are "moraines," a French word, as many glacier terms are ("glacier" itself a Savoy adaptation of *glacies,* the Latin word for ice). "A stupendous, ever-progressing, resistless plow," Twain wrote of a typical Alpine glacier, although if he were writing today he might call it a bulldozer or an earthmover. A glacier, in motion, shoves rocks and pebbles before it, some of which it has itself scraped off

* Ice can lose substance by evaporation even when temperatures are so cold that it can't lose it by melting; think of shirts on a clothesline on a winter day, frozen but dry.

the bed and crushed with its great weight. (A cubic foot of ice weighs nearly 60 pounds.) The glacier pushes the rocky debris at its foot up into "terminal" moraines which can be 200 or 300 feet high. To Twain, a terminal moraine that stretched all the way across a gorge looked like a "long, sharp roof."

Glaciers also produce "lateral" moraines, by pushing rocky debris off to either side, as well as by collecting boulders that have plummeted off mountain peaks. "A moraine is an ugly thing to assault head-first," Twain wrote of the 100-foot-high lateral one on the Mer de Glace he had just finished assaulting. "At a distance it looks like an endless grave of fine sand, accurately shaped and smoothed; but close by, it is found to be made mainly of rough boulders of all sizes, from that of a man's head to that of a cottage."

It is the "medial" moraines, however, that illustrate best the fact that glaciers are not static masses of ice adorning mountains. "How does the debris range itself upon the glacier in stripes some hundreds of yards from its edge," Tyndall asks, "leaving the space between them and the edge clear of rubbish?" When two glaciers meet, he answers himself, they lay their two lateral moraines together to form a single medial moraine, which continues on downhill. Like "a dark wall separating two white rivers," A. E. H. Tutton, author of the 1927 book *The High Alps: A Natural History of Ice and Snow,* said of a medial moraine he regarded on the Aar glacier. (After studying the Alps for 23 summers, Tutton concluded that "no more beautiful nor interesting natural phenomena can be conceived than those displayed by solidified water in its various manifestations.") Most mountain glaciers have tributaries, and many have many tributaries, so on a trunk glacier you often see a whole series of medial moraines, resembling, Tutton wrote, "great cart roads on the ice."

These moraines can be of "very different colours," he noted, "as the rocks composing them will probably come from widely separated parts of the glacier amphitheatre, and even two sides of the same medial moraine may differ in appearance." The left half of the medial moraine on the Aar is composed of dark mica schist while the right half is white granite, Tutton wrote, "the two halves thus standing out quite clearly in colour and general appearance from each other. An extreme case is the Baltora Glacier in the Hindu Koosh, which has no less than fifteen different moraines of very different colours."

Where Tyndall and the boy crossed the Mer de Glace, medial moraines stood 20 or 30 feet above the general level of the glacier, but most of the added elevation came from ice, not stone. "A great spine of ice . . . ran along the back of the glacier"; the rocky topping is "superficial." Before the reader can even imagine such a thing, Tyndall explains that the ice did not swell upward to produce the spine; it merely failed to melt as much as the glacier around it since the rocky debris was thick enough to protect it from the sun's rays.

• Third piece of evidence for glacier movement: the riven, broken surface. On their way home after a "hard day's work upon the Glacier du Géant," Tyndall and his climbing partner hear an explosion "under our feet, as if coming from the body of the glacier. . . . Somewhat startled, we look enquiringly over the ice. The sound is repeated, several shots being fired in quick succession. They seem sometimes to our right, sometimes to our left, giving the impression that the glacier is breaking all round us. Still nothing is to be seen." They spend an hour in a "strict search" before, cleverly tracing some rising air bubbles in a meltpool on the glacier's surface, they find a crack at the bottom of the pool. They trace the crack a long distance from the pool on either side, although it is almost too feeble to be seen at times, and "at no place is it wide enough to admit a knife-blade."

Contemplating the minuscule crack, Tyndall muses, "It is difficult to believe that the formidable fissures among which you and I have so often trodden with awe . . . the great and gaping chasms on and above the ice-falls of the Géant and the Talèfre . . . could commence in this small way. We are thus taught in an instructive and impressive way that appearances suggestive of very violent action may really be produced by processes so slow as to require observations to detect them."

The crack and explosion announced the birth of a crevasse (another French word, meaning in that language any crack, even in skin, but in English used most often for cracks in ice). Such a crack, Tutton wrote, "gradually widens until it becomes a broad chasm, shimmering in the sunshine with a blue light, delicately pale near the surface of the glacier, but deepening to a dark Prussian blue or indigo in the depths." Usually V-shaped and usually much deeper than it is wide, a crevasse appears where the stiff crust of some part of a glacier is pulled apart under tension. For instance, when ice passes over a large bump on the bed or when the valley forces the ice into a sharp turn, the ice splits to relieve the stress. The sides of a glacier are nearly always crevassed because the friction of the valley walls retards the bulk of the ice. Oddly, or so it must seem to the novice observer, these "marginal," or lateral, crevasses point *upstream* rather than downstream, at about a 45 degree angle, "giving the erroneous impression that the sides of the glacier are moving faster than the center," according to Tutton. The impression can be corrected by keeping in mind Tyndall's maxim that the break occurs *across* the line of tension; thus an up-angled crevasse indicates a down-angled pull, which makes sense if there is drag at the edges. The scenario is complicated somewhat by the fact that, over time, as the up-angled marginal crevasses are pulled along by the swifter central flow, they rotate *downhill.*

Crevasses also open up *across* the glacier because ice is being pulled *down* the glacier. Some of these "transverse" crevasses appear where the rockbed suddenly steepens and the ice at the bottom of the glacier speeds up, to as much as ten times its previous rate. As it passes over the brow of the dropoff, "the glacier breaks its back," Tyndall tells his young charge, and as fresh ice keeps on passing over the brow, the glacier keeps on breaking its back. The glacier certainly looks broken: the ice there is fractured, jumbled, "confused." It looks particularly confused if it's being pulled from directions other than the downhill one, perhaps because the bed is uneven or the channel widens, in which case crevasses cut across crevasses, creating checkerboard patterns of cracks. Ice blocks defined by the checkerboarding may become, as the cuts deepen around them, the "rude sculpture" Tyndall remarked on as the Mer de Glace descended a steep gorge.

This sculpture—the characteristic feature of an icefall—takes the form of "towers, pyramids, pinnacles, and grotesque or beautiful shapes of every possible kind and variety," Tutton noted. At least one shape, possibly grotesque, provoked the natives of Chamonix to name the towers "seracs" for their resemblance to "an inferior kind of curdy cheese" (Tyndall's footnote). Tall seracs are often called "gendarmes" (policemen), Tutton pointed out, and are particularly dangerous to climbers because they often topple over when their bases melt in the sunlight.

Once crevasses are open, they don't always stay open. At the base of an icefall, for instance, where the slope angle is abruptly reduced, the ice comes under pressure instead of tension. Tyndall asks the boy to demonstrate this. "Lay bare your arm and stretch it straight. Make two ink dots half an inch or an inch apart, exactly opposite the elbow. Bend your arm, the dots approach each other, and are finally brought together. Let the two dots represent the two sides of a crevasse at the bottom of an icefall; the bending of the arm resembles the bending of the ice, and the closing up of the dots resembles the closing of the fissures."

• Fourth line of evidence for glacial movement: corpse transport. "Here at the head of the Grand Plateau, and at the foot of the final slope of Mont Blanc, I should show you a great crevasse, into which three guides were poured by an avalanche in the year 1820." (It is of course a different crevasse Tyndall points to in 1871, yet a crevasse in the same place.) The crevasse was so deep in 1820 that the guides could not be rescued, and gradually it filled with snow and closed over them. Decades passed, during which scientists learned more about glaciers, enough so that one of them, English geologist J. D. Forbes, had the temerity to predict not only that the guides would reappear but when—35 or 40 years after they vanished—and where—at the foot of the glacier.

On August 12, 1861, another (luckier) mountain guide discovered, at the edge of a crevasse at the foot of the Bossons Glacier (according to Stephen d'Arve's *Histoire du Mont Blanc,* as condensed and quoted by Twain), several dreary objects: three human skulls, or parts thereof; a jaw "furnished with fine white teeth"; tufts of black and blond hair; a forearm (flexible); a hand, all fingers intact, blood visible on the ring finger; a left foot, "the flesh white and fresh"; portions of waistcoats, hats (one straw, one felt), and hobnailed shoes; a pigeon's wing (one of the guides had planned to free a cageful of pigeons on reaching the summit of Mont Blanc); a boiled leg of mutton, which, though odorless when extracted, soon was not.

Two guides who had survived the avalanche of 1820, elderly by 1861, identified the objects as belonging to their long-lost colleagues, with one of the old men clasping the resurrected hand, grateful for the chance to touch again the flesh of his brave friend before he too quit this world. A bloodstained green veil emerged from the ice the next year, Twain quoted d'Arve, as did a second arm, whose "extended fingers seemed to express an eloquent welcome to the long-lost light of day."

The disinterment by glacial wastage was, according to Tutton, "a wonderful [!] confirmation of the work of Forbes, and caused great interest at the time." In 41 years the bodies had traveled 9,250 feet down the mountain, or about 225 feet per year. "A dull, slow journey," Twain considered it, "which a rolling stone could make in a few seconds," but a journey nevertheless.

An even earlier trip, made by a scientist's hut, provided the hut's owner, a Swiss professor named Hugi, the chance to make what were probably the first measurements of glacier motion. Seven years after the three guides fell in one crevasse and 34 years before they emerged from another, Hugi had built a small hut on a medial moraine of the Unteraar with a view to studying that glacier. When he came back three years later the hut was 330 feet farther downhill; he made a note of it. Thirteen years after Hugi, Agassiz built another hut on a lateral moraine of the same glacier and began his more systematic investigations of glacier motion. He made his hut *out* of the moraine, using the projection of a giant boulder for a roof, smaller stones for two of the sides, and a space under a second large boulder for the kitchen area. (Ice for chilling the wine was in the backyard.)

Over the years that followed, Agassiz and the staff he'd assembled at his "hotel" on the Unteraar, as well as Tyndall and Forbes, who were working diligently on the Mer de Glace and other glaciers, determined, among many things, that:

1. The middle of a glacier moves faster than the sides. Deceived by the upstream slant of the marginal crevasses, Agassiz thought at first the opposite was true. Then he and colleagues drove stakes ten feet deep into the glacier in a straight line across it (something curious locals had probably been doing for years, using boulders laid on the surface instead), and when Agassiz returned a year later he found the straight line had become a curve. The line was higher upstream at the sides because of drag on the sides (and, it was later determined, because of the greater thickness of the center).

On the Mer de Glace, Forbes demonstrated pretty much the same thing, and what's more, according to Tyndall, he showed that it isn't necessary to wait a year or a week or even a day to determine the motion of a glacier. Using a carefully adjusted theodolite, a small telescope that determines horizontal and vertical angles, Forbes checked the position of the stakes against background features and detected changes hour to hour, during which time the ice appeared absolutely immobile to the naked eye.

2. The underside of a glacier moves more slowly than the top. After planting stakes in an exposed wall of ice at three levels, top, bottom, and somewhere in between, Tyndall determined that the top stakes moved more than twice as far in an hour as the bottom ones, and he credited the difference to friction from the rockbed: "the upper portions of the ice slide over the lower ones."

3. The higher reaches of a glacier move faster than the lower ones. When Tyndall set out three lines of stakes across the Morteratsch glacier, "high up," "lower down," and "lower still," then measured their downhill progress, he found that in 100 hours the distance traveled by the first line was almost double that of the third. "The upper portions of the Morteratsch glacier are advancing on the lower ones." This inspired him to address a much-discussed question: "whether a glacier is competent to scoop out or deepen a valley through which it moves." Some scientists, observing that the snout of the Morteratsch was "sensibly quiescent," concluded that there had not been any scooping in the past. "But those who contended for the power of glaciers to excavate valleys never stated . . . that it was the snout of the glacier which did the work." (Snout, foot, toe, even butt, these projecting parts of the body have all been used at some time or other to denote the front end of a glacier.) The swifter ice in the upper part of the Morteratsch could presumably do the work.

4. ***In a sinuous glacier, the fastest motion is on the side toward the outside curve.*** With the usual stakes set in lines across the usual glacier (Mer de Glace), Tyndall determined that the point of swiftest motion switches from one side to the other of the central axis according to the way the valley is curved, creating a line of maximum motion more deeply sinuous than the valley itself. "Can it be then that this 'water-rock,' as ice is sometimes called, acts in this respect also like water?"

5. ***A glacier moves more slowly in winter than in summer.*** "I will not . . . ask you to visit the Alps in mid-winter, but if you allow me, I will be your deputy to the mountains," Tyndall tells his young charge. Just after Christmas in the "inclement" winter of 1859 ("snow near Chamonix so deep that the road fences are entirely effaced"), he and four colleagues, "wrapped at intervals by whirling snow-wreaths," drove two lines of stakes into the Mer de Glace near the famous inn, and the next morning—"spangles innumerable . . . filling all the air"—measured the stakes' progress downhill. They found it to be "half the summer motion," presumably because there was less melt.

Thus did Alpine hiker-scientists show conclusively not only that glaciers moved but that they moved variously, within their own masses, some parts passing over others, or pulling apart from others, ice spreading out, turning corners, thinning and thickening, speeding up or slowing down according to the seasons or where it lay on the mountain slope. In their search for answers, the investigators even crawled *inside* glaciers, descending into vertical "moulins" (from the French for grinding mills), deep shafts in the ice created by meltwater streams on the surface flowing into cracks and eroding the sides, "the crack thus becoming the starting point of a funnel of unseen depth," Tyndall reports, "into which the water leaps with the sound of thunder." On the Unteraar, Agassiz had himself lowered into a deep moulin, discovered ice-cold water at the bottom when his feet got wet, and as he was being hauled up narrowly avoided being stabbed by sharp-pointed ice stalactites protruding from the dripping walls.

Yet all their findings left unanswered the question of *how* it is that glaciers move, by what mechanisms they make their way downhill. What is it in the nature of "water-rock" that allows it to *flow*?

Scientists proposed, among other things, that it was water freezing in fissures and expanding "with resistless force" that propelled glaciers downhill (Agassiz briefly held this view); that glaciers moved by sliding over their beds; and that glacial ice was not rigid but like softened wax, "a heap of coagulated matter" pressing downward. Tyndall came to accept Forbes's argument that glacier ice behaves as an "imperfect fluid": "We saw the branch glaciers coming down . . . , welding themselves together, pushing through Trélaporte, and afterwards moving through the sinuous valley of the Mer de Glace. These appearances alone . . . were sufficient to suggest the idea that glacier ice, however hard and brittle it may appear, is really a viscous substance, resembling treacle, or honey, or tar, or lava."

The question would be hotly debated for another three-quarters of a century before there was consensus among scientists about how glaciers move (which turned out to be more like molten iron than water or wax), and even then there were unanswered questions, some of which are still being addressed today.

"Here, my friend, our labours close," Tyndall concludes. "It has been a true pleasure to me to have you at my side so long. . . . You have been steadfast and industrious throughout, using in all possible cases your own muscles instead of relying upon mine. . . . And should we not meet again, the memory of these days will still unite us. Give me your hand.

"Good bye."

CHAPTER SEVEN

SURGING GLACIERS

The sheerly steadied stubborn tons of it

—Brad Leithauser, *"Glacier"*

THE TYPICAL SPEED of a mountain glacier (known also as an "alpine" or "alpine-valley" glacier even when it is not in the Alps) is much, much slower than the ooze rate of a slug crossing a spinach leaf: 1/10th or 1/20th of an inch a minute, three to six feet a day, 100 to 700 yards a year. A not untypical speed of a continental ice sheet—glaciers that instead of flowing down a mountain valley spread out like pancake batter in all directions and cover large areas (10 million acres at least) and which exist today only in Greenland and Antarctica—is a mere fraction of that. "Glacial" in English signifies movement that is not only slow but extremely slow, often maddeningly, exasperatingly slow—obdurate, sluggish, impassive, plodding. When Mark Twain, eschewing a mule, tried using the Gorner glacier in the Swiss Alps as a means of quick, cheap transportation to Zermatt, having read in his guidebook that the glacier was "moving all the time," he was "outraged" at its piddly progress.

Stepping onto the glacier, "[I] chose a good position to view the scenery as we passed along," he wrote. "I stood there some time enjoying the trip, but at last it occurred to me that we did not seem to be gaining any on the scenery." Consulting his guidebook again, Twain learned to his disgust that the Gorner travels at less than an inch a day. "Time required to go by glacier," he calculated, "*a little over five hundred years!*" As for his baggage, which as a cost-saving strategy Twain had placed "on the shoreward parts, to go as slow freight," since according to his guidebook the center of the glacier moved the fastest, *it* couldn't be expected to arrive in Zermatt until around the beginning of the 25th century!

A mule can always outrun ice, yet there are some glaciers that move at more than stereotypically glacial rates; by the standards of their species they are sprinters, grasshoppers among slugs. Some poke along for decades then have sudden bursts of speed, during which they flow at up to 100 times their usual rate, before returning to their former indolent pace. Some fairly sprint all the time, cramming the sea at their bases with chunks of broken ice. Some have sections that sprint while the rest of the mass remains dilatory. (To be considered a glacier and not just a patch of ice and snow, a glacier should move, at least a little; some advance only a quarter of an inch a year. Since a glacier moves downhill by gravity, it needs to be thick and heavy enough to offset the strength of the ice—that is, the tendency of its crystals to interlock—as well as the friction effect of the rock underneath; one suggestion of a minimum thickness is 60 feet. Also, to be considered a glacier, an ice patch should in some scientists' opinion cover a minimum surface area; some suggest 25 acres as a lower limit, some a hectare.)

Glaciers move in two main ways (the long, heated debate over mechanisms of motion were resolved to most glaciologists' satisfaction shortly after World War II). They move by (1) sliding over their beds and (2) deforming internally, or "creeping." It was the creeping that took a long time to sort out. Metals and minerals creep under stress when they're molten, or heated to near their melting points, and glaciologists hadn't been thinking in terms of metal and minerals but of water and those imperfect fluids, tar, honey, and treacle. A glacier is by nature near its melting point already, and the stress that the weight of overlying ice puts on crystals beneath it causes layers of atoms within those crystals to line up parallel to the glacier surface. On a slope, under the force of gravity, the layers start slipping over each other, like playing cards in a deck. This is creep, and all glaciers do it, slowly and continuously. The thicker, warmer, and steeper the glacier, the faster the creep.*

Not all glaciers slide over their beds, however. Sliding requires lubrication, and often in very cold places glaciers can't produce even a thin film of meltwater underneath themselves. They are frozen to their beds and move only by creep. (All the ice in a "warm," or temperate, glacier is about 32°F, except possibly on top next to wintry air, while in a "polar" glacier the ice is well below 32°F, except perhaps on top next to summery air.) Still, most glaciers move in both ways, some sliding more

* Creep is what caused a beam of ice hung in a cold room with a weight on the bottom to stretch taffy-like to more than twice its length (see Introduction). Even at much smaller levels of stress, ice will creep; a piece of ice in your refrigerator will, under the action of gravity, flow "like honey," according to one scientist, if you could observe it for a long enough time.

and creeping less and vice versa. According to one study, 75 percent of the flow of the Athabasca Glacier in the Canadian Rockies comes from sliding over the bed, while no more than a few miles away in the Saskatchewan Glacier only 20 percent does.*

In 1993 and 1994, Alaska's huge Bering Glacier fairly raced over its bed. The acceleration opened surface crevasses 50 miles upglacier and, according to ice-penetrating radar, snow-buried crevasses ten miles higher than that. What had been a low, gently inclining terminus turned into a steep cliff of ice which plowed forward and mowed down vegetation, overran islands, and littered several lakes with ice rubble. Toward the end of the event, the glacier released enough water, one journalist calculated, that if distributed evenly around the world it would provide every person with three glassfuls to drink (if no one minded a little sediment). In under two years, the glacier moved about five miles closer to the sea, then stopped dead in its tracks and started retreating, fast.

Only in the last few decades has such odd behavior been recognized as a normal (if small) part of the total glacial repertoire, even though glaciers have been doing it for centuries, maybe millennia. In 1906 Alaska's Variegated Glacier was seen advancing rapidly, but no one knew what to make of it or any of the half dozen or so other lunges reported afterward. Then one day during the International Geophysical Year (IGY)† 1957–58, hydrologist Austin Post was flying by small plane up the Muldrow Glacier on the northern slope of Mount McKinley (now Denali), taking photographs as part of an American Geographical Society project to map the small glaciers of Alaska. With a camera mounted in a hole in the side of the plane's fuselage, he focused on some startling sights: fringes of ice, vestiges of the glacier's former surface, clinging to

* When a glacier that's sliding reaches a small obstacle on the bed, a bump or a pebble, the ice sometimes gets around the object by changing phases. On the upstream side of the bump, the ice liquefies because increased pressure from the weight of the glacier against the resistant object lowers the ice's melting point. Then a very thin layer of meltwater flows around the bump, and on the downstream side where the pressure is relieved and the melting point raised, it turns back to ice. This is called "regelation."

Regelation is also responsible for that miraculous-seeming research feat in which a metal wire goes all the way through a block of ice yet leaves it whole instead of in two pieces (see Introduction). A wire is looped around a block of ice and hung with a weight. Pressure on the ice directly under the wire induced by the weight causes that ice to melt; the wire moves downward through the melt; the melt refreezes above the wire where the pressure is relieved, sealing the slit shut. The weight falls to the floor with a clunk.

† The first International Geophysical Year, which actually lasted a year and a half, was a cooperative effort by scientists around the world to study in a systematic way, using new technologies (including rockets), various aspects of the earth and its surroundings, glaciers included.

mountain walls 400 to 500 feet above the present surface (Post had already heard "excited tales" from park officials about the upper Muldrow having "collapsed" or "fallen in"); extensive crevassing; and a fat, bulging terminus. Obviously, a lot of ice had been "dumped" and had "slid down the mountain" in a very short time. Post realized that he was in the rare and privileged position of catching a glacier in the act of "sudden, exceptional advance," what he and others would soon call "surging."

Later, by examining other people's aerial photographs (including those of mountaineer Bradford Washburn) as well as his own, some of the photographs decades old, Post found evidence that surges had taken place on 204 glaciers in south central Alaska and the Yukon, including the Muldrow and the Bering. A key piece of photographic evidence of surges past was "looped moraines." Some medial moraines had not gone straight downhill parallel to the sides of the glacier, like A. E. H. Tutton's great cart roads, but had instead been pulled out in places into "large, bulb-like loops," which sometimes stretched nearly the width of the glaciers. Tellingly, nearly identical loops appeared at regular intervals on the moraines. Post surmised that each loop was produced by a tributary glacier flowing into the main glacier while it was moving very, very slowly, then the main glacier speeding up and carrying the loop downslope with it. When the main glacier came almost to a stop, a new loop formed where the tributary emptied into it, and later when the main glacier speeded up, or surged, it carried that loop downhill, and so on, until several of the loops merged into a debris pile.

Using these "really dandy" markers, Post says, plus other, more subtle evidence for rapid advances, he concluded not only that surges had occurred on certain glaciers in the past but that they had occurred repeatedly, and what's more that they had occurred *regularly.* By estimating the distance between loops, he could get a rough idea of how much time had passed between surges on a particular glacier and work out its surging cycle. Each glacier turned out to have its individual cycle, surging anywhere from once every 15 years to once every 100, "almost like a clock."

On the Bering Glacier, Post found "folding moraines." Because of the Bering's great size, its tributaries as they enter the slow-moving main glacier produce smaller irregularities in the medial moraines than appear on smaller glaciers. As the ice carrying these irregularities during a surge emerges from the mountain valley and enters the coastal plain (an area four times wider), it spreads out, and the folds are stretched sideways at the same time that they are compressed by the force of ice flowing in behind them; the effect is like an accordion being closed, according to Post. He also found zigzagging folds in the marginal moraines, "kinks" caused by drag at the edges, each kink probably laid down by a separate surge.

Why, though, would a glacier speed up, or surge? Post wondered, and so did just about everybody else. Why, after crawling along on its belly like an ordinary well-behaved glacier for 30 years, or 15, or 50, would it zip downslope for a year or two, then go back to crawling or even coming nearly to a stop? ("Zip" is a relative term; the human eye can't detect the difference in speed; it still perceives an hour hand.) The thousands of other glaciers in Alaska and the Yukon do not surge. No glaciers in the Alps are known to surge, none in Scandinavia except on the Arctic island of Spitzbergen, none in the American or Canadian Rockies, and none in the Southern Alps of New Zealand. However, some in the Andes, Greenland, Iceland, Russia, China, and the high mountains of central Asia *do* surge; the Kutiah Glacier in the Karakoram Range in northern Pakistan once "galloped" (a popular term for how surging glaciers move) at an average speed of 369 feet a day for nearly three months. On Spitzbergen, during a single surge in the late 1930s, the Bråsvellbreen Glacier spilled forward 12 miles.

So what is it that glaciers in Spitzbergen, Pakistan, Chile, and Russia share that two glaciers near each other in Alaska (the surging Susitna and the nonsurging Barnard, for example) do not? They don't share size, shape, complexity, bedrock type, or latitude but come in all varieties of these conditions. Some observers proposed that climate change, earthquakes, heavy snowfalls, or avalanches might be responsible, but once it was appreciated that surges are cyclic, it seemed reasonable to assume that erratic events don't act as triggers. In their book *Glacier Ice,* Post and Edward R. LaChapelle, professor emeritus of geophysics at the University of Washington, suggest that some external factor, "such as unusual bedrock roughness, permeability, or temperature," makes glaciers susceptible to surging.

Post observed that during their quiescent periods surge-type glaciers build "reservoirs" of ice, usually in their middle or upper reaches, where ice and snow pile up, while the lower parts, deprived of their full ration of ice from above, thin and flatten out through melting and evaporation. The glacier slope becomes steeper and steeper until, at some point, the ice in the reservoir abruptly starts to move downhill as a wave (often visible as a bulge or ridge) toward the starved lower parts, opening up crevasses as it goes. The ice moves fast, 10 to 100 times faster than the ice in the glacier does ordinarily, and when it reaches the end, it pushes the snout forward. "This is different from normal glacial advance," Bruce Molnia, chief of international polar programs for the U.S. Geological Survey, explains, "where the upper reaches thicken because of precipitation and flow slowly downhill. During surges, they're stealing material from upglacier and rapidly displacing it downglacier. They are thinning the glacier [by] extending it."

No extra ice is involved in this exchange; the level of the reservoir area just falls (the "collapse" mentioned by the excited rangers on the Muldrow), and the level in the terminus rises accordingly. Surging is thus a process of catching up. "Procrastinators on the grand scale," one scientist says of glaciers that surge. But why would glaciers hoard ice? And why is the hoarded ice released from storage when it is? And why does the ice, once released, move so uncharacteristically fast? Few scientists have had the chance to observe glaciers as they surge, since most of the surging ones are in remote areas, and surges aren't common events. But in 1973, alerted that Alaska's Variegated Glacier, whose 1906 advance had so puzzled observers, has a cycle of roughly 17 years, glaciologists from the California Institute of Technology, the University of Washington, and the University of Alaska descended on it and studied it before, during, and after a surge. For nearly a decade they drilled boreholes in the Variegated and watched water levels rise and fall in the holes; injected fluorescent dye in it and watched to see where the dye emerged; took water samples from crevasses and checked for sediment; made seismic readings of "icequakes" (caused by abrupt, severe movement of the ice); inserted temperature probes; and set out those old reliables, vertical stakes, to measure movement.

To their surprise, before its main surge, the Variegated went through several "mini-surges," some only a few hours long. The glacier "behaved like an athlete warming up for an event by taking deep breaths," glaciologist Robert P. Sharp wrote in his book *Living Ice: Understanding Glaciers and Glaciation,* "doing calisthenics, and engaging in short sprints to flex muscles and get in trim for the main effort." It even had a mini-surge afterward, "similar to a warming-down exercise."

The glaciologists found that during the mini-surges, as well as during two pulses of the main surge—in other words, whenever the Variegated speeded up—water pressure and volume underneath the glacier went up too. Mountain glaciers have what amounts to plumbing systems, passageways like household pipes in and under the ice for carrying away surface water. Meltwater and rain collect on the top to form rivulets, and the rivulets trickle into crevasses and cracks and form moulins, and the water descends through the channels to the bottom. Once on the bottom, the water passes down the length of the glacier through what is usually a single tunnel and exits at the end, into streams, lakes, and rivers-to-be like the Aveiron.

In surging glaciers, the plumbing cannot be normal, the glaciologists reasoned. They discovered that 95 percent of the Variegated's movement during the surge came from sliding over the bed; clearly, something was aiding and abetting the sliding. They concentrated on water, an obvious suspect, and after their investigations drew up a working hypothesis. Instead of a single tunnel, or at most a few tunnels, there exists under the reservoir part of a surging glacier a more "distributed," or spread-out, arrangement, with many tunnels linking "cavities" in the ice. Although the cavities are fairly large, they are connected to each other by passageways so narrow—less than an inch across—that as the reservoir above them thickens, the flow of water through them is severely restricted, or "throttled." As a result, water volume and pressure build up under the glacier and start lifting it off its bed, or "floating" it. With a film of water under its belly, some of the glacier slides rapidly downhill, all the way to the bottom, and beyond. "Almost like hydroplaning," said an Alaskan scientist who was on the scene recently when the great Bering surged.

cold blue chasms
And seracs that shone like frozen saltgreen waves.

—Earle Birney, *"David"*

At 2,300 square miles, the Bering is the largest glacier in North America, the largest glacier in the world outside Antarctica and Greenland, larger than the state of Rhode Island. It was named after a Danish explorer who passed by it in 1741 while sailing through the Gulf of Alaska but who apparently didn't notice it. The first explorer who described the Bering probably did so because (it can be said in hindsight) the glacier was in surge. In 1837, British naval officer Edward Belcher peered through his telescope at the "very peculiar outline of ridge in profile" which his draftsman happened to be sketching and realized "it was actually one mass of four-sided truncated pyramids" (the reader will recognize these as seracs). The whole ridge was "similarly composed," Belcher wrote, "and as the rays played on those near the beach, the brilliant illumination distinctly showed them to be ice." (Mr.

Bering might be forgiven his oversight; whenever his namesake glacier is not in surge, the part of it that is visible from the gulf is not broken up into four-sided chunks which the sun's rays could light up brilliantly but is covered with a camouflaging mantle of dark, rocky debris and vegetation. "A wasted, stagnant glacier," Sharp points out, "is not a pretty sight.")

Also realized in hindsight, from cores of dated trees found in terminal moraines, the Bering probably surged around 1900 and 1920 as well, and, from evidence of folds on old photographs, in 1940. Eyewitnesses testify to a "spectacular" advance around 1960. "It tore up the whole lower glacier," Post says. "There were over 1,000 square miles of shattered ice," an area larger than all of the glaciers in the Alps combined. In 1965 the Bering surged again, but less strongly; that might have been a delayed pulse of the 1960 surge. So, in round numbers, that makes it 1840, 1900, 1920, 1940, and 1960, with 1860 and 1880 inferred; Post confidently set the cycle at 20 years. On the basis of 27 sets of folded moraines, he reckoned that the glacier could have been keeping to that timetable for a thousand years.

By 1993, however, the Bering was long overdue for another surge, and colleagues were razzing Post. This "procrastinator on a grand scale" was decidedly procrastinating. Then, in early June, two New York geologists, P. Jay Fleisher, chair of the Earth Sciences Department at the State University of New York at Oneonta, and Ernie H. Muller, professor emeritus of Syracuse University, were flying over the Bering about eight miles from its snout. (Aircraft can do for today's glacial investigators what only sturdy legs and alpenstocks could do for their 19th-century counterparts.) They were interested less in the ice than in the sediments in lakes that the melted ice drains into. Since most of the groundwater used by New York state comes from deposits laid down in waterways by the retreating North American ice sheet during the last ice age, the geologists were hoping to understand those ancient deposits better by looking at the ones a living glacier leaves behind when it retreats. And the Bering had been retreating, having lost 40 square miles since the turn of the last century.

While flying over the glacier, the geologists noticed some abnormally large crevasses, hundreds of feet wide and up to a mile long, at the mouth of the valley, just before the glacier enters a "piedmont lobe" (the broad plain where the Bering accordion-pleats its folds; Fleisher likens the shape of a piedmont glacier to the profile of a mushroom: the stem is the narrow channel it flows through between mountains, and the cap is the much wider lobe it fans out into). Upstream of the crevasses, which were clearly not the product of a glacier merely crawling along, they noticed a marked thickening of the ice, a bulge about 500 feet high. In the weeks that followed, they observed the bulge passing into the lobe, producing as it went a series of cracked domes (probably in places where there were large bumps on the bed). Its spread into the wider lobe opened up crevasses that cut across ones already formed there by the rapid slide of the ice downhill, until the entire lobe was a mass of four-sided, truncated pyramids, a domain of shattered ice.

Once the leading wave had passed ("you can visualize it like a wave in water as the crest goes by," Post says), a fringe of ice remained along the margins, testament to the wave's height. The surge started not at the top but near the middle and propagated both up and down the glacier. The descending ice pushed the glacier's snout forward and obliterated most of the geologists' study subjects, including deposits from the previous surge, on which young alder trees and alpine plants had taken root (it crushed them), and outlet lakes, which it invaded and filled with ice rubble. Long before the obliteration, though, the New York geologists had turned their attention from the lake deposits to the ice. "How many lifetimes do you have to live," says an excited Fleisher, "before a glacier decides to surge *while you are there?*"

Other scientists came to survey the surge scene. Dennis Trabant of the U.S. Geological Survey had a time-lapse camera set up in front of the ice front to document its advance, until a grizzly bear vandalized the camera. (A diverted river wiped out another camera.) In Fairbanks, the Geophysical Institute of the University of Alaska collected images of the glacier over many weeks using SAR (Synthetic Aperture Radar) from a European satellite, then spliced some images together into a very short film so that a person could for the first time "see" a glacier in surge, by following the shifting shadows of hummocks and crevasses. Five months of motion were compressed into a single minute of film, and the resultant sight of ice making its way around glacial features suggests the twitching sway of a hula skirt.

In real time, glacial motion was still elusive, however. "Frustrating," Bruce Molnia called it. "You could stand there an hour or more and not really detect the thing moving, but in the morning you'd come back and all the stakes would have fallen over. We concluded that the glacier only moved when it wasn't being watched, at night." Fleisher and Muller resorted to hanging a pointer from one of the many low plates of ice that the snout pushed ahead of itself like a cowcatcher, laying a ruler under the pointer, and checking to see how the pointer's position changed vis-à-vis the ruler: the best estimate of velocity they got was $\frac{1}{12}$th of an inch per minute.

In May 1994, when the Bering was still in surge, I got a chance to fly over it myself, with a bush pilot who was delivering a moped and meat and mail to isolated settlements along the coast. On one of his runs several months before, the pilot, Steve Ranney, had seen early signs of a surge on Bering, what looked to him like "a set of rolling hills" on the eastern side of the glacier. "Just about every week we could see the hills moving forward," he reported. None of us in the plane could see very far upglacier—the sky was overcast, the ceiling low, the air thickened by drizzle and mist—but we did get a good look at the lobe. The devastation was so complete that bombing came to mind: nothing was spared; the entire surface was torn up and ripped apart. There were thousands of banged-up seracs, some crammed together, many tilted, their tops lopped off at different heights and angles, like a mouthful of big, crooked molars.

Beyond the lobe, in the sediment-filled water of Vitus Lake (which bears Bering's first name), lay toppled seracs, now icebergs, their weathered, rounded tops barely visible, like so many bottom-feeding hippos. The water between the bergs was almost solid with pulverized ice.

Two months later, in late July, the Bering had a "jökulhlaup" (not a French word this time but an Icelandic one, signifying "glacial burst," a flash flood, which in Iceland is nearly always caused by volcanic heat under a glacier that produces so much melt that water floods the surrounding countryside and encroaches on farmhouses). With noise and "plumes of spray, almost like an explosion," Molnia reported, water poured from beneath the glacier into adjacent lakes, entering from below the surface, causing them to "fountain," or the sediment to rise in mounds. Fleisher and Muller spotted chunks of ice "as big as houses" lying where the force of the jökulhlaup had flung them. The quantity of water released in the outburst was, according to Molnia, "hundreds of thousands, no *millions* of cubic meters of water." Presumably, the linked-cavity system, which trapped water, had collapsed and been replaced by the tunnel system, which drained water, and drained it with increasing efficiency, since the rapid flow would have enlarged the so-called pipes by melting their walls.

Within days the flood stopped, and so did the surge (at different times in different parts). Without water to float the glacier, the Bering settled back onto its bed and friction took over. Until the next surge or mini-surge, the glacier would have to rely for its motion almost entirely on creep. It wouldn't flow even as fast as the typical draggy mountain glacier (or slug). "It's like a traffic signal," Post explains. "When the light turns red, the glacier stops. When the light turns green, off it goes! Except the light is red nine-tenths of the time."

The fastest speed the Bering was clocked at when the light was green was 300 feet in a single day. Even the glacier's retreat was rapid, its snout pulling back an impressive 98 feet in 92 days. A surge "hurts" a glacier, one glaciologist points out. All those jagged pinnacles and deep, wide-open crevasses present a far greater surface area to the agents of melting than does the smooth profile of a nonsurging glacier in retreat. Also impressively rapid, at least to the New York geologists, was the rate with which the glacier and its meltwater eroded adjacent land and distributed the products of the erosion. "This revises our thinking about how fast landforms in front of the retreating Laurentide [North American] ice sheet were created," Fleisher says. "What we previously thought would have taken decades could have taken days."

What triggered the jökulhlaup? (The questions keep on coming.) Where did the water come from in sufficient amounts to send the glacier glissading downhill? How was the surge front—the wave of thickened ice—controlled by the water trapped under or inside it? An alternate theory to the linked-cavity theory was proposed: instead of a hard-rock bed, the bed under a surging glacier is soft, made up of ground-up rock, or till, the product of glacial erosion, and high water volume forces the rock particles apart, like wet sand, causing the till to lose strength, and the glacier to move, by pushing its way through the slurry.

"Surging is still one of the enigmas of glaciers," Post concludes. "We haven't come close to figuring it out." Not long ago, European glaciologists recorded "pulses of movement" in two Alpine glaciers—one of them the Unteraar, where Hugi and Agassiz built their famous huts—which in some ways resembled the Variegated's mini-surges. Also, all glaciers, the great majority of which are nonsurging, show temporal variations in motion: slower in winter than in spring and summer when there's plenty of meltwater. A glacier generally flows faster in the afternoon than in the morning when the chill of night is still upon it, suppressing melt. Some scientists suggest, as Charles F. Raymond, professor of geophysics at the University of Washington and an investigator of the Variegated, explains it, that surging in glaciers is "an extreme end member in a continuum of possible pulsating flow behavior."

On that continuum, some people would put Jakobshaven Isbrae, a tidewater glacier in Greenland which, driven by gravity down steep slopes, slides so fast day in and day out, year after year, dumping millions of tons of icebergs per day into the ocean, that they consider it to be in "continuous surge" (although by definition surging is noncontinuous, or cyclic). Also on that continuum, some people would put the "ice streams" in the large Antarctic ice sheets, sections that move much faster than the glaciers they're imbedded in. Which introduces a whole new set of questions . . .

CHAPTER EIGHT

WEST ANTARCTIC ICE SHEET

Cataclysms from the winepress of the glaciers

—Amy Clampitt, *"Rain at Bellagio"*

OF ALL THE DIRE THINGS people suggest could happen because of greenhouse warming, perhaps the direst is the collapse of the West Antarctic ice sheet (WAIS). The WAIS ("WAYSS," if you say it out loud) covers an area the size of France, Spain, Germany, and Italy combined and contains over seven million cubic miles of ice, and if all of that ice slid or melted into the ocean, sea level around the world would rise nearly 20 feet. Twenty feet is enough to inundate most of Florida, the Netherlands, and Bangladesh; put large parts of Manhattan and other coastal cities under water (half the population of the United States lives in a coastal area); and flood countless farms and settlements in river deltas. In 1978 J. H. Mercer of the Institute of Polar Studies at Ohio State University (now the Byrd Polar Research Center) warned that if consumption of fossil fuels kept rising at the rate it was then, the WAIS could begin to "deglaciate" in as little as 50 years.

Antarctica has two ice sheets, the WAIS and the older and much larger East Antarctic ice sheet (EAIS), separated by the Transantarctic Mountains. If the EAIS melted, sea levels would rise 200 feet—a truly dire event, since the majority of the world's population lives within 200 vertical feet of sea level. Yet scientists aren't worried so much about the EAIS because most of it is grounded on rock that is well above sea level. The WAIS, on the other hand, rests on rock that is in places thousands of feet *below* sea level. This came as something of a shock when it was first discovered, in the 1950s. "Our first progress report to Washington so astonished the officials that they asked their chief scientist to verify it," says Charles R. Bentley, A. P. Crary professor emeritus of geophysics at the University of Wisconsin at Madison. (That chief scientist happened to be the namesake of Bentley's chair, A. P. Crary, the first human being to set foot on both the North and South Poles). During IGY 1957–58, Bentley took part in an American program in which researchers traveled around Antarctica measuring the thickness of the ice cover by setting off small explosive charges in shot holes they hand-drilled in the ice, then recording the time it took for the sound waves to reach the bed and bounce back to the surface. Since there are mountains on both sides of the WAIS, the researchers assumed there would be mountains under it as well and that the ice sheet would be thin. "We thought maybe 1,000 meters [3,300 feet]," Bentley says. "It turned out to be 3,000 meters [9,900 feet]." Instead of mountains for a bed, they found an archipelago, with a deep central rift.

WAIS is, then, a marine ice sheet, the last large one left on Earth. The others collapsed toward the end of the last ice age. This raises the question: Is WAIS unstable by nature? The portion of the North American ice sheet that covered Hudson's Bay during the ice age lost its substance very fast, many times faster than the rest of the ice sheet, producing several jumps in sea level. In a warming climate, would WAIS do the same?

Glaciologists couldn't say one way or the other. They didn't know enough about WAIS to make a prediction. "We understand the forces *on* the ice," explains Robert Bindschadler, physical scientist at the National Aeronautics and Space Administration's (NASA) Goddard Space Flight Center, "but not the dynamics *of* the ice." In 1983, to be able to gauge the threat of a collapsing WAIS "and not say 'let our grandchildren worry about it,' " as Bentley describes the situation, he, Bindschadler, and several other glaciologists decided to find out about the ice sheet by studying its ice streams.

Ice streams are sort of glaciers-within-glaciers, parts of an ice sheet that flow faster, usually much faster, than the ice sheet as a whole. Just as mountain glaciers flow between walls of rock, ice streams flow between walls of ice. Hermann Engelhardt, geophysicist at the California Institute of Technology, calls them "transmitter belts" because most of the ice that drains out of an ice sheet's interior to the sea drains out through them. "When something happens," he says, "it happens through ice streams."

("Weird and wonderful"—Bentley)

As important as ice streams are, and as large as they are—in some cases hundreds of miles long—they are mysterious entities. "Why are ice streams moving so fast?" Engelhardt asks. Their slopes are shallow so their speed can't be explained on the basis of slope, and their thickness is about the same as in the rest of the ice sheet so their speed can't be explained on the basis of the weight of the ice, either. "Why are they *where* they are?" Engelhardt also asks. And, the ultimate question, in philosophy as in science, "Why do they exist?"

To answer such questions, as well as others, including, "Are they shrinking?" scientists from the universities of Wisconsin, Alaska, Kansas, Nebraska, and Chicago, Ohio State University, the California Institute of Technology, the U.S. Geological Survey, and NASA began focusing their research attention in 1983 on six ice streams that flow into the Ross Sea, which they labeled A through F. They experimented with different ways of finding out what the ice streams are doing. They spread black plastic sheets over the white glacial surface in various places and took photos from the air of the sheets' changing shapes and positions. They erected poles and tracked the poles' movements by Global Positioning System (GPS). They measured annual layers of snowfall in ice cores; got readings from strain gauges they buried in the ice; took radar and electrical-resistivity soundings of the ice sheet from both the air and the ground; mapped clusters of crevasses—their patterns as distinctive as human fingerprints—on satellite images and compared them with crevasse patterns on earlier images; and drilled holes all the way through the ice sheet to sample the bed the ice was resting upon.

("So many surprises so soon!"—Bentley)

After a decade of such work, they had learned, among other things, that Ice Stream B, the largest of the six, moved at a speed of over 1,200 feet a year, while ice on either side of it crept along at only one or two feet a year (they consider that ice stagnant). They learned that B was losing much more ice than it was gaining—for every 35 cubic feet of ice it accumulated as snow in its cachement area, it discharged 50 cubic feet of ice at its mouth—and was thus out of balance. They learned that in some places B rested upon a layer of thick, water-soaked till and that the water pressure in the till was 99 percent of what would be needed to float the ice. Bindschadler suggested that B might be moving in part by deforming this "very very very sloppy, mucky stuff," although in most areas it probably just slid over the slippery top of it. Having a well-lubricated bed was apparently what allowed B to move quickly despite its low "driving force," or combination of slope, weight, and gravity.

Within the till, scientists discovered fish teeth, diatom fossils, sponge spicules, and other marine sediments, an indication that at least once in the relatively recent past WAIS was not there and open ocean was. "It means that the ice sheet is unstable in interglacial times," Barclay Kamb, physicist at the California Institute of Technology, explains, "and it could disappear at any time. A small trigger could set it off, even without a change in climate."

As for Ice Stream C, B's neighbor to the north, it turned out to be inactive. Its existence was discovered only when radar detected the presence of crevasses 90 feet below the surface of the ice sheet, and scientists deduced that they were the old margins of an ice stream. Judging by the depth of the crevasses (greater depth means more snow fallen, more time passed since the crevasses formed), C was active until about 135 years ago. "A crack opens in a second," a young geologist muses, "and spends the next century getting buried."

("It's so difficult to get information from the ice. This is no bed of roses." —Kamb)

From such findings came more questions. Why would Ice Stream C shut down anyway? Did it drain so fast that it couldn't hang on to its water supply and froze to its bed? But why would a fast-moving ice stream freeze to its bed? (Engelhardt suggests this scenario: heat stoked by the rapid motion of C caused the underside of the ice there to melt, thereby removing a layer of friction-warmed ice and bringing cold ice closer to the bottom of the ice sheet, eventually so close that the ice affected the till, first stiffening then freezing it, and the freezing process "sucked water off the bottom like a sponge.") Or was C's meltwater "captured" by B? Was it a case of water piracy, the melt generated by C diverted somehow to B? Could B become so unbalanced in time that it too would shut down?

And another question: Why wasn't B moving even faster than it was, considering that it had such a gloppy bed? *What was holding B back?* Ice shelves (the parts of glaciers that, when they reach land's end, keep on flowing out over the sea as floating, horizontal extensions)? "Sticky spots" (rough, protruding places on the bed)? The margins?

("I wonder that it's so complicated we'll never figure it out."—Bentley)

I happened to be on hand during the 1993–94 Antarctic research season when scientists set up a field camp on WAIS near one of the margins of Ice Stream B. "Four groups of the top glaciologists in the world," Engelhardt said with characteristic ebullience, "all in one spot, working on the same theme. What an accumulation of authority and experience!" Besides Bentley, Engelhardt, and Kamb, on hand were Ian M. Whillans, professor of geological sciences at Ohio State University, and Keith A. Echelmeyer, geophysicist at the University of Alaska at Fairbanks. (Echelmeyer's colleague at Fairbanks, physicist William D. Harrison, was back at the main base, McMurdo, laid low by a vile stomach bug.) The field camp, called OutB, contained dark-green World War II canvas huts; bright yellow tepee-like tents (designed by polar explorer Robert Falcon Scott); sleds (after a design by Norwegian explorer Fridtjof Nansen), Ski-Doos, and Sno-Cats, plus outhouses made of Styrofoam and plywood, all plunked down on terrain so uninterruptedly flat and featureless and white that those working on it were sometimes stirred by mere shifts in the size and tint of shadows.

Two of the groups were going to study the margins close up. They wanted to know if B's ice walls could be providing enough friction to slow the ice streaming between them. "That's an extremely unusual proposition for glaciers," Kamb pointed out. B's margins are only a couple of miles wide, which is very narrow considering that they separate ice that's moving hundreds of times faster on one side than the other. Not surprisingly, those margins are very riven, very broken up. "Go a quarter mile from the edge," Bentley advised, "and you're in the middle of the most godawful mess you ever saw." "Half ice, half air," Echelmeyer said of them. Nevertheless, Echelmeyer and Engelhardt and several students would be going a mile from the edge of the margin called the "Dragon" (as in, "How much is it draggin' the ice?"). "It's the most extreme project in Antarctica," Engelhardt said proudly. "We had trouble getting it approved." Three or four people at a time would be skiing in, with artificial skins under their skis for traction and all ropes taut. They'd be hauling a sled with a booster heater and hose on it, linked to a larger heater back at the Dragon's edge. Using boiling snowmelt as a drill, they would bore five holes in the ice, about 1,000 feet deep, at regular intervals to about the center of the margin, then hang strings of temperature probes inside each and throw slush down to freeze the strings in place (slush freezes in about three minutes there). Later, they would come back and read the gauges on the probes. Just as a rubber band heats up when you pull on it, ice heats up under the strain of shearing—fast ice tugging on slow ice. If temperatures turned out to be much higher inside the Dragon than outside it, the intrepid glaciologists could conclude that the sides of the ice stream *are* draggin' the ice and estimate by how much.

("I was thrown for a loop with that."—Whillans)

Meanwhile, Whillans and his students would be driving around the "Unicorn"—the island of stagnant ice the camp was on—and patiently planting nine-foot-long aluminum poles in it. They would stop every

20 minutes or so, at "any old place," according to Whillans; stick a pole into the snow there; get the pole's latitude, longitude, and elevation with GPS; then move on. Since GPS testifies to one spot only, they would have to plant a great many of these poles, 647 in all, to get the information they needed to make a contour map of the Unicorn. They would also take readings from 50 poles they had planted the year before in a double row across a puzzling feature called the Rather Strange Hook which Whillans had spotted—very faintly shadowed—on a French satellite image two years earlier. It was a mile-wide, 20-mile-long, slightly elevated line with a sharp bend on one end; Whillans wanted to know if it, unlike the rest of the Unicorn, was still actively moving. "That might tell us something about ice-stream dynamics," he explained. Initial readings suggested that the Hook was an edge between two slopes, one steep and one gentle, resembling a terrace, "like if you put a house on a steep slope for a good view and it's flat there," Whillans said. "I couldn't *think* how it could get like that. It shook me to my socks."

("A very bizarre proposition!"—Kamb)

Meanwhile, Bentley would be driving around the Unicorn with his own students, directing radar waves down into the ice. Radar can detect such things as ice's internal temperature, he explained, as well as its crystalline structure and its age (that is, the portion of it that is old, formed during the last ice age, and therefore small grained), all characteristics that can affect how the ice flows. Radar could also reveal whether or not the Unicorn is frozen to its bed; a wet bed will give a much stronger echo than a dry one. "Radar doesn't see frozen ground."

("Either I'm stupid or it's a weird glacier."—Whillans)

Over Christmas, the glaciologists kept working—they had brought with them two tiny fake Christmas trees and a guitar and a harmonica for playing carols—after which they flew home and spent months analyzing the data they got. The Dragon's hot-water-drill-hole probes revealed that the temperature of ice inside the Dragon is 3° to 5°F higher than ice outside it, leading to the conclusion that the sides of the ice stream *are* helping to hold it back. "Without friction at the margins," Engelhardt explained, "the ice stream could take off and totally disappear." (The bed of the ice stream is so large that it is now thought to be helping hold B back too, despite its slipperiness. Glaciologists continued to search for sticky spots.) Another finding was that the Dragon is moving sideways at the expense of the Unicorn, gobbling up its ice and thus widening Ice Stream B, by 23 feet a year.

To everybody's surprise, Bentley's bunch found in the stagnant Unicorn a great many buried crevasses, which means that the Unicorn had been fast moving (that is, part of an ice stream) until about a century ago. "There was no hint at the surface of this history," Bentley declared. One of his graduate students, Ted S. Clarke, determined that the crevasses don't lie at equal depths in the Unicorn but at increasingly shallow ones the closer they get to the Dragon. "That seems to show a gradual shift of the margin," he surmised, "a migration rather than a jump." The Bentley group also found what appeared to be upside-down crevasses in the Unicorn, cracks that opened from the bottom of the ice sheet rather than the top and grew upward instead of downward. "It's something that ordinarily happens only in ice shelves," Bentley explained. "They've never been seen before in stable, inland ice." "Bentley's as confused as I am," Whillans remarked. "Maybe it's something spooky," he added hopefully.

Information gleaned from the poles that Whillans had set in lines across the Hook turned out to be "dullsville," he admitted. The feature that had tantalizingly suggested something "active and dramatic" was neither, probably a scar left from the period when B first formed, an old margin "which hasn't been buried by snowfall yet."

("B has proved all our original ideas wrong."—Whillans)

In the Antarctic field seasons that followed, these and other glaciologists went to B, to C, UpC, D, OutD, E, and F. They braved entry into other jumbled-up margins ("hair-raising work," says Gordon Hamilton, one of Whillans's students, now assistant professor of geology at the University of Maine at Orono, who surveyed the margins of D and E by walking into them; "your next step could be your last; you come back to camp every day with a racing heart"); onto other islands of stationary ice; across domes of ice. They did more hot-water ice drilling, more pole planting, and more remote sensing by satellite. Some climbed the slopes of the Transantarctic Mountains and others dug into the seabed under the Ross Ice Shelf, all in an effort to better understand the present state of the West Antarctic Ice Sheet and predict its future.

After the second decade of work on ice streams, there seemed to be a sense among the glaciologists that the collapse of WAIS would be neither rapid nor soon. Almost certainly not in a hundred years, probably not in several hundred years, but maybe, even probably, in a few thousand years. "There's no evidence at the moment of rapid retreat," Bentley says. "Overall the ice sheet is approximately stable. But that doesn't mean it's going to stay that way. In a few thousand years there's a reasonable chance the ice will disappear." The investigators have been impressed by how large some of the changes they've observed on the ice streams are and in how short a time the changes took place. Ice Stream B narrowed over the last century, and is now widening again. The ice coming out of its mouth has slowed 50 percent in a mere three decades, and it is now roughly in balance. Ice Stream C might have shut down in a single decade. Its upper part is now moving, and thickening. "Presumably we are in for a revitalization of C," Whillans noted. "All of this is evidence," Engelhardt summarizes, "that the whole system is very dynamic."

Some changes could come from forces intrinsic to the ice sheet. "It's doing it by itself," says Bentley. For instance, the earth's crust underneath WAIS is unusually thin because of the rift in the rocks (now called the Bentley Subglacial Trench), which permits a greater amount of heat from the earth's interior than would usually get through to reach and warm and melt and lubricate the bottom of the ice. "Another clue to why the ice sheet is so active," Engelhardt says. He compares ice streams to a surge system, only "on a huge scale. Ice streams have a life cycle—they're very active, slow, stop, and build to a point where they are active again—except on the order of thousands of years." "There may be no such thing as a typical ice stream," Bindschadler notes. D and E are active now while B was hyperactive and C hypoactive. "Ice streams may not be permanent features of the ice sheet but part of an evolutionary cycle."

If the cycle is being influenced by global warming, the warming is probably—for the time being—natural, not manmade. Engelhardt explains that at the end of the last ice age, 10,000 to 12,000 years ago, "there was an enormous increase in temperature, and the ice is still responding to that." It takes a very long time for a temperature difference at the surface of a thick ice sheet to register at the bottom of it, since ice is such a poor conductor. Engelhardt offers a formula: for half of the effect of a change in surface temperature to be felt 1,000 meters (3,300 feet) below the surface takes 7,000 years. Say there's a 6°F increase in air temperature. In 7,000 years the temperature of the ice in WAIS at a depth of 3,300 feet will be 3°F warmer. That won't be the end of the story, though. "The glacier is going to *keep* warming," Engelhardt points out, "always in the direction of more melting at the base. Maybe that's why ice streams are getting wider now."

At the peak of the last ice age about 20,000 years ago, WAIS was two-thirds larger than it is now. (Scientists conclude this from trimlines on mountains protruding from the ice sheet which are thousands of feet higher than the present surface of the ice sheet, as well as from scrapes and debris characteristic of glacier action they found farther out on the seabed than the present edge of the ice shelf.) As WAIS shrank to its present size, global sea levels rose 20 to 25 feet, and none of that rise can be blamed on human-induced warming. "The WAIS is changing *now*," Whillans and Richard B. Alley, geophysicist at Pennsylvania State University, once wrote, "even before there could have been important anthropogenic effects from industrial gases." Computer models suggest that WAIS oscillates between extreme states—large ice sheet, small or no ice sheet—and the swing now is in the direction of small. "Diminishment is not ambiguous," Bentley says. "It's already in the cards. The near-term future is already determined." (If WAIS should disappear, estimates are that at the current rate of snowfall it would take 10,000 years for it to grow back to its present size.)

All of which doesn't mean that our own greenhouse-gas contribution to global warming won't have some effect. "We may be enhancing what's going on naturally," Engelhardt states. "If there's an instability, we may be pushing [the ice sheet] over the edge." A few decades ago, Terry Hughes at the University of Maine suggested that the "weak underbelly" of WAIS, the place where its collapse would begin, is the area of the Thwaites and Pine Island glaciers, where the ice sheet flows into the Amundsen Sea. Recent observations have shown very large changes taking place there, with glaciers accelerating, grounding lines retreating, and ice thinning (at a rate double what it was in the 1990s). "It's a far cry from collapse," Bentley notes, "but it may be the first sign that something has already started, and it's happening fast."

Meanwhile, on the Antarctic Peninsula, where temperatures have risen 3.5° to 7°F in the last half century (as they have not on the rest of the continent), sections of the Larsen Ice Shelf went to pieces in 1995 and 2002, probably because the water in melt ponds leached into crevasses and, with its greater density, widened them. Where the shelf disintegrated, glaciers that had been feeding it speeded up dramatically. For a long time, glaciologists have been debating whether or not ice shelves act as brakes on glacier flow by providing back pressure, and what happened after the Larsen breakup suggests—though does not prove—that shelves are indeed "buttressing," as Mercer termed them. The speedup of some Amundsen Sea glaciers suggests the same. The ice shelves there have been reduced by warmer seawater melting their undersides, which caused them to float well above bedrock and, presumably, lose their buttressing capacity. Glaciologists—cautiously optimistic about WAIS's future after decades of research on the Ross Sea ice streams, which they concluded are probably not destabilizing WAIS by rapidly draining its inland stores of ice as they had feared—are now worrying about the integrity of the ice shelf the ice streams pour into. Recent satellite images show the Ross Ice Shelf has huge cracks in it.

Even if the whole West Antarctic ice sheet isn't expected to disintegrate any day now, sections of it could. "It is quite possible that small parts could collapse or melt or thin in as little as a few centuries," Hamilton says. "If 10 percent of the ice sheet goes, that would still raise sea level 20 inches or so, and 20 inches would have a huge impact on society. It's not just sea level you have to worry about but storm surges and other extreme events on top of that. Florida's geography would be different. The low-lying Rhine delta would be gone. Some South Pacific islands would disappear. Eight to 20 inches of sea-level rise in 100 or 200 years—these levels are definitely within the realm of possibility."

("None of what we learn is wasted. Everything lives."—Bentley)

The glaciologists continue to investigate.*

* A year after his 18th trip to Antarctica, at age 57, Ian Whillans died, and in honor of his pioneering work in glaciology, B has been renamed Whillans Ice Stream.

CHAPTER NINE

CORING

What the ice gets, the ice keeps.

—Sir Ernest Shackleton
(as reported by *Endurance*
captain Frank Worsley)

THE TASKS THAT Lonnie G. Thompson has performed in the name of glaciology include driving across sand dunes in 95°F heat for 88 straight hours in the fumey wake of a convoy of ancient Tibetan trucks; combing an Andean settlement for a potato farmer or alpaca herder capable of rounding up several dozen plow horses and pack burros; scouting a volcano from a hot-air balloon; and doing field tests while hail pellets from thunderstorms boiling up out of an Amazonian rain forest pricked his skin "like bumblebee stings" and air charged with electricity made his lank hair stick straight out.

The glaciers Thompson is most interested in are not in the polar regions. They are in the tropics and subtropics, in Peru, Ecuador, Tibet, China, Tanzania, Bolivia. They are on mountains and high plateaus—in low latitudes glaciers exist only at high altitudes—and most are harder to reach than Antarctic ice sheets. "You can't fly or drive to them," Thompson says. On a typical expedition, he and several other scientists and technicians land at an airport somewhere; drive hired trucks to the dead end of some unpaved road; load about six tons of equipment, including a portable hyperbaric chamber in case somebody gets altitude sickness, onto porters' backs and their own backs or onto snowmobiles or yaks or burros; trudge up a steep slope; spend three to six weeks doing hard physical labor several thousand feet above sea level in burning sunshine; then trudge back down carrying four extra tons. The extra tonnage is ice.

The ice is in the form of cylinders, each of them three to six feet long and about three inches wide, the width of the drill bit used to extract them. Each cylinder is a short section of one very long cylinder drilled vertically out of a glacier, then numbered with a blunt-tipped #1 lead pencil, wrapped in plastic, stored in a snow pit, and transported by the reverse route off the glacier down the unpaved road to the airport and flown, still frozen, to a cold room at the Byrd Polar Research Center at Ohio State University, where Thompson is a Distinguished University Professor. (The holes left behind when the cylinders were removed from the ice close up on their own, the ice walls creeping into the empty space. To keep the ice walls from creeping in before the drilling is finished, the crew dumps a fluid down the hole that has about the same density as the ice.)

The cylinders are ice cores, or, as CRREL geologist Tony Gow once referred to them, "file cabinets of glacial time turned upside down." Other people have called them libraries, archives, repositories, books, and calendars, all ways of saying that the ice keeps a record of past times in an organized way (annual layers, oldest at the bottom) which can be "read" by scientists. What the scientists most want to read about is climate. They want to know what the atmosphere was like at the time the snow that turned into the ice that makes up the glacier fell, even if the snow fell a very long time ago. A snow cover may *look* pristine and uniformly blank and just like other snow covers, but it is not. As snow forms and after it falls, it develops a complex chemical and physical identity, based on the temperature of the atmosphere at the time and incorporating many of the elements in that atmosphere, including sea salts, volcanic ash, desert dust, nitrates (from auroras and car exhaust), lead (from smelting during the Roman Empire and car exhaust), pollen, radioactive debris, extraterrestrial particles (nickel, platinum), ammonia (from terrestrial fires), methane (from swamp gas and burning of fossil fuels), carbon dioxide, microbes, DDT. And once the snow has turned to ice, that identity is preserved, for thousands of years, or as long as the ice survives.

In the half century since scientists realized there was climatic history locked up in ice and decided to do something about it, they have developed sophisticated ways of extracting not only the cores (early ice drills were essentially rock drills, little more than pipes with sharpened edges at the bottom, shoved downward through snow and ice by gasoline motors) but also information from the cores. There are now at least 50 tests that can be run on the ice, including ones so sensitive they can detect substances in parts per billion. The sensitivity is necessary because, despite the fact that the ice isn't pristine, it's "still a rather clean material," Richard Alley points out in his book *The Two-Mile Time*

Machine (two miles being the length of an ice core taken recently from the Greenland ice cap). "One nose drip," he writes, "has more of some contaminants than do armloads of ice cores."

In the codified language of the test results, scientists can read tales of droughts and monsoons, forest fires, powerful windstorms, sudden warmings, gradual coolings, dwindlings of snowfall, increases in solar activity, buildups of greenhouse gases, eruptions of volcanoes, the spread of wetlands, the beginning of the Industrial Revolution, the introduction of low-lead gas. Embedded in the ice are bubble-like inclusions of fossil atmosphere—actual atmosphere—former air bubbles that dissolved under the stress of the weight of the ice above them, then once the core was brought to the surface and freed of the confining pressure, were born again as bubbles, containing air that Caesar might have breathed, or Lady Murasaki, or cave painters, or a saber-toothed tiger in heat—and that air too can be analyzed, and compared with our own.

To serve as a calendar or file cabinet, an ice core has to be dated. But how does a person know when the snow that formed the ice that now lies 1,000 feet deep in the glacier actually fell? Or 2,000 feet deep? Gow likes to tell how as a geology student studying rock stratigraphy in New Zealand in the late 1950s he was recruited to drill Antarctic ice and study *its* stratigraphy. "I couldn't understand how you could take this plain white material [he'd never stood on snow before or even seen it fall] and count years in it, one by one. I wondered, 'What have I gotten myself into?' " (What he had got himself into was several decades of analyzing ice cores, from the North Pole to the South Pole and many places in between.) Counting years in the upper part of a core, the part nearest the surface of a glacier, is usually a cinch. All a glaciologist—or you or I—has to do is look.

If you roll a cylinder of ice onto a clear table with fluorescent light shining through from below, you'll see a series of dark and light bands in it, like tree rings. (Tree rings keep records from past climates, too, but the record stops at 5,000 years ago in the case of living trees and 12,000 years ago in the case of dead ones, and the information is limited to growing conditions, wetness, sunshine, etc. Deep-seabed and lakebed sediments also keep climate records, some of them very long, representing millions of years, but their layers become exceedingly compressed and are often disturbed by waves and worms. "For long-term history," Lonnie Thompson declares, "nothing beats ice.")

One light band or cluster of bands represents one summer. The band is light because crystals that form in warm temperatures tend to have recrystallized a couple of times and are thus coarse and loosely packed, with air around them to scatter light. One dark band represents one winter; snow grains in winter are usually small, and the strong winds of winter pack them densely together, leaving few air cavities. One dark band plus one light band equals a year in the life of the glacier.

"Each time a summer layer went by," Alley wrote, describing the experience of dating a Greenland core by eye, "I marked it on the book and also made a tally on a separate sheet. We thus could keep track of the passing of history: This snow fell the year I was born, that snow when Lincoln spoke at Gettysburg, and so on."

The farther down in the core the annual layers are, however, the more stretched-out and thin they are, since the ice is compressed and flows outward to the sea under the combined weight of all the ice and snow above it, until at some depth even an expert observer can't tell winter from summer or one year from two or five. In central Greenland, three feet of snow on the surface are transformed by compaction into one foot of ice 200 feet below the surface. That foot of ice is squeezed and spread into a half foot of ice by the time it is halfway through the ice sheet, according to Alley, and into a quarter foot of ice by the time it is three-quarters of the way through. "When buried seven-eighths of the way through the ice sheet," he writes, "the [ice] layer has lost seven-eighths of its original thickness and is only one-eighth foot thick, and so on." Which means that an ice core seven-eighths of the way through the glacier will contain eight times as many annual layers as a core of the same length near the top. In very long cores, the bottom 10 percent may hold 90 percent of the total annual layers.

Characteristics other than grain size distinguish summer from winter layers, and tests for them allow scientists to count years without seeing them. One such test, called the electrical conductivity measurement (ECM), involves running two electrodes over the outside of an ice core and noting where the current passing between them, through the ice, is strongest. A strong current means plenty of acids in the ice, and since the atmosphere is more acidic in summer than in winter (direct sunlight oxidizes airborne chemicals), that signals a summer layer.

Temperature, an obvious seasonal difference, can be determined by comparing the weight of oxygen atoms in various layers. Most oxygen atoms in water that falls as snow or rain have an atomic weight of 16, eight protons plus eight neutrons, but a small proportion, two-tenths of one percent, have a couple of extra neutrons and an atomic weight of 18. Since water molecules containing the heavier ^{18}O atoms are more likely to condense from vapor into snow than the lighter ones, and since as air gets colder much of the moisture in it condenses out, the amount of ^{18}O compared to ^{16}O in glacial ice is a good indicator of how cold the air was when the snow formed, and thus whether the season was summer or winter (in very cold places like Greenland, snow falls in summer as well as winter). Willi Dansgaard, a geochemist at the University of Copenhagen who was one of the first to recognize this phenomenon, proposed that the ratio between the two isotopes of oxygen be used as a stand-in for atmospheric temperature at the time the snow fell. In an ice core, a summer band of ice can be distinguished from a winter band by its lower ^{18}O to ^{16}O ratio.

Over the long term, protracted stretches of warm climate, such as the interglacial period we are in right now, can be told from predominantly cold periods—ice ages or glaciations—by their lower overall ratios of heavy-to-light oxygen isotopes.

Scientists can also count annual layers using dust. Dust particles are blown onto the ice from land that's free of ice, often land quite far away; in Greenland, much of the dust probably came from East Asia, in Antarctica from South America. Lasers can pick up reflections from dust particles in layers of ice thinner than a hundredth of an inch. According to Alley, the dust appears as "faint, grayish, ghostly bands" when seen in the beam of a fiberoptic lamp. "These are the late-winter dusty layers." Dustiness is a sign of winter because winds tend to be stronger in winter than in summer, due to the greater temperature contrasts then between the tropics and the poles. The same reasoning holds true for ice ages, only more so. "In ice from the cold, dry, windy ice age," Alley reports, "the bands are so strong that they can be counted from across the room, without any special lighting."

Uncertainties are involved in all counting methods (is this one summer layer or two summer layers close together?), so scientists rely on several and check the findings against each other and the seabed record. They also look for markers that can be linked to known events in history, such as volcanic eruptions and thermonuclear explosions. "Here's Pompeii, 79 A.D.," Gow says to himself as he comes across an ash-darkened, sulfuric-acid-rich band in an ice core, then makes the appropriate time-line adjustments. "We take everything and clean it up."

One of the first ice cores ever drilled came from beneath Camp Century in northwest Greenland, an underground city carved entirely out of snow. (Built by the American military in 1959, the camp was powered by a nuclear reactor, had 21 tunnels, a main tunnel "street" 1,200 feet long, a movie theater, chapel, infirmary, skating rink, and dorms where 100 men were able to live for two years, walking around coatless and taking steam baths while temperatures above ground registered −54°F.) The Camp Century ice core was nearly a mile long, extended all the way to bedrock, and represented about 110,000 years: the ice at its base dated from the beginning of the last ice age. When it was analyzed, the core revealed that in the coldest part of the last ice age, one-half to two-thirds *less* snow fell on Greenland than falls on it today, probably because most of the North Atlantic was sealed by an ice lid then, and since little moisture escaped from the surface of the frozen sea, the atmosphere became very dry.

The same core also revealed that during the transition from the ice age to our current interglacial period, after warming had already begun, Greenland went through a 1,300-year cold snap, a "final poop" of the ice age, as Gow puts it, during which temperatures fell more than 12°F. The cold snap had already been identified in Europe from other records and was named for, of all things, a flower, Younger Dryas, an eight-petaled mountain posy, *Dryas octopetala,* which disappeared in the European highlands during the initial warming but reappeared during the cold snap. Then when the snap ended and warm temperatures settled in for the long haul, it disappeared again. What the ice core added to the story was that the Younger Dryas ended abruptly, in a mere 20 to 50 years.

This is the major, surprising message to come out of the study of ice cores: that very large climate changes can happen very fast. The Camp Century core, along with a core taken 15 years later from another place in Greenland called Dye 3, showed that during a long stretch of the last ice age, between 75,000 and 11,500 years ago, the climate in the vicinity of Greenland "flip-flopped." It "flickered," it oscillated, it "yo-yoed." Instead of being relentlessly, unrelievedly cold for all those years, as most of us envision an ice age being, the climate warmed several times by 9° to 14°F and stayed warmed for anywhere from 500 to 2,000 years before slipping back into full glacial cold. In the Dye 3 core, Willi Dansgaard and Hans Oeschger of the University of Bern, Switzerland, identified ten such warm periods during the last glaciation, all of which arrived quickly, in as short a time as two or three years.

One problem with both cores, however, was that the oldest ice, the ice closest to the bed, the "really interesting ice," as Alley described it, was hard to read. The cores had been drilled from the flanks of the ice sheet, where rapid flow over an uneven bed could have distorted the layers, and summer meltwater percolating downward through the pack could have garbled them. In addition, neither core extended through the last interglacial, the warm period from 115,000 to 135,000 years ago which Europeans call Eemian and North Americans call Sangamon, and which climatologists on both sides of the Atlantic are particularly keen to learn about since what happened during it might help predict what will happen during our own interglacial, the Holocene, which started about 10,000 years ago.

So, to get a longer and better-resolved core, glaciologists decided to go to the summit of the Greenland ice cap, from which place the ice should flow like pancake batter in all directions, slowly and evenly, and the air should be too cold for the ice to have done much melting. In 1992, two groups (with acronyms that make them sound like a weightlifting duo), GRIP and GISP2, set up separate drill camps within 20 miles of each other. GRIP (the Greenland Ice Core Project) was made up mostly of European scientists, including Oeschger and Dansgaard, who chose as their drill site the very peak of the ice cap, where the ice is 10,000 feet thick, the thickest of any place in the northern hemisphere, and nearly stationary, or so they expected.

GISP2 (the Greenland Ice Sheet Project Two, "2" because earlier coring projects in Greenland were collectively known as GISP1), was made up mostly of Americans, including Richard Alley, Tony Gow, and Paul Mayewski, then at the University of New Hampshire, who was chief scientist of the project, and they set up their drill rig west of the ice peak, close enough that distortion of layers from ice flow would be minimal, or so they hoped.

Using electromechanical ice drills, the Europeans took a core four inches in diameter, the Americans a core five and a fifth inches in diameter. The Europeans poured diesel oil mixed with Freon down their borehole to keep the ice walls from closing in; the Americans used butyl acetate, a smelly organic salt used as a sweetener in chewing gum. Both teams worked only in summer and both wanted very much to get to the bottom first, although the competition was friendly and they called their cylinders "sister cores." Both lost parts of their cores to fracturing. Mayewski explains: "There's a layer in all cores about 1,300 to 4,000 feet down that's extremely strained. If you touch it, it will shatter." "The stuff starts to wafer," Bruce Koci, designer of the American ice drill, adds, "like coarse poker chips."

The reason for the wafering is that the weight on the ice at those depths is over a ton per square inch, Alley writes, "and the air in the bubbles has been compressed so that it pushes back with this pressure." (Bubbles can be squeezed to the size of pinpricks.) "Cores brought to the surface will pop, snap, and break as the air expands and fractures the ice." Below the brittle-ice layers, however, the increased weight of the overburden of ice causes the bubbles to dissolve into the ice; the molecules of air are forced by pressure to go inside ice crystals. These cages of ice with bits of air in them are called clathrates. "As clathrates replace bubbles," Alley reports, "the ice cores become as clear and easy to handle as plexiglass tubes."

In 1992, GRIP hit bedrock or, rather, silty ice just above it. The next summer GISP2 hit bedrock and kept on going, drilling out a five-foot-long core of rock as well. Both GRIP and GISP2 ice cores were about 10,000 feet long and spanned 150,000 years, into the Eemian/Sangamon interglacial, as had been hoped. Annual resolution appeared to be very good; GISP2 scientists could count layers by eye down to about 11,000 years. Dating by ^{18}O, ECM, and dust matched each other well. "In 2,000 years, we were only 12 years off," Gow confides. In the top 9,000 feet of both cores, representing about 95,000 years, the information turned out to be "astonishingly" similar, as one participant expressed it, confirming not only the ten warm periods in the last ice age that Dansgaard and Oeschger had discovered in earlier cores but adding 15 more, including one that lasted a mere decade.

With their finer resolution, the new cores showed an even greater temperature drop on entry into the Younger Dryas than the earlier ones had: 27°F! Winters grew longer by several months and the atmosphere became stormier. Three times as much sea salt and up to seven times more large dust particles fell on the ice sheet than fell in the first thousand warm years that followed. "It was a weird time indeed," Alley observes. Weird, too, was the speed with which the Younger Dryas ended, even more extreme than the other cores had shown. Suddenly, the climate became very warm and wet, and snowfall over Greenland doubled in only three years, with most of the increased snowfall falling in a single year.

"I cannot insist that the climate changed in one year," Alley avows (it is possible to misidentify a year if, for instance, a drill happens to hit a snowdrift), "but it certainly looks that way. And I can insist that the change was fast—not over a century, not even over a human generation, but maybe over a congressional term or even less."

The Greenland ice-core findings led Alley to compare the climate to a drunk. "When left alone, it sits; when forced to move, it staggers."

That is, climate tends to be stable unless it is "pushed," in which case "it often jumps suddenly to very different conditions, rather than changing gradually." Things that might provide a push include changes in the Earth's orbit, changes in ocean circulation, the drifting of continents, and the size of the continental ice sheets themselves.

> *Logically, benevolence surrounds us.*
> *In fire or ice, we would not be born.*
>
> —John Updike, *"Ode to Healing"*

As the GRIP and GISP2 cores opened onto the Eemian, toward the bottom of the ice cap, however, they started looking less like sisters and more like warring cousins. According to the GRIP scientists' reading of the deepest part of their core, the climate during the last interglacial behaved quite differently from the way ours has so far. While our climate has been dependably stable for 10,000 years, the climate during the Eemian was interrupted repeatedly by "cool events," as Dansgaard calls them, during which temperatures fell 18° to 27°F, to nearly ice-age levels. One cool event came on in less than a decade.

This was scary stuff. If such extreme flip-flopping could occur in one interglacial, why not another—our own precious Holocene, for example? The Eemian is generally considered to be an analogue for the Holocene, although not a precise one since the Eemian was a few degrees warmer overall. (In fact, as GRIP participant David Peel of the British Antarctic Survey pointed out, England's climate was so mild during the Eemian that hippopotamuses waded in the Thames and lions prowled the river's banks.) Could the temperatures we're enjoying now drop several degrees in under a decade? What would trigger such a drop? The climate question thus became "politically charged," as one GRIP member put it, to wit: Could we human beings set off the rapid temperature swings characteristic of the Eemian by heating our atmosphere with greenhouse gases to Eemian levels?

"The [ice-core] results bring rapid climate change to our doorstep," wrote GRIP participant J. W. C. White of the University of Colorado at Boulder. It is one thing to find instability during an ice age—we can always "take comfort" in the belief that the large ice sheets and vast extents of sea ice that existed back then brought the changes about—but quite another thing, and not at all comforting, to find a similar instability during an interglacial. "Adaptation—the peaceful shifting of food growing areas, coastal populations and so on—seemed possible, if difficult, when abrupt changes meant a few degrees in a century," White pointed out. But 10 to 18 degrees in less than a decade?

Although flip-flopping interglacials are "extremely difficult to explain," Dansgaard admitted, the fact that our present climate has been "strangely stable" for 10,000 years is likewise hard to explain. "Perhaps the most pressing question," he mused, "is why similar oscillations do not persist today." Perhaps oscillation during interglacial periods is normal and steadiness abnormal. Perhaps we have just been lucky in our abnormality.

When the full GISP2 results were in, the arduous business of setting up duplicate drilling operations within an hour's snowmobile ride of each other seemed justified, a case of smart thinking. "Two cores are better than one," went the headline over a commentary on the American findings. In the bottom 900 feet or so of the core, researchers found not flip-flopping but folds. "Moderate to severe distortion of layers," reported Gow, who, with Alley, did most of the analyzing. "Annual layers became wavy, wavy layers became folded, folded layers became overturned. Some layers were inclined 25°!" Some of the layers weren't simply stretched out horizontally but pushed to higher levels (perhaps because of raised areas in the bed, or stiff ice embedded in softer, flowing ice, or both) so that—since depth is equated with age—a cool event might appear at an earlier position in the chronology. "In rocks we see magnificent folding, outcropping, faulting, and shearing," Gow points out. "Ice is a rock. Why shouldn't this go on at the bottom of an ice sheet?"

The consensus among the GISP2 participants was that the bottom 10 percent of the ice in both summit cores was "discontinuous." But the GRIP group wasn't at all sure. "The test, of course, would be to drill another borehole on the divide," G. S. Boulton, glaciologist at the University of Edinburgh, concluded. In 1996, a group of European scientists started doing just that, drilling a core about 200 miles north of the Greenland summit, in a project called North GRIP. Over the next seven years, they extracted a core whose deepest ice they found was not distorted by folding or "ice mixing." However, it *was* melted on the bottom, because of geothermal heat, and didn't encompass the whole Eemian as they had hoped but only the tail end of it—and wouldn't be answering their key questions. The tail end did, however, provide a good record of the Eemian's transition into the ice age. A slow, steady cooldown was interrupted by a temperature rise of 9°F followed by a return to cooling, all in only 50 years, a short human lifetime. This suggests to Colorado's

J. W. C. White, a member of the drilling team, that significant, abrupt changes are possible at any time in the climate cycle without some obvious precipitating cause. "There was no smoking-gun event."

The bottom of the North GRIP core turned out to be pink, a "watermelon" pink. Red protozoa known to live in snow probably provided the tint. Buried by thousands of years of snowfall, along with particles of windblown material they could subsist on, the protozoa were presumably getting the liquid water they needed to live on from the melted ice.

A million years of history are waiting for us in that ice.

—British scientist in sci-fi movie *The Thing from Another World,* on seeing a giant body buried—eyes open—in polar ice

The longest ice core ever drilled was one taken in the late 1990s by a joint Russian-French-American team from a Russian station in East Antarctica called Vostok. Vostok has been called "the pole of cold"—it's the coldest spot on Earth (mean temperature −67°F), the spot where the coldest temperature ever recorded was reached, on July 21, 1983 (−129°F), and the spot where the ice is thickest of any place on Earth (12,280 feet). The core was more than two miles long and covered 420,000 years, or four full glacial-interglacial cycles. A shorter core had been taken in Vostok in the early 1980s, and Claude Lorius, then director of the Glacier Laboratory in Grenoble, France, made an important discovery in it. As the temperature in a particular layer of ice went up, he observed, so did the level of carbon dioxide (CO_2) in that layer's fossil air. Temperature and CO_2 rose and fell together, for the whole 160,000 years represented by the core, with methane (CH_4), another greenhouse gas, keeping pace with them. Lorius also found that the CO_2 levels were much higher during interglacials than they were during glacial periods, leading him to surmise that the CO_2 was acting as an "amplifier" to warming, at the very least. If certain factors heated the atmosphere one degree, the presence of large amounts of CO_2 would heat it up another degree. This finding in an Antarctic ice core kicked off the international debate on global warming. From then on, one of the missions of drilling ice cores became to find out what natural greenhouse warming is, or was, as a baseline for judging the human role in aggravating it.

In the later, longer, Vostok core—the "granddaddy" of all ice cores, as Jerry McManus of Woods Hole refers to it—CO_2 and methane stayed in lockstep with temperature for the entire 420,000 years. That connection lent support to the idea that greenhouse gases contribute to shifts between ice ages and interglacials. The longest of all ice cores wasn't quite as long as it could have been, however. Radar has revealed the presence of a huge lake under Vostok, 3,900 to 5,400 square miles in area and 1,600 feet deep, formed by the melting of the underside of the ice sheet by geothermal heat. (Radar has also found more than 75 other, smaller lakes under Antarctic ice.) The bottom 1,200 feet or so of the core, where glacial ice flowed over the lakes frozen surface, was "all scrambled eggs," according to Todd Sowers, research associate at Pennsylvania State University and a member of the drill team. "You couldn't make anything out of them." Under that scrambled layer was one made up not of glacial but lake ice (bigger crystals). In it, biologist John Priscu of Montana State University found remnants of bacteria, as well as the same level of dissolved organic carbon that exists in a typical American lake. If Lake Vostok has bacteria living in it, he concluded, they should have plenty to eat. Not wanting to contaminate the water with microbes from their drill before they could find out what microbes are in it naturally, the drillers stopped short. To reach the lake, they'll have to wait for engineers to figure out a way to sterilize a drill that has passed through more than two miles of ice.

Seeking a longer, more legible time line, drillers went after yet another core, at Dome C on the East Antarctic plateau, where ice rests on bedrock, not water. By 2004, a consortium of European scientists called EPICA (for European Project for Ice Coring in Antarctica) had succeeded in pulling out the "great granddaddy" of ice cores, as McManus calls *this* one. It encompassed at least 900,000 years and covered eight ice-age cycles, which makes it the oldest (although not the longest; Vostok still has that honor) ice core ever. One thing the core has shown is that the current levels of greenhouse gases in our atmosphere are about 30 percent higher than *at any time in nearly a million years.* "How our climate will respond" to these artificially high levels "is anybody's guess," White comments. "To me, it changes the boundary conditions. It shuffles the cards and puts a couple of jokers in the deck."

From the height of a glacier I beheld half a world, the earthly width.

—Joseph Brodsky, *"May 24, 1980"*

To get a truly global picture of past climates, Lonnie Thompson argued, ice cores should be collected from the tropics and subtropics as well as the poles. Ninety-eight percent of the glaciers in the world are in the polar regions, but 70 percent of the people in the world live in the tropics, he noted, and ice cores from those latitudes could shed light on matters of concern to them—monsoons, El Niño, volcanic eruptions—by showing how those things fluctuated in the past. Thompson's argument left many other glaciologists cold; several told him he was wasting his time. "I took my exam for an MBA in case I needed another career," he says. The first core he drilled came from the 18,670-foot-high Quelccaya ice cap in the Andes of southern Peru, within a few hundred miles of the equator. Although he had seen a few aerial photographs of Quelccaya in an atlas, he couldn't find anybody who'd actually been there. Finally he managed to track down a Canadian mountain climber who had *glimpsed* the ice cap from a neighboring mountain and talked him into joining a scouting expedition. "We rented horses and found our way," Thompson says. A year later, he went back with a crew to drill cores. "We were naive how this could be done," he says. A Peruvian Air Force helicopter was supposed to carry 1,500 pounds of equipment up onto the ice cap, but it foundered in the thin, turbulent high-altitude air. "It kept dropping, falling," Thompson recounts. He had been warned (by Dansgaard) that the places where he wanted to do his drilling were too high for both the drill and the drillers. "No one had worked at those elevations before," he admits.

Four years later, though, he was back at Quelccaya with a lighter drill, made of a fiberglass-like material with a Teflon-coated barrel, as well as a lighter power source, several solar panels he had tested on the roof of a parking garage in Columbus, Ohio, using ice he got from a company that makes ice cubes. Still, it took 40 packhorses and 17 porters two trips to carry all equipment and supplies up to the drill site. The dating of the cores proved to be simple because, although Quelccaya gets about ten feet of snow a year (almost all of it from thunderstorms brewed in the rainforest thousands of feet below the peak), the snow falls there only seven months of the year. The rest of the year—the dry season—dust falls. Thompson took a stunning photo of a margin of the ice cap in which the dust bands are so conspicuous and regularly spaced between wide bands of snow that it looks like a tiered wedding cake.

Next, Thompson went to Tibet. He and half a dozen other Americans and 30 Chinese from the Lanzhou Institute of Glaciology and Geocryology in China drilled three long cores from the 17,500-foot-high Dunde ice cap on the Tibetan highlands. (All cores were split down the middle so each country got half.) From travel journals and glaciology references of the 1920s, Thompson had known of Dunde's existence, but it took three months for the drillers to find it. The ice cap was frozen to the bed, "the first hint that the ice was very old," Thompson says. (Quelccaya's ice at the bed had been *melting,* losing a layer off the bottom each year and adding another layer of ice at the top, and proved to be only 1,500 years old.) Dunde's ice at the bed turned out to be 12,000 years old, dating from the last ice age! Nobody expected there to be ice-age ice in the subtropics, Thompson says. It had generally been believed that while continental ice sheets were growing in the north, the climate of the tropics stayed balmy. Dunde ice cores showed "the whole world cooled."

During the Little Ice Age (LIA), too, the whole world cooled, judging by ice cores from both Dunde and Quelccaya. The Little Ice Age was a roughly 400-year period from about 1500 (some people say 1350) to 1880 (some say 1900) when temperatures in northern Europe fell 1° or 2°F. Glaciers in the Alps advanced. The Thames froze (people sailed iceboats on it). Harvests failed. Poor people became poorer and hungrier. In her novel *Orlando,* Virginia Woolf conveys the spirit, if not the facts, of the "Great Frost" that resulted: "Birds froze in mid-air and fell like stones to the ground. . . . it was no uncommon sight to come upon a whole herd of swine frozen immovable upon the road. The severity of the frost was so extraordinary that a kind of petrifaction sometimes ensued." What the ice cores showed was that the Little Ice Age was not confined to Europe but a global phenomenon.

(Although the LIA is sometimes called a "neoglaciation," it is considered a mere blip on the essentially smooth temperature line of our interglacial, not a sign of its instability. When Agassiz and Tyndall began their questing jaunts on the glaciers of central Europe, the glaciers were probably bigger than they had been at any time since the end of the last ice age, but by the time Tyndall gave his lecture to his youthful audience, he was able to report that "for the last fifteen or sixteen years the glaciers of the Alps have been steadily shrinking; so that it is no uncommon thing to see the marginal rocks laid bare for a height of fifty, sixty, eighty, or even one hundred feet above the present glacier." The Little Ice Age, which had been largely responsible for the glaciers' buildup, was drawing to a close. By 1894, concerned scientists formed an international group based in Zurich to monitor the state of the glaciers because, according to Swiss glaciologist Wilfried Haeberli, "they were afraid the Little Ice Age would come *back,* and they had suffered terribly during it. They thought that glaciers oscillated like sunspots, advancing and retreating, and they made very precise maps of areas that were free of ice at the time but which they expected to be covered by ice later on, when the glaciers advanced as they inevitably would. Since then, only six of those areas have been covered.")

Ice-core experts have long ago concluded that Thompson is not wasting his time. In 32 years, he has led 49 coring expeditions to rugged, remote, and dangerous places and never lost an ice core or had one melt away. In the year 2000 alone, he and his team drilled cores on, among other places, Sajama, the highest mountain in Bolivia; Dasuopu, on the Tibetan plateau, the highest plateau in the world; and on Mount Kilimanjaro, the highest mountain in Africa. On Sajama, he was able for the first time to date ice cores using carbon 14. He found five species of insects in the ice core, perfectly preserved by it, "even their eyes." The bugs at the bottom were 25,000 years old, therefore so was the ice in which they were embedded. They had probably been flying over a lake in the valley, gotten caught in a thunderstorm, been carried by updrafts to the tops of clouds, been frozen to death, and when the snow fell on Sajama, the bugs fell with it.

Dasuopu, at 23,500 feet above sea level, is the highest place where ice cores have ever been drilled as well as the coldest place—mean temperature 3°F—they've ever been drilled outside the polar regions. Thompson expected to find the oldest ice in the world there, maybe a million years old. Because the layers were very thin, he reasoned, the time lines could be very long. The ice at the bottom, though, turned out to be only 9,000 years old. "There was no glaciation-stage ice in the icecap!" Thompson exclaims. The ice had *started* forming only when the coldest period *ended!* What was going on?

As it happened, the youth of Dasuopu's ice turned out to be at least as interesting as its old age would have been. It showed, according to Thompson, "glaciation on earth is not synchronous." Although the entire earth cooled during the ice age, not all cool places grew ice. He reviewed the ages of the oldest ice found at each drilling site and realized that the oldest ice grew successively younger as the locations went from south to north. For example, at Sajama in Bolivia, at 18°S, the oldest ice was 25,000 years old; at Huascarán in Peru, at 9°S, it was 19,000 years old; and at Dasuopu in Tibet, 28°N, it was 9,000 years old. What accounted for the "northward migration of 'the age of glacier formation,' " Thompson concluded, was the "northward migration of the axis of their major moisture supply." "*Moisture* is what drives glacier formation in the tropics," he explains. Cold is necessary "but you can't build ice unless you have water."

The water comes from a narrow band of vapor which circles the

Earth's midsection, feeding monsoons and snowfall (it's called the Intertropical Convergence Zone). It doesn't stay put but shifts north and south according to where the greatest amount of sunlight is delivered, and that's determined by "precession," the wobble in the tilt of the Earth's axis. In the northern hemisphere, the wobble isn't the main determinant of glaciations, but in the tropics it could well be, judging by what's in the ice cores. "This is a fundamental new piece of climate information," Thompson says. "It may be the new paradigm."

These days the average latitude of the vapor band is about 5°N, which suggests that glaciers south of the equator should be retreating. They *are* retreating. But north of the equator, where glaciers are getting plenty of snow and should be advancing, they are also retreating. Thompson concludes that the response of the glaciers to the increase in snowfall is being overridden by an increase in the temperature of the atmosphere due to a buildup of greenhouse gases. "What this tells me," he says, "is that the *human* driving system is going outside the *natural* driving system." According to what the ice cores taken from glaciers in Asia show, the last 50 years have been the warmest in the last 12,000 years there, and "*that's* outside the range of natural variability."

Thompson feels a sense of urgency about drilling more ice cores while there's readable ice to drill. "These archives of our past could literally melt away," he says. "It's possible that in the not too distant future the only tropical ice cores for study will be in our freezer." He has been back nine times to Quelccaya to remap it and each time found it markedly smaller. Its largest outlet glacier, Qori Kalis, retreated three times faster between 1983 and 1991 than it did between 1963 and 1983, five times faster from 1993 to 1995, eight times faster between 1995 and 1998, 33 times faster between 1998 and 2000, and 44 times faster between 2000 and 2001. In a valley that was covered by ice a little more than 25 years ago, three new lakes have appeared. The ice cliff with the wedding-cake dust bands no longer exists. "All gone," Thompson says, with a sweeping move of his arm.

In January 2000, "as soon as the millennium people got off the top," he and several colleagues climbed Mount Kilimanjaro and drilled six cores to bedrock. The information in the cores revealed that the late twentieth century was the warmest period on the African mountain in 11,000 years. Since 1912, when Kilimanjaro was first surveyed, it has lost 82 percent of its ice, in thickness and extent. When Thompson went back in 2001, he found that in the intervening year the summit had thinned by three and a half vertical feet. At that rate, he predicts, there will be no ice left on the mountain by 2020, maybe even by 2015.

The loss will be greater than a couple of square miles of ice out of a world total of hundreds of thousands.

> "Then they began to climb and they were going to the east it seemed, and then it darkened and they were in a storm, the rain so thick it seemed like flying through a waterfall, and they were out and Compie turned his head and grinned and pointed and there, ahead, all he could see, as wide as all the world, great, high, and unbelievably white in the sun, was the square top of Kilimanjaro. And then he knew that there was where he was going."*

"What is a snowcap worth to us?" Alley asks. "I don't know about you, but I like the snows of Kilimanjaro."

* Ernest Hemingway, "The Snows of Kilimanjaro."

CHAPTER TEN

ON GLACIERS

How determined we all are to be frozen
and how anxious to be among the ice.

—Lt. James Fitzjames, before setting
off with the Franklin expedition in search
of a Northwest Passage

THERE IS A POLE at the South Pole, a real pole with red and white barbershop stripes spiraling up to a mirrored globe a little bigger than a soccer ball which gives those who stand beside it looking down a view of the heavenly sphere with their own heads at the center. Having an actual pole at the pole, in a familiar, homey guise, helps a visitor handle the concepts of spin axis and zenith, although the first one installed there was meant to boost morale and maybe catch the attention of pilots of supply planes flying overhead so they'd know where to make a drop. Near the pole is a geodesic dome, half buried in snow, headquarters of the Amundsen-Scott research station, named for the leaders of the first team to reach the South Pole, which was Norwegian, and the second, which was British, but run by Americans since 1956. Inside the dome when I was there, besides laboratories, offices, refrigerators that kept food from freezing, machine shops, and a United States post office that sold liquor and souvenir T-shirts as well as stamps, there was a library with thousands of books, among them first-person accounts by the early Antarctic overland explorers—Amundsen, Scott, Shackleton, Mawson, Byrd, and Cherry-Garrard.

Even without reading these accounts, a person at the South Pole nowadays can get an inkling of what it must have been like in 1911 and 1912 to trek from the edge of the frozen Ross Sea over the Great Ice Barrier (now the Ross Ice Shelf) and up the glaciers that flow between the Transantarctic Mountains and across the polar plateau to the pole for the first time, and the second, as you stand beyond the last of the station's motley collection of huts and buildings on stilts and look out over an utterly empty flat white plate of a surface toward the horizon ("monotonous," Robert Falcon Scott declared). Or when, on a sunny summer day the video monitor in the station dining room reads "Temperature –60 With Wind Chill" and you step outside dressed like a Mongolian herdsman. Or as you regard the flags of the twelve nations that were signatory to the Antarctic Treaty of 1959 flapping on their high poles near the short barbershop pole and are reminded of the black flag that Roald Amundsen tied to a piece of sledge runner at his final camp before reaching the pole, which broke Scott's heart when he saw it.

"Great God! this is an awful place," Scott wrote in his journal when he and four companions reached the pole and "put up our poor slighted Union Jack." It was not, however, as awful a place as some of the ones that the men had labored over already and would labor over again as they tried to retrace their route to base camp. Seen in hindsight—even Scott's hindsight—many things contributed to making a "wreck" of the expedition—Scott named bad weather, particularly the prolonged blizzard at the end; a late start because of the failure of pony transport, which exposed them to extreme cold toward the last; and the illness of two of the party. Other people mentioned the decision not to rely entirely on dogs for transport, the decision to add a fifth man to the pole party when there were provisions for four, the decision even when the surviving men were weak and hungry to keep hauling 35 pounds of rock specimens collected off a moraine on the Beardmore Glacier. In addition, there were a couple of characteristics of the glaciers themselves that helped to delay and ultimately defeat the Englishmen: crevasses that functioned as booby traps and surfaces that seemed in some ways more appropriate to a beach than a glacier.

"The crevasse, Nature's pitfall—that grim trap for the unwary—no hunter could conceal his snare so perfectly," Scott wrote at one point, "—the light rippled snow bridge gives no hint or sign of the hidden danger, its position unguessable till man or beast is floundering, clawing and struggling for foothold on the brink." It was beasts that floundered during a cache-laying trip that several of Scott's men made over the Ice Barrier, using dogs. Two dog teams were "trotting side by side," Scott wrote, pulling their loaded sledges, when the middle dogs of one of the teams vanished. "In a moment the whole team were sinking—two by two we lost sight of them, each pair struggling for foothold. . . . The dogs hung in their harness in the abyss, suspended between the sledge and the leading dog . . . howling dismally." As it dangled, one dog kept trying to climb out, clawing at the walls on either side of the crack, and two dogs bit each other whenever they swung within range. Two other dogs had fallen out of their harnesses entirely and could be seen "indistinctly" on a collapsed snow bridge "far below." With thirteen wailing dogs trapped in it, the cold crevasse became "an inferno."

The men saved the beasts, working from a sledge pulled across the crack, with Scott retrieving the last two after having himself lowered on a rope onto the collapsed snow bridge 65 feet down. The air inside the crack, he reported, was warmer than outside. There were "curious curved splinters" of ice sticking out of the wall high on one side, he also reported. The crack narrowed below the snow bridge, so that if someone broke through the bridge he wouldn't fall far but—and Scott seemed to find this reassuring—be wedged in. Meanwhile, one of the men examining the crack from above, Apsley Cherry-Garrard, was so impressed by its size and steepness that he was reminded of "the cliff of Dover. . . . It is a great wonder the whole sledge did not drop through."

A few days before the dog incident, Scott had written a memo to himself to get "for next year," meaning when they were trying for the pole, "a stout male bamboo shod with a spike to sound for crevasses." (Female bamboos are evidently too willowy.) Whether or not Scott did get a spiked bamboo, once on the march he and the men ran into a great many crevasses and fell into more than a few of them. "We have all been half-way down," he remarked after a day in heavy fog on a badly crevassed region of the Beardmore Glacier—merely half-way down because they were roped together. Since the Beardmore funnels ice off the polar plateau through valleys to the Ross Ice Shelf, the ice in it gets severely broken up. (When I flew over the glacier in an airplane, the crevassed sections looked to me from a distance like choppy seas, on closer approach like deeply plowed earth, and from right above like mountain ranges, jagged and chaotically aligned, the valleys steep sided, narrow, and apparently bottomless, impossible to cross.)

Scott's men were on the Beardmore on Christmas Day, and after having eaten a celebratory treat of raisins and chocolate, one of them, W. Lashly, fell into a crevasse which had been hidden by a glasslike crust of *névé,* and Scott—the dutiful scientist—recorded that the crevasse with Lashly in it was "50 feet deep and eight feet across, in form U, showing that the word 'unfathomable' can rarely be applied." It was Lashly's 44th birthday, yet after being rescued he showed no sign of regret at having to spend a good part of the double holiday in an icy pit. "Hard as nails," Scott concluded.

Most of this falling into crevasses and getting "jerked about abominably" was merely "trying," according to Scott, but on the men's return from the pole, during which they faced "800 miles of solid dragging," one man's plunge proved disastrous not only to himself but to some degree to the whole expedition. While walking on a "good hard surface," Scott wrote, he and Edgar Evans fell together into a crevasse. Later that day Evans became "rather dull and incapable," apparently having suffered a concussion. His condition worsened over the days that followed. One day the group took a wrong turn, in "horrible light, which made everything look fantastic," and ended in the worst "ice mess" Scott had ever seen. "There were times when it seemed almost impossible to find a way out of the awful turmoil in which we found ourselves. . . . We could not manage our ski and pulled on foot, falling into crevasses every minute." Irregular cracks gave way to "huge chasms, closely packed and most difficult to cross. It was heavy work, but we had grown desperate." Evans stumbled on but by then had "nearly broken down in brain, we think." Two weeks after he fell into the crevasse, Evans collapsed in the snow, went into a coma, and died that night, at the foot of the Beardmore Glacier, the first of the five men to go.

As important in delaying the expedition as glacial surfaces that gave way under their feet were surfaces that hung on to their sledge runners. Before leaving base camp, one of Scott's men had given a lecture on "types of surface" to be found on glaciers: "Rippled, snow stool, glass house, coral reef, honeycomb, ploughshare, bastions, piecrust." Once the men had departed, however, Scott's descriptions of surfaces ran more to "awful," "wretched," and "terrible bad." They were pulling the sledges themselves—"heavy collar work," Scott called it—ponies and trucks having failed as transport early on, and the surface crystals were dry as sand. Because of the extreme cold, there was little liquid lubrication on the crystals; any melt caused by the friction of the runners quickly refroze. Most of the time the "sandy crystals," as Scott referred to them, were loose snow, but some of the time they were frost or even rime ice; "a beard of sharp branching crystals," he once complained, kept the sledges from gliding off ridges "even on a down grade."

In his journal, Scott spelled it out:

February 19: "A really terrible surface. . . . It has been like pulling over desert sand, not the least glide in the world."

March 3. "The surface . . . is coated with a thin layer of woolly crystals, formed by radiation no doubt. These are too firmly fixed to be removed by the wind and cause impossible friction on the runners. God help us, we can't keep up this pulling."

March 4: "Things look *very* black indeed . . . surface covered with sandy frost-rime."

March 12: "The surface remains awful."

On March 21, when they were only 11 miles from their next cache of food and fuel, a severe blizzard hit and confined them to their tent. By March 29, they were out of food, out of fuel, out of strength, and out of hope. "The end cannot be far," Scott wrote. Eight months later, a relief party of men from base camp found the frozen bodies of three men inside the tent.* They wrapped them *in* the tent, built a cairn of snow blocks above it with a cross on top made from parts of two skis, and left the bodies to lie "without change or bodily decay," the surgeon in the party wrote, "with the most fitting tomb in the world above them."

By contrast, Roald Amundsen's trip to the pole has been compared to a winter picnic, "ski racing writ large." He and four other Norwegians, all veteran skiers, one a champion, made the 870-mile run from the edge of the Ice Barrier to the pole and back in 99 days, covering as many as 50 miles in a single day on the return and having so much food to eat that it was like "living among the fleshpots of Egypt," Amundsen wrote in his book, *The South Pole.* Still, the accursed sandy crystals often created "wretched going" for the Norwegians, too. "A sledge journey through the Sahara could not have offered a worse surface to move over," Amundsen declared.

As for crevasses, most of the Norwegians would probably have ended up in them if they hadn't been wearing skis, according to Amundsen. "Many a time we traversed stretches of surface so cleft and disturbed that it would have been an impossibility to get over them on foot." The route they took up mountain glaciers between the Great Ice Barrier and the polar plateau was even steeper than the one the Scott party took up the Beardmore, and it was likewise riven by cracks and given to "insidious pitfalls." In a region filled with monster crevasses, "some hundreds of feet wide and possibly thousands of feet deep," Amundsen reported, they saw to one side of their scouting track a landscape so disturbed it "looked as if a battle had been fought here, and the ammunition had been great blocks of ice. They lay pell-mell, one on top of another, in all

* Another man, Lawrence Oates, with a bad foot, had earlier walked to his death in a blizzard rather than delay the party.

directions, and evoked a picture of violent confusion. Thank God we were not here while this was going on, I thought to myself, as I stood looking out over this battlefield, it must have been a spectacle like doomsday." They managed to find a passage through another chasm- and hummock-filled route, "if, indeed, it deserved the name of a passage. It was a bridge so narrow that it scarcely allowed room for the width of the sledge; a fearful abyss on each side. The crossing of this place reminded me of the tight-rope walker going over Niagara."

Once they reached the polar plateau, the crevasses petered out and so did the waves of wind-hardened snow called "sastrugi" (after the Russian word for the uneven surface of roughly planed wood), which had caused many falls on their skis. Then one day, which "passed without any occurrence worth mentioning," in brilliant weather, Amundsen's team reached 90°S. "It was a vast plain of the same character in every direction, mile after mile" and reminded at least one of the party of home. "Here it's as flat as the lake at Morgedal," Olav Bjaaland confided to his own diary that night, "and the skiing is good."

I think of Amundsen, enormously bit
By arch-dark flurries on the ice plateaus

—Richard Wilbur,
"On the Eyes of an SS Officer"

On the very day that Amundsen skied over the pole, December 14, 1911, the Australian explorer Douglas Mawson was crossing a glacial highland on the Antarctic coast with two sledge-driving companions, B. E. S. Ninnis and Xavier Mertz. The three had managed to get through an area with crevasses "like huge cauldrons," Mawson wrote in his book *The Home of the Blizzard,* and were on a smooth snowfield where they didn't expect any. Mertz was traveling in front, singing his student songs, followed by Mawson, then Ninnis. Mawson heard "a faint, plaintive whine from one of the dogs" but didn't think anything of it until he noticed an alarmed expression on Mertz's face. He looked behind him and saw a hole about eleven feet across with two pairs of sledge tracks leading up to it on the far side, one of the pairs being Mawson's own, and a single pair leading away from it on the near side—Mawson's alone.

"I leaned over and shouted into the dark depths below," Mawson reported. "No sound came back but the moaning of a dog, caught on a shelf just visible one hundred and fifty feet below. . . . For three hours we called unceasingly but no answering sound came back. The dog had ceased to moan and lay without a movement. A chill draught was blowing out of the abyss. . . . Our fellow, comrade, chum, in a woeful instant, buried in the bowels of the awful glacier."

Stunned and at first heedless of the danger to themselves, Mawson and Mertz kept going, toward a glacier that had earlier been named for their now-buried chum. "During descents I sat on the sledge and we slid over long crevassed slopes in a wild fashion," Mawson wrote, "almost with a languid feeling that the next one would probably swallow us up." Nearly all their food had gone down with Ninnis's sledge, and before long they began eating their dogs, the weakest first. They soon became ill, without knowing why. They had eaten the dogs' livers, which (it's now known) store toxic amounts of vitamin A. (So do polar bear livers, and Arctic explorers who ate them often became ill, too.) Their hair fell out in large tufts, nails grew loose, toes blackened, skin peeled off. One day Mertz said to Mawson, " 'Just a moment,' and, reaching over, lifted from my ear a perfect skin-cast. I was able to do the same for him." Mertz grew lethargic, weak, depressed, chilled, and one day delirious. He died, presumably of hypervitaminosis A, leaving Mawson by himself with little to eat, no dogs to pull his sledge, and his own symptoms: "raw patches all over the body, festering fingers and inflamed nostrils."

Alone, ill equipped, and sick, Mawson trudged on toward a glacier named earlier for this second now-deceased chum. Then, on January 17:

"Broke through lid of crevasse but caught myself at thighs, got out, turned fifty yards to the north . . . a few minutes later found myself dangling fourteen feet below on end of rope in crevasse—sledge creeping to mouth—had time to say to myself, 'so this is the end,' expecting the sledge every moment to crash on my head and all to go to the unseen bottom. . . . The width of the crevasse was about six feet, so I hung freely in space, turning slowly round.

"A great effort brought a knot in the rope within my grasp, and, after a moment's rest, I was able to draw myself up and reach another, and at length, hauled myself on to the overhanging snow-lid into which the rope had cut. Then, when I was carefully climbing out on to the surface, a further section of the lid gave way, precipitating me once more to the full length of the rope.

"Exhausted, weak and chilled (for my hands were bare and pounds of snow had got inside my clothing) I hung with the firm conviction that all was over except the passing. Below was a black chasm; it would be but the work of a moment to slip from the harness, then all the pain and toil would be over. It was a rare situation, a rare temptation—a chance to quit small things for great—to pass from the petty exploration of a planet to the contemplation of vaster worlds beyond. . . .

"My strength was fast ebbing; in a few minutes it would be too late. It was the occasion for a supreme attempt. . . . The struggle occupied some time, but by a miracle I rose slowly to the surface. This time I emerged feet first, still holding on to the rope, and pushed myself out, extended at full length, on the snow—on solid ground. Then came the reaction, and I could do nothing for quite an hour."

The "ground" did not stay solid. As Mawson plodded onward, he fell often into "space" but managed to get out with a ladder he fashioned from alpine rope and attached to the front of the sledge he was pulling. At night "loud booming noises, sharp cracks and muffled growls issued from the neighbouring crevasses . . . and I concluded that the ice was in rapid motion." He trudged over the ice sheet for a week and a half, until at last he happened upon a snow cairn where comrades had left a bag of food and a note saying that Amundsen had reached the pole and Scott was "remaining another year in Antarctica" (they couldn't have known it would be forever). A couple of days later, Mawson stumbled into a cavern he and his men had carved out of a glacier six months before, a "haven within the ice" which they had named Aladdin's Cave because it was "a truly magical world of glassy facets and scintillating crystals" on walls of solid ice. There he found—brought by men on the ship that he hoped was at that very moment lying at anchor waiting to pick him up although he was long overdue—three oranges and a pineapple.

More than a week later, after an immobilizing blizzard, Mawson got within a mile and a half of the base hut, but as he looked out over ice cliffs toward the sea, he saw "a speck on the north-west horizon." It was the ship, already departed, and there wouldn't be another until the next summer. (As it happened, six men had stayed behind on the ice in case he, Ninnis, and Mertz showed up, and they and their food stores kept Mawson alive through the long Antarctic winter.)

A half century later, when explorers started using motorized vehicles instead of skis and dogs to get around Antarctica, it was pretty much the same as far as crevasses were concerned. The transport may have been more modern and bulkier, but the crevasses could still booby-trap them. Sir Vivian Fuchs, leader of the British Commonwealth Trans-Antarctic Expedition, which in 1957–58 went from one Antarctic coast to another by way of the South Pole in Sno-Cats, Weasels, and Muskegs, those workhorses of the northern winter-road builders, said of crevasses he encountered on a scouting trip for the journey: they "could not have been more diabolical than if they had been traps deliberately set." Some were big enough to swallow a double-decker bus.

The men began roping the vehicles together, "like climbers on a mountain." They checked for snow bridges, using ice chisels mounted on long wooden poles. "We plunged them downward, butt end first," Fuchs wrote. "If they struck a crevasse bridge, it would reverberate loudly, and these we came to call 'boomers.' Then, with the chisel end, we would cut a hole large enough to thrust one's head, and sometimes shoulders, through the lid to see the width and direction of the crevasse. Hanging head down over a bottomless pit, with sloping blue-white sides disappearing into the depths, can be somewhat alarming—especially when you know that very soon you will be driving a three- or four-ton vehicle, together with heavy sledges, over the precarious snow bridge above the dark abyss."

Barely 30 miles into their 2,000-mile journey, a snow bridge gave way under a Sno-Cat, leaving Fuchs and another man "suspended in mid-air over an impressive chasm." To get themselves out, they had to crawl over a pontoon and ladder-like track "as it hung in space"; to get the Sno-Cat out, two other Sno-Cats plus two Weasels and a Muskeg had to pull on it. The men started traveling at night, in the hope that colder temperatures would strengthen the snow bridges, and employed "crevasse-bridging units," 14-foot-long pieces of metal which they'd lay along the edges of the cracks, as reinforcement, so a vehicle that fell in had something firm to climb out on. That must have presented some spectacle. One Sno-Cat, Fuchs wrote, "heaved and wallowed its way to the surface . . . like some monster rising from the deep."

Meanwhile, on the other side of the continent, Sir Edmund Hillary was leading a similar caravan of tracked vehicles, planting caches that Fuchs was to pick up on his outward leg from the pole. Although Hillary had encountered crevasses on Mount Everest, those in Antarctica were of a different order. In his book *Nothing Venture, Nothing Win,* he wrote about finding a punched-out snow bridge, then creeping up to the edge of it and looking down: "To my amazement I saw . . . a huge void large enough to absorb a house. It was a most unpleasant sight with sheer ice walls dropping away to vast depths and enough room to put a hundred Fergusons [tractors]."

One time Hillary almost made it into the void himself. He had been driving through an area with many ice hummocks—"familiar signs of a crevasse area"—when "suddenly there was a thud beneath me, and my tractor tipped steeply backwards. The bridge had gone and I was going with it! For a few awful moments we teetered on the lip of the crevasse, with the tracks clawing at the edge and the nose high in the air. Then the extra bit of power told and we seemed literally to climb up the wall of the crevasse and thump to a level keel on the far side . . . If the motor had stalled, I wouldn't have had a chance." The conqueror of the heights jumped out the door of his tractor, "somewhat shaken" by what he had nearly experienced of the depths.

As things worked out, Hillary reached the pole from his side of the continent before Fuchs did from his, which made Hillary's "the first vehicle party to travel overland to the South Pole." (His was also the first party of any kind to make it to the pole in the 45 years since Scott had stood dejectedly there.) Yet Hillary had his doubts about the worthiness of this, his second first. "We showed that if you were enthusiastic enough and had good mechanics you could get a farm tractor to the South Pole—which doesn't sound much to risk your life for," he declared. "But we had produced no scientific data about the ice and little information about its properties." Fuchs's group did get some information about the ice. Taking seismic shots every 30 miles across the polar plateau, the explorers determined that the ice at the pole is about 8,000 feet thick, over a rock basin 50 miles wide with sides a mile high.

Sweetness & light, ice & fire

—Richard Wilbur, *"&"*

The barbershop pole is ceremonial; nearby is a second pole with a marker on it that shows precisely where the true geographic South Pole is. Every January 1, Americans working at the Scott-Amundsen base take down the existing marker pole and put up a new one 33 feet away. Thirty-three feet is how far the ice in the ice cap moves every year as it flows northwest toward the Weddell Sea, taking poles (and the motley collection of buildings) with it. The year I was there, Larry Hothem, physical scientist at the U.S. Geological Survey, had calculated that by the year 2017 the pole would be underneath the station's geodesic dome or, rather, the ice under the dome would have slid across the pole. The part of the dome that would be right over the pole on January 1, 2017, is the medical center. You too can be an Antarctic pioneer, Hothem suggested. "Come down, arrange to get sick during that month, and you can be lying in bed as the bed slides toward the pole, and you will be at the southernmost center of the world."

CHAPTER ELEVEN

ICEBERGS I

It is not its air but our own awe
That freezes us.

—W. S. Merwin,
"The Iceberg"

THE MOST FAMOUS PIECE of ice in the history of the world is the one the steamship *Titanic* hit on April 14, 1912, during its maiden voyage across the Atlantic. The night was calm, cloudless, moonless, and cold enough that the watch on the bridge noticed "whiskers" around the deck lights from ice crystals in the air. Two lookouts in the crow's nest were scanning the sea for icebergs; that very day six other ships had telegraphed messages warning of ice in the area. Shortly before midnight, one of the lookouts saw in the darkness off the bow a darker shape; he banged a warning bell three times and spoke three words into the phone to the bridge: "Iceberg right ahead." The first officer shoved the wheel hard to starboard and shouted at the crew to stop the engines and then reverse them, but it was too late. The rest is history, oft-told history. The impact of the iceberg grazing the luxury liner was "not very great," according to Walter Lord, author of *A Night to Remember* and *The Night Lives On*. "Some people never did wake up." Those people whom the impact did cause to wake up, or who while already awake noticed a jolt, Lord noted, described it variously as being:

"A little like coming alongside a dock wall rather heavily" (quartermaster on the afterbridge). "Like a heavy wave striking the ship" (Canadian chemical manufacturer). Reminiscent of the "sloppy landing" of a Zurich lake ferry (Swiss girl). As though the ship had rolled "over a thousand marbles" (woman about to turn off her bed light).

Stewards in the dining saloon, who noticed the silverware rattling, also sensed "a faint grinding jar from somewhere deep inside the ship." One young woman assumed the men in the next cabin were having yet another rowdy pillow fight. Several firemen, hearing a thud, then a grinding, tearing noise, believed the ship had gone aground on the Grand Banks. A woman who felt nothing did hear an "unpleasant ripping sound as of someone tearing a long, long strip of calico."

Few of the more than 2,000 people on board actually saw the iceberg as it went by, but the quartermaster, one of the few, said it resembled "a windjammer, sails set, passing along the starboard side." A passenger who had leapt out of bed when he felt a bump and had run over to the porthole observed "a wall of ice gliding by" (in other starboard cabins, passengers with their portholes open found chunks of ice on their floors). "Round, with one big point sticking up on one side," an American woman described the iceberg, which she probably saw after it had passed. "The Rock of Gibraltar as seen from Europe Point," another said, adding "only much smaller."

Some passengers in the third-class recreation space, where several tons of loose ice landed, threw scraps of it at each other, and in steerage, men played soccer with ice chunks. Here and there people picked up samples—variously the size of a pocket watch, a teacup, and a small washbasin, according to Lord—and showed them around as curiosities. One crewman threw a chunk of ice—conveying what message?—onto the bunk of another crewman, his son-in-law.

Below deck, there was no such frivolity, however. The collision had opened up holes in the first six of the ship's sixteen "watertight" compartments, which had automatic closing doors, the innovative safety feature that had led some people to call the ship "unsinkable." Seawater poured into the compartments so fast that a couple of firemen barely made it out. As the water rose in one compartment, it would flow over a bulkhead into the next one, from there into the next, and on down the line.

Everyone knows the story: the slow assembling on deck of passengers in all manner of deshabille and habille—nightgowns, furs—the redeeming of jewels and money from the purser; the shortage of lifeboats—space for only 1,178 of the 2,201 people on board—the lowering of some boats without all of the places filled; the radio operator who had earlier brushed off an ice warning desperately sending out SOS's; the captain's "women and children first" order observed, occasionally in the breach; the lights in the saloons continuing to blaze; the band playing gamely on; the decks tilting; two men changing into formal attire for the finale; the passenger ship *Californian,* ten miles away, standing still through the whole ordeal; then at 2:20 a.m., fewer than 2 1/2 hours after the wall of ice had scraped by with a modest jolt and a tearing sound, the

Titanic breaking in two and all 46,000 tons of it sliding into the water, carrying tennis racquets, and golf clubs, stocks, bonds, potted palms, 15,000 bottles of ale and stout, 800 cases of shelled walnuts, 30,000 eggs, five grand pianos, a case of gloves, a jeweled copy of the *Rubaiyat,* an automobile, a telephone switchboard, a bracelet with "Amy" written on it in diamonds, shuffleboard sticks, a silver duck press, and an ice-making machine (!) 13,000 feet down; the gulp the rent ship made as it vanished; the howling from people released into the freezing water and soon to perish.

Meanwhile, the liner *Carpathia* was steaming full speed to the spot from fifty miles away, its captain risking iceberg encounters himself. (The ship avoided one iceberg only because, according to Lord, the second officer spotted "a tiny shaft of light glistening two points off the port bow, revealed by, of all things, the reflected light of a star.") One passenger on the *Carpathia,* peering anxiously out her porthole, believed what she was seeing was a rock on shore, although she did wonder how that could be "in mid-ocean four days out of New York." One woman likened a grayish iceberg she saw from the *Carpathia* to the frame of a dirigible fallen onto the sea. When dawn came and revealed to the captain what the ship had passed through unscathed, he was astonished; behind him lay "a plain of packed ice" with several "towering monsters" of icebergs in it "150 to 200 feet high."

By the time the *Carpathia* reached the *Titanic*'s last radioed position, there was no sign of it, only its lifeboats, difficult to distinguish at first from small icebergs, scattered over a four-mile area. The *Carpathia* picked up survivors in the lifeboats; there were none in the water. Several days later another ship, this one loaded with coffins, shrouds, embalming fluid, and tons of ice, arrived to collect bodies, which were floating in a line 50 miles long, some so chewed up from being banged by ice floes that their faces were unrecognizable. These were buried at sea; the rest were laid in coffins packed with ice and taken to Halifax, Nova Scotia. Autopsies on several of them showed death came not by drowning but from the shock of the cold water, which the *Titanic*'s crew had measured at 28°F, the freezing point of seawater. One man who jumped in and was picked up by a lifeboat described the water as "a thousand knives."

Anyone who reads accounts of the sinking-by-iceberg cannot help thinking, "Oh, if only . . ." Among the many "if only's" mentioned by Lord, any one of which might have averted the tragedy and saved 1,502 lives, were, besides the obvious ones—if only the ship had been going more slowly, if only the ship had carried more lifeboats, if only the watertight bulkheads had been one deck higher, if only the radio had had a more powerful signal, if only the *Californian* had responded when the flares went up (its captain had never sailed in ice before, and neither evidently had one of the officers, who mistook the first white patches that appeared on the water for a school of porpoises crossing the bow)—were several "if only's" that concerned the iceberg itself. If only ice conditions had been normal—the ice was unusually heavy because of a stormy winter. If only the air hadn't been "flat calm," a breeze would have pushed waves against the iceberg and produced a white foam at its base, which lookouts might have spotted in time. If only the moon had been out that night to light up the berg. Most important, if only the officers had paid greater heed that day to the six messages warning of ice which, when considered together, indicated a large belt of ice running for 78 miles across the path of the ship, in which case the captain might have posted more lookouts and reduced speed, giving the ship time to complete a turn away from the iceberg when the lookout spotted it.

The iceberg that sent the largest passenger liner in the world to a premature grave and caused a society to question the very idea of progress was not a particularly large one, certainly not the largest around. The lookout who spotted it later testified that it was "a little higher than forecastle head but not as high as crow's nest." Another estimated that it was 50 feet high, another 100 feet high. After the berg disappeared behind the ship's stern, it was never seen again, or at least not that anybody knew for sure. For years, the chief candidate was a long, low berg with a knob at the end whose photo had been taken by a person on the recovery ship days after the sinking, by which time the berg would almost certainly have moved on. Another candidate was an iceberg with a steep central peak and two smaller peaks flanking it that the chief steward on the liner *Prinze Adelbert* took a photo of the morning the *Titanic* went down, a few miles south of where it sank but without knowing that it had sunk. What caught his attention was a line of "red paint" along the base. Skeptics note that icebergs often have stripes of sediment on them, which can be orangish. As likely as not, the perpetrator berg made its escape without anybody's getting a snapshot of it.

Nevertheless, the iceberg took on a certain celebrity, a spooky, menacing persona: the ocean beast that sank a ship! In Richard Brown's *Voyage of the Iceberg: The Story of the Iceberg That Sank the Titanic,* the iceberg—or the Iceberg, as he always refers to it—is the central character, around which various other dramas unfold. He traces its natural history from birth to death, which he was able to do without going too far wrong because most icebergs in the North Atlantic have a similar life history. Composed of snow that fell a thousand years before Christ onto the Greenland ice cap, he wrote, the Iceberg broke away from the Jakobshaven glacier, one of dozens of glaciers along Greenland's west coast which funnel ice from inland to the sea. The Jakobshaven was moving toward the sea at the rate of 65 feet a day (probably the fastest of any nonsurging glacier), producing an iceberg a minute, 20 to 30 million tons of ice a day, and five cubic miles of ice a year.

Somewhere in that tonnage in September 1910 is the Iceberg, white on top but "black as jet" underneath, Brown wrote, from coal dust the ice scraped off the mountains as it went. At first, currents carry it north, not south, up the west coast of Greenland; then the Iceberg goes across Baffin Bay to the coast of Ellesmere Island, where a cold current from the North Pole takes hold of it and carries it southward, past Baffin, Devon, Philpott, and Bylot islands to Davis Strait. On its way, it gets stuck on the seabed several times, is jostled free by a gale, rolls, and as the bottom turns into the top becomes rounded "like a big white whale," only with "a streak of frozen coal dust across its shoulder like a black ribbon." Accompanied by pack ice and fellow icebergs, the Iceberg is visited by seals and sealers, nibbled by crustaceans, carved by waves, lifted and dropped by tides. A crevasse opens up in its surface, a slab breaks off. It runs aground yet again, shakes loose, runs aground, then slips into the tepid waters of the shipping lanes of the Grand Banks, melting as it goes, "making fizzing noises as the bubbles in it pop."

By this time it "has the winged look of bergs that have reached the Grand Banks, and rolled often along the way: a tall, central spike of ice flanked by two smaller ones, like the back and arms of a chair." In the sea-ice cover around it are many other icebergs with more mundane destinies, so many bergs that "they look like sails of a fishing fleet." After the Iceberg has its midnight rendezvous with the great steamship, it drifts on, farther south, until it reaches the Gulf Stream, where the water temperature is many degrees higher, and there, after a journey of a year and a half, it succumbs. "Clear as glass" and glowing "a brilliant blue in the Sargasso water," the Iceberg is "scarcely strong enough to scratch a whaleboat."

Brown's book came out in 1983, 71 years after the *Titanic* sank; at the time of the sinking, little was known about the life cycle of icebergs. "Do you know what an iceberg is composed of?" William Alden Smith, the senator heading the American inquiry into the disaster, asked an English officer, hoping to learn, Lord surmised, whether the iceberg was carrying "more familiar lethal material like rocks" that might explain its destructive power. "It was hard to believe," Lord explained, "that a mere piece of ice, however big, could rip apart a steel hull." The English officer answered the question in a manner that conveyed disdain for what many in his country considered the American senator's unseemly effort to expose British incompetence: "Ice, I suppose, sir."

It wasn't long after the *Titanic* went down that there was talk of bringing it up. Proposals for ways to do that included pumping the hull full of molten wax which when hardened would make it buoyant; injecting it with thousands of Ping-Pong balls, causing it to float; and encasing it in ice. "Then, like an ordinary cube in a drink," Lord noted, "the ice would rise to the surface, bringing the *Titanic* with it." The liner stayed where it was, though. It was only in 1985 that anybody was able to locate it, 15 miles east of where it made its last distress calls. Several exploration teams in submersibles and robots were able to examine, among many things, the long gash the iceberg supposedly inflicted on the hull. Instead of a gash they found six unconnected narrow slits, each no wider than a person's hand in one comparison, in places no wider than a person's finger, covering a total area equal to that of a person's body or, in another comparison, to two sidewalk squares. Instead of slashing the ship's side, as had been assumed, the iceberg had applied so much pressure to it that the rivets popped and the metal plates buckled, perhaps because the rivets were defective and the steel low grade and brittle. The slits were 20 feet below the waterline, where pressure was so great that water rushed through them, narrow as they were. The fact that the openings didn't extend all the way to the stern suggested to Lord that the iceberg suffered in the collision, too. "It [the iceberg] is generally pictured as a great natural force," he wrote, "impervious to the assault of mere man, but the collision could have caused it to lose part of its mass, and spare the rear of the ship."

Today only three *Titanic* survivors are still alive, two in England and one in the United States. But the night lives on, and every now and then a tabloid newspaper discovers a new survivor. Among headlines like "Man shoots out sore tooth—with a pistol!" could be read not long ago "TITANIC BABY FOUND ALIVE! Sailors find infant girl floating in old life preserver. She's dressed in 1912 clothes & crying for her mother. Scientists say she passed through a time warp." Or: "TITANIC SUR-

VIVOR FOUND ON ICEBERG. She thinks it's April 15, 1912—and her dress is still wet!" Fishermen who found the woman reported her as "suffering from severe exposure" after almost 80 years on an iceberg.

A few *Titanic* survivors who lived in unwarped time until their turn came chose to have their ashes scattered over the spot where their ship, shipmates, and belongings went down. Several years ago, at the annual laying (or tossing out an airplane window) of a floral wreath on that spot by the Titanic Historical Society, the ashes of a woman who was 12 years old when she entered a lifeboat and had lived a mostly quiet life ever after were tossed out by members of the International Iceberg Patrol (IIP), whose pilots and scientists chipped in to purchase their own memorial wreath. The IIP, part of the U.S. Coast Guard, keeps track of icebergs in the shipping lanes of the North Atlantic, an area equal in size to a quarter of the continental United States, and lets ships know by radio and satellite of the icebergs' positions. It has been doing this every year since the sinking of the *Titanic,* except during the two world wars when forces other than icebergs doomed ships in the North Atlantic and left bodies bobbing around on its surface.

One thousand icebergs
in the sunlight

—George Bowering, *"Did"*

In 1914, when several maritime nations set up the ice patrol following a conference on safety at sea, it was known as the International Ice and Derelict Destruction Patrol. In those days, abandoned wrecks of wooden ships were drifting around the North Atlantic, too; many of them had been drifting for years, and often crews of working ships couldn't see them any better than they could icebergs, and the ships would hit them and become derelict themselves. The ice patrols destroyed the derelicts by blowing them up, but they couldn't destroy the icebergs, although they tried to. Over the years the IIP has shot bullets at icebergs, dropped bombs on them, set off mines inside them, and blasted them with underwater explosives, all of which efforts were, as former commandant of the U.S. Coast Guard Admiral W. J. "Iceberg" Smith put it, "spectacularly unsuccessful." The patrol eventually figured out that it would take more than 1,900 tons of TNT to break apart one average-sized iceberg and all the heat generated by the burning of 2.4 million gallons of gasoline to melt it. Likewise unsuccessful was the trick the patrol tried of dumping bags full of carbon black on an iceberg and spreading it around with big brooms to darken the surface so that sunlight would melt it faster; the melting washed off the carbon. Nowadays the IIP just waits for icebergs to disappear in their own time and concentrates on spotting them and telling mariners where they are or are likely soon to be.

Before World War II the patrol did the spotting mostly from ships, after World War II mostly from airplanes. During the "ice season," which usually runs from the beginning of March through the end of June but may start in February and last through July and into August, C-130 turboprops fly search patterns over half a million square miles of ocean off the east coast of Canada, covering the equivalent distance north and south as from, say, Miami to Washington, D.C. Aboard are three observers, or "ice picks," who aim side-scanning radar over a 27-mile-wide swath of ocean on either side of the airplane. Whenever the radar screen registers a target, the picks try to confirm by sight that it's an iceberg and not, say, a fishing trawler. If they can't see the target because of fog or clouds, they direct forward-looking radar at it to find out how it's responding to waves. Targets that sit stolidly in the water are probably icebergs. Targets that do a lot of rolling, pitching, and yawing are probably trawlers. Icebergs have deep keels; fishing trawlers do not.

The ice picks note the positions and sizes of all icebergs, then radio them to IIP headquarters (on whose office walls, lest anyone forget, are framed front pages from April 16, 1912, of the *St. Louis Post-Dispatch*—headline: "1302 Lives Lost When 'Titanic' Sank; 868 Saved"—and the *New York Times*—headline: "Titanic Sinks Four Hours After Hitting Iceberg; . . . Probably 1250 Perish; Ismay Safe, Mrs. Astor Maybe, Noted Names Missing"). There the positions are fed into a computer model, along with wind and current measurements plus sightings made by other interested parties, including some on ships passing through the area, in lighthouses, on aircraft checking for violations of Canadian fishing regulations, even an occasional transatlantic airline captain gazing down at the Atlantic from 33,000 feet. The model comes up with likely drift paths for the bergs over the next 12 hours. "It's important to know not only where an iceberg is but where it will be tonight or in tomorrow's fog," Donald Murphy, an oceanographer and the IIP's chief scientist, explains. Twice a day the IIP issues "ice bulletins" to ships planning to enter the shipping lanes, or already in them, along with a chart showing the southern and eastern "Limit of All Known Ice."

Close to the limit's edge, the chart displays individual icebergs, "to give the limits some credibility," Murphy says. Each iceberg is a triangle and each "growler," or small iceberg, a sort of decapitated triangle. (Growlers got their name from the sound they make when sea swells set them riding up and down and, on one of their aggravated downward swings, they plunge deeper into the water, sucking and growling.) Away from the limit's edge, however, the IIP notes on the chart simply "Area of Many Bergs." "It's a 'There Be Dragons' approach," Murphy explains. "We don't plot all icebergs because we don't want a mariner to think he can zigzag around them like with a street map." The chart shows the limits of *sea* ice, too, as the Canadian Ice Service determines it. In spring the southern edge of the sea-ice pack may define the southern border of all known ice, but in summer the pack is likely to have retreated 400 or 500 miles north, even as icebergs continue to drift south.

In the model there's an "error circle" around each iceberg, which grows larger and larger as the days go by and it's not sighted again. Sighting an iceberg again, or knowing that you are sighting it again, isn't as easy as you might think. In the Arctic, where water temperatures are close to freezing, icebergs last a long time, but once they reach the waters of the Grand Banks they deteriorate fast, not only shrinking but changing shape. "Icebergs normally change their appearance so radically," an IIP publication points out, "that it is impossible to positively identify them after only a few days." For years, the patrol assigned ships to follow individual icebergs until they melted, in the belief that "only continuous contact could ensure that the iceberg being tracked remained the same piece of ice." It also tried tagging them with color, dropping mayonnaise jars full of red dye onto them from airplanes, and shooting test tubes of dye by bow and arrow at them from boat decks. It tried putting transponders on them by wrapping a nooselike rope around them with a transponder hooked to it; by parachuting transponders onto their surfaces; and by firing a large steel dart "like a little rocket," says Murphy, deep into the flank of an iceberg, with a transponder trailing behind at the end of a tether. But the dyes washed off as the bergs melted and rolled around, and the transponders got lost as the bergs slipped their nooses or broke apart. "We gave up," Murphy says. "We are just going to have to go looking for the icebergs and ask ourselves, 'Is it *reasonable* that this is the one the *QE2* saw three days ago?' "

The Iceberg hit the *Titanic* at 41°46′N, the latitude of Providence, Rhode Island. Occasionally, icebergs drift south to 40° N, the latitude of Philadelphia. In 1991, the IIP tracked one weathered berg to 38°45′N, the latitude of Washington, D.C.; they spotted it floating 1,200 miles east of the capital. The farthest south that humans have reported sighting an iceberg is 30°20′N, the latitude of New Orleans, Louisiana, and Jacksonville, Florida.

The IIP chooses to track only those icebergs that drift south of 48°N, the latitude of, say, St. Narcisse de Rimouski in Quebec, a few miles north of the northern tip of Maine. That's the northern boundary of the main shipping lanes. Above 48°N, there are more icebergs but also fewer ships. "People think an iceberg patrol is a polar operation," Donald Murphy points out, "but it isn't. It's a temperate-water operation." (The Canadian Ice Service spots icebergs along with sea ice in Canadian waters up to the tip of Labrador, 60°N.) Ships that travel in the Atlantic north of 48°N, to catch fish or go to and from Iceland, usually have experience ice-wise and know what they're doing. Nevertheless, in January 1959, the steamship *Hanshedtoft,* returning from *its* maiden voyage between Denmark and Greenland and equipped with the latest electronic aids, hit an iceberg 40 miles south of Greenland and sank without a trace—except for a single life buoy that washed ashore on the coast of Iceland—taking all 95 people on board down with it.

The number of icebergs that drift south of 48°N varies widely year to year. In 1984, 2,002 icebergs did, the most the IIP has recorded. In 1966, none did. "There was no ice season," Murphy declares. In 1999, the patrol reported seeing only a dozen icebergs. "It was a little spooky. We worried that we were missing something," Murphy says. During an average iceberg season, of the many thousands that are calved each year by glaciers in Greenland and the Canadian High Arctic islands—a region so productive of icebergs that an old-time whaler called it an "ice manufactory"; one current estimate of output is 300 *billion* cubic meters of iceberg ice a year—roughly 600 sizable icebergs manage to get south of 48°N. The rest are grounded in shallow water or stuck in bays farther north and end their lives there.

Growlers aren't counted because they're so hard to see the IIP is sure to miss some, but they are displayed on the chart if they show up near the ice edge. A growler, in the IIP's official classification of iceberg sizes, when it compares them to boats, is as big as a "dory." A bergybit (what Canadians call a large growler) is the size of a "survival craft." A small iceberg is equal to a "long liner," a medium iceberg to a "supply vessel," and a large iceberg to a "drilling rig." For those of us who grew up well inland, and whose idea of a survival craft is how to make your own candle out of old newspaper strips, other comparisons are more helpful. The U.S. Navy after World War II likened a growler to a grand piano, a bergy bit to a small cottage, and an iceberg to a ship. On its charts, the IIP keeps it simple, displaying only icebergs and, near the limits of all known ice, growlers, as triangles and half triangles.

As the Iceberg showed, an iceberg doesn't have to be enormous to make an impression on a ship. An iceberg with a sail—the part above the waterline—that's only three feet high can weigh 100 tons, since its weight includes the keel, the part below the waterline. Murphy says, "If your ship weighs 100,000 or 200,000 tons and is going 15 to 18 knots and hits a piece of ice that weighs 1,000 tons, it can hurt. Glacial ice is very hard." And many iceberg sails are a lot higher than three feet, some as high as 200 to 300 feet. The tallest sail the IIP ever spotted was 550 feet high; in a photo taken of it from a ship's deck, it looks like one of those steep-sided, domed mountains in old Chinese paintings.

> A solemn, unsmiling, sanctimonious old iceberg that looked like he was waiting for a vacancy in the Trinity.
>
> —Mark Twain, letter to the *Alta Californian*

Besides sizes, the IIP classifies icebergs by their shapes. Not that mariners care about an iceberg's shape, but knowing what it is sometimes helps the IIP predict how fast the berg will deteriorate. "If an iceberg is peaky," Murphy points out, "there's less mass."

Tabular icebergs have horizontal tops, steep sides, and a ratio of length to height of at least 5:1. *Blocky* icebergs are much the same, except their length-to-height ratio is less than 5:1. (*Wedged* icebergs are a subclass of blocky, with a flat but tilted top.) *Drydock* icebergs have one or more large U-shaped slots in the middle of the top reaching to or below the waterline, leaving columns or pinnacles of ice on either side which, in late stages, can be "horned" or "winged." "Domed" icebergs are solid, with smooth, rounded tops. "Pinnacled" icebergs have large spires in the middle or else several spires forming a pyramid; they're also known as "picturesque Greenland type."

Even "picturesque Greenland" doesn't begin to suggest the richness of iceberg shapes. Usually irregular to begin with, icebergs split, flake, erode, roll, explode, melt, and morph into other configurations. "The ice-modification process is so complex that icebergs could be any shape you want," says Gregory Crocker, former director of ice engineering at C-CORE, previously known as the Centre for Cold Oceans Resources Engineering, in St. John's, Newfoundland, "except maybe a toothpick."

"Caterpillars, sleeping bags, pork chops, man-of-wars, slippers,"

Deborah Diemand, a scientist who studied icebergs for C-CORE, says. "Oh, lots of slippers." Author Richard Brown mentions "backs of dinosaurs, giant scallop shells, bows of ocean-liners" plus the more generic "tent-shaped, conch-shaped, gable-shaped, and fluted bergs." People have seen the Sydney Opera House in icebergs, crosses, dromedaries sitting down, bowler hats, palaces. Many see cathedrals. From his ship Shackleton admired an iceberg that "exactly resembled a large two-funnel liner, complete in silhouette except for smoke." On Scott's last expedition, Herbert Ponting took a photo of what came to be called the Castle Berg, "commonly felt to be the most wonderful iceberg ever reported," the caption in a book of Ponting's photographs stated, "nature having carved it into the semblance of a medieval castle."

Their life is death for us.

—Randall Jarrell,
"The Iceberg"

On a clear, dark, starlit night—the sort that prevailed on the Grand Banks on April 14, 1912—a ship's lookout will probably fail to see an iceberg when it's a quarter of a mile away. If the lookout has binoculars (which those on the *Titanic* did not; nobody knows why not), or if there is a breeze pushing a surf up against the side of the berg and producing "an occasional light spot," the IIP notes, a lookout may be able to see the iceberg when it's a mile away. With a full moon, he can see it when it is three miles off.

Yet nights are foggy as often as clear on the Grand Banks, where the cold Labrador Current meets the warm Gulf Stream and the temperature difference of the two water masses can be more than 30°F. A lookout can't see an iceberg at "any appreciable distance ahead" if there's heavy fog, according to the IIP. If by chance he can see it, "it may first appear as a luminous white mass when the sun is shining." ("It's shocking how white they are when you fly over," Donald Murphy notes. "They stand out dramatically against the gray and foggy background.") Although shipboard radar can detect large icebergs, it often misses smaller ones. Ice reflects radar weakly, returning signals only 1/60th as strongly as does a metal ship with the same cross section. Besides, the odd shapes of most small icebergs scatter beams in many directions.

Mariners sometimes infer the presence of icebergs in the vicinity if they see tiny pieces of ice floating near the ship, since chances are the pieces came from a larger, deteriorating iceberg to windward. Or they may hear the cracking noises of icebergs breaking up, or the popping sounds icebergs make as they melt and compressed air in their bubbles escapes. Air pressure inside bubbles of glacier ice can be 20 times greater than in a standard automobile tire, according to glaciologist Sue A. Ferguson. In his book *The Antarctic Circumpolar Ocean,* Sir George Deacon mentioned a "continuous" underwater noise that issues from melting icebergs "which, at the common listening frequencies, sounds like frying fat." Nevertheless, listening for such sounds is not considered a reliable way of detecting icebergs when they can't be seen.

"The only sure sign of an iceberg," the IIP concludes, "is to see it."

In the 80 years that the IIP has been looking for icebergs, and finding 80,000 of them, not a single person has died or piece of property been lost because of a collision between an iceberg and a ship outside the Limits of Known Ice that the patrol has so carefully set. *Inside* the limits is another matter, however. Some captains want to go through the ice zone no matter what, and some, if they're heading into St. John's, have no choice. Some choose to work their way through the ice because it's quicker than going around and they have confidence in their radar and it's daytime and the weather is fair, and for them, Murphy says, "time is money. It all depends on how much risk they're willing to take." In 1993, two captains chose to take the risk, probably to their regret. In February, the *OOCL Challenge* was doing 18½ knots while going from Montreal to the United Kingdom and struck a growler which made a 30-foot gash in its bow and opened cracks in its hull; and in April, the *Omikronventure,* carrying 600,000 barrels of oil from Norway to New York, hit a growler, which opened cracks in its hull and a 20-foot-long hole in its forepeak. Most ship captains, though, stay clear of the ice zone. "You can see their track going completely around it," Murphy says.

To respond here more fully than the English officer did to the American senator's "born-fool question" about composition, an iceberg is made up of frozen fresh water, ice that was formerly part of a glacier, ice that had once been snow. Although usually found in the sea, an iceberg does not come from the sea. Some glaciers terminate on land but most flow to the edge of land and keep on going, out onto a body of water, occasionally a river, sometimes a lake, usually the ocean. Some glaciers form tongues of ice which float on the water and are worked at by tides and waves and buoyancy forces, until one day the forward part of the tongue breaks off, probably at the point where there was a crevasse, and drifts away, a separate, free-floating chunk of ice. Sometimes the front edge of the glacier is a cliff grounded on the seabed, and water works away at it too until the upper part, deprived of support, falls away. (If you've seen that happen on TV, which is the only place I've seen it happen, the collapse of the glacier's face might remind you of the demolition of a large building by explosives, the structure giving way and falling in awesome slow motion.)

This breaking away—bodies of glacial ice giving birth to smaller bodies of ice in water—is called "calving" and can be a noisy and messy affair, the parturition accompanied by roars and rumbles and agitated waters and a great deal of ice debris surrounding the new offspring. When calved under water, the newly released ice pieces often shoot right up out of it. (Naturalist John Muir wrote in *Travels in Alaska* about a Hoona[h] Indian who was reluctant to approach the front of a glacier because, he told Muir, many of his tribe had been lost "by the sudden rising of bergs from the bottom.") When calved from an ice cliff, icebergs may plunge into the water and *then* shoot up out of it, "sometimes with a roll like a giant porpoise," according to one animamorphic account, or with "the ponderous grace of a great blue whale," according to another. In his book *The Arctic Regions,* American artist William Bradford told of seeing the birth of an iceberg starting at the moment the first roar issued from the mother glacier. "The roaring continued, and in a few seconds more a section of the glacier's front . . . was in motion, swaying backwards and forwards, gently at first, and then with an increased thundering . . . plunged bodily into the sea." The ice submerged, then rose "until its summit was at least two hundred feet above the water; while from its rugged and uneven sides poured a thousand streams and cascades, which, combined with the swaying motion of a new-born berg, gave it the appearance of some mighty sea monster shaking his shaggy head."

A calving can produce what has been called a "tsunami-like wave." One long and light-filled summer's eve in June 1993, Frances Smith and two women friends were standing across the Copper River from the Childs Glacier near Cordova, Alaska, taking photos of the glacier's impressive 300-foot-high face. They were standing at least a quarter of a mile from the glacier and at least 50 feet back of the riverbank, listening to "popping sounds" the glacier made as small pieces of ice broke from it, when, without any other warning, as the next day's edition of the *Anchorage Daily News* noted, a piece of ice "half the size of a football field" separated from the glacier face and fell into the river, not with a popping sound but with a giant roar, sending a wall of water 15 to 20 feet high in the women's direction. The women ran but the wall of water caught up with them. It carried Smith into the woods; "I tumbled, tossed, and turned," she said. "It was such a massive wave, with ice chunks, sand, stones, and sticks. . . . I kept thinking, 'Where are those trees?' " The next thing she knew, she was lying in a clump of alder brush, her pelvis broken in two places. She could see "chunks of ice all around" and "this thing slithering toward me. It was a salmon."

> *O solemn, floating field,*
> *Are you aware an iceberg takes repose*
> *With you, and when it wakes may pasture on your snows?*
>
> —Elizabeth Bishop, *"The Imaginary Iceberg"*

Icebergs and sea ice often coexist, although their relationship can be a bit contrary. Icebergs are 3-D entities while sea ice, formed as lake ice is by the freezing of water from the surface downward, is mostly 2-D—a floating field, maybe solemn. Even at the North Pole, sea ice is rarely thicker than nine feet (except where there's ridging) while icebergs are often hundreds of feet thick. In places where they coexist, you can usually spot the icebergs because they tend to stick up much higher than the sea-ice floes, like, to use *Carpathia* witnesses' descriptions, rocks or boats or collapsed dirigibles. While sea ice responds mostly to wind, icebergs, with their deeper keels, respond mostly to currents. These differences produce some bizarre sights. Icebergs can be moving in one direction at the very same time that sea ice is moving in another, even opposite, direction. When American explorer Frederick Cook was returning from his putative discovery of the North Pole, he and several companions were forced to take shelter on a series of floating pieces of ice south of Ellesmere Island, and one day in tossing seas, they switched from a low-lying if strong floe of sea ice to a more elevated iceberg. "What a relief to be raised above the crumbling pack-ice and to watch from safety the thundering of the elements!" Cook wrote. Once the party was settled on the "secure mass of crystal," Cook took time to observe its movements. He found them to be complex and "at variance with the pack-ice. It [the iceberg] ploughed up miles of sea-ice, crushing and throwing it aside. After several hours of this kind of navigation . . . the berg suddenly, without any apparent reason, took a course at right angles to the wind, and deliberately pushed out of the pack into the seething seas."

Icebergs are generally harder than sea ice, too, since they're made of fresh water (salts soften ice). "I've seen icebergs cut through pack ice like a knife through butter," Mary Williams, director general of the Institute for Ocean Technology in St. John's, Newfoundland, says. Lieutenant Sherard Osborn, commander of the ship *Pioneer,* part of a British flotilla looking for the lost Franklin expedition in 1850, wrote of seeing "the pressure of the pack expending itself on a chain of bergs. . . . The unequal contest between floe and iceberg," he recounted, "exhibited itself there in a fearful manner, for the former pressing onward against the huge grounded masses was torn into shreds and thrown back piece-meal, layer on layer of many feet in elevation, as if it was mere shreds of some flimsy material, instead of solid, hard ice, every cubic yard of which weighed nearly a ton." Solid and hard as the sea-ice floes were, the icebergs were more solid, harder still.

Only a small portion of Arctic icebergs are tabular but most Antarctic icebergs are, with a length-to-height ratio above water that can be as much as 100:1. They are calved from the huge, slow-moving ice shelves that drain glaciers along the coasts of Antarctica—the Ross Ice Shelf, the Weddell, the Amery—as well as from smaller and thinner glacial tongues. The tabular bergs I saw in Antarctica came from the Ross Ice Shelf but didn't get very far before becoming grounded or frozen into sea ice. Immobilized, sticking up above sea-ice or water level and displaying the long, low, simple lines of the shelves they broke from, the icebergs looked like so much furniture, tables, of course, but also sofas, highboys, chaises, beds. To Frank Worsley, on Shackleton's 1914 expedition, tabular bergs, though, resembled "huge warehouses and grain elevators." Even simple tabular bergs, though, are eventually transformed by sun and wind and water into more elaborate, fanciful shapes.

It's easy when looking at icebergs to understand why they seem to many people to be more than compressed snow. Spiritual entities, perhaps. Suffused with what often seems to be an inner light, displaying curves of a chaste voluptuousness, materializing like ghosts out of the mists, moving as if of their own volition, unpredictably, emitting sounds and morphing shapes, looming dangerously. A critic commenting on the work of photographer Lynn Davis, who has taken thousands of pictures of icebergs, describes them as being "embodiments of something profound and pure beyond the self, beyond man and close to God." Australian explorer Douglas Mawson came across this vision while trying to work his ship through "an ice-strewn sea" off Antarctica in late 1911:

"Majestic tabular bergs whose crevices exhaled a vaporous azure; lofty spires, radiant turrets, and splendid castles; honeycombed masses illumined by pale green light within whose fairy labyrinths the water washed and gurgled. Seals and penguins on magic gondolas were the silent denizens of this dreamy Venice."

Iceberg colors contribute to their aura. Bergs are usually white, from light-scattering bubbles inside and sometimes snow on top. When explorer Joseph Banks came upon an iceberg off Labrador one hazy but cloudless and moonless night in 1766, he noted: "The ice itself appears like a body of whitish light, the waves dashing against it appear much more luminous. The whole is not quite unlike the gleaming of the Aurora Borealis."

Icebergs can also be deep blue, if the ice is old, pure, and dense, with few bubbles. "No art can reproduce such colours as the deep blue of the iceberg," Scott wrote at the very moment that two of the expedition's party were doing their best to reproduce the scenes before them. Icebergs can be almost any shade between blue and white—sky blue, cerulean, robin's-egg, ultramarine—or both blue and white interspersed. Bradford saw one iceberg with "streaks a few inches in width, the colour of Prussian blue, which appeared to extend clear through the mass, presenting an agreeable relief to the otherwise dazzling whiteness."

Besides the innate colors of icebergs, "With borrowed beams they shine," as poet William Cowper wrote. The highest iceberg Douglas Mawson ever saw was in the shape of a "turreted castle—a keep of the icy solitudes!" whose peak, "in the waning light" of a summer evening, he wrote, "was tinged with the palest lilac." In the low-angled light of a northern sunset, Isaac R. Hayes, surgeon on Kane's ship *Advance,* described several icebergs he saw off Greenland as "masses of solid flame or burnished metal."

Another hue seen in icebergs, although not commonly, had been thought by some scientists to be a product of waning light, too; the reddened light of a low sun's passing through the intrinsic blue of ice would, they reasoned, produce green. Probably not the green recalled by Coleridge's old seaman in the "Rime of the Ancient Mariner": "And ice, mast-high, came floating by. / As green as emerald." Rather, the green of "gemstone jade," as Eliot Porter described the color of a grounded iceberg he photographed near Livingston Island in Antarctica in the 1970s. Or "bottle-green," as some scientists called the color of a berg they spotted in the South Shetland Islands near the Antarctic Peninsula in 1976. Those scientists also called the color "an oddity not readily explained." Some on board thought it was most likely "enclosed fine rock material," iron and copper maybe, another that it was organic material, another that it was fine sediment stirred up from the seabed and frozen into the iceberg, changing the way light passed through it.

Sightings of green icebergs are so infrequent that the source of the color wasn't pinned down until the summer of 1988 when an Australian research ship happened upon a 500-foot-long green berg grounded north of Mawson Station in Antarctica. It was two-toned, green on top and blue-white on the bottom. The scientists took samples and eventually determined that the green part was sea ice that had frozen to the underside of an ice shelf and trapped microorganisms—which are known to absorb blue light as well as red—inside. Then it broke from the ice shelf and one day rolled over to expose its yellow-green underbelly. The reason there aren't many bottle-green icebergs drifting around the seas, offering enchantment to passing mariners, is that conditions have to be just right (low temperatures, lots of organic material) for a very long time. No icebergs in the Arctic form under these strict conditions, and few in the Antarctic do, either.

Chunks of ice floating in the water around the main green berg were pale green, the scientists noted, but only if they were at least three feet long. Under three feet, they were colorless. Pure blue ice is also colorless in short pieces; for blue and green tints to show up in ice, even faintly, there evidently needs to be long light paths.

On Frederick Cook's trek toward the North Pole, he observed another effect that small marine organisms can have on icebergs. One time he saw distant lights winking against the black of the ocean, like "lighthouse signals, or the mast lanterns on passing ships. They flashed and suddenly faded," he wrote, "these strange will-o'-the wisps of the Arctic sea. In a moment I realized that the lights were caused by distant icebergs crashing against one another. On the bergs as on the surface of the sea, as it happened now, were coatings of a teeming germ life, the same which causes phosphorescence in the trail of an ocean ship. The effect was indescribably weird."

CHAPTER TWELVE

ICEBERGS II

In shadowy silent distance grew the Iceberg too.

—Thomas Hardy,
"The Convergence of the Twain"

AS BIG AS ARCTIC ICEBERGS GET, Antarctic icebergs get bigger. Much, much bigger. An Arctic iceberg can be the size of a cottage or a city block. An Antarctic iceberg can be the size of an American state. "How can you compare a whole state with a cottage?" the International Ice Patrol's Donald Murphy asks. An Arctic berg is considered large if it's 200 meters (656 feet) long, newsworthy if it's over 1,000 meters (3,280 feet) long. In autumn 1987 the Joint Ice Center of the U.S. Navy and National Oceanic and Atmospheric Administration (NOAA) began tracking an Antarctic iceberg that was 96 *miles* long and 22 miles wide, or about the size of Long Island, New York. It broke away from the Ross Ice Shelf (itself the size of Texas) and was so huge that all the people in the world could stand on top of it at once and have almost 40 square inches of standing room each, so huge that if all its 287 cubic miles of fresh water were melted and evenly shared, each person in the world would get two glasses of water every day for the next 1,977 years.

The center (renamed the National Ice Center after the addition of a Coast Guard employee) doesn't usually track icebergs. In the south there's no need to track them because shipping routes don't pass close to Antarctica, and in the north the IIP does the tracking. The center's mission is to monitor the extent of sea ice around the world by using images from polar orbiting satellites, mostly for the benefit of scientists. But when an iceberg shows up that's over 17 miles long (any shorter and it can't be distinguished on satellite pictures from a large floe of sea ice), or when it has a distinctive shape such as the one that looked to the center's former technical director Don Barnett like a tiger, they keep an eye on it.

The iceberg that was as big as Long Island, and happened to be shaped like Long Island, the technicians found while taking a routine look at satellite images of the Ross Ice Shelf and noticing that a crack in it that had been visible for the last 20 years now had water in it. They named the new iceberg B-9, B for the quadrant of the continent it was in. Barnett has regrets about that. "If we had had any foresight, we would have named it like a hurricane," he says. "Some of us tried calling it 'Ralph,' a pet name, but it didn't stick. We used to name each iceberg for the person who found it until a guy named Harris found one, and 'Harrisberg' didn't sound too good."

A pet name might have been fitting because B-9 became a popular iceberg. Oceanographer Stanley S. Jacobs, a senior staff associate at Columbia University's Lamont-Doherty Earth Observatory, kept his eye on it, too. He decided to use the berg as "an enormous scientific instrument," a substitute for manmade current drifters, which are drogues dangling from buoys. Since radar soundings showed that the ice shelf extended downward 800 feet below the waterline, B-9 would also extend down that far, and a keel that deep, Jacobs reasoned, could function as a drogue, catching and being pushed around by deep currents. By following B-9's wanderings in the southern seas, he could study the strength and direction of deep currents there, about which, he admitted, scientists knew "surprisingly little."

The navy dropped a satellite transmitter on top of B-9, and using that the center kept track. "Once B-9 was set adrift, every nuance of its motion had a story to tell," Jacobs wrote. First, it floated west in the company of 25 smaller bergs that were calved at the same time. Having come from the same ice shelf, the others had keels as deep as B-9's, but their undersides were not much larger than their sides, while B-9's underside was *50 times* larger than its sides. That made a difference in how it moved and where it went. The smaller bergs were propelled mostly by the push of ocean currents on their large side walls, but with B-9 the effect of the push was overwhelmed by another force, the friction of water flowing across its vast underside. B-9 therefore responded more strongly to deep currents than its siblings whose ice was just as thick. As B-9 drifted with the others along the edge of the ice shelf, it began lagging behind—"dawdling," Jacobs called it—due to its greater drag. Then it veered away from the coast as well as from the other bergs and headed northwest, through the Ross Sea, for the next seven months.

After that, B-9 did "something extraordinary," according to Jacobs. It went back to the ice shelf it came from and "butted" into it. The tipoff that it had made contact with the shelf was that it changed directions in a single day, from south to west. The contact didn't knock off any new icebergs, although Jacobs looked for some since iceberg strikes on shelves are considered by some scientists to be a major trigger for the calving of more icebergs. Gigantic as B-9 was, it apparently didn't trigger anything. When it butted the ice shelf, it was moving at only 0.14 miles an hour and, Jacobs notes, "even Detroit bumpers can handle that speed on smaller projectiles."

Once it rebounded, B-9 followed the coastline again, rotating as it went, then for some reason—a "final puzzle"—broke out of the coastal current six months later and headed northwest again, but at triple speed, even as its smaller counterparts were still hugging the coastline and floes of sea ice were taking still another, unrelated path across the Ross Sea. When B-9 was less than two years old, it rounded a bank and a cape and broke into three parts, perhaps because it ran aground, although that's hard to tell from a satellite.

When the Joint Ice Center first noticed B-9 in 1987, it might have been the largest iceberg ever seen by humans. In 1956, people on the icebreaker USS *Glacier,* when it was 150 miles west of Scott Island in Antarctica, had reported seeing an iceberg that was 208 miles long and 60 miles wide, bigger than the state of Massachusetts, but the sighting is disputed. "Shipboard estimates lead to exaggerated fish stories," Jacobs points out. Since B-9 was born, however, satellites have registered the birth of two even larger icebergs. In 1998 an iceberg the size of Delaware, 90 miles long and 30 miles wide, broke from the Ronne Ice Shelf and drifted off; it was named A-138. In March 2000, an iceberg nearly as big as Connecticut, 185 miles long and 23 miles wide, split from the Ross Ice Shelf; it was called B-15. B-15 is considered by many to be the champion of champion icebergs. After the split, it swung around and knocked another piece of ice off the ice shelf; Matthew Lazzara, a meteorologist at the Antarctic Meteorological Research Center at the University of Wisconsin, estimates that B-15 was a quarter of a mile thick and contained 500 to 1,000 trillion gallons of fresh water, enough if melted to dump four to eight inches of rain on the entire land surface of the Earth. In his view, the calving of B-15 was part of a natural cycle of ice shelf buildup and breakoff and not an effect of global warming. "The ice shelves have been growing and growing for a long time and are overdue to calve off," he says.

To people not interested in deep-ocean drogues, the movements and breakups of large icebergs must seem a dreadful waste. *What?* Enough fresh water to give every person in the world a couple of glasses of water a day for a couple of thousand years allowed just to melt into the salt-filled sea? In olden days, whalers, sealers, and explorers often took advantage of these handy reservoirs of fresh water before they vanished. While off the coast of Antarctica in January 1773, Captain James Cook noted in his log "an important and triumphant experiment, the taking in of loose ice from round a berg for water—arduous and freezing, as well as picturesque work," he wrote. To him, it was "the most expeditious way of watering I ever met with." In one day, the ship picked up 15 tons of iceberg pieces.

But these were small pickings compared to what John Isaacs of the Scripps Institution of Oceanography in California had in mind when he proposed that a severe drought in California in the early 1950s be relieved with icebergs—whole icebergs, large Antarctic icebergs, towed 10,000 miles north to California, where their meltwater could be spread on parched California farmland. Nobody paid much attention, except to have a good laugh, but over the next couple of decades, in the course of which the need for fresh water around the world increased while the supply decreased, some people began to see the possibilities. Why *not* icebergs? Studies were made, companies formed, papers written, and after another calamitous drought in California in 1977, during which there was talk of carting in snow from Buffalo, New York—182 million railroad carloads of it at a cost of $437 billion—a mixed group of climatologists, glaciologists, entrepreneurs, oceanographers, inventors, and one prince gathered in Ames, Iowa, a place that had never seen an iceberg, for the First International Congress on Iceberg Utilization. "The laughter has ceased," said one participant, in all seriousness.

Attendees were reminded that it had been done before. Shortly after Isaacs made his proposal he received a letter from the director of a Chilean technological institute informing him that between 1890 and 1900 South American ships had towed and sailed small icebergs from the southern coast of Chile northward along the coast to Santiago and beyond, to a site near Lima, Peru, 2,500 miles away

Cautiously optimistic, they got down to particulars, including which icebergs would be right for the job. Suggestions were that any "towbergs" be tabular to reduce chances of their rolling en route; have a high length-to-width ratio, to minimize drag; be very large so they wouldn't lose so much ice that the end product wouldn't be worth the ride; and have only 5/6ths of their mass below sea level, to make floating easier. Pick "a berg that maintains a white luster, looking like plaster of paris," one economist advised, ". . . since it contains vast quantities of air bubbles."

"The icebergs are there," Willy F. Weeks, then at CRREL, and W. J. Campbell, member of the Ice Dynamics Project of the U.S. Geological Survey, concluded in their paper, "Icebergs as a Fresh-Water Source: An Appraisal." "The problem is only in arranging suitable transportation," by which the geologists meant supertugs, since dragging a miles-long iceberg through thousands of miles of ocean would take powerful engines and large propellers.

One proposal for how to connect bergs to tugs involved chipping a groove in an iceberg and laying a wire in it with reinforced plastic cushions underneath to keep the wire from digging its way through the ice. Another was to put the berg inside a harness of cables, with hitching posts for ropes embedded at each "corner" of the berg. Several bergs might be towed together, in a train, each berg linked to one in front and one in back by a rope running from stakes in the top, the whole line of bergs pushed by a power plant at the rear.

Or the icebergs could make the trip on their own. Prince Mohamad Al-Faisal of Saudi Arabia had already formed a company, Iceberg Transportation International, which had explored the possibility of towing icebergs from Antarctica to the port of Jidda, as a way of getting fresh water more cheaply than by desalinating seawater. He and his technical adviser came up with a design for a self-propelling iceberg with underwater paddle wheels mounted on two sides and units on top to power the paddle wheels. When one paddle wheel rotated faster than the other, the berg would turn. To reduce melting, the iceberg could be sprayed with urethane foams and underlaid and overlaid with polyester films.

Other ideas for insulating a berg during its long ocean journey north included tying a plastic skirt around it at the waterline and wrapping it entirely in an 18-inch-thick sheet of plastic (a project Christo might want to take on). Even so, an iceberg could lose a third to a half of its mass en route, mostly from water erosion. To reduce friction with the water, the iceberg should be towed slowly, at a knot or two an hour, at which rate a trip from Antarctica to Jidda might take six months to a year.

The best part of the iceberg-towing scheme, according to Weeks and Campbell, is that iceberg ice would no longer be "completely wasted as regards man's needs." Every year Antarctic glaciers cast into the sea an estimated 625 cubic miles of ice, or an amount equal to two-thirds of all the fresh water used annually by all the people in the world. (Antarctic icebergs are not only much bigger than Arctic icebergs but there are also a lot more of them.) Once the iceberg water has irrigated crops (or whatever), it would return to the sea. "The end result is the same," Weeks and Campbell wrote, "only the path is different," although admittedly the path the water would take is a long and "tortuous" one.

None of it came to pass. Economics didn't favor the scheme; for one thing, the price of the oil needed to power the supertugs went up. Today there are no mountains of white ice rising behind palm trees in the harbor of Jidda or catching the surf off the parched coast of California, no plastic-wrapped blocks of compressed snows paddling their way through the tropical oceans.

The gray shape with the paleolithic face
Was still the master of the longitudes.

—Ned Pratt, *"The Titanic"*

Even as people were talking in Iowa about towing Antarctic icebergs north one day, in the north some towing of icebergs was already going on, although of much smaller bergs, over much shorter distances, and for a different purpose entirely. The towing (plus some vigorous nudging) was being done to keep icebergs from hitting exploratory oil-drilling rigs in the North Atlantic. In the late 1960s, oil companies started looking for oil and gas deposits in waters off the east coast of Newfoundland, as well as in the Beaufort and Chukchi seas north of Alaska and northwest Canada, all waters that have ice in them a good part of the year. The companies hadn't had experience with ice before and didn't know what it might do to their drill rigs, tankers, and underwater pipelines if it hit them, although they probably had some inkling, since almost everybody knows the story of the *Titanic.*

(Brian T. Hill, supervisor of the ice tank at the Institute for Ocean Technology [IOT] in St. John's, has put together a list from historical records of ships besides the *Titanic* that have collided with icebergs, and in the North Atlantic alone he has found more than 600, from the oldest documented hit, in July 1686 when the sailing ship *Happy Return* "struck ice on way to Hudson Bay for North West Fur Co. [and] sank" [as Hill's notes read] to the most recent, in June 2004 when the shrimp trawler *Solberg* collided in fog with a 75-foot-long iceberg which "heavily dented and dished" its bow and dumped tons of ice onto its deck. [The weight of iceberg ice piled on decks can by itself sink ships.] Not surprisingly, Hill found good correlation between the numbers of icebergs in the North Atlantic in a given year and the number of ships struck that year, until about 1920 when the iceberg patrol was in full operation and collisions started tailing off.)

The ice most oil companies drilling in the North Atlantic are most concerned about is iceberg ice, since by the time sea ice gets as far south as the Grand Banks—if it does—it tends to be loose and not exerting pressure on anything. Also, of course, iceberg ice is usually harder than sea ice. But in order to know how to design ice-resistant structures and operate drilling rigs in ice-rich waters, oil companies needed a lot more information than they already had about iceberg strength, iceberg behavior, movements, etc. They hired their own "ice experts" (Esso alone had several); Canadian government agencies funded ice researchers; and over the next four decades people tracked icebergs with radar, made plastic models of icebergs and set them adrift in ice tanks, towed icebergs into instrumented cliffs, and cut beams from icebergs and subjected them to pressure, tension, and sharp blows.

"We couldn't just tell the companies 'enormous,'" IOT's Mary Williams says. "They needed numbers. Otherwise, to cover their own uncertainties, they'd have to build their platforms really big and really strong and that's really expensive."

Say you're the operator of a drill rig and an iceberg has been spotted 25 miles from it, mightn't you be interested in knowing how long it would take for the berg to get to you? A group at C-CORE, Peter Ball, Herbert S. Gaskill, and Robert J. Lopez, made 19,000 hourly observations of the changes in position of icebergs at drill sites in the Labrador Sea and concluded that the average speed of all bergs is about seven miles a day. So the iceberg spotted 25 miles from your rig and moving in your direction will reach you in a few days, right? Wrong. "Bergs do not go in a straight line," the group stated. They "wiggle in the drift track." Sometimes they do more than wiggle. In another C-CORE study, senior research engineer Mona El-Tahan and two associates observed icebergs that were within a mere 18 miles of each other going in "different (even opposite) directions simultaneously." What's more, their manners of movement differed from those of others close by: while several were making straight beelines through the water, others not far away were describing large U's and still others were making loops.

Some of the looping—done in a clockwise direction—was caused by tidal currents and some, the larger loops, by the rotary motion of eddies. In places where strong winds blew the icebergs into calm waters, the Coriolis effect made them spiral. The conclusion to be drawn is that icebergs could take two to three times as long to reach a drill rig as their

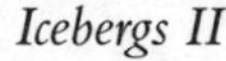

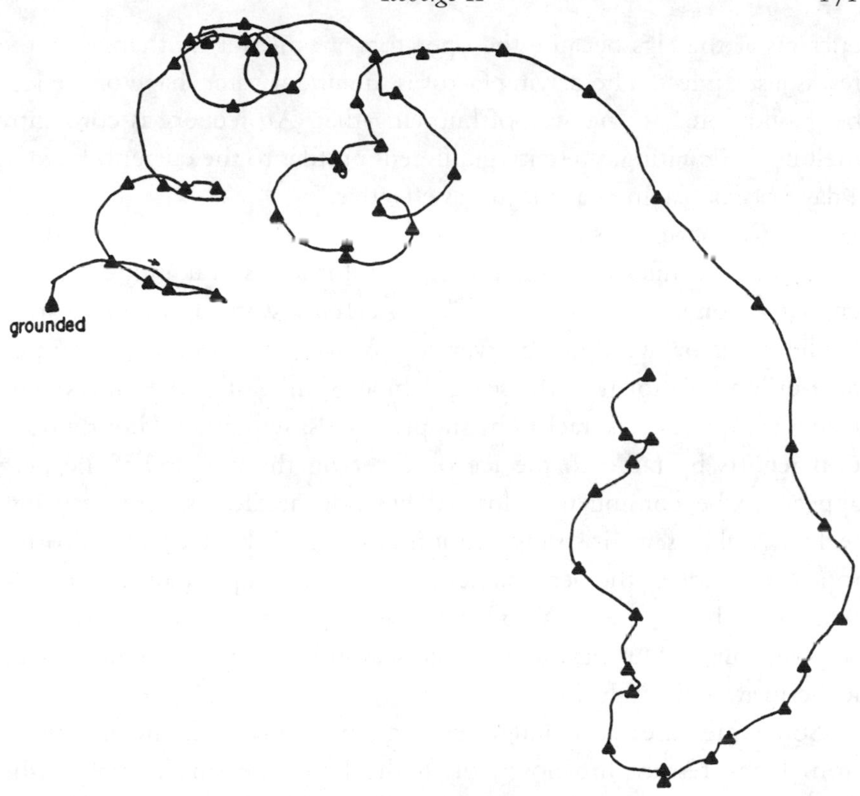

An iceberg's meandering path

average speeds through the water would suggest—if they reach the drill rig at all.

The limits of ice that the IIP concentrates on are usually south and east of the Grand Banks, but most of the oil rigs are on the Grand Banks, so to monitor icebergs there companies hire a private airline to fly small planes over the area and map targets with radar. "We eliminate anything that's moving with any kind of speed," Pip Rudkin, manager of environmental services for Provincial Airlines, explains. "Anything faster than two knots isn't an iceberg." Anything slower than two knots could also be a fishing boat, however, so they go back and examine the target with a rotating gyrostabilized, telescope-fitted television camera which can pick up an image as far away as 40 miles. If light is low, they use infrared sensors. "Icebergs don't give a thermal signature," Rudkin explains. "Boats do."

Using the map and a computer model, they forecast the movement of all "icebergs of interest" over the next two or three days. The forecasts are "not too bad but not good either," Rudkin admits. "You know the currents at the rigs because the operators are measuring them, but the rest is just a guess. The driving factor is the underwater shape of the iceberg, and you have no way of knowing that. An iceberg is constantly melting and rotating, presenting different profiles to the current. Beyond a day or two, playing darts is just as effective."

What rig operators do in practice is set up a safety zone around the rig, maybe 25 miles in all directions, and if it looks as if an iceberg will enter the zone, "we take action," says Gerard Watson, a combination "radio operator–weather observer–ice observer" on Hibernia, a fixed-base drilling platform in the sea 200 miles southeast of St. John's. The iceberg in question is tracked by supply vessels, which send hourly position reports by radio to the ice observer on the rig, and if the berg appears to be coming too close, they'll tow it. Towing is "very low tech," Rudkin says. It's slow-motion lassoing. A boat drops a floating rope at one side of the berg, circles the berg, picks up the line, and pulls. It tows the berg far enough off trajectory that it will clear the rig by at least five miles. During the 2000 ice season, boats towed almost 1,000 icebergs away from drill rigs.

Sometimes they just nudge icebergs, by blasting them with water from high-pressure fire hoses, or, if the bergs are small, "propwash" them, by backing up to them, putting the engines full ahead, and letting the wash push the bergs out of the way. The small icebergs, "say the size of a garbage can," Watson notes, are the ones that operators of the semi-submersible rigs are most concerned about. "If you're in a 20-foot sea and a small piece of ice keeps rubbing against one of the legs, it won't take long before you've got a hole in the leg. People ask us, 'Why not blow the big ones up?' and I tell them that with a big iceberg you have one problem and if you blow it up you have 100 or 1,000 problems."

When the icebergs are large, operators have to move the rig. They don't take chances; "any spilled oil could go to Ireland," a scientist points out. One year in prime iceberg season, the six-week period from mid-April to early June, drill rigs in the Grand Banks were off their drill sites 75 percent of the time. It takes about two days to disconnect a floating rig and two more days to reconnect it after the berg has gone by, and in those four days the companies will lose a couple of million dollars in revenue. Ice observer Watson remembers one "wild night" on a semi-submersible rig when the crew had hoisted seven of the eight anchors but hadn't had time to pull the last one up before the iceberg showed up, and "we were still on the anchor swinging off to one side just as the ice passed right over the drill hole!" Although there have been other near misses as well, in all the years of exploration and operation, only one iceberg has hit a drill rig, and it was a bergy bit that Provincial's Rudkin says "looked as if all it did was scrape a bit of paint off." A year later, nevertheless, when the drill ship went in for maintenance, mechanics found many fractured ribs in the hull—"just from that little thing!" says Rudkin, impressed.

Hence! Seek your home; no longer rashly dare
The darts of Phoebus, and a softer air;
Lest ye regret, too late, your native coast

—William Cowper, *"On the Ice Islands Seen Floating in the Germanic Ocean"*

Even as an iceberg is bearing down on an oil rig, it is deteriorating. In the month of September, when water temperatures are at their highest on the Grand Banks, up to 60 percent of the icebergs there can melt completely in a single day. Over its lifetime, however, much of an iceberg's loss of mass comes from breakup. "Icebergs are under a lot of stress," Jack Clark, former president of C-CORE, explains. "That's why they don't end up in Bermuda. That's why we don't see 60-million-ton icebergs in the Grand Banks. They destroy themselves." Ocean waves erode notches in the ice at the waterline, producing an unsupported, overhanging slab of ice, which falls off. (Once calved, icebergs themselves calve.) "It's like an avalanche," the IIP's Murphy says. The breaking away leaves large "rams," or ledges, of ice under water, which sets the stage for yet another breakup. Even as buoyant forces push upward on the rams, the weight of the sail pushes downward at the center, and the berg splits in two. It's the same sort of stress you'd get if you pressed your hand down hard on the center of a floating slab of Styrofoam.

During a study of iceberg grounding, Jim Lever of CRREL took before-and-after photos of an iceberg that split in two. The iceberg weighed eight million tons and was named Gladys. (People on C-CORE expeditions did name icebergs like hurricanes for a time, or as hurricanes used to be named, for females only.) In a 3-D model Lever made of Gladys you can see that although her sail wasn't large, only 550 feet long—her underwater rams were—1,640 feet end to end. As Gladys split, she spun with such high energy that she dug a pit in the seabed. In Lever's "before" photo, you can see a single tabular shape; in his "after" photo, the table has been turned into a pair of slant-top desks.

Another stress on bergs is thermal: warm on the outside, cold on the inside. Before starting a study on iceberg splitting, Deborah Diemand, then at C-CORE, decided to measure temperature gradients in icebergs. " 'Don't waste your time and money,' people told me," she says. " 'Of *course* icebergs are all 32 degrees,' they said. 'Icebergs are *known* to be at 32 degrees.' It was so obvious to everybody. The water is warm—40 degrees—so the icebergs must be warmed. To me it seemed blindingly obvious that icebergs *weren't* all 32 degrees. How about the temperatures of those glaciers they came from?"

In a borrowed rubber dinghy, Diemand would pull up next to an iceberg, climb up onto it, drill a couple of holes several feet deep, and put thermistors in the holes. She managed to take the temperatures of 20 icebergs. All were 32°F on the outside, as she expected, and all were 5°F to −4°F on the inside, as she also expected. "Icebergs are produced up north in places where the average annual temperature is 5 degrees to minus 13," she explains, "and when they fall in the water and then loaf around Baffin Bay—that's a serious cold body of water!—they're not going to melt right away. Ice is a good insulator, and in very large pieces icebergs should stay the same temperature inside as when they left the glacier. After several years the icebergs move toward Labrador and start to warm up, but very, very slowly. You can't hasten the warming of ice."

As they warm, the icebergs melt on the outside. Inside they're still 5° to −4°F, but because some of the outside is gone, the cold inside has moved closer to the surface. "Early in the season, the change from 32 degrees to minus 4 takes place through 100 feet of ice," Diemand points out, "but later it takes place through only three feet or maybe a foot and a half of ice." The stress of the steep difference in temperature causes the outside layers to break away, or spall. "They often come away like orange peels," C-CORE's Clark says.

One day Diemand was standing on an iceberg, drilling a hole in it, when the iceberg rolled. It was "a nowhere sort of day, in May," she relates, "with an overcast sky. The berg was ratty, all covered with rocks and gravel. I remember one of the rocks was the size and shape of a coffin. I could hear the ice creaking and cracking and snapping and carrying on. I climbed from the low end toward the high end which had a hammerhead knob on it. I heard *'Mumph'* and the knob fell off. The iceberg started going up and down, up and down, up and down, in one plane. Then one end went deep in the water and took me with it. I was convinced I was dead meat. I thought of my high school yearbook and the quote after my name, 'You should always be a little bit improbable.' I don't know why I thought of it. I hated high school."

Diemand got a cut on her head where she'd hit it with her own ice pick and came up bleeding, but she was lucky. When a large iceberg rolls, it creates currents that can suck you down, Greg Crocker reports. It can collapse on top of you. "I don't mind the small ones," he says. "They're unstable and hard to stay on but if they roll you just jump off, like jumping off the end of a wharf. But I won't go on big ones. Nobody does anymore. They're too dangerous." When icebergs melt or have parts fall off, they make adjustments to get the center of gravity under the new center of buoyancy, which has been compared to the shifting around a person does to get comfortable in an old armchair.

Once it rolls, a berg rocks. It keeps on rocking until the energy released by the overthrow is absorbed by friction with the water. The bigger the iceberg, the more potential energy it has to get rid of and the longer it goes on rocking. Like bridges and tall buildings that the wind sets to swinging, each iceberg has its own frequency at which it oscillates, based on its size, shape, and stability.

Some icebergs explode in breakup. A photograph Greg Crocker took from the air of a stretch of ocean after an iceberg had come violently apart shows a few small white ice remnants surrounded by a swirl of thousands of smaller remnants; it looks like a star-studded nebula. Most of the ice pieces were too small to threaten a ship, Crocker says; some of them were probably "no bigger than a cornflake." But not all. "For the pieces which break from the large Islands are more dangerous then the Islands themselves," Captain James Cook observed. ("Ice islands" was the term for icebergs back then.) Small icebergs tend to respond more to surface waves and winds than large icebergs and thus to move faster. Growlers have been timed traveling at nine miles an hour. "That doesn't sound fast if you're driving a car," Crocker notes, "but if the car hits a brick wall [or ship or drilling rig], it is."

> Let anyone who has witnessed the crash of even so small an object as a ship, when run into by another having only a barely perceptible movement, reflect on the terrific momentum of an iceberg, some mile or two square, and from 1000 to 2000 feet in thickness, when, borne onwards by a current of only half a mile per hour, it runs on a submarine bank
>
> —Charles Darwin, 1855

Charles Darwin had opinions even about icebergs. In a book of his collected papers, appearing between an article titled "Bucket Ropes for

Wells" and one called "Does Sea-Water Kill Seeds?" is a journal article "On the Power of Icebergs to Make Rectilinear, Uniformly-Directed Grooves Across a Submarine Undulatory Surface." In it, Darwin addressed the claim made by some geologists that long scratch marks found on exposed rocks on land were created not by ancient glaciers but by ancient *icebergs* when the land was below water. How, Darwin wondered, could a floating body scratch an undulatory seabed? On reflection he decided that while a glacier "must press on its bed with the whole immense weight of the superincumbent ice," an iceberg will press on it with "only the weight of as much ice as is forced up above the natural level of the floating mass," and therefore it could scratch a wavy surface.

One hundred and twenty years later, oil-company employees beaming side-scan radar over the Grand Banks while searching for oil-well sites discovered large grooves on the seabed, crisscrossing each other like giant slug tracks, and deduced that the grooves had been made by icebergs dragging their keels across the bed. This mattered to oil companies because bergs that can scrape across the seafloor can also scrape across anything laid across the floor, including pipelines, cables, and wellheads. Investigators descended in a miniature submarine and looked at the grooves. "The forces at work must be colossal," Christopher Woodworth-Lynas, senior researcher at C-CORE, reported. Grooves were on average 80 feet wide and four feet deep. The largest of all was in Hudson Strait: a third of a mile wide, 80 feet deep, and about six miles long. None of today's Arctic icebergs could carve out such a trench. The iceberg that made it had to have been around at least 10,000 years ago, at the end of the last ice age. Most grooves found in the Grand Banks probably date from that time; they haven't been erased in 10,000 years because below 160 feet waves don't have much effect on the seabed.

One way of dealing with the scouring problem would be to reduce the draft of an iceberg that threatens a pipeline. One company's scheme is proprietary; a spokesman says only, "We call it a 'weapon of mass reduction.' "

> solid, like earth—impermanent, like cloud
>
> —Charles A. Lindbergh, *The Spirit of St. Louis,*
> of ice "cakes" he flew over on his way to Paris

The reason more icebergs end up around oil rigs and in shipping lanes in some years than in others seems to have less to do with the icebergs themselves (although more are thought to calve in some years than others) and more to do with their 2-D cousin, the sea ice. Scientists have determined that when sea ice is extensive near the beginning of winter in the High Arctic, it blocks channels and keeps icebergs from moving into them and getting stuck there for good. Sea ice can keep icebergs as far offshore as 40 miles; then, in spring, when the ice breaks up and drifts southward, it carries the icebergs with it. A good indicator of what kind of iceberg season it's going to be in the Grand Banks in June, say, may be how much sea ice is in the strait between Greenland and Baffin Island in January.

The average number of icebergs making it to the Grand Banks probably won't change even with continued global warming. Although higher temperatures could cause the glaciers to calve more icebergs, they would also reduce the extent of sea ice. The effects would offset each other. Less sea ice would mean more grounding and melting of icebergs in the north, therefore fewer icebergs (of the greater number produced) making it south.

hope i do
not meet any icebergs how
would one talk an iceberg out
of it
archy

—Don Marquis, *archyology: archy repels an attack of whales*

When the tanker *Exxon Valdez* ran aground on Bligh Reef in Prince William Sound in 1989 and dumped 11 million gallons of crude oil into the water and over 1,300 miles of Alaskan shoreline and onto a quarter million birds and several thousand sea mammals and billions of fish eggs, the captain had been trying to avoid icebergs. He had switched shipping lanes to be clear of them and hadn't switched back. The icebergs came from the Columbia Glacier, one of several glaciers that release icebergs into the sound, and were probably small ones, growlers, even small growlers. Icebergs in the temperate clime of southern Alaska, unlike icebergs in the Arctic and Antarctic, don't last long. Most melt within a few hours, Nancy R. Lethcoe points out in her book *Glaciers of Prince William Sound, Alaska,* with "only a very few" surviving a day or more. Many are "bottom bergs," dark blue or black, blue from having been near the base of the glacier where air in bubbles got squeezed out by the weight of overlying ice, black from the rocks and gravel the ice picked up from the earth on its way to the water. Icebergs carrying many rocks sit very low in the water, making them particularly hard to spot.

On the Pacific side of North America, there is no IIP equivalent. It's not that the glaciers of southern Alaska don't produce a great many icebergs; a single active one can calve 50 billion pounds of icebergs a day, Sue A. Ferguson noted in her book *Glaciers of North America.* One year, when visiting the Alaskan glacier named for him, John Muir observed that it calved at the rate of three to 22 icebergs an hour ("On one rising tide, six hours, there were sixty bergs discharged, . . . and on one succeeding falling tide, six hours, sixty-nine"), with an average of one iceberg calved every five minutes. When his group approached the bay into which the Muir Glacier calved, by canoe, they found it so crammed with bergs that their guide, a Stickeen Indian, warned they "might be entering a skookum-house (jail) of ice, from which there might be no escape."

The reason there is no IIP equivalent on the western side of the continent is that the North Pacific is almost iceberg free. Few icebergs make it out of the Gulf of Alaska; most spend a large part of their short lives trapped in fjords or stranded on beaches and moraines at low tide, and when they do float away they don't usually get very far before breaking up or melting. Also, there are fewer ships on the Pacific side; the North Atlantic has the busiest shipping lanes in the world.

Still, where there are ships and icebergs, the twain shall sometimes meet. Just before dawn in January 1994, the oil tanker *Overseas Ohio* was heading in the direction opposite to the one the *Exxon Valdez* had been, from Bligh *into* Valdez Narrows, when it struck an iceberg. The hit opened a puncture hole in its bulbous bow, from which there poured into the waters of the sound . . . more water, ballast water. A company spokesman said the berg was probably an "underwater iceberg," maybe one of the dark, stone-laden ones.

> The fleet of icebergs set forth on their voyages with the upspringing breeze; and on the . . . shattered crystal walls of the glaciers, common white light and rainbow light began to burn.
>
> —John Muir, *Travels in Alaska*

A year after the *Overseas Ohio* hit an iceberg, I signed up for a ride on a sightseeing boat so I could get a look at the Columbia Glacier from the water. It was the boat's first outing of the season, and the air was cold and damp. The captain tried to work us as close as possible to the glacier, but there were so many icebergs in the water he had to stop well short. The ice was mostly "brash," defined as ice fragments under six feet wide, "the wreckage of other forms of ice." We were told that the Columbia is in fast retreat, and its front had pulled so far back it had lost contact with the ridge it was grounded on, and it floated now in deeper water back of the ridge and was therefore unstable and releasing huge numbers of icebergs. The larger ones got hung up behind the ridge, where waves and tides beat them up into smaller pieces, which floated out. The ice around the boat was low, clumpy, white, soft-looking stuff that covered the surface loosely, like crushed ice floating in a soft drink; to me it looked like something the boat could easily push through, but the captain wasn't convinced. Sticking up from the white field here and there were some larger ice chunks, of a beautiful shade of aquamarine. Regarding them was like taking a deep breath of clean air. The color was most saturated next to the water and in holes and cracks. A seal was draped across the top of one of them. Another one had a dark bird perched on it. Muir, "feeling sad" over his "weary failure to explore" a particular glacier because of all the icebergs in the water in front of it, had been consoled when an ousel, a waterbird, flew straight from shore toward him, circled his head three times "with a happy salute," then "alighted on the topmost jag of a stranded iceberg, and began to nod and bow as though he were on one of his favorite rocks in the middle of . . . sunny California."

Our steward was very enthusiastic about the bergs. He fished a chunk of one out of the water and showed it around. It was about the size of a manhole cover. "Great ice!" he said and poured green food dye over it. The green slid across the surface and collected in depressions between crystals, defining their boundaries. With the butt of his hand, he hit the ice a blow, and several crystals separated out of it. Each was about ten inches long. "It took hundreds of years to make this ice," he declared. He picked up one crystal and placed it carefully back into the space it came out of. "See," he said, delighted, "the pieces fit together, like pieces of a puzzle." He told us he had taken some iceberg ice to his sister's wedding. "I put it in the champagne punch, and it fizzled and popped. The rest"—he gestured vaguely toward some place on the ship where manmade ice cubes were presumably kept—"is garbage."

We'd rather have the iceberg than the ship,
Although it meant the end of travel.

—Elizabeth Bishop, *"The Imaginary Iceberg"*

A man in the second-class smoking room of the *Titanic* when the ship hit the Iceberg joked that he'd like some iceberg ice for his drink. Perhaps he didn't know what lay ahead for him; perhaps he guessed and was being bravely lighthearted. Only a dozen men in second class made it to New York; chances are he was not one of them.

CHAPTER THIRTEEN

SEA ICE I

The sheeted deep

—Frederick A. Cook,
My Attainment of the Pole

"RETURNING TO CAMP THAT NIGHT," Frederick Cook wrote while crossing sea ice on his try for the North Pole, "we surprised our stomachs by a little frozen musk ox tenderloin and tallow. . . . Then we retired. Ice was our pillow. Ice was our bed. A dome of snow above us held off the descending liquid air of frost. . . . Beneath my ears I heard the noise of the moving, grinding, crashing pack. It sounded terrifyingly like a distant thunder of guns. I could not sleep. Sick anxiety filled me. Could we cross the dreadful river [broad crack in the ice] on the morrow? Would the ice freeze? Or might the black space not hopelessly widen during the night? I lay awake, shivering with cold. I felt within me the blank loneliness of the thousands of desolate miles about me."

Cook, his men, and his sled dogs had reached what he called the Big Lead, a channel of water that lay between the ice that was attached to shore and the ice beyond it that was not. With the setting sun, the temperature had fallen rapidly, Cook noted, "and the wind was just strong enough to sweep off the heated vapors" rising from the thin ice, under which conditions, he understood, "new ice forms rapidly." Still, he worried. "Long before the suppressed incandescent night changed to the prism sparkle of day we were out seeking a way over the miles of insecure young ice separating us from the central Pack. . . . I knew, as I gently placed my foot upon the thin yellowish surface, that at any moment I might sink into an icy grave."

To spread their weight, the men wore snowshoes and took long, gliding steps with their legs well apart. "Stealthily, as though we were trying to filch some victory, we crept forward," Cook wrote. "None of us spoke during the dangerous crossing. . . . We covered the two miles safely, yet our snail-like progress seemed to cover many anxious years. . . . When the crossing was accomplished . . . I could have cheered with joy."

Explorers who managed to cross frozen polar seas discovered more than new shores, new wildlife, and new routes to the East. Forced to deal with sea ice at almost every turn, by pushing through it in ships or slogging over it on sledges or foot, they came to know it intimately—though often to their sorrow. Whether they wanted the ice to give way (if they were in a ship) or not (if they were standing on it), they couldn't help but notice something of its behavior, properties, distribution, life cycle. Many of the explorers* had a strong scientific interest in their frigid environment, took measurements, collected specimens; Cook wasn't the most keen, but he described what he saw. As he placed his foot gently on the newly frozen ice of the Big Lead, he observed that it was "mottled and tawny colored, like the skin of a great constrictor" (probably from dirt trapped between crystals) and as flexible as "a sheet of rubber" (from saltwater trapped between crystals). "We rocked on the heaving ice," he wrote, "as a boat on waves of water."

and the breadth of the waters is straitened.

—Job 37:10

Sea ice is frozen ocean water, the ice that icebergs sometimes cut through like knives through butter. Because it has dissolved salts in it, seawater freezes at a lower temperature than fresh water does, typically 3°F lower, or at about 29°F. In the early years of exploration, many seafarers believed that seawater was too salty to freeze—that what they saw floating around on polar seas was actually freshwater ice that had formed on land, in rivers, and been disgorged into the sea. In 1772, when Captain James Cook was searching for a southern continent, which no one had yet seen but which philosophers believed existed as a balance to the northern continents, and his ship was stopped by sea ice just short of 70°S, he assumed, on the basis of "received opinion," that "there must be land in the Neighbourhood of this ice." Although Cook "had been attentive to the ice, its appearance and movement, since he first encountered it," J. C. Beaglehole wrote in his biography of Cook, ". . . he was still prepared to admit—wrongly, though in accord with the philosophers—that sea ice invariably implied land."

* As well as a few whalers and sealers, most notably William Scoresby Jr., a whaler-explorer-scientist and author of the remarkable *An Account of the Arctic Regions with a History and Description of the Northern Whale Fishery.*

Some years later, when Captain Cook was searching for a Northwest Passage from the Pacific side of North America and his ship was stopped by ice at 70° N ("The ice had beaten him," Beaglehole wrote, in the north as it had in the south), he "meditated on the nature of the enemy. . . . The enemy was ice." Cook figured that the ice was too thick to have come from rivers, which were so shallow that "there is hardly water for a boat," and, besides, the ice wasn't carrying anything that originated on land. Yet "even now, with all his experience," Beaglehole pointed out, "he would not assert, or guess, that the sea froze." Cook decided what he was seeing was frozen snow. Snow fell on top of the ocean, froze, and served as an icy foundation on which more snow fell and froze.

When ice forms from seawater, the crystals reject impurities, including salts, and most of the salts end up being rejected out the bottom of the ice as it freezes. But not all of them. About a fifth of the salts stay in the ice, in pockets between crystals. The pockets form because the very salty layer (brine) is too salty for the crystals to convert to ice, so they extend the freshwater tips of their constituent platelets into the ordinary, less salty seawater below, which they *can* convert to ice, and thereby they grow longer. The underside of sea ice, then, unlike the smooth underside of a layer of freshwater ice, has a structure made up of many delicate platelet tips, 15 to 30 of them per square inch; in a schematic drawing, they look like the pointed teeth of a comb. Some of the brine from the layer just under the ice gets drawn up into the grooves between the tips and is held there by capillary action. The platelets widen as they lengthen, and here and there ice bridges form between them, creating the pockets, which are long and narrow and contain brine. As the ice cover gets thicker, the pockets line up in vertical strings. "Without this entrapment mechanism," Gary A. Maykut, research professor at the University of Washington in Seattle, concludes, "there would be little to distinguish sea ice from freshwater ice."

Because of the entrapment, sea ice has different properties, a different consistency, and even a different look from freshwater ice. Freshwater ice, 20th-century explorer Vilhjalmur Stefansson wrote, has a "glasslike quality" when it freezes beyond the slush stage: "A quarter-inch-thick layer is strong enough so you can handle it almost like a pane of window glass." With sea ice, the slush stage persists longer (the ice being closer to its melting point, with some of it not yet frozen), and a piece of it cannot be handled like glass until it is more than two inches thick, "perhaps three." Even a "piece four inches thick, if you allow it to drop on any hard surface from a height of three or four feet, will splash like a chunk of ice-cream, instead of falling like a sheet of glass." On four inches of sea ice, Stefansson pointed out, "a man does not venture to walk upright" as he would on freshwater ice. "He will crawl, or walk on snowshoes, or, best of all, on skis." Because of this difference, critical to people who might want to traverse ice, "the expression 'young ice' is seldom used for fresh water ice, constantly for salt."

The brine doesn't stay in the pockets, however; it migrates through the ice. The bridges between platelets melt or break due to stresses on the ice, and strings of pockets turn into tubes or channels. The brine near the top of the ice cover, closer to the air, is usually colder and therefore heavier than brine toward the bottom of the ice and drains slowly downward through the channels, by gravity. When it reaches the bottom, it is pushed out, into the water, in plumes or streamers. Since the brine is as cold as the ice, which can be very cold, when it hits the water it can cause the water around it to supercool, then freeze, into "ice stalactites," or hollow tubes through which more brine flows. In the Arctic ice stalactites can be several feet long, in the Antarctic up to 20 feet long.

In summer, the migration of brine speeds up, as snowmelt percolates through the cover and flushes out the channels. One day, Australian explorer Douglas Mawson and his men were off the Antarctic coast digging "ice-shafts" in sea ice (which the morning light had turned into "a lilac plateau," he wrote), so they could compare its structure and temperature to those of glacier ice and lake ice, when he noticed that as soon as a shaft got to be three feet deep "brine may begin to trickle into the hole, and this increases in amount until the worker is in a puddle. The leakage takes place," he noted, "if not along cracks, through capillary channels, which are everywhere present in sea ice."

Farther out, the waves will be mouthing icecakes

—Sylvia Plath, *"A Winter Ship"*

A sea-ice cover starts out as individual platelets and needles of frazil suspended in water. As temperatures fall and more particles form, they produce a soupy layer on the surface called "grease ice" (Mawson described the effect as an "oily lustre"). If the sea is calm, grease ice solidifies into nilas, a layer up to four inches thick riddled with saltwater pockets, which render it elastic, easily bent by waves. "New ice often looks like someone rumpled a rug," says Richard Glenn, an ice geologist in Barrow, Alaska (whose Eskimo mother's name in Inupiat means "snowflake"). Under pressure the thin ice breaks, and the detached pieces push against each other, so that rectangular strips from each override and underride the other in alternating fashion like fingers of two hands, creating a pattern that is striking in its regularity.

Slightly thicker ice is called by its tint, "dark gray" and, up to 12 inches thick, "light gray ice."

Waves and swells often break up the sheets of young ice, then herd the pieces, along with grease ice and slush, into assemblages which freeze together into pancakes. These pancakes are usually round with raised rims from being jostled by other pancakes as they rotate. Grease ice freezes between them and welds the cakes into floes. The floes freeze together into fields. An ice field, according to a post–World War II U.S. Navy publication, is more than five miles long. A "giant" ice floe is 3,000 feet to five miles long, or "the size of a small city." A medium floe is 600 to 3,000 feet long, "the size of a golf course," a small floe 30 to 600 feet long, "the size of a city block." An *ice* block is six to 30 feet across, the size of a volleyball court, and a piece of brash is at most six feet long, the size of a pool-table top.

Floes grow smaller as well as larger, breaking apart or being ground down as they knock into or rub against other floes. Ice meeting ice "makes much ice," Stefansson wrote. The ice may be "a soft slush or fragments the size of your fist, the size of a kitchen range, or of a house." These may freeze onto other fragments to form larger pieces again. "Sea ice is always changing," Willy Weeks, one of the pioneer investigators of sea ice, points out. "Temperatures change, properties change, positions change. What's here today is gone tomorrow."

And the sundry high shoals of ice

—A. R. Ammons, *"Conserving the Magnitude of Uselessness"*

Sea ice covers about 7 percent of the world's oceans at any one time—"at any one time" because the extent of the ice changes according to the seasons. When the sea ice is at its maximum extent at one pole at winter's end, it tends to be at its minimum extent at the other pole at summer's end. During the long polar nights, the covers expand; during the long polar days (or "double days of joy," as Frederick Cook called them), they shrink. In the Antarctic, sea-ice extent goes from almost 8 million square miles in winter to only a little over half a million square miles in summer. The difference melts. Most Antarctic sea ice doesn't last more than a year. In the Arctic, sea-ice extent is less at its maximum than in the Antarctic but more at its minimum; half the sea ice lasts more than a year. (In winter it covers almost 5.5 million square miles, in summer over 2.5 million.)

The difference in survival times is caused by asymmetry at the poles. The Antarctic is land surrounded by water, the Arctic water surrounded by land. Sea ice rings the Antarctic continent and, unconfined by outlying land, spreads away from it, northward, into warmer waters. The Arctic Ocean, by contrast, is a "Mediterranean sea," Roger Colony, research scientist at the International Arctic Research Center in Fairbanks, points out. "It's almost enclosed, with only a few connections." The main connection, the channel through which 95 percent of the ice that leaves the Arctic Ocean leaves, is a strait between Greenland and Spitzbergen. Only 10 percent of the ice in the Arctic Ocean does leave it every year, though, to melt in the Atlantic. Some of the rest melts in place during those summer double days, mostly around the edges, but that leaves a lot of ice hanging around the Arctic Ocean until the following summer, and sometimes the summer after that, and often the summer after that.

This ice is called "multiyear," "perennial," "permanent," or just "old." In the Arctic, it's seven years old on average, but some of it is probably decades old. Ice that has not yet been through a summer is called "first-year ice," "winter ice," or "ice of the season," as Eskimos of northwest Alaska called it, according to Richard K. Nelson, author of *Hunters of the Northern Ice.* If you stood on an ice cover made up of both first-year and multiyear ice, you could probably tell which is which. Multiyear ice tends to be bluish, first-year grayish (or "an ugly green," Glenn says). Multiyear is thicker, 10 to 16 feet thick compared to 3 to 7 feet thick for first-year, and therefore sticks up higher above the waterline. Multiyear ice often has a glare surface (which "gives poor footing for men and dogs," Stefansson wrote), first-year ice a sticky one ("therefore always has snow adhering to it"). Multiyear ice splinters more easily than first-year ice "when you peck at it with your hunting knife," Stefansson noted, and its melt rarely tastes salty. Because of the salt drainage, he claimed, "ice which is 2 years old is probably fresher than average river or spring water."

A sea-ice cover, then, can be a mishmash of ages, sizes, stages, degrees of saltiness, thicknesses, and types. "Now I can see why the Eskimos have 200 words for ice," Richard Weber, a member of the 1986 Steger International Polar Expedition to the North Pole, declared. The usual phrase you hear is that Eskimos have 200 words for *snow.* ("Why don't we have that many words for love?" Margaret Atwood supposedly asked once.) People don't usually talk about how many words there are for ice, but Nelson lists dozens for sea ice alone, which an Eskimo "informant" shared with him (he himself wasn't a linguist), some with such fine distinctions in English translation as "rough ice which consists largely of pieces of ice which have been pushed into a vertical position" and "ice which is about to begin piling." I picked up a list of Inupiat words for ice and snow that the North Slope Borough in Barrow gave visitors to the annual bowhead-whale hunt and showed it to Kenny Toovak, a former whale hunter and construction supervisor for ice-engineering projects, whom I found waiting for his wife at a senior citizens' center. Did people actually use all the words for ice on the list? I wanted to know. There were 76. After Toovak gave the list a hard, blistering look, he said he'd get back to me. Very early the next morning he appeared at my dormitory door with comments.

Some words, including "*sikuliagruaq*" ("thick ice," greater than three feet, in the borough's definition) and "*agiuppak*" ("a smooth wall of ice along the edge of landfast ice formed by other moving ice"), you heard people use, he said, but others, like "*aluksraq*" ("young ice punched by seals forming a seal blowhole"), not much. He laughed at two separate words on the list; they were, like "cap" and "hat," the same thing! When I pointed to "*nunagvaq*" ("ice once used by walrus") and asked how anyone could tell, he frowned at my thickheadedness. The ice was *brown,* of course; walruses defecated all over it. Several times when evaluating some word he would look out the window with such a fierce expression that I looked too, half expecting to see the ice in question gliding by over the ground.

When I showed Richard Glenn the list, he added "*sugainnuq*" and defined it as "huge, moving ice that threatens integrity of the lead edge; could be large piece of pack ice or agglomeration of multiyear ice and first-year ice." He translated the often-used "*sagrat*" as "ice floes of random sizes, beltways of ice with water on both sides." On seeing "*qinu*" (slush), he was prompted to tell the story of a hunter who was paddling his kayak with a single paddle until he came to *qinu* and had to start using two.

In *The Language Instinct,* Steven Pinker, Johnstone professor of psychology at Harvard University, contended that the 200-words-for-snow claim is a myth, an "anthropological canard." Eskimos don't have a hundred words for snow, he wrote, "or forty-eight, or even nine. One dictionary puts the figure at two. Counting generously, experts can come up with about a dozen, but by such standards English would not be far behind." He quoted anthropologists who dubbed the claim the "Great Eskimo Vocabulary Hoax" and traced the hoax to a more modest claim dealing with roots of words for snow which had been greatly amplified in a misguided attempt by some people to show that nonliterate cultures, too, can be sophisticated. Even if the claim were true, the anthropologists argued, it would be a boring truth because why wouldn't people who have a lot to do with a particular substance in their everyday lives concoct a variety of names for it? In the case of sea ice, Eskimos have clearly concocted many more than a dozen words, although I never could pin down what proportion of them are in common use.

In the sunless cold of the lingering night
into marble statues grow!

—John Greenleaf Whittier,
"The Frost Spirit"

Sea ice makes up two-thirds of the earth's permanent ice cover by area but only one-tenth of one percent of it by volume. "Sea ice is just this little thin veneer," Gary Maykut points out. A mere sheet on the deep, a lid upon the great salt flood. Even in extremely cold air, sea ice doesn't grow to be more than about 12 feet thick. As with lake ice, the more it grows, the slower it grows, since the heat that water must give up in order to freeze to the bottom of the ice cover has to pass through a greater and greater thickness of insulating ice to reach cold air and dissipate. Still, some sea ice gets to be thicker than 12 feet, a whole lot thicker, 20, 30, 40, even 60 feet thick. It does so not by growing but by building. One day I visited an eighth-grade class at the Eben Hopson Middle School in Barrow, and the teacher, Mrs. Moye, asked her kids to draw a picture of sea ice for me. Erik Edwardson drew this:

(Note two climbing polar bears, "one mother and a big baby.")

Nora Itta drew:

Brandon Fishel drew this:

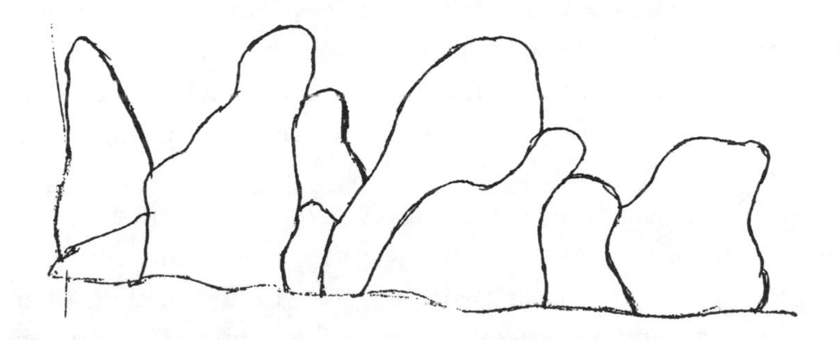

None of the drawings was supposed to be of an iceberg. There are no glaciers in northwest Alaska to throw off icebergs. The pictures were all of sea ice as the students routinely see it along Barrow's shoreline. "There's always ice rubble in Barrow," Richard Glenn says. Onshore winds drive floating pack ice against the landfast ice, and it buckles, fractures, and piles up.

Sometimes the winds drive ice clear onto land, producing ice overrides (*ivu*). In 1937, a fire burned the hospital in Barrow to the ground, after which "nature showed what can be done with frozen water," Charles Brower wrote in his book *Fifty Years Below Zero.* "Suddenly ten million tons of sea ice, driven by a southwest gale, came surging in to finish the rest of Barrow. They thundered against the anchored ice along the shore. They shoved it across the sandspits and up the slopes beyond. They squeezed and pressured it into fantastic masses which towered seventy-five feet high. All in twenty minutes! Then, when there seemed no hope left for any wooden structure, the pressure ceased as abruptly as it had started and the fleeing populace, not yet adjusted to loss by fire, came back to survey its loss by ice."

According to Inupiat oral history, heavy ice once hit a seaside cliff with a village on top, tore away part of the bank, and destroyed a house "so suddenly that one of the inmates, a large, stout man, was unable to escape through the trap-door and was crushed to death."

Out in the pack, too, floes bang into and scrape against one another, creating ridges and rubble and hummocks. (A "pack" is defined as a region of floating ice.) "In order to move, ice in the landlocked Arctic Ocean has to push something else out of the way," Willy Weeks explains, "shove it up, down, or sideways." For instance, a thin or fast-moving piece of ice will ride up and over a thicker or slower-moving piece of ice, producing a piece that is the sum of their thicknesses. When the *Advance,* a ship commanded by American explorer Elisha Kent Kane, was crossing Melville Bay on Greenland's west coast, giant floes that by his estimation weighed over two million tons were slamming together and flying about. "Tables of white marble were thrust into the air, as if by invisible machinery," he wrote, landing on and breaking one another, producing "a chaotic mass of crushed marble."

From aboard the *Endurance,* Shackleton's men were witness to the process of ridge building. Based on their journals, Frank Lansing wrote this description of it: "Whenever two thick floes came together, their edges butted and ground against one another for a time. Then, when neither of them showed signs of yielding, they rose, slowly and often quiveringly, driven by the implacable power behind them. Sometimes they would stop abruptly as the unseen force affecting the ice appeared mysteriously to lose interest. More frequently, though, the two floes—often 10 feet thick or more—would continue to rise, tenting up, until one or both of them broke and toppled over, creating a pressure ridge."

So common is all this shoving around in the Arctic that 40 percent of the mass of sea ice there is thought to be concentrated in ridges and hummocks. That's including of course the ice keels, which stick down three or four times as far below sea level as ridges and hummocks do above it. In the Antarctic, where the ice is generally thinner and looser, the ridges aren't as high.

By contemplating the profiles of pressure ridges, you can usually tell whether the ice in them is multiyear or first year. The freshly broken ice in first-year ridges, according to Vilhjalmur Stefansson, "may well be compared with the masses of rock in a quarry just after the blast," or if the ice is thinner, to "the broken-bottle glass on top of a stone wall." Yet after thaws, "all the sharp outlines are softened so that at the end of the first summer they are no more jagged than a typical mountain range. At the end of 2 or 3 years they resemble the rolling hills of a western prairie."

Almost all Arctic sea ice is deformed in some way or other because, except where it's attached to land or grounded to the bed, it's always on the move, being propelled this way and that by winds and currents and tides and the Coriolis effect (mostly winds). "All the world on which we traveled was in motion," Frederick Cook wrote while trekking toward the pole. "We moved, but we took our landscape with us. Our footing was seemingly a solid stable ice crust, which was, however, constantly shifting." Even when ice floes are tightly packed together, Richard Glenn points out, "and everything is butting up against something else, and there's no water in sight, the ice is still in motion, not locked. A given piece will be actively deforming the ice next to it." Tom Curtin of the U.S. Navy compares the movements of sea ice to plate tectonics. "The pack is a floating plate that's being stressed," he says, "like the floating crust of the continents."

In their movement, floes diverge as well as converge. They pull away from each other or crack open, exposing pools or lines of water. A linear crack is called a lead—"ugly black snake," one explorer called a lead that kept him from going where he wanted to. If the line is narrower than the ice is thick, it's just a crack. If it's wide enough for a ship to pass through, some people call it a lane. A pool of water, up to hundreds of miles square, like a lagoon, is called by the Russian word "polynya." Polynyas often show up in the same places year after year, probably because strong winds show up in the same places year after year and sweep ice from the surface.

A sea-ice cover is never complete; there's always open water somewhere. About a fifth of the area in the Arctic Basin is unfrozen in summer; in winter about half of one percent is. The temperature difference between suddenly exposed seawater and winter air can be as great as 40°F. "Frost-smoke from opening cracks was showing in all directions," Shackleton wrote of steam coming off leads on a day when air temperatures in the Weddell Sea were falling rapidly. "It had the appearance in one place of a great prairie fire, rising from the surface and getting higher as it drifted off before the wind in heavy, dark, rolling masses."

The amount of heat lost to the air by water in a lead, or by water in a lead only lightly iced over, can be as much as 100 times greater than that lost from a thick layer of ice right beside it. The heat flux from a few patches of open water in a sea-ice cover—a mere one percent of the total area—can equal that from all the rest of the cover combined.

"The opening of leads in the ice ahead of you is not a serious matter in winter," Stefansson assured those of his readers who were likely to be going sledging, "since after a night's wait it should freeze over." All you have to do when faced with an uncrossable water barrier is stand around and cool your heels. Once when Frederick Cook was stopped by a lead he heard a "peculiar noise, like that of a crying child. . . . It came seemingly from everywhere," he wrote, "intermittently, in successive crying spells. In an effort to locate the cry, I searched diligently along the lead. I came to a spot where two tiny pieces of ice served as a mouthpiece. About every fifteen seconds there were two or three sharp, successive cries. With the ice-axe I detached one. The cries stopped; but other cries were heard further along the line. . . . Returning later to the lead, to watch the seas breathe, the cry seemed stilled. The thin ice-sheets were cemented together."

On another occasion, on the "ebony mirrored surface" of a newly refrozen lead, Cook found "a profusion of fantastic frost-crystals arranged . . . in bunches resembling white and saffron-colored flowers." That's where you usually find frost flowers in the sea, on a newly refrozen lead, particularly in springtime when there's lots of moisture in the air yet the air is still cold. Will Steger, leader of a 1986 expedition to the North Pole, came across what looked like "miniature white maple leaves or fern fronds" while skiing across a firmly refrozen lead. "They shimmered with the slightest breeze," he wrote, "acting as prisms that set the surface aglow, twinkling in every color of the spectrum."

Open water absorbs about 90 percent of the sunlight that hits it, or four to seven times more radiation than the ice on either side of it. This means that water is reflecting only 10 percent of the light coming in at the same time that bare ice nearby is reflecting about 60 percent of the light and snow about 80 percent. "That's one of the great things about sea ice," Don Perovich, research geophysicist at CRREL, points out. "Here's the lightest, most reflecting thing in the world right next to the darkest, least reflecting thing in the world." As sunlight warms the water, the water melts the edges of the ice, which widens the lead and exposes more water to the warming effect of sunlight.

A curdled sea

—Fridtjof Nansen,
Farthest North

The first look Fridtjof Nansen got of sea ice was when he was on a sealing ship on his way from the west coast of Norway to the east coast of Greenland, where he planned to disembark and cross the ice cap for the first time ever. (There wasn't any ice on the west coast of Norway when he left because of the warming effects of the Gulf Stream.) As the sealer passed from the Norwegian Sea to the Greenland Sea, "suddenly something huge and white loomed out of the darkness," Nansen wrote, "and grew in size and whiteness, a marvellous whiteness in contrast to the inky sea, on the dark waves of which it rocked and swayed. This was the first floe gliding by us. Soon more came, gleaming far ahead, rustling by us with a strange rippling sound, and disappearing again far behind. Then I saw a singular light in the northern sky, brightest down at the horizon, but stretching far up towards the zenith . . . and as I looked I heard a curious murmur to the north like that of breakers on a rocky coast, but more rustling and crisper in sound . . . I felt instinctively that I stood on the threshold of a new world. . . . The light was the reflection which the white masses of ice always threw up when the air is thick, as it was that night, and the sound came from the sea breaking over the floes while they collided and grated one against the other."

The light Nansen saw was "ice blink," light reflected off sea ice onto the underside of low clouds. When he can't see ice, a person can infer its presence by reading the "sky map," if the sky is overcast. The counterpart of ice blink is "water sky," the black of open water darkening the clouds overhead. A "mottled sky" indicates there's water both frozen and unfrozen below the clouds, according to Mawson, "an ice-strewn but navigable sea."

The sound that Nansen heard, "like that of breakers on a rocky coast" but crisper, is but one of an immense variety of noises that sea ice has been known to make, and by no means the oddest. In the pack, there were "the basic noises, the grunting and whining of the floes, along with an occasional thud as a heavy block collapsed," Alfred Lansing wrote, conveying the impressions of men aboard the *Endurance.* "But in addition, the pack under compression seemed to have an almost limitless repertoire of other sounds, many of which seemed strangely unrelated to the noise of ice undergoing pressure. Sometimes there was a sound like a gigantic train with squeaky axles being shunted roughly about with a great deal of bumping and clattering. At the same time a huge ship's whistle blew, mingling with the crowing of roosters, the roar of a distant surf, the soft throb of an engine far away, and the moaning cries of an old woman."

Once I asked a biologist what effect the booms of supersonic airplanes have on Arctic wildlife, and he told me that the animals aren't disturbed; ice is so noisy they are used to booms. " 'The eternal polar silence' writes the poet in his London attic," Stefansson pointed out. "But Shackleton's men, as quoted in his book 'South,' now and again commence their diary entries with the words 'din, *Din, DIN.*' Robert Service some distance south of the arctic circle in a small house in the city of Dawson, wrote much of the arctic silence. But we of the far north never forget the boom and screech and roar of the polar pack."

The shapes of ice blocks produced by the movement of the pack are as varied and odd as the sounds, judging by on-scene accounts. Explorer Sir Ranulph Fiennes wrote of seeing, by the light of a full moon, giant blocks that looked "like elephants or skyscrapers or demons." On his trek north, Will Steger found many "an ice sculpture garden. . . . The diversity of sea-ice forms and colors was incredible," he noted. "The ice piled and shattered in an endless array of beautiful and ominous shapes." Among the shapes seen by another member of the same expedition, Brent Boddy, were "dorsal fins on whales, giant ice cream cones, statues of warriors, and lots and lots of modern art."

When he was on board the *Fram,* Nansen called the sea-ice statuary a "shifting pageant of loveliness," but when he left the ship to try to reach the pole on foot, accompanied by one of his men, Hjalmar Johansen, and a pack of dogs, he seemed to have found the piled-up ice blocks somewhat less than lovely:

"Began our march at three yesterday morning," he wrote on April 5. "The ice, however, was bad, with lanes and ridges, so that our progress was but little. These lanes, with rubble thrown up on each side, are our despair. It is like driving over a tract of rocks. . . . This continual lifting of the sledges over every irregularity is enough to tire out giants."

July 11: "Our hearts fail us when we see the ice lying before us like an impenetrable maze. . . . There are moments when it seems impossible that any creature not possessed of wings can get farther, and one longingly follows the flight of a passing gull, and thinks how far away one would soon be could one borrow its wings."

"There's a misperception that going to the North Pole is like going across *pond* ice," says Geoff Carroll, area wildlife biologist for the State of Alaska and a member of the Steger expedition. "For us, the whole way was one pressure ridge after another, all jumbled up, huge blocks of ice. Some days we worked real hard to make one mile. The Iditarod* is like

* Alaskan dog-sled race, considered grueling.

a *biking trip* compared to what we did. Their dogs pull lightly loaded sleds over a nice, groomed trail; we pushed a wheelbarrow full of rocks through a boulder field."

The ice clamps and will not open,
For a year it has not opened

—Gwendolyn MacEwen,
"Afterworlds"

Men on ships often had similar thoughts about ice's power to thwart their plans. "Desolate, hummocky pack," Scott wrote irritably when ice kept the *Discovery* from making headway through the Weddell Sea. On Christmas Day 1910, he complained that the scene at Cape Crozier, where a solid sheet of pack ice held the ship tight, was "altogether too Christmassy. Ice surrounds us," he reported. "The white haze of ice blink is pervading. We are captured. We do practically nothing under sail to push through, and could do little under steam, and at each step forward the possibility of advance seems to lessen."

Scott's crew had just, "by hard fighting," worked their ship toward an opening they saw in the ice ahead—"a long lane of open water"—and, believing it to be the end of their troubles, had been jubilant, yet they soon discovered ("alas!") that at the end of the opening was more heavy bay ice. "What an exasperating game this is!" Scott declared. On that occasion hard fighting consisted of a lot of stopping, backing the ship up, and starting it going forward again. Really hard fighting could involve such things as sawing away the ice in front of the bow, chopping it with axes, and blasting it with powder. (Captain James Cook is said to have had penguin grease smeared on his ship's hull to ease its passage through ice.) Shackleton described using his ship to ram open a lane:

"When the way was barred by a floe of moderate thickness," he wrote—his idea of moderate being up to three feet thick—"we would drive the ship at half speed against it, stopping the engines just before the impact. At the first blow the *Endurance* would cut a V shaped nick in the face of the floe." After backing the ship up 200 or 300 yards, the men would drive it "full speed into the V, taking care to hit the centre accurately. The operation would be repeated until a short dock was cut, into which the ship, acting as a large wedge, was driven. At about the fourth attempt, if it was to succeed at all, the floe would yield. A black, sinuous line, as though pen-drawn on white paper, would appear ahead, broadening as the eye traced it back to the ship. Presently it would be broad enough to receive her, and we would forge ahead."

Or not. One time the men tried to open a channel for the ship with "ice-chisels, prickers, saws, and picks," Shackleton wrote. "The men cut away the young ice before the bows and pulled it aside with great energy. But . . . every opening we made froze up again quickly owing to the unseasonably low temperature." He explained: "The young ice was elastic and prevented the ship delivering a strong, splitting blow to the floe, while at the same time it held the older ice against any movement. . . . The task was beyond our powers." It was at this point that Shackleton became resigned to the possibility that the *Endurance* would be spending the winter "in the inhospitable arms of the pack."

In the Arctic, a great many more ships ended up in the inhospitable arms of the pack, and some for more than a winter. In 1829, the steamer *Victory,* commanded by John Ross, was gripped by ice west of Baffin Island and stayed gripped for more than three years.

First winter: "The prison door was shut upon us for the first time," Ross wrote, "leaving us feeling that we were as helpless as hopeless captives, for many a long and weary month to come."

First summer: "Whatever value voyages of discovery may have in these countries, they are certainly purchased at a high price in time. We might have circumnavigated the globe in the same period."

Late summer (when the ice had slackened): "Under sail! . . . What seaman could feel this as we did, when this creature which used to carry us buoyantly over the ocean, had been during an entire year immovable as the ice and rocks around it, helpless, disobedient, dead." The *Victory* managed to get three miles away before ice tightened around it, again.

Second winter: "Our winter prison was before us; and all that we had now to do was . . . set up our amphibious house."

Second summer: "It is difficult to convey to my readers the impression produced by this stationary condition of a sea thus impracticably frozen. . . . 'Till the rocks melt with the sun' is held an impossible event, in one of the songs of my native land [England], and I believe we began at last to think that it would never melt those ice rocks; which even at this late period of the year continued to beset us . . . Oh! for a fire to melt these refractory masses, was our hourly wish."

Third winter: "The thickness of the ice round the ship was such as to prevent all hopes of her liberation."

Beginning of third summer: "We drank a parting glass to our poor ship, and having seen every man out I took my own adieu of the *Vic-*

tory. . . . It was like the last parting with an old friend; and I did not pass the point where she ceased to be visible without stopping to take a sketch of this melancholy desert, rendered more melancholy by the solitary, abandoned, helpless home of our past years, fixed in immovable ice, till time should perform on her his usual work."

For another year, Ross and his men sledged and boated over pack ice until, near Cape York, a ship's sail appeared in the distance. Some of the "despairers" among the men swore it was nothing but an iceberg.

During the long search for the Northwest Passage, as well as during the long search for Sir John Franklin, who went looking for a Northwest Passage in 1845 and never returned, other ships were "glued in" (as Elisha Kent Kane put it), mostly in the bays, sounds, and straits of the Canadian archipelago. Franklin's two ships, the *Erebus* and the *Terror,* were last seen, by whalers, in June of 1845, tethered to an iceberg. On the basis of artifacts that eventually turned up in Eskimo camps—a silver pencil case, a piece of embroidered vest—plus skeletons on an island with bits of wool on them plus a note left in a cairn by an explorer looking for Franklin, historians have speculated that the two ships had been heading south toward King William Island in Victoria Strait in the summer of 1846 when they ran into an ice stream flowing out of the Beaufort Sea.

"Driven by the prevailing winds from the northwest, this ponderous frozen stream, awesome in its power, squeezes its way between the bleak islands, as it seeks warmer waters," Pierre Berton wrote of that perennial iceflow in his book *The Arctic Grail.* "Like a floating glacier up to one hundred feet thick, unbroken by any lane or channel, the moving pack is impenetrable." The *Erebus* and *Terror* were beset in September, "imprisoned in that frozen river that moved south at the frustratingly slow speed of one and a half miles a month," and a year later still were. "There was no way out of the floating trap. It had been the coldest winter in living memory. The ice had not melted; its progress [southward] was too slow."

For a long time, seamen had been searching for a North*east* as well as a North*west* Passage, a route between the Atlantic and Pacific by way of water north of Siberia in the Barents, Kara, Laptev, East Siberian, and Chukchi seas. The first person to make it through the Northeast Passage (or *a* Northeast Passage; there are three possible ones, or four if you consider going over the North Pole an option, just as there is no single Northwest Passage but several possible ones) was Baron Nils Adolf Erik Nordenskiöld, a Swede born in Finland, who once tried to reach the North Pole with reindeer pulling his sleds. He was reputedly more interested in opening up a trade route along the Siberian coast as an outlet for furs and other goods from the hinterland than the glory of setting a record. "The romantic and incredible and heroic do not lard the log of the *Vega* [Nordenskiöld's ship]," Jeannette Mirsky wrote in her book *To the Arctic!* "There are no crises, no hairbreadth escapes."

There was, however, ice. It had been a summer of light ice, and in only two months the *Vega* had nearly completed the passage when, near the northern entrance to the Bering Strait—"the very threshhold of the northern portal of the Pacific," wrote the anonymous author of *The Realm of the Ice King* (1883), where the ice was "unusually bad"—the ship was stopped short. Very short. *The Realm* quoted a tantalized Nordenskiöld: "When we were beset, there was ice-free water some few minutes further east. A single hour's steaming at full speed would probably have sufficed to accomplish this distance." Not until ten months later when the ice relaxed its grip was the *Vega* able to close the distance. Two days later, its cannon sounded a "salute in that strait where the Old and the New World seem to shake hands."

Less than a decade earlier, in 1871, a party of Austrians had tried to gain easy entrance to a Northeast Passage by following one branch of the Gulf Stream north, past the island of Novaya Zemlya, in the expectation that the stream's warmth would keep the water there ice-free. The water turned out to be ice-rich, however, and their ship, the *Tegetthoff,* became mired in it. "Little as both officers and men then suspected it, *they were destined never again to see their vessel in open water,*" *The Realm* noted. As the ice was borne on currents northward, so were they. "We were drifting into unknown regions," Army Lieutenant Julius Payer wrote in his journal, "utterly uncertain of the end."

Summer passed, winter passed, and the ice and the ship kept moving north, farther and farther from any known land. Spring came, summer again, and the crewmen "dug and sawed to get the *Tegetthoff* free," Payer reported, "to no avail. They bored through the ice to see how deeply they were caught. After going to a depth of 27 feet, they still struck ice." They gave up. "We were mere insects, who dwell on the leaf of a tree and care not to know its edges." At the end of the second summer spent in the clutches of the ice, they spotted, "not by our action, but through the happy caprice of the floe," a cluster of islands. "Land, Land, Land at last!" Payer wrote. They named the islands Franz Josef Land, for their emperor.

The Austrians' notion about the Gulf Stream was but one manifestation of a centuries-old belief that there exists, beyond a wall of ice or land circling the earth at high northern latitudes, an ocean that is free of ice. In that view, the North Pole is under water. A map that cartographer Gerardus Mercator drew in 1569 shows a ring of land north of Europe and North America and Asia with four large gulfs in it, through which water could flow to and from this polar sea. (Mercator was apparently influenced by reports from voyagers who sailed north during a time of comparative warmth, the Climatic Optimum, before the Little Ice Age, without encountering much ice.) Even after whalers who had been there dismissed the idea of an open polar sea as fantasy, the concept still had a grip on the imagination of some Arctic explorers and influenced their choices.

Lieutenant Edward Parry dragged two 1,480-pound amphibious boat-sledges—good for use in open sea—on his 1827 try for the North Pole; their weight helped defeat his quest. Kane extended his search for Franklin into the northern part of the Canadian archipelago, on the assumption that Franklin would have headed that way to gain access to an open sea, through which he could then sail to the Bering Strait. Kane claimed to have found such a sea, when what he probably saw was a polynya, or a mirage.

> The dense pack had come, and hardly a square foot of space showed amongst the blocks. . . . The ominous sound arising from thousands of faces rubbing together . . . spoke of a force all-powerful, in whose grip puny ships might be locked for years and the less fortunate receive their last embrace.
>
> —Douglas Mawson, *Home of the Blizzard*

Of all the ships that received their last embrace from sea ice, the best known is Shackleton's *Endurance*. Shackleton's goal on his 1915 expedition had been to cross land ice, to lead the first party across Antarctica from one coast to the other, but the expedition ended on sea ice. The ship had been drifting with the pack in the Weddell Sea for months when a severe blizzard increased pressure in the ice and broke up the "island" floe the ship was embedded in, thereby exposing it to "attacks" from other ice. As he wrote in his book *South:*

October 26: "The ship was bending like a bow under titanic pressure. Almost like a living creature, she resisted the forces that would crush her; but it was a one-sided battle. Millions of tons of ice pressed inexorably upon the little ship that had dared the challenge of the Antarctic."

Suddenly, eight penguins emerged from a crack in the ice 100 yards away, walked toward the ship, and "after a few ordinary calls proceeded to utter weird cries that sounded like a dirge for the ship," a sound none of the men had ever heard penguins make before.

October 27: "At 5 p.m. I ordered all hands on to the ice. . . . The floes, with the force of millions of tons of moving ice behind them, were simply annihilating the ship."

October 28: "The whole of the after part of the ship had been crushed concertina fashion . . . and the wardroom was three-quarters full of ice. . . . Only six of the cabins had not been pierced by floes and blocks of ice."

October 29: "The ship is still afloat, with the spurs of the pack driven through her and holding her up."

November 21: "It was with a feeling almost of relief that the end came."

From a camp they had set up on an ice floe a mile and a half away, Shackleton and his men watched the *Endurance* "struggling in her death-agony," as one of the men reported. " 'She's going, boys!' " Shackleton called. "She then gave one quick dive and the ice closed over her forever. . . . 'She's gone, boys.' "

"There she goes, there she goes," the watch on the *Jeannette* yelled as the ship dropped through ice in the Siberian Sea. It was June 1880, and Captain George Washington DeLong of the U.S. Navy had been trying to reach the North Pole by a Pacific route, going north through the Bering Strait and along the east coast of Wrangel Island, on the premise that the warmth of the Japanese current would open a way through the ice pack. But ice took hold of the ship south of Wrangel and didn't let go for two years. "People beset in the pack before always drifted somewhere to some land," DeLong wrote in his journal, "but we are drifting about like a modern Flying Dutchman, never getting anywhere, but always restless and on the move. . . . Thirty three people are wearing out their lives and souls like men doomed to imprisonment for life." After the sinking, 24 of the 33 survived the slog by foot and boat across ice-filled seas to a village on a Siberian river delta, but DeLong was not among them.

Three years later, relics from the *Jeannette*—a pair of oilskin breeches with a crew member's name on them, a list of ship's provisions signed by DeLong—washed up on the southwest coast of Greenland. Hearing of them, a Norwegian professor concluded that they had made the long journey on an ice floe carried by currents across the polar sea. Hearing of the professor's interpretation of events, Fridtjof Nansen decided to duplicate the relics' journey, to freeze a ship into ice north of Asia and let the currents carry it across the Arctic Ocean to Greenland, perhaps by way of the North Pole. "If a floe could drift right across the unknown region," he reasoned, "that drift might also be enlisted in the service of exploration."

Nansen had a ship built with rounded sides, like a bowl, or half an egg, so that when the ice pressed on it, it wouldn't be crushed but instead squeezed up, "like an eel out of the embraces of the ice. The ship will simply be hoisted up," he predicted. "Henceforth the current will be our motive power."

> I laugh at the ice. We are living as it were in an impregnable castle.
>
> —Fridtjof Nansen, *Farthest North*

On Christmas Day 1893, not long after Nansen's ship, the *Fram* ("Forward" in Norwegian), had been hoisted up as he intended, he took a moonlit walk over the sea ice beside it. "They will be thinking much of us just now at home," he mused, "and giving many a pitying sigh over all the hardships we are enduring in this cold, cheerless, icy region. But I am afraid their compassion would cool if they could look in upon us, hear the merriment that goes on, and see all our comforts and good cheer. . . . I myself have certainly never lived a more sybaritic life. . . . Just listen to today's dinner menu": Oxtail soup, fish pudding with potatoes and melted butter, roast of reindeer with peas, French beans, potatoes, cranberry jam, cloudberries with cream, Ringnes bock beer, coffee, pineapple preserve, gingerbread, vanilla cakes, coconut macaroons, figs, almonds, and raisins.

"Was this the sort of dinner for men who are to be hardened against the horrors of the Arctic night?" Nansen asked. The following day, it was the "same luxurious living . . . a dinner of four courses," followed by "shooting darts at a target for cigarettes."

The *Fram* took three years to make it to Greenland, the same amount of time it took the relics. Although ice threatened the ship several times—"the ship is getting violent shocks," Nansen wrote during one of those times; "I stand gazing out at the welter of ice-masses that resemble giant snakes writhing and twisting their great bodies out there under the quiet starry sky"—the ship survived the threats. Among the things that the men learned as it was carried through uncharted territory was that the Arctic Ocean is not shallow—as many scientists, including Nansen, had believed, the existing islands being the remains of "an extensive tract of land" which had formerly covered the basin—but deep; in one place they got a sounding of 14,000 feet. Also, the ship's journey, along with the sledge trip that Nansen and Johansen made from the ship to within 240 miles of the North Pole, swept away any remaining beliefs among scientists that there is in the farthest north an ocean without ice on it.

CHAPTER FOURTEEN

SEA ICE II

Ice, and again ice.

—Fridtjof Nansen,
Farthest North

IN EARLY MAY 1937, a Soviet N-170 airplane landed on sea ice a dozen or so miles west of the North Pole, deposited four men and a load of supplies on it, stuck around for a few weeks, then flew off, never to return. In late February the following year, a Soviet icebreaker picked the same four men off the same ice floe, a thousand miles away in the southern Greenland sea. In the nine months between dropoff and pickup, whither the floe went, so did the men. That was a time when the Soviets were taking a particularly keen interest in the Arctic, sending icebreakers to push their way through the Northeast Passage (one, the *Sibiryakov,* was the first vessel to make it all the way through without having to winter on the way, and another, the *Chelyuskin,* made it through but was shoved back by ice and crushed before it reached the Bering Strait); and sending aviators on record-breaking flights over the North Pole, around the Soviet Arctic, nonstop between Moscow and North America. "There was national pride," Roger Colony says, "similar to our pride in our space program."

The reason the four men chose to take up residence on a slab of sea ice was to unravel the "secrets and mysteries of the Central Polar Basin," as their leader, Ivan Papanin, wrote in his book *Life on an Ice Floe.* When they first set up their camp on the floe, it was ten feet thick (undoubtedly multiyear); large enough to accommodate a couple of landing strips; surrounded completely by other floes; and "tough," as Papanin described it. When they were plucked from the floe 274 days later, leaving a Soviet flag flapping from its highest point, it was only 100 feet long and 30 feet wide; full of cracks; "very rampageous"; and surrounded by water and brash, "a little island; there are even waves lapping its edges."

Floe life for the "Papaninites," as they called themselves, consisted of working and eating in "ice buildings" they made out of bricks of snow mixed with water and sleeping in tents of fur and down. For power, they used a windmill and a bicycle generator. Instead of "the classical explorers' food" (pemmican) or polar bear and seal they shot themselves ("no time for hunting"), they ate what they brought from Russia: caviar, borscht, noodle pudding, cranberry jelly, Siberian salmon, barley soup, and buckwheat porridge (often tainted, often sour tasting, often diluted with snow). They drank tea—"we could kiss the person who invented tea!"—five times a day.

In the evenings, sitting crosslegged "in Eastern fashion" in their tent under a portrait of Stalin, they listened over a wireless to broadcasts of opera and concerts from Paris and Stockholm and speeches from Red Square ("We hear the chiming of the clock on the Kremlin tower"). They received radio messages from their wives and war news from Spain ("Fascists are bombing Barcelona"). They read Gorky and Dreiser and Balzac, played chess, and listened to their collection of 68 gramophone records ("we particularly enjoyed Leonid Utesov's Jazz Band"). They wrote articles for *Pravda* and *Izvestia* and answered questions from a public that considered them celebrities in their isolation: What is your favorite sport and why? What do you expect of the Soviet cinema? "Maybe next they will ask," Papanin wrote, " 'What brand of eau-de-cologne do you use?' . . . 'What is your favourite color for pyjamas?' " One newspaper asked them to interview their dog. Once a month they washed their heads and necks and changed their underwear.

They worked 14 to 16 hours a day, taking depth soundings of the ocean under their floe and collecting water samples from a hole they made in the ice (and remade every time the hole froze over) to check for life-forms. They measured drift velocity and wind and magnetic variations and seismic shocks (when other floes hit theirs). They recorded the appearance of wildlife ("Birds at the North Pole!") and noted changes in their floe, including the coming and going of melt ponds. In summer, water from melted snow almost completely covered the surface of the ice, and they had to get about in a rubber "clipper-boat." "At times I forgot these were not deep polynyas," Papanin wrote, "but lakes with ice 3 metres thick underneath them."

During their long, meandering drift southward, they endured blizzards ("confined us to the tent, like badgers in a burrow"); freezing rain ("When we do go out the icicles drop from us"); illness (an "unlucky sausage"); gales ("our tent shook wildly"); intense cold ("while the jelly was cooking the barley soup froze"; "one dives into a sleeping-bag like lightning"); spoiled meat, from sunlight's penetrating the ice where the meat was buried ("Alas! electrical refrigerators seem to be necessary even in the Arctic"); saucepans that ice could not scrub clean; months without sunlight ("taking its toll of our health, appetite, and sleep"); threats to their floe ("a vast ridge of multi-colored ice has piled up opposite our tent"); and fatigue ("we were always tired").

Still, Papanin wrote, "It is surprising what one can get used to, living on an icefloe. . . . We have become so accustomed to it that we often forget the raging ocean beneath us and around us."

They came home to a heroes' welcome. All had been elected from afar to the Supreme Soviet of the USSR. They found themselves compared to the knights of old ("Legendary knights indeed!" Papanin scoffed. "I'm only 5 feet 3 1/2 inches tall!"). During their time away they had been not only bold, stalwart, conscientious, and resourceful but lucky; "throughout the entire region," Papanin wrote at midjourney, "there is no other icefield as level and undamaged" as theirs. They were sad to leave it: "It had proved so enduring and hospitable."

After World War II, the Soviets set up other research camps on other ice floes elsewhere in the Arctic (naming them North Pole II—the first having been I—III, IV, V, up to XXXI), and Americans started doing the same. Both countries wanted to find out as much about the Arctic as possible in order not to give the other country an advantage in the cold war. It was believed then that the next war would be fought in the Arctic, with missiles fired from submarines hidden in the ice fields. "If World War III should come," General H. H. (Hap) Arnold is reported to have said, "its strategic center will be the North Pole."

During the 1957–58 IGY, Americans set up a station called Alpha (code name "Ice") on an ice floe 500 miles north of Barrow, but within a year the ocean had taken "whale-sized mouthfuls" out of it, Tim Weeks and Ramona Maher reported in their book *Ice Island: Polar Science and the Arctic Research Laboratory,* reducing it eventually to a piece only 1,000 feet long. In 1960, Americans set up a research station on an ice floe near Banks Island in the Canadian High Arctic, but it kept cracking until, after only half a year, it became, according to Weeks and Maher, "too small for safety."

"While offering many of the amenities of land," U.S. Air Force Lieutenant Colonel Joseph O. Fletcher concluded, "pack ice has one immense disadvantage as a site for carrying out scientific observation, namely, it is everywhere subject to buckling and cracking." But another, more sturdy sort of ice had recently caught the attention of the American military. As Fletcher tells the story in *National Geographic,* a radar operator aboard a converted B-29 Superfortress on a routine flight over the frozen ocean 300 miles north of Barrow saw on his scope a "picture of land—an island rising from deep sea where no land should be." Pilots of other planes, when they heard of it, flew over the island, dubbed Target X, to see it for themselves. "They looked down on the robin's-egg blue of lakes," Fletcher wrote, "the steely glint of rushing streams, and a coast 20 to 40 feet high rising from the tumbled sea ice of the polar pack." They also looked down on rocks and patches of earth. The "land" mass covered an area of 200 square miles. "But, mysteriously, it was several miles from where it had first been plotted," Fletcher wrote. "There could be only one answer. Target X was a drifting island of ice!"

In time, other "ice islands," as they came to be called, were spotted from the air and traced to their common birthplace, an ice shelf on the northern edge of the northernmost island in the Canadian archipelago, Ellesmere Island. The reason the scientists knew the ice islands had come from there was that the islands had gentle swells and troughs on top laid out in parallel lines instead of the chaotic mix of ridges and cracks and rubble piles typical of sea-ice floes, and so did the ice shelf. (One explorer who sledged along the north coast of Ellesmere in the mid-1870s described the ice surface there as rising "in the form of a roller, with a second roller behind it, exactly as water rolls on a beach after a breeze of wind."*)

The ice shelf consists mostly of glacial ice coming off the land but also some landfast sea ice, both thickened with snow-ice on the top and sea ice on the bottom. As with Antarctic ice shelves, pieces of the Ellesmere shelf break off from time to time and float away to join the sea-ice floes; they are the Arctic version of Antarctica's tabular icebergs. They are much thicker than ordinary sea-ice floes, up to 20 times thicker, and much harder. Ice islands were probably what Arctic explorers who claimed to have found new land, which could never afterward be confirmed as land, were actually seeing: "Crocker Land," named by Peary in 1906 and on maps until 1914; Sannikov Land, "discovered" by a Russian of that name in 1811 and declared nonexistent in 1938 by other Russians; Takpuk Island in the Beaufort Sea, found and pho-

*At first, scientists were puzzled about what causes the rolls. Movement of the glacier? Pressure from the pack? Tides? Finally, they settled on offshore winds, which produce a similar striping in sand dunes.

tographed by Eskimos and never found again; Keenan Land; President's Land—all could have been ice islands.

Figuring that an ice island would be more enduring than an ordinary sea-ice floe, Lieutenant Colonel Fletcher decided to put a research station on one. He chose a nine-mile-long, kidney-shaped ice island called T-3 (it was the third target spotted), and in May 1952, when it was 150 miles from the North Pole, he and a handful of other men took possession of it. T-3, or "Fletcher's Ice Island," was occupied off and on for the next 27 years, during which time it drifted around the Beaufort Gyre, the revolving currents in the Beaufort Sea, before heading toward the North Atlantic and breakup. Even then, it resisted complete destruction. "I've been told," Fletcher says, "there are fragments that can still be found along the west coast of Greenland."

Five of the Soviet North Pole research camps were set up on ice islands, the rest on sea-ice floes. The Soviets tried to keep at least two stations going at once but eventually had to give all of them up because of hard economic times. In their many years of operation, the Soviet drift stations, and the American ones to a lesser extent, amassed a great deal of information about the Arctic Basin—meteorological, oceanographic, and geological. "They filled up a big blank space on the map," Fletcher says. Still, they couldn't fill it in completely. The Arctic Basin is vast, "about the same size as the lower 48 United States," Roger Colony points out (and about the same shape; "if you cut them out and set them on top of the basin, they would just about fit"), and there's tremendous small-scale variation within it. About 25 years ago, Americans started putting unmanned stations on ice floes instead of manned ones. In all, the University of Washington in Seattle has parachuted about 500 instrument packets onto multiyear ice floes, where they measure air temperature, atmospheric pressure, and wind speed as the ice moves around, and 25 times a day they send the information along to a passing satellite. "We try to keep 30 or 40 of these little weather stations going at once," says Colony, who headed the program.

> Ice. Ice. Ice. Nothing but ice. Boy, I'll tell you it's going to be a drunk night when we reach England.
>
> —Submarine sailor, quoted by William R. Anderson in *Nautilus 90 North*

Another way that men thought of investigating the Arctic was from underneath the sea ice. In 1914, both Vilhjalmur Stefansson and Sir Hubert Wilkins (the first man to land an airplane on sea ice) proposed sailing a submarine under the Arctic ice cover, and in 1931 Wilkins did something about it. "Wilkins believed," U.S. Navy Commander William R. Anderson wrote in his book *Nautilus 90 North,* "that there were enough leads and polynyas to enable a short-legged, battery-driven submarine to transit the ocean by puddle-jumping beneath the ice and recharging batteries on the surface in open water." The U.S. military offered Wilkins a submarine it was about to scrap, and he outfitted it with a device for drilling through ice, so that if there wasn't a hole in the ice when he needed it, he could make one. He also mounted runners on top of the sub; he assumed the underside of sea ice would be fairly smooth and therefore planned "to keep the submarine in a state of positive buoyancy," according to Anderson, "—actually pushing up against the ice."

Wilkins's sub made it to the southern edge of the ice pack between Greenland and Spitzbergen but because of an equipment problem had trouble submerging. Wilkins tried to "skid" it under the ice. "Frost formed on the inside hull," Anderson wrote, ". . . and the crew became less than enthusastic about the polar attempt." (Wilkins gave up.)

During World War II, Nazi submarines sometimes hid under ice at the fringes of the pack, Anderson pointed out. "After an attack on a convoy, they would dart under the edge of the ice, beyond reach of allied antisubmarine vessels." But these were brief and decidedly nonscientific forays. It wasn't until the late 1950s after the development of nuclear-powered submarines, which could dive deeper than conventional subs and so pass beneath the thickest Arctic ice (it was assumed) and stay under almost indefinitely without having to puddle-jump, that submarines ventured much beyond the ice margin. "By then it was clear to strategic planners that when the nuclear-powered, missile-firing submarine became a reality, the Arctic, dominating over three thousand miles of Soviet coastline, would be an ideal launching spot." Submarines could fire missiles from holes in the ice, then be concealed by the ice. In "the white, trackless Arctic, it would be . . . difficult to detect a missile-firing atomic submarine," Anderson noted, "and harder to kill it if detected, since the submarine could use the thick ice floes as a bomb shelter."

In 1958, two nuclear-powered U.S. submarines went where none had gone before, all the way to the North Pole, under ice. Anderson was commander of one of them, the *Nautilus,* named for Wilkins's submarine, which itself was probably named for Captain Nemo's submarine in Jules Verne's *20,000 Leagues Under the Sea.* This latest *Nautilus* was equipped with "ice detectors," or inverted fathometers, which could bounce sonar signals off the underside of the ice and thereby measure the sub's distance below it. (The sonar was so sensitive that one time in open water the crew got a reading of ice off the bow—which turned out to be "two sea gulls bobbing in the water.") The submarine also had a lamp to illuminate the bottom of the ice so it could be seen through a periscope. The only crew's drill added to the routine because of the ice was "a vertical, elevatorlike ascent, which we would have to know how to do if we surfaced through a crack or patch of water in the ice," Anderson explained. "Ordinarily, submarines surface with forward momentum." Anderson also figured that if they got lost or had a fire while submerged and needed to find an opening in ice in a hurry but couldn't, they could always "tilt the ship up at an angle and fire a salvo of six torpedoes into the underside of the ice. If that didn't open a hole, we still had nineteen more torpedoes to send behind them."

After a week and a half of sailing through open water, the *Nautilus* reached the southern margin of the pack between Greenland and Spitzbergen. "At long last, we had arrived at the edge of the unknown, the ice pack," Anderson declared. "It lay before us, stretching seemingly unbroken to the horizon, a trackless, colorless desert, highlighted by a kind of bright halo: . . . iceblink." (Despite the packs being trackless, the submarine, Anderson realized, would be gliding beneath the "footprints" of the great explorers, "Ross, Peary, Cook, Amundsen, Stefansson, Wilkins.") Not long after the sub dived under the ice, Anderson realized that the "underside of the ice is not smooth, as many have believed, but distinctly irregular." While one pen on the ice detector was tracing a line representing the water level and the other a line describing the underside of the ice—"in effect, [the pens] traced a profile of the shape and thickness of the ice"—the second pen would dip occasionally, "indicating the ice was, perhaps, fifty or more feet thick. We dismissed these dips as 'ghosts,' " Anderson recounted, "or imperfect returns from the sonar. In actuality, they . . . were deep-hanging pressure ridges, dangerous and deadly to our operation. . . . The ice was much thicker, the pressure ridges much deeper, than we thought."

At one point, to get a "firsthand" look at the underside of the ice, Anderson had the sub brought higher in the water and the periscope raised. "The water was grayish and not at all dark, as the sunlight filtered through the ice," he wrote. "I turned the field of the periscope up, bringing my eye, through magnification, within a few feet of the underside of the floes, which appeared to be scudding overhead like gray clouds. It was a fascinating but eerie experience. In fact, it was a little unnerving. . . . I put the scope back down. . . . It was much better watching the ice on sonar."

At another point, sonar picked out a large lead or polynya ahead, and Anderson decided to practice making an ascent through it. As the submarine rose slowly toward the opening, "I kept my eye glued into the periscope, expecting the upper glass to break water at any moment. A split second later I was startled to find not water overhead, but solid ice! I thought it must be a very thin sheet—so thin it had not registered on our sonar—and since there was nothing to do now to stop Nautilus' ascent, I waited for the periscope to shatter through the ice." It did not. "Instead, our vertical ascent terminated abruptly and a shudder swept through the ship." The ice bent the periscope. The crew tried again, using a second periscope; the ice bent it, too. Back to port they went to repair the scopes.

On their second try for the pole, they made it to 87°N, "much farther north than any other ship under its own power," but had to turn back, this time to repair their gyroscope. On their third try, they took a riskier route, up the Pacific side, where the seabed is higher and the ice thicker than on the Atlantic side. "It would be something like a small boy trying to squirm under a low-hanging fence," Anderson wrote, describing what the submarine needed to do, "with the big difference that there is nothing small about Nautilus—long as a city block and four thousand tons submerged." While below, they encountered "a mass of ice big enough to supply a hundred-pound block to every man, woman, and child in the United States," according to Anderson. Hitting it could have meant "slow death for those on board." They missed the mass "by an incredible five feet." Their fourth try was also from the Pacific side, this time with a closed-circuit TV set that had an upward-facing lens mounted on top of the sub so they could have a better view of the ice than they got through a periscope. Sixty-two under-ice hours later, the *Nautilus* reached 90°N, the first ship of any kind to make it to the North Pole by any means. The submarine couldn't surface at the pole, however, because the ice detector revealed a pressure ridge there that protruded 25 feet down into the water.

Word of *Nautilus*'s feat reached the men on the *Skate,* the other American nuclear-powered submarine traveling top-secret under the Arctic ice at the time. Its mission was to learn how to surface a submarine in ice-covered waters and thus increase the size of the navy's "useable ocean," but they had hoped to be first at the pole themselves.

"We felt as the ill-fated Robert Falcon Scott must have felt when he found to his crushing disappointment that Amundsen in 1911 had preceded him to the South Pole by a month," U.S. Navy Commander James F. Calvert admitted. They did make it to the pole, several months later, and looked for a lead or "skylight" to surface through (a skylight is a stretch of thin ice; ice with many large skylights in it is "friendly ice"). But "everywhere ice of 10 feet or more created a black ceiling for our icy world," Calvert wrote. "Hours went by." When a skylight did appear, its ice proved to be heavier than any they had broken through before, yet there they were, so they went ahead and crunched through it anyway. Once on top, they found a "wild and forbidding scene," with "the heaviest and ruggedest hummocks I had yet seen in the Arctic," Calvert wrote. He and several men stepped out, set up on the ice a table with a green cloth cover, planted a few flags, read the service for burial at sea, then scattered over the ice the ashes of Sir Hubert Wilkins, who had died four months earlier but who had long ago dreamed that a submarine could be where this one was that day. The *Skate* had set a record of its own: the first ship to reach the top of the top of the world.

All all and all the dry worlds lever,
Stage of the ice, the solid ocean

—Dylan Thomas,
"All all and all the dry worlds lever"

The solid ocean, thin as it is, greatly affects the liquid one, as well as the air above both. For one not insignificant thing, it dampens waves and tempers storms. "The ice acts as a filter," Gregory Crocker, former research scientist at C-CORE, explains. It scatters and absorbs the energy of incoming waves, attenuating them. "Gradually the swell subsided," Mawson noted as his ship headed from open water into the pack around Antarctica, "smoothed by the weight of the ice."

Since ice is a buffer against motion, if you changed the distribution of ice, you'd change the currents, David Battisti, professor of atmospheric sciences at the University of Washington, points out. "And that would change ecosystems in ways I couldn't begin to anticipate," he says, "and nobody else could."

For another, significant thing, sea ice keeps the water under it and air over it cooler in summer than they would otherwise be, by reflecting into space a large portion of the sun's rays that hit it. "Sea ice changes tremendously the whole disposition of how energy from the sun is used in the climate system," Battisti states. In winter, as an insulating physical barrier between water and air—water that's, say, 32°F and air that's, say, −32°F—sea ice keeps the atmosphere cooler than it would otherwise be. "The heat of the ocean has a much tougher time getting into the atmosphere than if the covers weren't there," Claire Parkinson, research scientist at NASA's Goddard Space Flight Center, notes. Without sea ice, according to Battisti, "the Arctic would be much warmer and more moist, and land at the edges would be warmer and foggier."

For yet another thing, sea ice affects ocean circulation and temperature by draining salt into the water. The saltier water is, the denser and heavier it is. In the Antarctic, the salty layers sink, pass under ice shelves, become supercooled (the pressure at that depth being too great for the water to freeze), and flow northward, as extremely cold bottom water, into other oceans of the world. "The ocean is cold because of the sea ice," polar oceanographer Miles McPhee concludes. "There's no other reason why the ocean is cold. If you take off that thin veneer of sea ice, it would probably be warm."

What will happen to the sea ice if global warming continues? And if something happens to the sea ice, what will that do to global warming? "Sea ice is a pretty sensitive indicator of climate change," Battisti notes, "since it's relatively tenuous. There's lots of heat just below the ice, and cold above it, and on average it's only ten feet thick. That's not a lot of buffer; it's pretty easy to melt ten feet." According to some climate models, greenhouse gases in the atmosphere will at least double over the next 50 years, in which case air temperatures around the world will go up 1° to 2°F. But in polar regions they'll go up 3° to 4°F, mostly because of feedback effects from the melting of the reflective, insulating ice cover. One model suggests that if the sea ice were to disappear completely, winter air temperatures in the Arctic would run 18° to 54°F higher than if it stayed in place, obviously a "dramatic" difference, as Richard E. Moritz, climatologist and research oceanographer at the University of Washington, points out. (In summer, the surface of sea ice is close to the melting point anyway, so air temperature changes would be more modest then.)

Where sea ice is involved, however, scientists don't have that much confidence in their predictions. "A lot of it is just guessing," McPhee says. "There isn't any certainty in this game. The system is too intertwined, too chaotic if you will." "Sea ice is unbelievably complex," Willy Weeks elaborates. "The diameter of nearly every kinky little brine

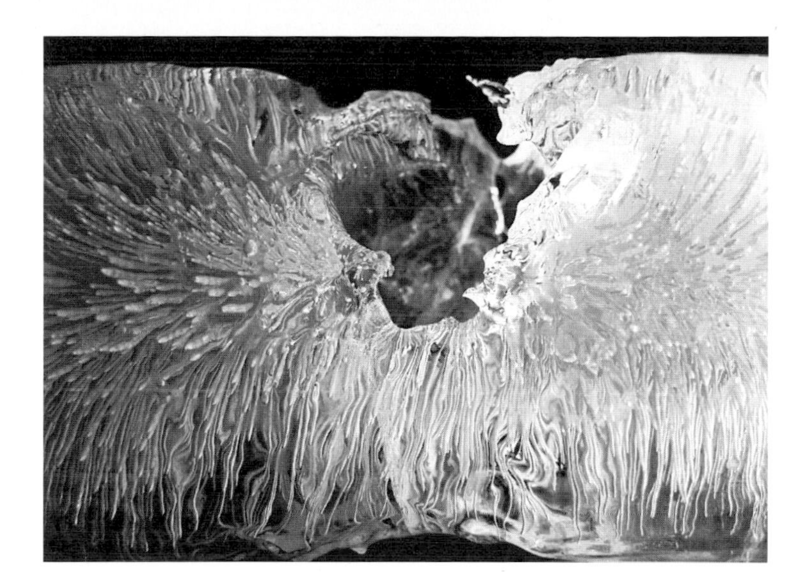

Brooklyn artist David Hirmes filled vases, balloons, and bowls with water, put them in the freezer, set each ice form that emerged on a dark plate in his darkened living room, shone flashlights on it, brought his digital camera to within inches of its surface, placed an ordinary magnifying glass in front of it, and clicked. Looking deep into the heart of the ice, as his camera did, you can see sprays of air bubbles, misty threads of interior cracks, and light glowing from the slick surface. Photos by David Hirmes.

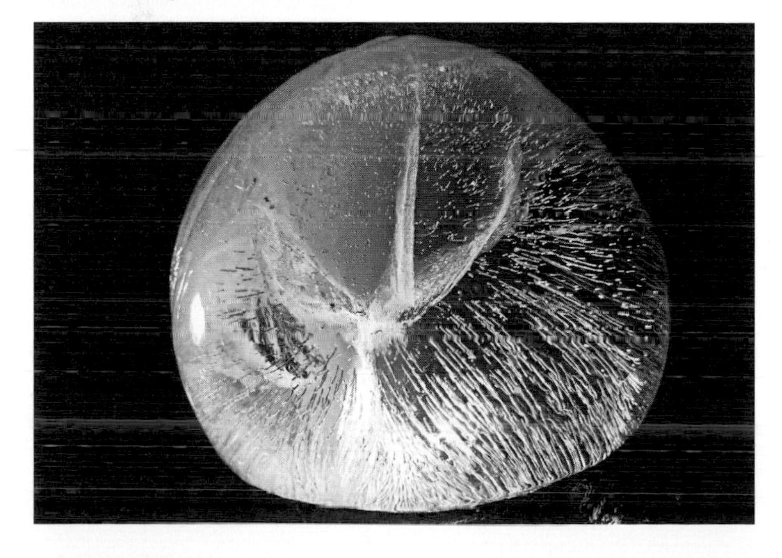

Opposite top left: Many ice fishermen choose to fish from ice huts equipped with sofas, stoves, and TV's, but some are happy to squat by a hole and drop their line. Photo by author.

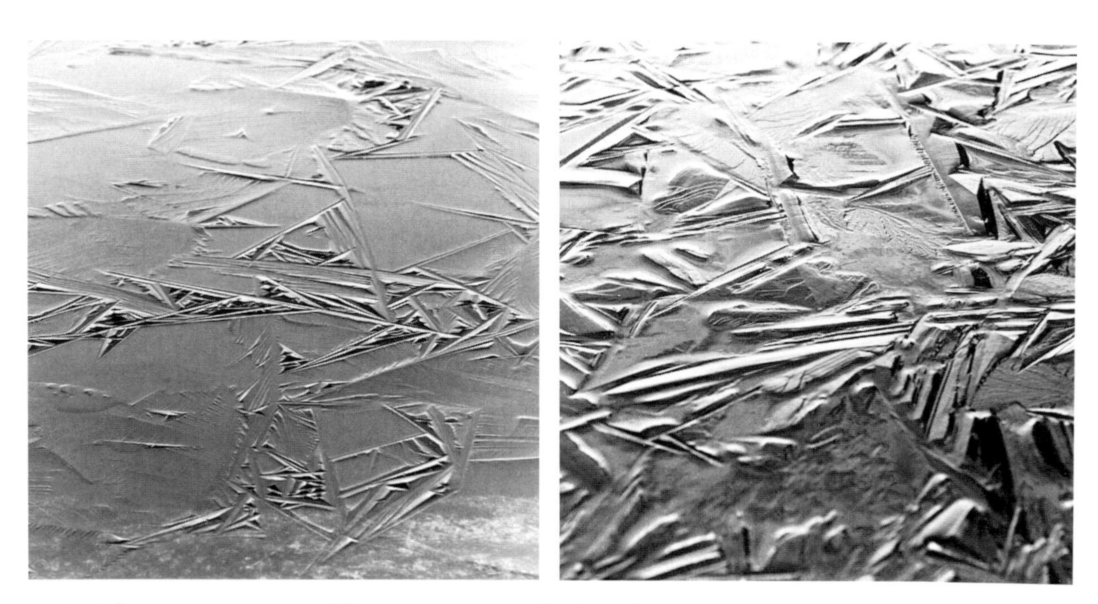

A rich, strange variety of forms appears in the first thin ice of a lake. Photos by author.

Slush holes on frozen lakes produce patterns that look like giant flowers, or in some cases spiders, stars, and centipedes. Photo by Charles Knight.

These chaps seem to be enjoying immensely an ice jam on a creek in Oil City, Pa., probably in the 1930s, even though such jams caused damaging floods. Photo courtesy of Larry Berlin.

In the far north, hardy men build thousands of miles of winter roads every year, mostly over frozen lakes, so trucks can carry supplies to isolated settlements. In the not-so-far north, too, men make ice roads, like these on Lake of the Woods, Ontario, which people often drive around on for fun. Photo by author.

One winter when the Hudson River froze almost solid past the Tappan Zee Bridge north of New York City, Coast Guard icebreakers had to free hundreds of barges trapped in the ice whose cargo of oil was desperately needed upriver. Photo by N. J. Fenwick from the USCGC *Sorrel.*

On Alaska's slow-moving Susitna Glacier, the dark lines of rocky moraines form loops and folds that are carried far down-glacier during "surges," or periods of speedier flow. Photo by Austin Post, U.S. Geological Survey.

Right: Ice cores, cylinders of ice containing long-term climate records, were ranged on the glacier from which they came by a member of the drill team, forming an "Icehenge." Photo by Bruce Koci.

On the margin of the Quelccaya ice cap in the Peruvian Andes, dust bands separate annual layers of snow, giving it the look of a wedding cake. Within decades the "cake" had completely vanished, melted away due to global warming. Photo by Lonnie G. Thompson.

"Drydock" iceberg off Canadian coast—not two bergs but one, its sails joined underwater by a keel. Photo courtesy of the Canadian Ice Service.

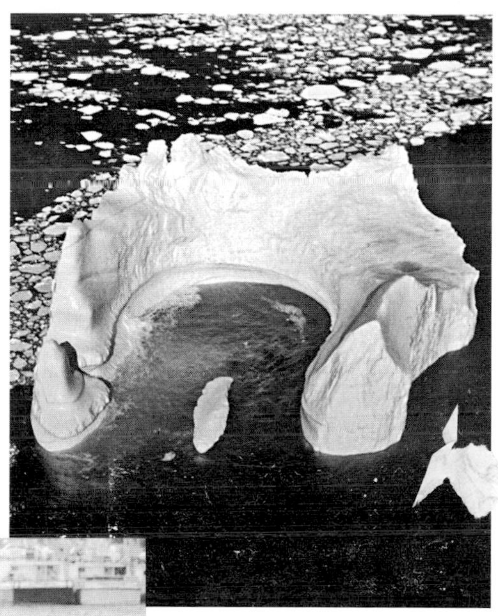

A weathered iceberg in the North Atlantic encloses its own inviting, yet probably treacherous, bay. Photo courtesy of the U.S. Coast Guard International Ice Patrol.

In 2004, almost a century after the *Titanic* met up with its iceberg, the shrimp trawler *Solberg* collided in fog with a 75-foot-long berg, which dented its bow and dumped a load of iceberg ice on its deck. Photo by Keith Gosse for *The Telegram,* St. John's, Newfoundland.

A 3D iceberg floats in the North Atlantic among 2D sea-ice floes. Icebergs sometimes cut through softer sea ice "like a knife through butter." Photo courtesy of the U.S. Coast Guard International Ice Patrol.

Right: The nuclear-powered attack submarine USS *Billfish* surfaces at the North Pole, with the sail-mounted diving planes on its conning tower in the vertical position for breaking through thick ice. Photo by Chuck Mussi, courtesy of the U.S. Navy.

Floes in the Gulf of St. Lawrence as seen from an icebreaker's helicopter: "The polar pack," wrote an explorer who'd sledded over it, "that great, moving crust of the earth." Photo by author.

"Pingos," conical hills with cores of ice, often resemble volcanoes because the ice growing inside splits the soil on top. This one is on Canada's northwest coast. Photo by J. Ross Mackay.

Above: Some earth artist could have created these look-alike stone circles arrayed on a plain on Spitsbergen, an island of Norway in the Arctic Ocean, but it was the repeated freezing and thawing of water, over eons, that made them. Photo by Bernard Hallet.

In winter, the ground under roads on either side of a bridge may retain some heat from milder times, even as the air under the bridge is freezing cold. Photo by author.

Top left: A wavy ribbon of ice, made up of many microribbons of ice fused together, forms when watery sap flows out of slits in the stem of a dock plant in northern India and freezes in the mountain air. Photo by Brian Swanson. *Top right:* Individual micro-ribbons of ice ("hair ice") curl around the twig of a red alder tree in British Columbia. Photo by Kathleen Jansen.

An otter pops up through spring ice on a lake in Yellowstone National Park; minutes later it caught a fish. Photo by Michael Quinton.

Polar bears are totally at home on sea ice; some never set foot, or belly, on land. Photo by Michio Hoshin/Minden Pictures.

From a perch of piled-up fast ice off Barrow, Alaska, Eskimo hunters scan the sea for bowhead whales, their sealskin boat ready to be slipped silently into the water the instant one appears. Photo by Bill Hess.

Emperor penguins manage to reproduce under the harshest circumstances of any vertebrate on earth, in the windy cold of a long Antarctic night, on a slab of sea ice. Photo by Gerald L. Kooyman.

Men court frostbite in ice-sitting contest in Chicago, Ill., 1933. The guy in front lasted only 26 hours. Photo in collection of Ripley's Believe It or Not!

An ice climber ascends a near-vertical column of ice on Frankenstein Cliff in N.H., adhering to the ice with only the axes in his hands and the toe points on his boots. Photo by George Hurley.

In 1916 in Greenland, Wegipsoo, in a classic kid's pose, lies on muskox fur; explorer Hviljalmur Stefansson touted caribou-skin underwear and polar-bear trousers. Photo courtesy of the American Museum of Natural History Library.

A Parisian takes inventive advantage of the slipperiness of ice, with a stunt called "Glace de la Concorde (1938)." Photographer unknown.

Right: Missie Hattie Atwater poses in her skating outfit, described as "Fanciful Fans" (Montreal 1870). Photo courtesy of the McCord Museum, Montreal.

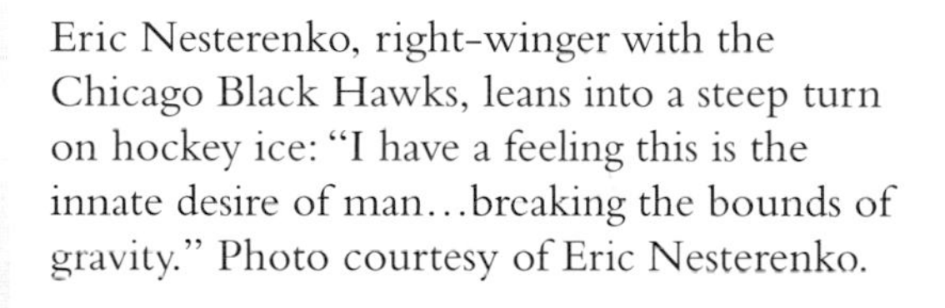

Eric Nesterenko, right-winger with the Chicago Black Hawks, leans into a steep turn on hockey ice: "I have a feeling this is the innate desire of man…breaking the bounds of gravity." Photo courtesy of Eric Nesterenko.

Iceboats on Lake Winnepasaukee, N.H. stand ready for a "long distance slide." Photo by author.

Designed by the mayor of Saranac Lake, N.Y., and built entirely of ice, an Irish castle features five towers, throne room, slide, and two "swaneagles." Photo by author.

"Joe Ice," credited with making ice sculptures cool, carved a 1950 Luscombe airplane once owned by the author. Photo by Joseph O'Donoghue.

Opposite bottom left: A boy sits by his family's traditional icebox, but most Amish use unplugged refrigerators with the coils ripped out or freezers with a shelf added for ice. Photo by author.

Opposite bottom right: From a sidewalk in Manhattan, Isabel Rivera sells "ice cones." She shaves a 50-pound block of ice with a "snow plane," piles the chips in a paper cup, and pours over it one of 12 flavors of syrup, with cherry the most popular. Photo by author.

Iceman Marion Weaver slides a 300-pound block of ice to his truck before setting out to deliver ice to Amish farmers in northeast Ohio, since they still use ice to preserve their food. Except for weddings, an Amish family doesn't need a full-size block, so the iceman breaks it into smaller ones; here he is carrying a mere 25-pounder. Photos by author.

For almost 50 years, Vermont farmer William Bentley photographed snowflakes through his microscope; at top are two of thousands. He also took photos of frost on windowpanes, many with a botanical motif, like the two at bottom. Courtesy of Dover Publications.

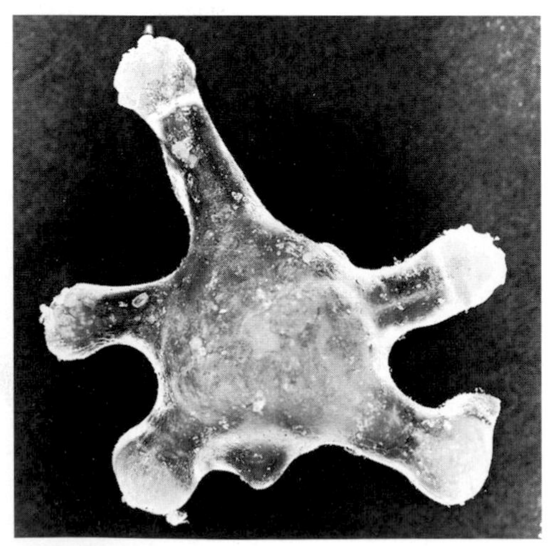

Left: A hailstone can develop appendages, by throwing water off while rotating, like mud is spun off a tire. Photo by Charles Knight.

Opposite bottom left: On a post in Scotland, rime ice grew *into* the wind, which kept it supplied with supercooled droplets. Copyright of the Royal Meteorological Society.

A frost crystal, plucked from a crevasse, rivals jewelry. Photo by author.

Below: This "bouquet" of hexagonal cups grew in a cave under an Antarctic glacier. Photo by Charles Knight.

Below: A crewmember on the Canadian CSS *Dawson* used a timeless method for getting rid of shipboard ice: hitting it with a sledgehammer. Photo courtesy of the Bedford Institute of Oceanography.

Saturn's rings are made up overwhelmingly of water ice, gazillions of particles from a fraction of an inch to 40 feet across. Photo courtesy of NASA.

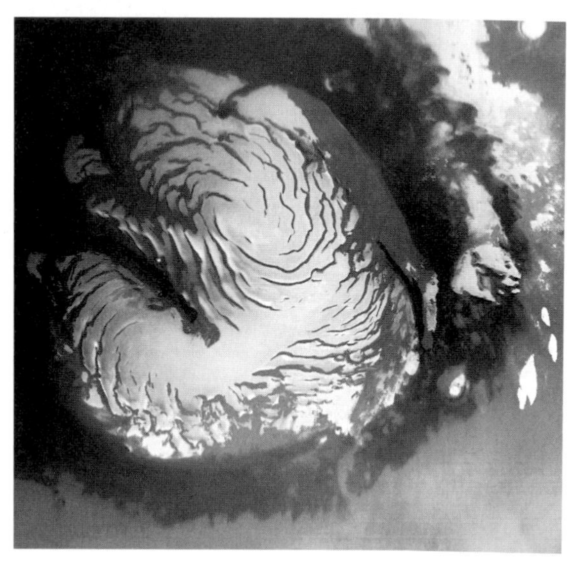

A cap of water ice covers Mars's north pole; the dark lines swirling through the white are probably exposed valleys. Photo courtesy of NASA.

Europa, one of Jupiter's moons, has chunks on its icy surface that seem to have floated around like icebergs in some slushy matrix, raising the question: is there a liquid ocean underneath? Photo courtesy of NASA.

pocket is different, every stinking centimeter in the ice is different. You're setting up for an impossible task."

Nevertheless, in September 1997, a little over a century after Nansen froze his ship into sea ice north of Siberia, 50 scientists from the United States, Russia, Canada, and Norway froze a Canadian icebreaker into sea ice 350 miles north of Alaska, in the middle of the Beaufort Sea, so that over the course of a single year they could investigate almost every centimeter of a representative piece of Arctic sea, and thereby improve the accuracy of their climate models. They called their project SHEBA, for Surface Heat Budget of the Arctic Ocean.

Their strategy was to place thousands of sensors, measuring everything from temperature of the air and stress in the ice to currents in the water, inside an imaginary column that extended 30 miles outward from the ship in all directions and 300 feet down into the water and upward through clouds all the way to the top of the atmosphere. The scientists were particularly interested in shedding light on two sensitive, interacting feedback mechanisms, one concerning clouds. Don Perovich of CRREL, SHEBA's chief scientist, explains that a cloud cover serves both as an umbrella, keeping short-wave solar radiation from reaching the ice, and as a blanket, keeping long-wave radiation from escaping from Earth into the atmosphere. As an umbrella, the clouds have a cooling effect; as a blanket, a warming one. Which effect dominates in summer, SHEBA wanted to know, the umbrella or the blanket? (In winter, the blanket does, because with 24-hour darkness there isn't short-wave radiation coming in.)

The other feedback mechanism they were interested in is "albedo." For most of the year, the usually snow-covered ice reflects a large portion of the radiation that reaches it, but in summer, when the snow melts and bares the ice and meltponds form and leads appear, the surface darkens and absorbs more sunlight, producing more melting and more darkening and more melting . . . How much more melting? SHEBA wanted to know. "I've seen the albedo change so fast in summer I could walk 30 feet over ice and have it go from 15 percent to 70 percent," Perovich reports. Melting means more evaporation and more clouds, and depending on whether the clouds act mainly as blankets or umbrellas, the feedback is positive or negative, the effect amplified or reduced.

The very first observation the scientists made as the icebreaker embedded its prow in a large floe was that the ice was quite thin. "We thought we would find some floes that were seven to eight feet thick," Perovich says, "but they were three feet thick on average." McPhee adds: "We were all sort of amazed at how thin the ice was. The center of the Beaufort Gyre is where Arctic ice is at its most compact and long-lived, and if it is thinning very rapidly *there* . . ." Tests he ran on the water under the ice showed its salt content to be 2½ parts per 1,000 less than it had been only 22 years earlier when a drift station passed that way. "This represents a net melt of at least six feet," he calculates, and all of it recent. "To effect this kind of freshening, you would have to divert the entire freshwater output of the Mackenzie River from its normal pattern into the center of the Beaufort Gyre for two or three years. In my view, this is a smoking gun," powerful indication that global warming is causing sea ice in the Arctic to melt.

(When sea ice melts, ocean levels don't change, as they do when glaciers melt, since sea ice is already present in the ocean and taking up space.)

After a year, when the scientists abandoned their chosen floe, it was even thinner than when they arrived. "All of us sea-ice experts thought the ice would be thicker in a year," Perovich says, "because it would grow in winter and reach equilibrium thickness. Well, all of us sea-ice experts were wrong." The ice did grow in winter by over a foot and a half, but the next summer it melted by almost *four feet*. "Boy, was there a lot of melting!" Perovich exclaims. "We found ourselves rooting for the ice to survive the melt season."

The blanket won, "at least for the pervasive low wet cloud conditions of the SHEBA summer," the scientists reported. Clouds blocked outgoing radiation more than they did incoming radiation and boosted warming at the surface. As for albedo, what proved important was the *timing* of the usual transitions—from dry ice surface to wet, wet to melting, snow cover to meltponds, small meltponds to large meltponds, and shallow meltponds to deep ponds with thinner and thinner ice beneath them.

SHEBA was never meant to detect global warming, only to understand the processes by which sea ice might contribute to it, and yet "when you're up there and this floe is melting out from under your feet," says Perovich, "human nature being what it is you start thinking about global warming." In recent years, he notes, "there's more of a sense among people in the sea-ice community that global warming is happening, although it's more of a visceral sense than anything we've been able to establish to ten significant figures." The fall-winter season of 1998, when SHEBA was shutting down, was a light ice year—that is, the southermost sea-ice edge was particularly far north—but even lighter years followed. The winter of 2002 set a record for smallest extent of sea-ice cover, and the winter of 2003 came close to breaking that record. "Is this a natural cycle," Perovich asks the tough question, "or a cycle being amplified by global warming?"

Safe and secure as winter ice

—Orpingalik (a Greenland Eskimo)
"My Breath"

Since the early 1970s, satellites have been keeping track of sea-ice edges in both the Arctic and the Antarctic. Recently, NASA announced that the extent of Arctic ice had decreased 3 percent per decade between 1973 and 2003. "That would jibe with global-warming studies," NASA's Claire Parkinson says. But she also says, "The satellite record is only a couple of decades long, not anywhere near as long as we would like, and ice covers can fluctuate a great deal [naturally]."

During that same 1973–2003 period, NASA also announced, the extent of *Antarctic* sea ice increased, by 1.3 percent per decade. That would jibe with global-warming studies, too, according to this scenario: warmer temperatures would have increased precipitation in the region, the precipitation would have freshened the surface water, and the lighter, fresher water would have stayed near the surface instead of sinking and mixing with and bringing up from below much warmer saltwater, thus rendering the surface cooler than usual and allowing more ice to form there. In the Arctic, where large rivers in Russia and Canada dump an enormous amount of fresh water into the ocean, there's already a fresh-water layer on top of the saltwater and little mixing.

Satellites, with their high overhead vantage points, are great at detecting sea-ice extent, as well as duration (ice seasons in the Arctic are getting shorter), but not ice thicknesses. Submarines, with their low vantage points and upward-looking sonar, are great at it. The best way of determining if there have been changes in the thickness of sea ice over time, besides drilling several million holes in the pack ice, is by looking at submarine records. The records are spotty, or rather spokey, limited to narrow tracks that submarines make when they're cruising around. Still, a fair number of those tracks have been made by now, and a comparison of ice profiles made during cruises years ago with ones made more recently suggests that sea ice is thinning all over the Arctic, and fast.

For instance, when scientists compared measurements of sea-ice drafts—thicknesses—taken by U.S. Navy submarines during cruises between 1958 and 1976 with ones taken between 1993 and 1997, they determined that in that short time the ice had thinned by *five feet*. Most of the thinning, they found, was in the central Arctic Ocean, the most perennially ice-covered part. Some of them have suggested that the thinning came from climate warming—indirectly. Higher air temperatures in the Arctic would have contributed to a change in wind strength in the Atlantic Ocean (related to the North Atlantic Oscillation), which weakened the Beaufort Gyre. Ordinarily, the gyre keeps ice circulating in the Beaufort Sea for years, during which time the ice gets thicker through the usual rafting and ridge building and hummocking as well as by pulling in ice from around the Canadian High Arctic islands. A weakened gyre, then, would allow ice to be moved out of the Beaufort into the Atlantic more rapidly than usual, not only whisking away ice that had already thickened but depriving thinner ice of the opportunity to thicken.

Speaking from an admittedly "unscientific standpoint," Roger Colony estimates the likelihood that there won't be any ice left in the Arctic Ocean in 30 years due to global warming at "fifty-fifty," adding, "and I'm a pretty conservative person. Nobody knows how to stop the warming of the earth. Once the ice starts to melt, and the ocean surface turns from white to dark, it keeps on melting. The rich get richer and the poor get poorer." Miles McPhee says, "I can't quite picture any scenario that would result in *no* ice. The heat balance is such that without solar input in winter, the ocean will still radiate so much heat it will almost have to form ice." Richard Moritz, SHEBA's director, declares, "I think sea ice will disappear in the summer in the next 50 years." Maybe ice sparkling on top of the ocean during the double days of joy will just be something to read about in explorers' books.

> The pack was coming in fast, and the signal, "Prepare to take the ice," flying from the Commodore's masthead. We did take it, . . . and in a few hours the "nip"* took place.
>
> —Lieutenant Sherard Osborn, *Stray Leaves from an Arctic Journal*

When 19th-century explorers "took the ice," it was in ships strengthened to resist the ice. The keel of Nansen's *Fram* "was of American elm, imported from Glasgow," Roland Huntford wrote in his book

*A nip is said to take place when ice forcibly presses against a beset ship; a beset ship is one surrounded by ice and unable to move.

Nansen. "The frames were of oak, grown to shape and seasoned for thirty years. They were double; each fifty centimeters [almost 20 inches] thick, and closely spaced. Weak point of any ship, the stem was made of three massive oak timbers, one inside the other, giving one and a quarter metres [four feet] of solid wood fore and aft, almost forty centimetres [15 inches] wide." Amundsen's ship *Gjoa,* the first to navigate a Northwest Passage, had a three-inch covering of oak planks laid over the hull of a fishing boat, and Shackleton's *Endurance* had a sheathing of greenheart—a very hard tropical wood—over a hull of fir and oak planks, in places two and a half feet thick.

Toward the end of the 19th century, ships began to be designed specifically for breaking ice, with some of the design features found in icebreakers today. A typical modern icebreaker is heavy, with large displacement and lots of power. It is short and wide—stubby—so the weight is compact. It has a strong bow, and the shape of the bow can be one of several developed for various ice regimes: rounded, concave, spoon shaped, "landing craft," cylindrical, and so on. In profile, the bow tends to slope inward from the top down, meeting the ice sheet at a small angle, to minimize the ice's resistance. For an icebreaker breaks ice not only by hitting it head-on (as the *Endurance* did in Shackleton's account) but by riding up onto it and letting the weight of the ship do the breaking. As one observer put it, an icebreaker is "in effect repeatedly running aground" (afloe), and the strain that this puts on a hull is lessened if the bow is inclined.

About 100 polar-class icebreakers are operating in the world today, mostly in the subarctic: the Baltic, the Gulf of Bothnia between Sweden and Finland, the Great Lakes, the Saint Lawrence Seaway. The only Arctic region where icebreakers are in regular use is the Russian Arctic, on the Northern Sea Route, the 2,500-mile-long part of the Northeast Passage that runs from Novaya Zemlya to the Bering Strait. In 1959, the Soviets built the first nuclear-powered icebreaker in the world, the *Lenin,* to use on the route. Since then, they have built several more, including the *Arktika,* the first surface ship to muscle its way through ice to the North Pole. "Because of their command economy," explains Lawson Brigham, deputy director of the U.S. Arctic Research Commission and former U.S. Coast Guard commander, "the Soviets were able to build these extraordinary ships without concern to the cost to the nation." Now, in a struggling market economy, "the trick is how to employ these ships," since the cost is too high for them to be used on the Northern Sea Route.

One trick is to take tourists to the North Pole. In 1990, about 50 people from several foreign countries paid $20,000 each for a nine-day ride from Russia to the North Pole on the nuclear-powered icebreaker *Rossiya.* According to one passenger, the way it broke ice up to 20 feet thick was "seemingly effortless." In August 2000, however, when the nuclear icebreaker *Tamal* reached the pole, the tourists and scientist-lecturers on board found . . . *no ice.* "Ages-Old Icecap at North Pole Is Now Liquid, Scientists Find" read the next day's headline in the *New York Times.* "The North Pole is melting" read the lead sentence of the article under it. "The thick ice that has for ages covered the Arctic Ocean at the pole has turned to water." The message of the article was that "an ice-free patch of ocean . . . at the very top of the world, something that has presumably never before been seen by humans, . . . is more evidence that global warming may be real and already affecting climate."

It wasn't the first time that people had seen water at the pole. When Amundsen flew over it in the airship *Norge* in 1926, for example, three days after Byrd had apparently flown over it in a Fokker airplane, he remarked that "the ice was much broken up at the Pole and a mass of small ice-floes was observable. It was quite different from the other ice we had passed over." A photo he took in the area shows small dark spots of water in spaces between floes, where floe edges didn't quite meet. Still, a hole in the ice a mile across? A lecturer on the *Tamal* cruise, paleontologist Malcolm C. McKenna of the American Museum of Natural History, reported that the ice on the way from Spitzbergen to the pole had been unusually thin, with intermittent open areas. "Some folks who pooh-pooh global warming might wake up if shown that even the pole is beginning to melt at least sometimes," the *Times* article quoted him, "as in the Eocene," a warm period 55 million years ago when the Arctic had tropical plants growing in it.

> Trackless, colorless, inhospitable desert of ice
>
> —Robert Peary, *The North Pole*

In 1993, I spent five days on a Canadian Coast Guard icebreaker in the Gulf of Saint Lawrence, where it did the old-fashioned thing, help beset ships get free of the ice. It was a particularly heavy ice year, with temperatures 7° to 11°F below normal, and the gulf was almost completely frozen over. The icebreaker was the *Sir William Alexander,* named for a Scottish baronet who once possessed royal land grants in today's Nova Scotia and New Brunswick; "he was a nobody," said Captain

Harvey Adams, who favored naming the icebreaker after a Canadian who had actually done something. A light-to-medium icebreaker, it was not the most powerful in the fleet but not the least, either. I'd been told that icebreaker captains work in their slippers, in the comfort of the heated bridge; Adams wore Birkenstock sandals. His job wasn't really cushy, though. Three times in the previous two weeks he had been up for 24 hours straight, helping ships get through the ice. "Twelve to 16 hours a day is normal," he said. Ninety percent of the ships that enter the gulf are non-Canadian, he explained, many of them small and underpowered, not equipped to handle ice. "Some captains have never seen ice, never felt cold, and they're nervous. We advise them to bring an ice pilot on board. Half do."

When I arrived in late March, eight ships were beset in the gulf. One was the *Medallion,* which belonged to a Danish company that named its other ships *Charm, Amulet, Scarab,* and *Talisman,* hoping for luck, presumably, but something unlucky happened to the *Charm,* I was told; it was wrecked in a storm. The *Medallion*'s last port was Newark, New Jersey, where it dropped off granite from Portugal. Now it was in the middle of the gulf, halfway between Newfoundland and New Brunswick, icebound. As the *Sir William* headed out of Sydney harbor in Nova Scotia, to go to the *Medallion*'s aid, I stood at the portside rail and looked down. The bow was noisily cleaving and knocking apart the ice, tearing it into blocks, which, as the hull brushed past them, plunged and tilted and rocked in the water. O the destruction! The *violence!* I was breathless with shock, something was happening that oughtn't to, some harm being done that should be stopped, as if somebody were breaking the best china, an intact thing being smashed to pieces! The ship shuddered as it collided with the ice; brown liquid boiled up around the broken blocks; and behind the icebreaker a shadowy track filled with bits and pieces of ice grew.

"It looks like a nice little ship, clean," Adams says as the *Sir William* approaches the *Medallion* after a couple of hours of inflicting severe damage on the ice. The ship has a cloverleaf painted on its side. A radio exchange between it and the icebreaker follows and goes, in part, like this:

Medallion: We're stuck, I'm afraid.

Sir William: Try backing up and give it a go. We're going through a big floe then wait for you.

M: [doubtful tone] All right, we'll try that.
M: We are still stuck.
SW: Okay, we are coming back.

The icebreaker's captain, first officer, third officer, and quartermaster talk among themselves on the bridge, sometimes to enlighten this observer. "The pressure came on," one says, in the way you or I might say, "The lights came on." "When pressure is really on," another explains, "the track closes up and ridges. It can shut like a zipper."

The *Sir William* backs up along the *Medallion*'s port side, staying 50 feet out, until it is half a length behind the ship, then starts forward. The *Medallion* moves into the new track.

M: Ja, we are moving.
M: We are stuck again.
SW: Keep your engines slow ahead.

To make a track for a ship to follow, an icebreaker usually backs up in its own track, so it can get a good run at the ice, then drives forward at full power. "Different people back up different amounts," one of the officers remarks, "a mile, a half mile."

The *Sir William* rides up onto the ice ahead, is stopped there, backs up, goes forward. "We don't baby an icebreaker," the captain points out. "We beat hell out of it. No mercy. Bow up, down, up, down. Everything is used to the max. There's lots of shaking and vibrating. Things break, things get worn. Icebreaking is probably the most stressful thing any ship can do, short of going on the rocks."

M: Ja, we were moving very slightly.
SW: Keep moving. I'll back up a little.
M: No. [Sigh] We are not moving.
SW: Okay, we'll back up.

The sky is growing dark. The icebreaker has its spotlights on, and snow can be seen falling through the beams.

SW: You'll be able to slide into the track, I think.
M: We have trouble getting into the track but we are still moving.
M: I'm afraid we are stuck again.

"Did the wind change?" one of the icebreaker officers wonders out loud. "No, the tide maybe . . . I don't know. The pressure came on again."

M: Well, we *are* stuck now.
SW: Mind backing up 50 feet and see if you can turn?

"The old darling," somebody on the bridge mutters. "Maybe this time."

M: Now we are moving ahead again.
SW: Okay, let me know if you start to pick up speed.
M: Well, I was moving, but we are stopped again.

From an officer: "Sometimes a ship only five minutes behind you hasn't a chance in hell of getting through. The ice isn't bobbing around in that water. You could walk over it, only five minutes after it was chewed up."

From the captain: "What have you got on the radar?" According to the Canadian Coast Guard *Ice Navigation* manual, ice makes a poor radar target beyond three or four miles. "Areas of open water and smooth floes have a very similar appearance."

M: We are stuck. We are not moving through the ice. Engines full ahead.
SW: Try backing up.
M: Will do.

"Where there's a northeast wind," the first officer says, "it will push the ice against shore and block this area tight as a drum. In a bad year, you could walk from here to Montreal!"

M: Ja, we are moving.
SW: Tell me when you are picking up speed.
M: We are stuck.
SW: Medallion, shut down for the night.
M: Okay, thank you very much and have a good night.

No sighs from Adams. I am amazed at how much patience icebreaking requires, how much optimism, resistance to frustration. "It'll let up," the captain says, "It always does. If not tomorrow, then the next day, that's the way I look at it."

The *Sir William* has its own helicopter, and once or twice a day the pilot takes an "ice pick," or ice-service specialist, aloft for a couple of hours so he can survey the ice in the area. Yves Sivret, a man with a remarkable nose, like an Atlantic puffin's, very thin and deep billed, is the icebreaker's ice pick. From above, he judges the ice by its shade (the darker, the younger), amount of ridging, and thickness as best he can determine it by eyeballing the edge of a piece that's been broken and turned on its side. From what he finds he'll make "eggs." Eggs are printed oval shapes scattered across maps of the gulf on which ice conditions for each area are noted by number code. On a map for March 20, when I was on the *Sir William,* almost all the eggs in the middle of the gulf had "9+" in a band at the top, meaning the water was more than $^{9}/_{10}$ths covered with ice. On the egg that characterized the ice at the spot where the *Medallion* was beset, you would know from the code that $^{2}/_{10}$ths of the ice cover there consisted of a giant floe of gray-white ice; $^{2}/_{10}$ths a giant floe of gray ice; and $^{4}/_{10}$ths a big—not giant—floe of gray-white ice.

Ice picks forward their eggs to the Ice Office in Halifax, which combines the numbers with reports from satellites, ships, airplanes, and shore stations to produce ice charts, in color, with the hardest ice the most intense color (magenta), the newest a cool green. Every day at 3 p.m., the Ice Office faxes the charts to ships in the gulf.

After breakfast the next day:

SW to *M:* Can you move, eh?
M: We can try.
SW: Wondering if you can move into that open water.
M: Ja, we are stuck in the ice.

"If I get too far ahead, he sticks," the captain points out. "If I don't go fast, I can't break the ice. You've got to weigh one against the other."

SW: Okay, *Medallion,* you can come full ahead now. You follow close as you can. We'll be making a lot of turns.
M: Okay, we will try to follow you.
SW: We'll go for that little patch of water. Don't know what good it will do when we get there.

At the approach of the *Sir William,* a large, dark, hooded seal slips from a floe into the water. "You don't see the pups at first because they are white," the captain comments. "Sometimes we are on top of a whole herd and the little fellows get jammed down between the ice and you can't help but run over them."

By now the ships are so close together the men on the *Sir William* can see spider plants in the windows of the *Medallion*. "Those boys have plants on the bridge!"

M: We are stuck again.

Meanwhile, calls are coming in by radio. The *Voyageur* is stuck, the *Scarab* is out there crying for help. "We don't break off if we're doing a job," the captain says. "Better to do one at a time."

M: I'm afraid we are stuck again.

An officer: "That's where he got pinched, Captain, just at the turn."

And so on. More tracks laid down by the *Sir William*, more starts and stops in the tracks by the *Medallion*. Captain Adams scans the area often with binoculars, looking for leads and water sky. Murres—black-and-white seabirds—flit across the icebreaker's bow. Snow falls, cutting visibility, slowing movement. The day ends at near midnight again, after offering a faint display of northern lights. The *Medallion* hasn't made much progress. But the men of the *Sir William* don't lose their optimism. One remarks, "It won't be long now said the monkey when he lost his tail."

After escorting the *Medallion* all the way to the Gaspé Peninsula, the *Sir William* heads north to pick up the *Scarab*, then leaves the *Scarab* to go after the *Lucien Paquin* . . . That is one lonesome sight, a small dark freighter stranded upon an immense white plate of frozen sea. It's hard to believe that the time could ever come when nobody will be stuck there again.

CHAPTER FIFTEEN

GROUND ICE I

In the bleak mid-winter
Frosty wind made moan
Earth stood hard as iron
Water like a stone

—Christina Rossetti,
"A Christmas Carol"

WHEN AIR TEMPERATURES fall and lakes, rivers, ponds, and seas start stiffening, developing their winter carapaces, the ground starts stiffening too. There is, after all, some moisture in nearly all soils, in the spaces around pebbles and cobbles and particles of silt, sand, and clay, in the channels that worms create as they work their way up toward the robin's beak.

Say you are walking across a field early some morning and hear a crunching sound under your boots and you wonder what it is you're smashing: if it isn't a broken whiskey bottle, it's probably "needle ice." Even though the ice is at the surface, you probably don't see it, since as it grows it pushes upward, like a bean sprout, and lifts with it bits of dirt, twigs, and leaves, which may stay on top, like camouflaging little hats. Once you know to look for it, though, the signs are clear. In wet ground after a cold night you catch sight of several deep, odd-looking holes scattered about; they are the former surface, low spots between elevated clumps of ice. While making the crunching sound, your boots will sink a half inch or so into the earth, which seems fragile, not iron-hard. With a pocketknife, you dig out a sample of this soil and find bundles of what look like coarse fibers in it, cold to the touch.

Jim Lacombe, geophysicist at CRREL, was bicycling on a path near his office one morning in late autumn when the ground gave way under a wheel. Investigating the spot, he found needle ice and dug out a large clump of it. Shortly afterward, I saw the clump in a cold room at CRREL, and it looked like a big dirty sponge, except with all the holes aligned vertically. Lacombe also dug out a solitary needle near the path and brought it in; about a foot long, it looked like a glass candlestick.

He knew what to look for, but if conditions are just right—night air frosty, ground sodden from recent thaw, soil fine-grained—needle ice can make such a conspicuous showing that even a person with her mind on higher things—birds, clouds, lunch—can't help taking note. While hiking in a Vermont pine woods one day in early spring, when snow was still on the ground, Boston art teacher Sandra Olansky saw beside a trail what she thought at first were spikes of old snow left after the rest melted. The spikes were about four inches high, "stood straight up off the earth, and looked like miniature cities of ice," she says. While hiking in Vermont on an autumn morning, Jeffrey May of Cambridge, Massachusetts, observed in a field of newly overturned, clayey soil what he described as "plant-like shoots of ice . . . in thin, parallel fragile bundles." He took a photo, sent it to *Weatherwise* magazine, and asked for an explanation.

He got one from Charles Knight at NCAR. Needle ice, not to be confused with needle crystals of snow, Knight wrote, is very common, although not commonly noticed. The needles May saw were a typical size, 3/4 inch high, but Knight has seen ones several feet high, near hot springs in cold climates.

How needle ice forms is difficult to explain, Knight admitted, then attempted to explain it, in part. "The ice surface must become very highly curved to penetrate between the soil particles," he wrote. Say there's a wet spot on the ground, and when exposed to cold air the water turns to ice. If the ground is made up of a very fine-grained soil such as silt or clay, the channels between the particles will likewise be very fine, maybe a hundredth of a thousandth of an inch across, and the ice surface won't be tightly curved enough to allow it to enter the channels and set off freezing in the water there. The ice surface will bulge downward at the channel entrances but not go through them. The water in the channels becomes supercooled and gives up heat through the ice above to the cold air. Then if those aforementioned conditions are just right—night air frosty, ground sodden, etc.—the water is drawn up through the channels to the base of the ice, where it freezes and pushes the ice upward. If conditions *stay* just right, the supercooled water keeps moving upward and freezing to the base of the existing ice and pushing it upward, creating needles, or columns, of ice.

But why does the water get "drawn up" anyway? Jeffrey May might have asked of *Weatherwise*. Bob Eaton, former research civil engineer at

CRREL, likens the process to "a thousand little straws sucking water from below." Knight sees it more like pistons at work (conceptual pistons; "they don't move"). What provides the power for the piston-like effect is the drive to equalize pressures on either side of the curved ice/water interface (which are unequal *because of* the curve). It is this effort to impose stability on the system that attracts the water to the ice. "The pistons create a kind of suction," Knight says, "which pulls the water up through each of the little channels in the soil and spreads it out along the base of the ice. They force the whole piece of ice up, away from the soil, as freezing at the base replenishes it."

Although on the small scale each new ice layer is laid on horizontally, the ice becomes a vertical feature. Or clump of features. Since soil isn't uniform, needle ice tends not to grow into a solid piece. "Suppose there is a leaf or a twig or a piece of gravel," Knight says. "The ice cannot grow above that big piece. The needly appearance comes from oddly shaped and interconnected ice columns growing around all these bigger obstructions."

Meanwhile, the water in the channels becomes more and more supercooled and the interface more and more curved until at some point the curve *is* tight enough for the ice surface to penetrate the channels from which it had been excluded and trigger freezing of the water there. With its water source thus frozen and unavailable, the needle stops growing. Researchers in Japan, where needle ice, *shimobashira,* is common, have demonstrated in lab experiments that the smaller soil channels are, the more supercooled the water must be before ice can trigger freezing inside them and halt the growth of needle ice. In other words—*the bottom line in real life*—the smaller the channels in moist soil are, the more likely it is that needle ice will form on the surface.

There is a limit; channels can be *too* tight for needle ice to form at all. Some Japanese farmers, in an effort to keep *shimobashira* from growing in their fields and loosening the soil and making it vulnerable to erosion, reportedly stomp on the soil to compact it. Even supercooled water can't climb through channels that are smashed shut by a determined farmer's foot.

Some scientists might argue with fine points of this scenario, but probably all would agree that it is the microdrama, of molecules of water being drawn through channels narrower than a hair to a curved interface too small to be seen that is ultimately responsible for the buckling of asphalt roadways and the cracking of foundations under multistory buildings. Needle ice is, on the very small scale, a *frost heave.*

hill and field
Harden, and summer's easy
Wheel-ruts lie congealed.
—Richard Wilbur,
"Wyeth's Milk Cans"

For a long time, it was assumed that frost heaves are caused by water expanding in soil channels when it freezes, as water does in other confined spaces, like household pipes and milk bottles. But, as Gregory Dash, physicist at the University of Washington, emphatically points out, frost heaves are "*not* the same as bursting milk bottles!" The main thing that causes soil to heave, or swell, is water being sucked through fine soil channels to scattered sites and being converted into discrete bodies of ice, which push soil particles apart to make room for themselves. At the surface, these discrete bodies form needles; inside the ground—and much more commonly—they form "lenses." While needles are usually vertical features, lenses are usually horizontal ones, lying parallel to the surface and the source of cold. An ice lens can be thinner than a pencil lead or thicker than a piano. Both lenses and needles are considered to be "segregation" ice, since they usually consist of pure ice.

You don't get frost heaves in sand, gravel, and other large-grained soils because with large grains you also get large channels, which the ice surface has no trouble entering. In that case, instead of being drawn to and enlarging ice bodies elsewhere, the water freezes inside the channels, cementing the soil particles together, like a pale glue. The cementing is incomplete if there's a fair amount of air in the soil along with the water, and the soil freezes in clumps, "like those awful, sticky balls of popcorn one gets at carnivals," Knight says. If the channels are filled with water and there's no air, the cementing is complete when freezing and the soil becomes like concrete. When you're walking over the ground in winter and it feels like a sidewalk under your shoes, it *is* like a sidewalk.

This "pore ice" doesn't create frost heaves, however. Segregation ice does. A growing lens of ice can put up to 20 megapascals of pressure on the overlying soil, or enough to lift a gravel truck. Even the slender bristles of needle ice, forced upward by the pressure of supercooled water rising in the channels, are capable of lifting a 33-pound rock. "It's amazing how much pressure you can generate in this way," Charles Knight says.

In the winter of 1994, New York City went through twice as many freeze-thaw cycles as it usually does, and road crews filled twice as many potholes in city streets as they usually do, a quarter million of them. Potholes aren't all caused by frost heaves—Hawaii gets potholes too—but in cold regions most of them are. What typically happens is that water in fine soil under an asphalt road freezes and produces an ice lens, to which supercooled water is drawn, and that water freezes in turn, enlarging the lens. The lens pushes the asphalt and gravel above it upward into a bump. When the lens melts, there's less solid at that place but more water. Truck wheels roll over the spot and press the asphalt into the softened material below it, and the flexing causes the asphalt to weaken and break up. The spot bulges down instead of up: a pothole!

The Army Corps of Engineers' most-oft-requested publication is *Pothole Primer,* and in it road builders are advised to spread a layer of gravel at least a foot thick between the pavement and the natural soil, to keep (freezable) water from collecting there. "The three most important things in road construction," Eaton intones, "are drainage, drainage, and drainage."

> December locks up all in Ice and Snow, and constipates the Pores of the Earth that it cannot be tilled.
>
> —John Fryer, *A New Account of East India and Persia. Being Nine Years' Travels, 1672–81*

Maybe you've heard tales about pioneer farmers in New England and the Midwest digging up and carting away boulders from their stony fields in summer and autumn, then the following spring, after the snow melts, gazing out upon the same fields and seeing more dadburned boulders lying there. Scientists now know that those boulders were "frost-pulled" upward. They "upfroze," probably in this way:

A rock the size of a soccerball is buried in fine, wet soil. The water in the soil starts to freeze, from the top down. It freezes onto the sides of the rock boulder before it freezes onto the small particles around the rock because the rock cools faster than the particles do, due to its larger size and its conduction properties. When the frozen part of the soil has gained a good eight- or ten-inch grip on the rock, it is able to overcome the holding force of the *unfrozen* soil and the pull of gravity so that when the soil heaves upward, it carries the rock with it. That leaves a small air space below the rock, which the rock doesn't settle back into once the soil thaws because dirt and pebbles have since tumbled into the space. Each time the soil refreezes, the rock is carried to a slightly higher level, and each time the soil thaws, it is kept from returning to its former level.

After several freezes and several thaws, the rock pushes up a little mound of dirt at the surface, then breaks through it and waits for the exasperated farmer to arrive.

Today's farmers don't have much of a problem with "growing stones," according to Bob Eaton. Those who did lived mostly in the 18th and early 19th centuries, and "eventually you run out of rocks" in the upper layers. But many other things upfreeze besides rocks. Frost pulling can lift fence posts, utility poles, and foundation pilings right out of the ground if they're not set in it deeply enough. Put a six-inch-long stick in fine soil that gets a lot of frost action, and in a couple of weeks you may see it has levitated four inches. Recently, in Brooklyn, New York, artist Carol Carey was raking her patio garden before spring planting when she noticed bits of glass, pebbles, and vinyl records that hadn't been there in the fall. Over the winter, many freezes and thaws had unearthed trash from earlier tenants. "There's no such thing as clean dirt," she says.

Wind and water and ice and life
have powdered our planet's obdurate skin

—John Updike, *"Ode to Fragmentation"*

Ice can not only lift rocks, it can break them apart. In contrast to frost heaving, most of the power for splitting rocks *is* water's expansion into ice. A tiny crack appears in a large slab of bedrock; water fills the crack; the water freezes; the growing volume of the ice exerts pressure on either side of the crack; the rock splits in two. By laboratory measurement, the maximum pressure generated by the expansion of water into ice is 31,000 pounds per square inch (although, as A. L. Washburn noted in his book *Geocryology,* that's "probably never attained in nature"). In some areas, entire hillsides are covered with jagged chunks of rock produced by this "frost wedging," some of the chunks as big as buses. The big chunks often get split into smaller ones, and those into still smaller ones. Frost wedging is thought to produce rockfalls in cold regions more effectively, according to Washburn, than anything "exclusive of earthquakes."

We love thee, we love thee, we love thee frozen land.

—Sir Cavendish Boyle, *"The Ode to Newfoundland"*

One of the characteristics that define a "cold region" is the fact that the ground there freezes. For a region to qualify, the ground has to freeze at least one foot down at least one year in ten. By that standard, almost half of the land area of the northern hemisphere can be considered a cold region. The ground freezes that deep that often in nearly all of Scandinavia and the former Soviet Union—no surprise—but also in most of Japan, Korea, and China; in much of Austria, Turkey, and the Balkan states; in all of Canada except the Pacific coast; and in more than half of the United States, including Iowa, Pennsylvania, Nebraska, and most of Illinois, Indiana, and Ohio. The boundary between where the ground freezes and where it does not corresponds roughly to the line where, during the coldest month of the year, the average temperature is no more than 32° F, which underscores the fact that, although surface conditions count (is there an insulating cover of moss? of snow?), and soil structure counts (Bob Eaton calls soil the "skeleton" around which ice accumulates, and an image arises of ice as flesh, covering the bones of soil with its mottled tissue), it is chiefly temperature that determines whether ground freezes, and how far down.

In New York City, where the mean daily temperature in winter months is 34° F, the ground under the littered crabgrass typically freezes a foot down. Farther north, in Hanover, New Hampshire, where the mean daily winter temperature is 22° F, the ground freezes about three feet down. In Minneapolis, where mean winter temperatures run about 16° F, the ground freezes four to eight feet down. Inhabitants of those places keep such numbers in mind when they put their water pipes in the ground, although now and then the ground freezes farther down than usual and water inside the pipes freezes too, expands, and cracks the pipes. Every October a plumber goes around to the 105 outdoor drinking fountains in New York City's Central Park and turns the water off, since all pipelines to the fountains pass of necessity through the freezing level; then six months later, in an act many New Yorkers welcome as a sign of spring's arrival, he turns the water back on.

It came of winter's giving ground
So that the freeze was coming out,
As when a set mind, blessed by doubt,
Relaxes into mother-wit.
Flowers, I said, will come of it.

—Richard Wilbur, *"April 5, 1974"*

When spring comes to Minneapolis and Hanover, the ice goes. Even in Moscow and Ottawa, that's all there is to it, the ice goes. Iron-hard ground gives way under boots. Water leaps in drinking fountains. Nevertheless, there are places in the world where, even as flowers bloom at the surface under summer sunshine, ice does not vanish from the ground. These are not all isolated places with extreme environmental conditions, either. "Permafrost" (what one scientist called ground that is permanently frozen, in a slip of the tongue so useful that the word entered the language) underlies *one-quarter of the entire land area on Earth.* It underlies half of Canada, 80 percent of Alaska, 60 percent of Russia, a quarter of China if you include Tibet, and about 10 percent of Scandinavia. It lies beneath some continental shelves; under Antarctica's Dry Valleys and some of its islands; below sections of the Rocky Mountains, the Himalayas, Alps, Andes, and Urals; even under the peak of Mauna Kea in Hawaii. Permafrost* is so widespread that Alan S. Judge, retired scientist with the Geological Survey of Canada, who has studied permafrost for 25 years, calls it "the other ice sheet."

Still, all the ice in permafrost everywhere combined adds up to less than one percent of all the freshwater ice in the world; most of the rest is in the first two ice sheets, in Antarctica and Greenland. Yet "its significance far transcends its quantity," A. L. Washburn noted. For it is permafrost and not, as you might suspect, cold air or short growing seasons or long gloomy winter nights or scant precipitation that is "the single most important feature affecting twentieth century man's development of the northlands," as Jack D. Ives, of the Institute of Arctic and Alpine Research of the University of Colorado, wrote. Anybody in the Arctic or subarctic who wants to build a house, drill for oil, dig a mine, lay a pipe, or bulldoze a road has to face the possibility that permafrost could sabotage the project.

In 1743, while posted to a remote northern station of the Hudson's Bay Company, James Isham made note: "In Dig'ing three or four foot downe . . . in the mids't of the summer you shall find . . . Ice and above six or Eight feet Downe itt's all hard Ice." His has to be one of the earliest written references to permafrost. Isham even figured out why the ice was there. "The shortness of the summer's is not Sufficient to thaw the Ice the severity of winter occation's," he wrote, "therefore it gathers more and more Every year, for which Reason the frost is never out of the ground, in these parts."

*Although by definition permafrost is any earth material that has been 32°F or colder for two or more summers, and although permafrost occasionally consists of frozen rock without any spaces for water to collect in and freeze, and although the water in the ground is sometimes so salty that it stays liquid at temperatures well below freezing, the great bulk of permafrost does contain ice.

"These parts" was the western side of Hudson Bay, which today is within the zone of "continuous" permafrost. Continuous means that almost anywhere a person digs down at any time of year he'll eventually run into permafrost. South of this zone (in the northern hemisphere, where most permafrost is) is a zone of *discontinuous* permafrost, meaning that if a person digs down he may eventually run into permafrost or he may not; there are breaks in the other ice sheet. Still farther south is a zone of *sporadic* permafrost, where permafrost occurs in patches. In mountains, too, permafrost occurs in patches. Mount Washington in New Hampshire reportedly has some permafrost on it, and so do Mount Fuji in Japan, Mount Jacques Cartier in Quebec, and the Beartooth Mountains of Wyoming.

Isham didn't note, and almost certainly didn't know, how thick the layer of hard ice he found was; he dug to it but probably not through it. In *The Klondike Fever,* Pierre Berton described how difficult it was for 19th-century gold miners to reach gold-bearing rock "ten, twenty, and even fifty feet below the surface" when there was permafrost in between. They couldn't pickax their way down or, as miners do now, wash the frozen soil out with high-pressure water sprays. "At first, the miners let the sun do the work," Berton recounted. "This was a long, laborious process; a few inches of thawed earth were scraped away each day, and an entire summer might pass by before the goal was attained. Soon, however, wood fires replaced the sun. The gold-seekers lit them by night, removed the ashes and the thawed earth in the morning, then lit a new fire, burning their way slowly down to form a shaft whose sides remained frozen as hard as granite."

The thickest permafrost in the United States is in Prudhoe Bay, Alaska, more than 2,000 feet as measured in a shaft drilled by oil workers. The thickest permafrost in Canada is either in the Beaufort Sea, under water, on a continental shelf that was exposed to cold air after sea levels fell during the last glaciation, or else around the rim of the Sverdrup Basin in the High Arctic islands. In both places, it's about 2,600 feet thick. The thickest permafrost anywhere on Earth is in Siberia, Yakutia by the upper Markha River, where it was measured in a tunnel dug by salt miners. It is 4,350 feet, or four-fifths of a mile, thick.

Still, no matter how thick permafrost gets, there's always an "active layer" of ground on top of it. The active layer thaws in summer and refreezes in winter, just as the ground in places with no permafrost does. The layer tends to be thin where the permafrost layer is thick and vice versa. In the far north, an active layer could be a few inches thick, while toward the southern end of the permafrost zone it could be six feet thick. Underneath the permafrost layer as well there is always unfrozen ground, in every season. At a certain depth, heat radiating upward from the Earth's interior overcomes the effect of cold moving downward from its exterior, producing a temperature that's above freezing. At four-fifths of a mile, the Yakutian permafrost may be as thick as permafrost can get.

> *the mastodon*
> *Hung in the warehouse of a glacier, upside down.*
>
> —Brad Leithauser, *Darlington's Fall*

Most permafrost is old, very old. Since ice is a pretty good insulator, the more permafrost there is for the latent heat of freezing to have to pass through before reaching cold air, the more delayed the addition of a new layer of ice at the bottom will be. Thus it takes a very long time for even very cold air to create a very thick slab of permafrost. Most of the permafrost in the world has been around not just for two summers but for thousands of summers, some of it for hundreds of thousands of summers. In Alaska and the Yukon there's permafrost believed to be 2 to 2½ million years old.

One clue to the hoary age of some permafrost is the fact that it is much thinner in parts of the Canadian High Arctic than it is in parts of Alaska and Siberia that have similar air temperatures today. During the last glaciation, most of Canada was covered by a thick sheet of ice while much of Alaska and Siberia had little or no ice. Apparently, thick permafrost grew in the ground only where there was no thick blanket of ice above the ground to protect it from ice-age air.

Another clue that permafrost could be ancient was the discovery in it of woolly mammoths, which haven't been seen alive for 10,000 years. (Catherine the Great said as much to Voltaire: "What proves, I think, that the world is a little older than our nurses tell us are the finds of bones of elephants long ago extinct embedded in the ground in northern Siberia.") For centuries, Siberian tribesmen have sold as ivory the mammoth tusks they found in riverbeds, into which the tusks had fallen when cliffs of permafrost eroded. Sometimes they discovered mammoth hides as well and occasionally even flesh; in the late 1970s, a Russian paleontologist found a six-month-old baby mammoth with an intact body, down to its chestnut-colored woolly fur. It breathed its last 40,000 years ago and had been lying in the deep freeze of permafrost ever since, along with probably millions of others of its kind.

CHAPTER SIXTEEN

GROUND ICE II

In the tangled lakes
of its eyes a mirror of ice

—John Haines,
"The Tundra"

YOU CAN'T SEE MOST PERMAFROST, hidden as it is below the active layer, but you can often tell where it is by all the water on top. Without permafrost, as Canadian botanist A. E. Porsild noted, "most of the arctic zone would be a lifeless desert"; it gets little rain or snow. Yet with permafrost, it is decidedly wet in summer, waterlogged and soupy, dotted with meltwater lakes and rivers. The permafrost acts as a rigid underground table, impervious to liquid, which keeps meltwater at the surface from draining downward, so that what water there is stays high, in the active layer. "A land where . . . half of every square mile is a shallow lake or a sluggish river," Gordon Speck wrote of the vast permafrosted tundra of northern Canada in his book *Samuel Hearne and the Northwest Passage.* You can also tell sometimes where permafrost is by the geometrical patterns on top—polygons, circles, cones, and stripes, "as if," according to Speck, "some giant mathematician had practiced his art with a world for a slate"—most of the patterns produced by permafrost and the freezes and thaws of the active layer above it.

• ***Polygons*** form when wet ground freezes in very cold air, contracts, and cracks like dried mud at the bottom of a puddle. The cracks meet to define four-, five-, six-, or seven-sided polygons, just as cracks in dried mud do, except that in frozen ground the polygons, are far larger, up to 500 feet across. In early spring, meltwater trickles into the cracks and, since the soil on either side of the cracks is still cold, quickly refreezes. Then in summer the soil warms and swells, producing ridges on either side of the ice-filled cracks, which turn into troughs. The next winter or the next, cracks form *within* the ice that fills the cracks, and those cracks fill with meltwater, which freezes and expands and widens the cracks, and so on, over many winters and springs and summers, until the cracks defining the polygons no longer hold thin veins of ice but large wedges of it.

• ***Circles.*** On an otherwise barren plain on the Norwegian island of West Spitzbergen, hundreds of miles above the Arctic Circle, are arrayed scores of what look like slightly collapsed car tires, with vaguely rounded centers that suggest hubcaps. The "tires" are made up of light-colored gravel, the "hubs" of dark, fine-grained dirt. The circles of stone are 10 to 20 feet across, close together, and so regular in shape and contour that when whalers first saw them, they thought they were manmade. Even to Bernard Hallet, geologist at the University of Washington in Seattle, who knows that nature made them, they are "spectacular." Their level of self-organization is "striking," he declares, "given that they originate from a featureless mixture of mineral material," what eons ago was probably a beach.

What happened first, Hallet deduced after studying the circles over several years, was that frost-pulling sorted the mix of stones on the ancient beach according to size. Since frost-pulling has more effect on large stone fragments than on small ones, eventually a layer of large stones ended up on the surface and a layer from which the large stones had been removed, by being spirited upward, lay underneath. It was made up of small particles called fines.

Next, the sorted stones organized themselves into circles (and, remarkably in Hallet's view, continued to preserve those circles over thousands of years despite the perturbations of wind and rain and crawling organisms). Being lighter and more buoyant than the large stones, the fines would well up during thaws and push the coarse gravel radially outward in all directions to form raised borders, which left a smooth mound of fines in the center. At the edge of the mound, the fines would descend, move toward the center then back up to the surface, describing a vertical circle, as in a mud boil. Meanwhile, the gravel rims were rolling, too, at the same rate but in the opposite direction. Where the gravel met the fines, a downwelling occurred, which Hallet compares to the downwelling in plate tectonics. The soil is pulled under, sharply, abruptly, into a steep-sided trough, "like an ocean trench."

All of this took place, and still takes place, in extreme slow motion; the soil may need more than a century to complete a single circuit, Hallet says. What actually drives the soil convection, however—what got the circular movement going and keeps it going—"remains elusive," he admits. He favors a "dirty" mechanism called "border forcing" in which larger stones in the rims settle during thaws and exert sideways pressure on the fines, causing them to move upward.

• *Cones.* Inuit call them *pingos,* meaning "conical hills," and now so do scientists. On tundra, which is where pingos are usually found, they can be arresting sights, projecting above the level ground and reminding those who see them of volcanoes or Mayan temples or medieval castles. In 1848, when Sir John Richardson sailed along the Mackenzie River in northwest Canada looking for the lost Franklin expedition, he made a sketch of an "eminence" with steep sides and three small points at the top which reminded him of "an artificial barrow." The Inuit have two names for that very pingo: *Aklisuktuk,* "the one that is growing," and *Pingorssarajuk,* "the poor thing that is getting to be a pingo." Both names are tantalizing to J. Ross Mackay, emeritus professor of geography at the University of British Columbia, who for over 40 years has been studying pingos along the western coast of Canada, which has nearly 1,500 of them, the greatest concentration in the world.

The reason the names are tantalizing to Mackay is that no living Inuit has actually seen the hill growing or getting to be a pingo, and those who did see it have been dead for at least 150 years, and the pingo began growing at least 100 years before that. Oral tradition has simply kept the pingo's youthful names frozen in time.

Pingos have cores of ice—*"mamelons de glace pure,"* as E. DeSainville said of them when he was traveling in western Canada in the 1890s and, presumably, looked into a partly eroded pingo. There are two kinds of pingos, known as open and closed (in the open, water moves more freely to the freezing front). Common in Alaska, Greenland, and Russia, the open kind forms when groundwater flows downhill, usually into a valley, and with gravity providing the pressure moves upward through the earth until it reaches a freezing layer, at which point it turns to ice and forces the earth up into a cone. The closed kind is common in Canada, but nobody fully understood how it formed until Mackay, with "incredible perseverance," as Hallet views it, worked out the intricate series of events required to produce one. (Mackay went so far as to drain a small lake to see if he could get a pingo to start growing there.)

In Mackay's script, the closed pingo starts with a lake, like those along Canada's west coast. Under the lake is sandy soil, unfrozen even in winter because the lake is deeper than the ice is thick. Around the lake, though, and underneath the unfrozen bed, there's permafrost. One spring, when the lake's normal outlet for overflow is blocked by snow or there's excess runoff, water flows out of the lake through networks of ice wedges around adjacent polygons. The outflow erodes the wedges into deep troughs, through which the water drains so quickly that in only a few days the lake is empty, except for a small residual pond in its deepest part. The following winter, the sandy lake bottom, now exposed to cold air, begins freezing, from the top down. The pressure of the water's expansion into ice expels *unfrozen* water from the pores in the sand under the frozen layer, which is becoming permafrost. Under pressure, that water migrates from areas of the lake where the frozen layer is thick to an area where it is thin ("It's easier to lift up ten feet of frozen ground than 100 feet," Mackay points out). There it freezes, expands, and pushes the thin permafrost up into a mound. *Pingorssarajuk!*

As more water migrates through the sand and the ice mound grows, the soil on top may split, producing a crater which adds to the impression that a pingo is a volcano. The pingo may even erupt! On one occasion two Russian scientists watched a pingo toss chunks of ice a distance of 75 feet.

Pingos can get to be 50, 60, even 150 feet high, and Inuit hunters used to stand on them and scout for caribou herds. The soil on pingos is different from what's on the surrounding tundra, Kaye Everett, professor of agronomy at Ohio State University, pointed out, "like prairie soil." Plant species are far more diverse than on tundra; some even rare. Ptarmigan appear on the little hills, and foxes, and ground squirrels, and grizzly bears, which come to eat the ground squirrels. "Pingos are unique little environments," Everett concluded, "quite unlike anything in the area."

The growth of some "poor" pingos leads to their decline. Expanding ice cores split the overlying soil, thereby exposing themselves to sunlight, and melting. Decay is variable; for some pingos, it could take thousands of years; ice cores melt slowly. *Aklisuktuk* is subsiding at a fraction of an inch a year; Mackay reports. Someday there'll be nothing left of it but the bottom of its earthen walls, like old volcano rims.

Even as these and other geometric figures (including "stripes," which are stone circles pulled out long on steep slopes by gravity, with stones in one stripe and fines in the next) embellish the surface of some cold regions, there may lie underneath others what scientists call, with uncharacteristic straightforwardness, "massive" ice. Massive ice is big bodies of ice in the ground. The ice bodies can be lenses (Washburn suggested that any lenses over six feet thick are massive) or buried ice (glaciers or frozen streams or snowdrifts covered with soil) or even pingo cores, just as long as there's plenty of ice and not much soil involved. The ice should weigh at least 2 1/2 times as much as any contained soil. The most massive of all massive ice is on the Jamal Peninsula of northern Siberia, which Washburn described as "tabular, monolithic, continuous" beds of solid ice. "No one knows the origin of these massive sheets," Everett said. "One theory holds that they are an extreme form of segregation ice [lenses]. Another holds that they are remnants of former ice sheets, which were stranded when the rest of the ice sheet melted." Buried beneath 20 feet of frozen earth, they are 100 to 130 feet thick and miles long, or taller than some hills, longer than some valleys.

For eight months of every year
there is frost unbearable,
and in these you shall not make mud
by pouring out water
but by lighting a fire.

—Herodotus, *Histories* (on Siberia)

Permafrost is not really permanent. (Scientists tend to say "perennially" frozen ground instead of "permanently" nowadays but won't give up the handy word "permafrost.") Several warm summers in a row can thaw the top of a permafrost slab and deepen the active layer or even detach the active layer from the permafrost so that between them lies a layer of perennially unfrozen ground, what the Russians call *talik,* or "melted soil." When that happens, winter temperatures no longer reach the permafrost and it melts all year long.

In addition, streams of meltwater can flow through networks of ice wedges in spring and summer and erode the ice into troughs, which then carry more eroding meltwater. Or wind and waves can rip grass and soil off shorelines and expose ice in the ground to more wind and waves. If there happens to be a considerable amount of ice in the ground, the thawing can change the whole look of the landscape, producing depressions, mud slides, rock slides, and toppled trees.

"You get everything from pits and ponds to great gaping holes," says Thomas E. Osterkamp, professor emeritus of physics at the Geophysical

Institute of the University of Alaska at Fairbanks and an expert on permafrost. "The surface turns rough, uneven." In Russia, he points out, there are thaw holes more than half a mile wide. Behind his own house in Fairbanks he found a depression as long as a football field, which he's certain is a former lakebed that formed when permafrost thawed. "It could even have been a pingo." He also discovered, in a woods near campus, only "a baseball throw away from student housing," a network of melted-out ice wedges, troughs that are 13 feet deep in places, with fallen trees inside them and hummocky ground between. He figures that the thawing took place sometime in the last century, and was natural.

But permafrost thaws unnaturally, too, and often with even more dramatic effects. "*Anything* which disturbs the delicate temperature balance protecting permafrost can bring about a horrendous change," Farley Mowat wrote in his 1970 book *The Siberians.* "A tracked vehicle grinding over summer tundra and breaking through the thin insulating layer of moss and lichen can create vast, heaving ditches which will endure for centuries." In the 1940s and 1950s, when oil and gas production got started in Alaska, "everybody drove any place they wanted to," Kaye Everett recalled. Trucks going off road compressed vegetation, the vegetation darkened, sunlight heated up the dark truck tracks, the heat penetrated downward to the top of the permafrost, the ice in the permafrost melted. Without ice to provide support, the surface sank. Tracks turned into ruts, ruts into canals. The routes became impassable.

"Even the simpler activities of man initiate such changes," Peter J. Williams, professor emeritus of geography at Carleton University in Ottawa, observed in his book *Pipelines and Permafrost: Science in a Cold Climate.* He gave an example of the simplest. "Ross Mackay tells of a path worn, in ten days, by the feet of an Eskimo's chained dog. The wearing down of the vegetation modified the exchanges of energy through the ground surface, and led to the thawing of a layer of the permafrost with a subsequent settlement of the ground by 38 centimetres [15 inches] in the next three years." A week and a half of nervous pacing by a single dog produced more than a foot of ground sink!

A couple of centuries ago, when the Russians opened up Siberia to mining, agriculture, and the fur trade, permafrost often let them, and their dwellings, down. "The log cabin of the early settler, with its fireplace, thawed the permafrost beneath the structure," Jack Ives wrote, "and, in extreme cases, groundwater was tapped which entered the house. The ultimate condition was a wrecked house filled with a mass of ice." (In the margin of the library book where I read this, someone had penciled in "yes yes.") Even today in the city of Fairbanks, Osterkamp points out, there are "dozens and dozens" of houses with windows and doors that won't open or close, torn wiring, broken plumbing, and cracked foundations because their presence has caused the permafrost underneath them to thaw. The warmest parts of houses sink the farthest; a person standing outside a listing house can usually tell where the kitchen is.

The permafrost in Fairbanks is discontinuous—here but not there, under a shady or north-facing spot but not one nearby. Most of the permafrost is probably within a degree or two of melting. When people in the area need a loan so they can build a new house, some banks require them to dig five holes in their lot, one for each of the house's four corners and one at the center, to check for "excess" ice. Excess means the ice takes up more space in the soil channels than the normal volume of air and water would, in which case the ice will have pushed the soil grains apart until they are no longer in contact with each other. The weight of a new house could, by causing a slight rise in soil temperature and thus the melting of some ice, bring the grains back into contact, unfortunately for the new house. Slumping earth means a slumping abode.

To learn how to deal with permafrost, the Russians set up the Eternal Frost Institute in 1930 in the city of Yakutsk, in the Siberian republic of Sakha. Permafrost underlies 90 percent of the ground surface in Sakha and is on average over 1,000 feet deep. When Farley Mowat visited the institute 40 years later, Pavel Melnikov, director at the time, spoke of progress. "Whenever planners or constructors come to Siberia to begin work on a new project we tell them: 'Be nice to the Eternal Frost Queen. Keep her well covered under nice thick blankets and she will let you do anything you wish. But if you strip off her clothes, you had better watch out!' "

One way to keep the queen decently clothed, the institute found, is to build new structures on refrigerated pads which, as long as they stay frozen, can be like concrete. Another way is to put pilings under structures so that heat emanating from their undersides can be released into the cold air instead of the cold ground. "All buildings [in Yakutsk] were raised above the permafrost a regulation four feet on concrete stilts," Colin Thubron observed in his 1999 book *In Siberia*. Small structures in Yakutsk evidently didn't rate stilts, however, and the Eternal Frost

Queen was not amused. "Between the tenements long streets of cottages dip and sway over the unstable earth," Thubron reported. "As their weight softens the permafrost, one side or another starts to sink, until their walls loosen to a wave of unaligned shutters."

Earth, earth, earth, thy cold is keen,
Earth grown old.

—Christina Rossetti, *"Advent"*

Permafrost is chiefly a product of cold; so what happens to it when there's a lot less cold? In its 2001 report, the Intergovernmental Panel on Climate Change, a United Nations–sponsored group charged with assessing the impact that future climate will have on the Earth, including the Earth's "cryosphere," or domain of snow and ice, predicted that if average global air temperatures rose 2° to 3.5°F, permafrost would underlie *a quarter less surface area* in the northern hemisphere than it does today, with zones of continuous as well as discontinuous permafrost moving "substantially" northward. Possible local effects, listed in the panel's earlier, interim report, include wetter soil (since "ground ice is generally concentrated in the upper few meters of permafrost, the very layers that will thaw first as permafrost degrades"); *drier* soil (since as the permafrost table is lowered, there's more unfrozen ground for water to drain into); landslides (since the active layer could separate from the permafrost and, with a lubricating layer at its base, take off downhill); more rocks falling off mountains (since ice had been bonding the rocks together); and receding shorelines (since unfrozen soil is more easily eroded than frozen soil).

But how does anybody know what's happening to permafrost? After all, it's nearly always buried, sometimes deeply buried. One way is just to drive around the countryside and look for gaping holes, which can be dated (by, for instance, examining trees that have fallen in). Another way to check on permafrost is by measuring the active layer; if it thickens, you can presume the permafrost thinned. Still another way of checking on changes in permafrost is by looking at satellite images for places where the boundaries of vegetation zones have shifted. Tundra is typical of continuous permafrost, Oleg A. Anisimov, geographer at Russia's State Hydrological Institute in Saint Petersburg and professor of geography at the University of Delaware, points out; boreal forest is typical of discontinuous or sporadic permafrost. "You can easily see the differences."

The only way of knowing for sure what's happening to permafrost, though, is by digging a deep hole in the ground and taking the ground's temperature. "The ground remembers," geologist Alan Judge says, "like an old cassette tape." The deeper you dig, the further back in time you go. Near the top of the hole, ground temperature will be close to air temperature, although a degree or two warmer, since in most places where there's permafrost there's also an insulating cover of snow in winter. If the mean annual temperature of the air stays the same over a period of years, the temperature of the ground will rise steadily with depth, as heat from the Earth's core exerts a greater and greater influence, until at the base of the permafrost layer it's 32°F. If you were to draw a temperature profile for the hole you just dug, it would be represented by a straight line.

If the mean annual air temperature should *rise* a degree, the line will begin to curve, as ground temperature reflects the air temperature. Gradually, over thousands of years, the curve will penetrate downward through the ground and the straight line reestablish itself above the curve, based on a new steady temperature. If, on the other hand, the mean annual air temperature should *fall* a degree, the line will curve in the other direction, then slowly this curve too will penetrate downward and the temperature line above it smooth out.

"You can relate the history of air temperature at a locale by the shape of the line," Judge concludes. "You know by how far down a curve developed when an increase or decrease of temperature must have occurred." "This is not a proxy," Osterkamp insists, "like every method we have of determining past temperatures—ice cores, tree rings, snow melt. This is a true measure of what has gone on at the surface." From temperatures taken in a series of deep boreholes drilled in northern Alaska, he was able to conclude that the air there has warmed 3.5° to 7°F since the mid- to late 1800s. The shape of the line told him that "some thin, discontinuous permafrost is definitely thawing."

Every borehole speaks only for itself, though. Keeping track of the health of permafrost worldwide would require a whole network of boreholes, Judge points out, but so far the distribution has been "wild, buckshot." "Very sparse," Osterkamp agrees.

Permafrost doesn't thaw overnight. Even when permafrost is already close to the melting point, a layer ten feet thick could take 20 to 40 years to thaw, and one 35 to 65 feet thick would take 100 years. Like other forms of ice, ground ice can absorb a lot of latent heat, and a lot of energy is required to melt it. Once the permafrost has melted, though, it cannot be replaced, probably for hundreds or thousands of years. "The loss of volume," the climate panel stated in its interim report, will be "irreversible."

As soon
Seek roses in December, ice in June; . . .
Or any other thing that's false

—Lord Byron,
"English Bards and Scotch Reviewers"

On a summer day in 1877, Edwin Swift Balch—age 21, recent Princeton graduate, future lawyer—was hiking in the White Mountains of New Hampshire when his guide surprised him with a question. "Would you like a piece of ice?" the man asked, then promised, "I can get you some presently." Balch answered "Certainly" but wondered where the guide would find ice as "there hadn't been any ice or snow on the surrounding mountains for months." The men made their way to the bottom of a boulder-filled ravine where the guide "climbed down under one of the biggest [boulders] and presently reappeared with a good sized lump of ice." Balch recalled, "I was much impressed."

Very much impressed. Although the amount of ice was paltry and the site a mere gap between boulders, the encounter was the beginning of a beautiful friendship. Off and on for the next 40 years, Balch traveled in search of holes in the ground that had perennial (lasting through the summer) ice in them, mostly mountain caves but also wells, mines, lava tubes, and the various basins, troughs, chasms, gullies, gorges, and glens he called "roofless" caves. At the turn of the 20th century he published a book listing 300 of these, in, among other places, Germany, France, England (Wordsworth wrote of one in Helvellyn, "It was a cove, a huge recess / That keeps, till June, December's snow"), Hungary, Serbia, Russia, Iceland (a volcanic crater, Balch noted, which "hurled out simultaneously into the air lumps of lava and ice"), Japan, the Himalayas, South America, and North America, where he found 50. To his chagrin, most people kept calling them "ice caves" (and still do), despite the fact that most people also call the hollows that swirling meltwater carves out of glaciers "ice caves." The two ice caves are not related. One is holes *in* ice, the other is holes with ice in them. Balch's interest was in the second kind of ice caves, and he named his book *Glacières or Freezing Caverns* for his two choices about what else to call them.

He studied the caverns and figured out how the ice inside could last through hot summers. In winter, cold, dense air sinks by gravity into a hole, where it displaces warmer air. The cold air lies below the level of the entrance, so it cannot rise and flow out. Then, in spring, meltwater at the surface drips into the hole through openings in the rock, and the low-lying cold air causes the dripwater to freeze into icicles (some "as big around as large barrels"), stalagmites (some resembling porcelain), pillars (with "all sorts of icy ornaments hanging about them in tufts and fringes"), and ice floors (very smooth).

The weirdness of ice lying in an open hole in the ground in midsummer has sometimes elicited spiritual responses. In the White Mountains of Arizona, Zuñi Indians worshipped before a roofed cleft in the rock where ice lingered until June or July, "much later than anywhere else in the neighborhood," Balch noted. That the region was arid and any water therefore precious caused the "element of mysticism" to develop among the Zuñis, who called the subterranean ice "the breath of the Gods."

Although some American pilgrims who headed west in wagon trains in the mid-1800s attributed the presence of any ice in the ground in summer to divine provision, they also made practical use of it: they dug it up and consumed it. The ice was beside the Oregon Trail, reported to be two wagon days east of South Pass in Wyoming. On June 26, 1849, Alonzo Delano (a distant relative of Franklin Delano Roosevelt's) was on his way from Illinois to California when, as he noted in his diary, "I followed the [wagon] train and overtook it about four o'clock in the afternoon, on the borders of a morass. . . . Some of the boys, thinking that water could be easily obtained, took a spade and going out on the wild grass, commenced digging. About a foot from the surface, instead of water, they struck a beautiful layer of ice, five or six inches in thickness. Many trains were passing at the time, and all stopped and supplied themselves with the clear, cooling element, and buckets were brought into use to supply ourselves with frozen water for our supplies. This natural icehouse is not only a great curiosity in itself, but from its peculiar situation, in this dry, barren, sandy plain, is justly entitled to be called the diamond of the desert."

The ice kept replenishing itself. On June 19, 1821, Amelia Hadley of Cleveland, Ohio, had been riding on a wagon train with her new husband when after passing two graves they came upon an ice spring "in a marshy swale, which is mirey . . . there is a solid cake of ice as clear as any I ever saw and more so cut a piece as large as a pail and took and rapt it in a blanket, to take along"; and precisely 30 years to the day later, on June 19, 1851, Harriet Talcott Buckingham, another Ohioan on her way to Oregon, told of finding ice in the "Ice spring 1/4 mile from the road. It lies under the sod & can be dug in large pieces & very clear too. It is singular that it did not melt, it was so warm. . . . Mrs S. got off the horse & I had to ride alone. Made 14 miles."

CHAPTER SEVENTEEN

PLANTS

behold the junipers shagged with ice,
the spruces rough in the distant glitter

—Wallace Stevens,
"The Snow Man"

IT ISN'T LOW TEMPERATURES that kills most plants in cold regions in winter. It's ice. The heaving, bucking, frozen-and-thawed and refrozen-and-rethawed earth hoists up more than the usual inanimate boulders and asphalt (and the occasional coffin); needle ice can yank young, living plants clear out of the ground, exposing their roots, sometimes slashing the roots with its sharp crystals. The more powerful heaving caused by ice lenses can uproot even mature plants, tipping trees and creating "drunken forests."

Then there are ice storms. Of birches "loaded with ice a sunny winter morning / After a rain," Robert Frost wrote, "once they are bowed / So low for long, they never right themselves." Easily made "limp" by the weight of ice, the birches live on, but other trees are not so pliant when ice hangs heavy on them; their limbs snap off.

There's also "rime" ice, which low clouds deposit on the surface of cold things, and which can encase a packet of evergreen needles or leaves so firmly that when winds knock off the ice, the needles and leaves come off with it. Many fir trees in the northern Appalachians, where fog is common in winter, are thought to perish this way.

On the ground, a blanket of ice can, if it covers plants long enough, kill by smothering; the gases the plants give off in respiration can't pass to the air outside. As for the ice of glaciers, it can overrun and crush trees and plants that have taken root on rubble left from the last surge.

goodbye, shrub, if ice breaks you down
goodbye

—A. R. Ammons, *"Configurations"*

Still, it's not the ice without that is the main threat to plants' survival in winter. It's the ice within. Plants are made up mostly of water, and when air temperatures drop, that water can freeze just as water does in nonbiological enclosures (ice-cube trays, and the like). Christina Walters, plant physiologist at the U.S. Department of Agriculture's National Seed Storage Laboratory in Colorado, describes the death of a plant cell by ice as viewed through a cryomicroscope. "You see a little spot, then the spot spreads and consumes the whole space. It happens very fast," she says. "There's nothing subtle about it. That cell is *dead.* It's a mess, a total maceration. You hear a crackling with it, too." The death rattle of a minuscule victim of freezing . . .

Once ice is inside a plant cell, the cell nearly always dies, and if enough cells die, so does the plant. You can see the effect in your garden when there's a frost before the end of the growing season. "It looks like someone went through with a blowtorch," Walters says. Leaves darken and droop ("black flags," one gardener calls them). Stems go from upright to supine in a matter of hours ("mush," the same gardener says). Yet many plants live through winters when temperatures fall far lower than the freezing point and last much longer than a one-night frost. A red maple in upper Michigan can stand bare-limbed in a matrix of subfreezing air for days or weeks and still greet the spring with plump buds, flowing sap, and perfect leaves. Seeds of many plants lie in iron-hard ground for months, yet as soon as the ground softens they start sprouting. According to Charles Guy, professor of plant physiology and biochemistry at the University of Florida at Gainesville, "Some seeds you could put at any temperature on earth and they'll survive. Whatever it is they do, it renders them impervious."

In an example of near imperviousness in trees, Dahurian larches (*Larix dahurica*) of northeastern Siberia make it all the way through what are probably the coldest winters outside Antarctica, with air temperatures dipping as low as −85°F. How do they manage? How does *any* overwintering plant manage? Unlike animals, plants don't have the option of flying south or finding a snug cave when cold weather comes on. Although early thinkers speculated otherwise, plants contain no "vital heat" as animals do that protects them from freezing. "They have to dig their heels in and take it on the chin," Walters notes. Yet as Marilyn Griffith, associate professor of biology at Waterloo University in Ontario, emphasizes, "plants are not helpless. They control their own freezing destiny. That's an idea it took people a long time to warm to."

Some plants control their freezing destiny by keeping ice from forming inside them, and others allow ice to form inside but in ways that won't kill them. The first kind try supercooling. Crop plants whose growing season may be a bit too short where they're cultivated—tomatoes, beans, potatoes, strawberries—and parts of plants that are particularly vulnerable—tree buds, young fruit—can often make it through brief, mild freezes if they can keep the watery parts liquid below 32° F. Citrus plants in Florida, doing just fine in chilly but above-freezing daytime temperatures, can survive nights when temperatures fall to 25° F as long as they can stay supercooled for the six to nine hours it takes for the sun to reappear and reheat them.

How plants manage to supercool, though, is a puzzle to scientists since they contain many natural nucleators that can trigger freezing: carbohydrates, proteins, rough edges. "They do it by being very careful," Truman Young, research ecologist at the University of California at Davis, suggests, "staying still and having smooth surfaces." Or they get rid of nucleators somehow. Or, as Marilyn Griffith proposes, the thick, inflexible walls in the parts of plants that tend to supercool confine the water so severely that it can't expand into ice. "If the volume can't change," she reasons, "the phase transition can't take place."

Supercooling is a limited strategy, however. Higher plants generally can't supercool more than a few degrees; it doesn't take much to set off freezing in a rooted organism exposed to the elements, a wind batting it around and agitating the water inside, frost depositing on its leaves. Besides, the water-carrying vessels in plants tend to be large, and, as J. Levitt pointed out in Meryman's book *Cryobiology,* the "ability of the water to remain supercooled varies inversely with the diameter of the capillary in which it occurs" (think of those narrow soil channels).

Supercooling is a risky strategy, too. If plants should manage to supercool to very cold temperatures and then freeze, the result would be disastrous. The colder the water is when freezing begins, the faster the freezing will be (since at very low temperatures even very small particles can nucleate ice), and the faster the freezing is, the more deadly it will be (with ice macerating the cells). The second kind of plants, those that control their freezing destiny by allowing ice to form in them, freeze slowly. Some plants—wheat, cabbage, broccoli—actually *encourage* ice to form early, while temperatures are still relatively high, thus ensuring that freezing will be slow. Griffith is studying winter rye to find out how it manages to thrive on the wide-open prairies of central Canada, where strong winds blow insulating snow covers off the ground in winter and temperatures fall as low as −40° F, and she found some very large proteins in the rye leaves, which she concluded were acting as nucleators, triggering ice formation at temperatures just below the freezing point. (Large proteins nucleate ice close to the freezing point because by their size they provide many sites for water molecules to hitch on to.)

Yet even before air temperatures reach the freezing point, while they are still 10° or 20°F above it, the overwintering plant has started making the adjustments it will need in order to survive ice formation. "Plants aren't stupid," Mike Thomashow, molecular geneticist at Michigan State University, points out. "If they waited for freezing temperatures to make changes, it would be too late." When he took plants that had been growing at 72°F and subjected them to an air temperature of 23°F for three days then thawed them, they turned "flaccid and water-soaked immediately" and in a few days were dead. But when he took other plants of the same species, exposed them to *39°F* air for three days, *then* subjected them to several days of 23°F temperatures followed by a thaw, they suffered no "obvious" injury and did not die. In only three days they had become "cold-hardy" or "frost-tolerant." They had "cold-acclimated." In later experiments, Thomashow showed that plants could begin to be frost-tolerant in as little as 24 hours.

> Hurrah! . . . It is a frost! The dahlias are all dead!
>
> —Robert Smith Surtees, *Handley Cross*

In nature, frost tolerance usually develops in stages over the autumn and winter. (Although the plants quit growing, they remain reactive.) The shortened days of autumn, as perceived by sensors in their leaves, trigger changes that allow the plants to survive the first light frost, and being exposed to that frost stimulates changes that help them to survive harsher frosts, and so on. The changes are anything but simple, though. Gaining maximum cold hardiness, according to Peter L. Steponkus, Liberty Hyde Bailey Professor of Crop and Soil Sciences at Cornell University, requires an "orchestration of many events": hormonal, structural, metabolic, genetic. Learning what those events are and how they're orchestrated is a far more difficult task than a home gardener could ever imagine. "Tangible advances are few and far between," Charles Guy says. He remembers a professor telling him when he entered the field 25 years ago that finding out how plants deal with low-temperature stress was likely to be just as complex as finding out what goes on in a cancer cell.

They take affliction in until it jells
To crystal ice between their frozen cells

—Richard Wilbur,
"Orchard Trees, January"

Let us consider what happens to the aforementioned red maple tree in northern Michigan. In late autumn, inside a branch, water temperatures are dropping, slowly, in concert with air temperatures. Once they reach 32°F, the water in the branch cools another tenth or couple of tenths of a degree, then starts freezing. It freezes first in the spaces between cells, since particles outside cells are more effective at nucleating ice at high temperatures than particles inside cells. Ice outside cells, unlike ice inside them, is usually innocuous. The crystals don't act like little daggers, as many people imagine, slashing cells and spilling their contents. If an ice crystal should poke a cell, the resilient membrane will just deflect under the pressure.

Once ice exists outside cells, water inside cells starts moving out through their membranes to join the ice.* It is estimated that by the time a plant has frozen to 14°F, more than 90 percent of the water that had been inside its cells will probably have left and converted to ice outside them. "Think of a plum becoming a prune," Guy says. "Or a grape turning into a raisin." What makes the drying-out possible is changes the plant makes in its cell membranes, particularly their fats. The membranes become more and more porous, eventually so porous that they offer almost no resistance to the outward passage of water, at the same time that they are acting as a barrier to ice that would go the other way, into the vulnerable cell.

Drying-out is good for a plant, up to a point. Less water in cells means less chance of deep supercooling and flash freezing and ice forming inside cells. When a fully acclimated plant dies, though, it's usually because of dehydration. "The plant needs to dry out but not too much," Christina Walters explains. "The secret of survival in overwintering plants is balance." During extended periods of cold, so much water may be pulled out of cells and changed into ice outside that their membranes become destabilized. Layers fuse, and sections get scrambled about.

*This is due to a difference between water and ice in their vapor pressures, or abundance of water molecules in a gaseous state above their surfaces, with water's vapor pressure being much higher than ice's at the same temperature. In an attempt to equalize vapor pressures, the water moves toward the ice.

Holes develop, and cell contents leak out of the holes. Ice may break through the barrier and start its butchery within the cells. "It's all yuck," reports Thomashow.

> *We think of the tree. If it never again has leaves,*
> *We'll know, we say, that this was the night it died.*
> *It is very far north, we admit, to have brought the peach.*
>
> —Robert Frost, *"There Are Roughly Zones"*

Every year, farmers around the world lose billions of dollars because of freezing injury to their crops. The cold may come on too fast, before the plants are hardened, or be very severe, or be interrupted by periods of warmth which reverse the process, dehardening the plants. Even in suitable growing areas, crops often fail because of "unpredictable and aberrant" temperature fluctuations, according to Peter Steponkus, and a great many crops are planted in not very suitable areas. ("Man has a tendency to want to grow what he sees elsewhere in his own back yard," Charles Guy notes. "Near the edge of a species' range, you find more and more freezing damage as it reaches the limit of cold temperatures it can stand." Guy also notes however that "fruit-bearing species are often best where the climate is beginning to be marginal. Maybe low temperatures make fruit a little bit sweeter.")

A small increase in frost tolerance could therefore have a large effect on agricultural production and profit, Steponkus concluded. But how can people make plants more frost-tolerant than nature already has? For centuries, they have been crossbreeding plants, setting new varieties out in fields and seeing which ones do best during harsh winters. They haven't had much success. Screening for cold-hardiness is difficult; unlike flower color, it's not a single-gene trait. Hundreds of genes may be involved. Also, when selecting for cold-hardiness, other desirable qualities such as flavor and size can be lost. "You'd have to be *starving* to eat a frost-tolerant orange," one orchard owner admits.

So scientists are now trying to cold-harden plants more directly, through genetic engineering. They have isolated over 100 genes so far that boost frost-tolerance in plants, including some they call COR genes (for *co*ld-*r*egulated). They found the genes by noticing which ones were mute or operating at low volume during warm periods yet became active during cold ones, then testing to see if the activation made a difference in the plants' ability to withstand subfreezing temperatures. When Mike Thomashow attached a promoter to a COR gene on *Ara-*

bidopsis thaliana, a small, rosette-flowered plant which, although inedible, is closely related to broccoli and cauliflower and almost as frost-tolerant as they, the gene expressed even at warm temperatures and increased the plant's freezing tolerance by 2° to 3.5°F. Since in nature the plant's freezing tolerance is about 11°F, Thomashow concluded that other genes must be involved as well, "a whole slew of them." He started looking for a "master-control gene" that could turn on "tens of other genes, or scores, or hundreds," with the idea that it could be used to boost cold-hardiness in plants that need it. He figured a master gene might turn on a plant's own COR genes earlier in the growing season if that's when the plant usually suffers freezing damage, or later in the growing season, or to higher levels than usual. "We could fool plants into thinking it's colder than it is," he says.

In practice, how would the fooling work? A master gene might be inserted into a crop plant with promoters hooked to it that would keep it turned on full-time (although this might result in impaired growth and stunted plants). Or the promoters could be ones that are themselves turned on by cold so that the master gene would be activated at the usual time but to a higher level; the plant's normal frost tolerance would simply be enhanced. Or—"this is theoretical," Thomashow says—the promoters could be turned on with some safe chemical whenever a farmer needed greater frost tolerance. Say the farmer knows a week in advance that a cold wave is coming through. He could spray the chemical on his crops, which had been genetically altered to include the promoters and master control gene. The spray would cause the promoters to turn on the master gene, which would turn on the plants' own cold-regulated genes so that they expressed at a higher level and helped the plants survive the cold wave. If the farmer needed more frost tolerance, he could just spray on more of the chemical; the master gene would probably work more like a rheostat than an on/off switch.

"Everything
Falls back to coldness,
Even the musky muscadines,
The melons, the vermilion pears
Of the leafless garden."

—Wallace Stevens, *"The Reader"*

Plants that might benefit from this genetic approach include:

Winter canola. As things stand now, winter canola can't be grown in Canada and the northern United States because winter temperatures there are just too cold. "It's an automatic hit," Mike Thomashow says. But winter grains give 20 to 30 percent higher yields than spring-sown ones, and farmers like to grow them. If a master control gene could boost canola's maximum freezing tolerance, its range could be expanded northward; there's a rule of thumb that for every degree of freezing tolerance a plant gains, its planting line can be moved 100 miles north.

Sour cherries. Michigan grows a lot of sour cherries, "but they have a real problem," Thomashow says. "The blossoms look fine after a spring freeze-thaw cycle, but the tree doesn't give cherries. Only the pistil goes down, not the whole plant. Maybe the pistil doesn't have enough of the right stuff. We'd like to turn on COR genes to supply it."

Potatoes. The common potato, *Solanum tuberosum,* doesn't acclimate well to cold but its close relative, *Solanum commersonii,* does. This suggests to Thomashow that the common potato had the capacity to resist freezing and lost it, although it may not have lost it entirely, in which case a master control gene might reactivate what remains. If the potatoes weren't at risk of being zapped by early or late frosts, they could be planted sooner and left in the ground later and thereby get the longer growing season they often need.

Eucalyptus trees. The forest industry would like to grow eucalyptus trees in the southeastern United States since the wood produces a high-quality paper, but all too often late-winter frosts damage the trees. They acclimate to cold but apparently *de-acclimate* too soon in the spring. Maybe a master gene could be used to keep the trees' own cold-regulated genes switched on during that critical time.

If plants don't have their own cold-regulatory genes—either never had them or completely lost them—adding a master control gene wouldn't help as there'd be nothing for it to boost. But Thomashow suspects that most plants, even many tropical and subtropical ones, *do* have cold-regulated genes, if incomplete or inactivated ones, since the plants are resistant to drought. (Some genes are turned on by *either* cold or drought—not unreasonable, Thomashow says, given that both involve a water deficit.) Perhaps someday geneticists will be able to design a frost-tolerant tomato or a frost-tolerant orange that doesn't taste vile. Perhaps they could produce hardier flowers. "My wife wants me to do something about impatiens," Thomashow declares.

Not long ago, Thomashow's lab found the gene he was looking for. Called "cold box factor," it is now undergoing field trials to see if it will boost the freezing tolerance of crops outside the lab—starting with canola.

What delicacie can in fields appeare,
Whil'st Flora'herselfe doth a freeze jerkin weare?

—John Donne, *"Ecclogue"*

There are other ways to help crop plants get through cold times besides manipulating their genes. One is to make supercooling easier for those plants that rely on it, by getting rid of ice nucleators. Not the ice nucleators *in* the plants, like the large proteins Griffiths found in the leaves of her winter rye, but the nucleators *on* them. In the early 1970s, Steven Lindow—now a professor at the University of California at Berkeley, then a graduate student at Wisconsin—discovered that *Pseudomonas syringae,* a bacterium so common that "most plants are exposed to a constant rain of them," he says, was acting as a nucleator and causing the leaves and stems and blossoms of plants it landed on to freeze. In *syringae*'s outer membrane is a protein containing a cylinder that binds water inside it into a formation that "looks like ice," John Bedbrook, vice president of DNA Plant Technology Corp., points out, "and for all intents and purposes is ice, a pre-made crystal." He finds it intriguing that "probably the highest quality agent for nucleating water should be a biological organism."

Commercial companies now sell a strain of bacteria that is closely related to *syringae* but doesn't have its ice-nucleating protein, and farmers spray it on their pear and apple trees in early spring, "just as the buds are bursting," Bedbrook says. The little ringers consume most of the available nutrients and keep the *syringae* from settling in and triggering ice formation.

Another strategy for keeping plants from freezing is hair of the dog: spraying ordinary water on them and letting the water freeze. In the 1980s, there were five major freezes in north-central Florida, during which almost a quarter of the citrus trees there died. During the 1983 freeze, Nick Faryna, a grower of tangerines, tangelos, clementines, oranges, and sunbursts in Umatilla, lost all his young trees and had to buckhorn (cut back scaffold limbs on) his mature ones, some of them survivors of freezes back in the 1890s. During the 1985 freeze, when temperatures dropped into the low teens and stayed there for over 15 hours, he lost even the buckhorned trees.

"Dead to the ground," he says. "We got killed out 100 percent. You work your whole life and maybe your parents' and grandparents' lives and one morning you wake up and it's all gone." Citrus isn't like cabbage or carrots, he explains, "which if you lose them you can replant and get going again in six months. Citrus growers may have to wait five to seven years to get back a commercial crop. For us, 28 degrees can be a death sentence."

The 1989 freeze lasted 48 hours—"unusual in Florida, it's usually a one-day event"—but the trees Faryna planted after the 1985 freeze did not die. This time he watered them. Neighbors had tried watering their trees during a freeze in 1962, the thinking being in those days that a blanket of ice would insulate the trees from cold air, but the weight of the ice destroyed the trees. "They were crushed under tons and tons of ice," Farinya says. He figures each gallon of sprayed water weighs 7½ pounds when it turns to ice. His neighbors had sprayed entire trees from above, but he sprayed only the trunk and the lower, scaffold limbs. Spraying low, he explains, keeps the bud union, which is essential to the plant's survival and grows only a few inches off the ground, from freezing. It also builds what he calls "ice bridges," rods of ice that extend "like stalactites" from the limbs to the ground and help support the ice burden. During the freeze of 1989, "the interior of the tree became one big icicle."

Insulation probably played some role in saving Faryna's citrus trees, but it didn't play the main one. "Ice is *not* a wool jacket," Larry Parsons, professor of horticulture at the University of Florida, says emphatically. The bud unions were warmed by the latent heat given off when the water changed into ice and by the high temperature of the water Faryna used for spraying. So long as he kept spraying, the trees' surface temperature stayed close to 32°F even though the air temperature was in the low 20s and high teens. As for the fruit, Faryna was able to save some—if it was encased in ice.

Nowadays almost half of the 850,000 acres of citrus groves in Florida are watered by microsprinklers during freezes, according to Parsons. Faryna sprays only if there's a "doomsday forecast of a killer freeze," he says, since he can't afford to spray every night that frost is predicted, which could be 20 nights a year. "It's a last-ditch, Hail Mary effort to save the grove." Prayer figures in because sometimes the watering strategy backfires. "If there's a wind, you get evaporative cooling," Faryna explains, "in which case you could end up worse than if you did nothing. Every gallon of water that evaporates loses *seven times* the energy that's gained from freezing!"

> Ice plant or houseleek: . . . protect[s] the house from conflagration and LIGHTNING
>
> —*Oxford Dictionary of Plant Lore*

Another way of keeping frost from killing marginally hardy plants is to cover them. *The Harrowsmith Northern Gardener* advises its readers to keep on hand a supply of "portable" crop covers—bedsheets, garbage bags, newspapers, mulch—since frost "could come at almost any time" to their plots. "A single layer of plastic will protect plants from very light frost, while a double layer will keep plants from harm to about 27° F." Below 27° F, "cover the plastic with blankets." Poet Donald Hall used to pull holey blankets out of the closet whenever he heard that the first deep cold of the winter was on its way, although he confessed that "if the zucchini expire we will feel only gratitude." French market gardeners go so far as to put *cloches,* or glass bell jars, over individual plants when there's a forecast of *gelée blanche.* The covers are not meant to keep plants growing into winter but only into the milder days that follow a night or two of frost. Growers of ornamental ferns erect "ice roofs" over them, according to Faryna. They build a little structure over the plants, lay an open-weave cloth across that, and sprinkle the cloth with water, which hardens into a thin layer of ice. "They make themselves a little igloo."

Dirt and snow can provide cover of a sort for some dormant plants. "I know a guy in Minnesota who buries his long-stemmed tree roses," Charles Guy says. "Every fall he lays the stems down on the ground—they can be six or seven feet long—and covers them up with dirt. They freeze in the soil, which might get as cold as 20° F under a thick snow cover, but that's a lot warmer than the air in Minnesota!"

Gardening much farther north than that, people run out of options. "For most Northerners, gardening is an unpredictable adventure," *Northern Gardener* concludes. "An Inuvik gardener lamented that she put her seedlings outdoors in early July, and they were promptly snowed upon, hailed upon and then rolled upon by dogs ecstatic to find a patch of cool soft soil. On Mackenzie River permafrost, some bounteous crops have been produced, but further east or north only one crop is really reliable—bean sprouts, grown indoors."

Perhaps the most charming example of a wild plant controlling its freezing destiny—making its own icy tarpaulin, not to mention its own water-preserving mucilaginous goo—is the *Lobelia keniensis* in East Africa, which grows on the slopes of Mount Kenya. It has overlapping leaves in its rosette, which gets as big as a basketball when it's ready to reproduce, and a spike of a flower with a structure like a Chinese finger puzzle. It can live to be over 100 years old and may flower for the first time when it is 60.

At the altitude where the lobelia grows, 11,000 to 14,000 feet above sea level, temperatures fall below freezing almost every night of the year, but at the latitude where it grows, on the equator, temperatures climb high during the day. A half hour before sunup the air could be 25°F and an hour after sunup 100°F. With such temperature swings, there's no question of the lobelia getting cold-acclimated. So what's a giant of a plant to do to keep from freezing overnight? Truman Young of the University of California at Davis, who spent a decade studying the plant, found out.

Between its overlapping leaf bases, the lobelia traps not only rainwater but also water it squeezes out of its own leaves—"kind of a rare and difficult thing for plants to do," Young says—up to three quarts of water a day. After sunset, when temperatures fall precipitously, the surface of the trapped water freezes. Under the ice, the bulk of the water stays liquid, for the same reasons that water does in lakes and oceans under *their* lids of ice: because the ice partially insulates the water from the cold air, and the latent heat released during crystallization warms it. Surrounded thus by liquid water and close-packed leaves, the heart of the lobelia, the precious leaf bud, does not freeze.

One refinement is that the coldest water, the layer right under the ice, doesn't sink as it ordinarily would but stays high and gradually turns to ice itself while the warmer water stays low, near the precious leaf bud. What keeps the water from turning over and bringing the iciest water down is pectin, a water-soluble substance used in making jellies, which the lobelia itself secretes.

The plant is appreciated for more than its ingenuity. As mountain climbers trudge up and down the slopes of Mount Kenya, their boots sometimes jostle a lobelia beside the path, sending water splashing to the ground—and ice cubes popping out of the rosette. The ice cubes, formed in the crannies between tightly packed leaves, are shaped like crescents and about an inch long. Young finds them "elegant." He also finds them refreshing. "I suck on them," he says. Because of ice's habit of excluding solutes, they contain no pectin. Despite the fact that they're available only at breakfast time, before the sun has ignited the mountain air, the ice cubes have earned the plant the name "gin-and-tonic lobelia."

CHAPTER EIGHTEEN

ANIMALS I

Mother Nature has a remarkable power of producing life everywhere—even this ice is a fruitful soil for her.

—Fridtjof Nansen, *Farthest North*

SOMEONE ONCE CALLED ME a "pagophile," and quite correctly, too. A pagophile is a lover of ice (*pago-*, from the Greek for "ice, frost"—it was after all a Greek, Pytheas of Massillia, who made the first recorded sighting of sea ice, while sailing north of England in 350 B.C.; "lungs of the sea," he called it—and *-phile,* from the Greek for "loving"). You could say that an Inuit hunter who greets the freezing-over of the bay beside his village with cries of pleasure is a pagophile, by virtue of his need, and Wayne Gretzky was one, by virtue of his craft. But the way the word is used these days, by some biologists at least, it applies to marine mammals that spend a major part of their lives on and around sea ice. In the Arctic, pagophiles include several ice seals, a few ice whales, and one ice bear.

The bowhead is one of the ice whales. (Others are the beluga and the narwhal.) The bowhead seems to have, if not love, then a liking for ice, or at least a serious ongoing relationship with it. According to one contributor to the publication *The Bowhead Whale,* the bowhead "lives year round within sight of ice and can easily swim beneath it," which makes it, in his judgment, "an outlier on the curve of mammalian evolution." An environment dominated by ice would indeed seem to be marginal habitat for a gigantic, warm-blooded, air-breathing mammal, and that's probably the main reason why the bowhead moved into it and stayed in it. There were no other large resident whales in that apparently hostile niche, no giant competitors for the tidbits of food it strains from the water with strings hanging from the roof of its mouth. Its only natural predator, the killer whale, can't follow the bowhead very far under ice, in part perhaps because the killer has a large dorsal fin, which could scrape on the underside of the ice, while the bowhead, in apparent adaptation to its partially roofed icy environment, does not.

Along the coast of northwest Alaska, bowheads migrate with the ice, heading north in spring as the sea ice retreats and south in autumn as it advances. They spend winter in the Bering Sea and summer in the Beaufort Sea and swim between the two through the Chukchi Sea. While going through the Chukchi, they swim more or less parallel with the coast, through openings in the ice, some of the openings miles wide, others mere cracks. "It's not like some *trout stream* running up the coast," Craig George, biologist at the North Slope Borough Department of Wildlife Management, says of the bowhead routes. "It's just a bunch of busted-up ice."

Waiting at the edge of the shorefast ice for the bowheads to swim by each spring are other creatures with an attraction to ice: human beings. Standing atop ice ridges in their fur-trimmed snow-white parkas, they scan the sea for low, dark, rounded mounds—whale backs—and mists shooting high off the water—whale blows. "If the light is right," George says, "you can see the blows several miles away, like light bulbs." For a couple of thousand years at least, Eskimos in Alaska have been hunting bowheads from the floe edge, the boundary where the landfast ice meets the pack ice. Recently, the journal of the Alaska North Slope Borough, *Uiñiq* (*The Open Lead*), chronicled a modern bowhead hunt by following the actions of a single crew from Hollywood, a suburb of Barrow, 300 miles north of the Arctic Circle. The crew had five men in it, including the captain, Jonathan Aiken (Kunuk).

Written by *Uiñiq* editor Bill Hess, the account begins in early April, when bowheads are spotted moving up the coast toward Barrow "in heavy numbers." Kunuk's men get busy "cutting trail" to the camp they plan to set up at the floe edge. The floe edge is five miles from the land's edge, and much of the ice in those five miles is rough and jumbled, pushed up into piles by the pressure of the pack. Using picks, axes, and shovels, the men chop, dig, and crush ice to make roads for their snowmobiles and sledges. It's "hard labor." Word comes that a man on another crew has broken his collarbone hacking at the ice. While Kunuk's crew is cutting trail, so are dozens of other crews. They'll share roads for part of the way out, then branch to their separate campsites, which will be ranged along 25 miles of floe edge.

It takes Kunuk's men two weeks to hack out the trail, after which they head for their campsite, on snowmobiles, pulling sledges behind them. They chose this campsite because it has a high ice ridge and therefore probably a deep keel reaching the seabed, which would make the ice less likely than ice elsewhere to be torn loose and carried out to sea; it has a large stretch of flat ice back of the ridge on which a tent can be pitched; and it is on a projecting point of ice, and bowheads are thought to migrate point to point.

Once they reach the floe edge, the men do more chopping, to make a ramp for their boat, a slope down which it can be slipped into the water, noiselessly, without a plop the whales can hear. "Noise drives the whales away," the Barrow Whaling Captains Association's code of conduct informs visitors to the hunt. The crews are advised to sweep and scrape snow off the ice between their tent and their boat to minimize crunching sounds, since walking on snow is noisier than walking on ice.

After eight hours of working on the ramp, the men lay an umiak, a sealskin boat, at the bottom of it, right at the water's edge, with a block of ice under the stern so the boat won't freeze to the ramp its entire length. Next to the ramp, in a line along the ice edge, they place rectangular blocks of ice and snow, like stones along a crenellated wall, so passing whales won't see what's stowed behind it: harpoon guns, floats, long-handled lances, coiled line—the tools for killing and landing them.

"Hunting means waiting," *Uiñiq* continues. The hunters sleep with their clothes on, sometimes with their gloves on, on caribou skins or blankets they can quickly leap off of. There may be only a minute and a half—which is how long it takes for a whale to blow three times—between the time the lookout spots a bowhead to the time they get their last chance to hit it. One day, the lookout spots several bowheads in a lead, and the men paddle out in the umiak toward one of them. When they get to within striking distance of the whale, it dives. They wait for it to come up to breathe. "Finally, it is obvious that it is not going to." In the morning the northeast wind has turned into a northwest one. "Ice from the pack grows from slivers and bumps on the horizon into great floes and bergs bearing down on the campsite." The incoming ice could close the lead the whales have been migrating through. "At worst, large bergs could come crashing through campsites, breaking and crushing ice, and throwing anything and anyone standing in the wrong place into the sea."

Reluctantly, the men pack their gear onto sledges and move back to "safe ice," or ice grounded to the seabed. They pass the time waiting for the wind to change by playing pinochle in their tent, drinking coffee and Pepsi, snacking on "Eskimo doughnuts," smoking, and eating frozen-fish *quaq* and caribou that their wives bring out. "Still, everyone longs for that day when the water opens back up, the whales return, and there is some oily, steaming, *uunaalik* [boiled fresh whale skin with a layer of blubber attached, a delicacy said to taste like hazelnuts or hard-boiled eggs] to eat. Day after day," though, the wind blows from the northwest, the west, the southwest. The lead stays closed with "the pack ice and the shorefast ice welded together into one, massive, solid, and very rugged sheet." The crew learns that a man at a nearby camp has killed a polar bear that wandered in looking for food.

Finally, the wind starts blowing from the northeast and opens a new lead, which is, however, several miles farther out than the last lead, with "extremely rough rubble" filling those miles. Kunuk's men must chop a new road through the ice to a new camp on the new floe edge. That takes another 16 hours, then the wind shifts to the southwest, pushes the pack ice toward shore, and closes the new lead. "Tired, and discouraged," they head to safe ice again. Meanwhile, a "good long snowmachine drive to the Southwest," where ice conditions are apparently better, a crew has struck a whale, and two men in Kunuk's crew drive over to help bring it in. By the time they get back, a new lead has opened up, and they and the others spend four days chopping ice for 16 hours a day to cut a trail out to it, then two days later the wind shifts again, and again they have to move back to safe ice.

It's now the end of May. The Hollywood crew has been on the ice for over a month. Few bowheads are coming through, and most of the other hunters have left the ice. The ice edge is growing slushy and weak. Word comes that another crew has struck a whale and needs help landing it, so they all drive over. This whale is "fat" and almost sixty feet long. Facing it, Captain Joash "Ivik" Tukle "offers a prayer of thanks," then "lifts up his hands and shouts for joy." Three pairs of holes are chopped through the ice, one in back of the other leading away from the edge, and ropes have been strung between the holes and slits in the whale's skin, in a sort of block-and-tackle arrangement. Men and women from other crews and from town stand on the ice and "pull and pull" on the ropes, trying to get the whale out of the water and up onto the ice, but every time they get it up, the ice breaks under the whale and it slips back into the water.

All in all, it takes several dozen people 24 hours to work the whale up onto the platform of ice and another 30 hours to butcher it there. Leaving only some bones with bits of tissue clinging to them on the ice, where polar bears can reach them, they load the meat, blubber, organs, etc., on sledges, then haul the load over softening ice trails back to Barrow, where it is divided among the captain, his crew, and every person that helped bring in the whale. For Kunuk and his men, "it was a very tough spring." They didn't toss a single harpoon. However, another crew had an even tougher time of it. The ice they were on broke loose and took them out to sea. "One problem when you're on a big piece of ice," Geoff Carroll, wildlife biologist for the state of Alaska, explains, "is that you don't feel anything when it moves. The ice is the earth to you. It turns, you turn." Some hunters set a compass on a flat patch of ice and check it from time to time to see if the needle moves; "if it moves *at all,*" Carroll says, "they get out of there. Even a small pivot, and the ice is gone." In the old days, floating out to sea might have meant the end of the hunters, but now a helicopter picks them up, along with their snowmobiles.

The International Whaling Commission declared a ban on hunting bowheads in 1976, when the number remaining in the western Arctic was estimated to be 1,500 at most, but the Eskimos protested that the estimate was too low. They contended that bowheads migrate for long distances under the ice, where they can't be seen to be counted. Scientists were assuming that if leads freeze over or are forced shut by drifting pack ice, the whales will quit migrating on that route. Over the next decade, observers (including Craig George and Geoff Carroll), in partnership with the Eskimos, counted every passing bowhead they could see from ice ridges north of Point Barrow, while engineers listened for whale calls on hydrophones they dangled over the ice edge. During one four-day period, the counters-by-eye recorded 3 bowheads swimming through a mostly refrozen lead while the counters-by-ear registered *339* whales swimming through (their calls, according to one engineer, a "constant bantering between group members, much like a flock of geese in the spring"). The next year, during the same four days that the observers on the ice ridge counted 117 passing bowheads, the engineers at the ice edge counted 665.

Bowheads do migrate under ice, then. But how do the whales manage to breathe beneath the ice? Sea ice off Point Barrow was six inches to a foot thick, enough to support "the combined weight of three men and a heavily loaded sled and snow machine," George and Carroll pointed out. As it happens, when out on sea ice the two of them had come across many "ice hummocks," raised areas about three feet wide with "ragged central openings," and they concluded that the bowheads had made them as openings to breathe through and had used their heads to do it (as elder Eskimos had observed). When bowheads were feeding in open water, the scientists had noticed that the skin around many of the whales' blowholes was "roughened and abraded" and surmised that the wounds were "ice-inflicted." They figured the whales were placing their blowholes up against the underside of the ice and, using "fluke and flipper strokes and/or their own buoyant force," pushing upward until the ice gave way. Connective tissue around the blowholes can be up to nine inches thick, they noted, and appears "to be specialized for absorbing the impact of breaking ice."

Craig George tells how he was standing one day near an ice hummock when suddenly a whale's lower jaw shot through it and protruded three feet into the air. The whale was presumably attracted to the hummock by the skylight effect, more light coming through it than the ice around it. (Why abrade your blowhole more than you have to?)

After years of counting by land, sea, and airplane, the census takers concluded that at least twice as many bowheads were migrating up the coast as had been thought, and the Eskimo hunt was allowed to resume, with quotas.

For about three centuries, commercial whalers from Europe and the United States took bowheads from ships, in the Atlantic as well as the Pacific. They weren't interested in the meat but in the long, horny strings (baleen) in the whales' mouths, mostly for use as corset stays. The captains had to know a lot about ice, too, particularly since as they killed more and more whales they were forced to take their ships farther and farther into the ice fields to find them. In the Pacific, whalers hunted not only from ships but along the floe edge, once they saw how well the Eskimos were doing at it. One such whaler, Charles D. Brower, tells the harrowing story in his book *Fifty Years Below Zero* of being marooned on drifting sea ice with 31 other onshore whalers by a captain who ordered them out onto the ice to scout a way ahead, then abandoned them 100 miles from land. They had no boat, almost no food, thin clothes ("mostly fawnskin"), and only a couple of pairs of spare boots among them. "Ice stretched in every direction," Brower wrote, "—miles of it. The whole world had turned into drifting ice."

For nearly two weeks, the men walked across the sea ice, hoping to sight a ship. When on the twelfth day those who were still alive—riding on a cake of ice—spotted a whaling ship in the distance, they were too weak to signal it. In the crow's nest of the ship there happened to be a Siberian Eskimo who was gazing through binoculars at what he thought was a herd of walrus on an ice floe. He remarked on it to a ship's officer, who "turned curious glasses on where we were huddled together on the ice," Brower wrote. By the time the ship picked up the 16 survivors, half were unconscious (and eventually four had to have their frostbitten feet amputated). "Twelve days?" the captain exclaimed when he learned about their ordeal on the ice. "My God!"

The reason the officer turned "curious glasses" on the ice floe with "walruses" on it was that in the region where the ship found itself, in the Beaufort Sea more than 100 miles east of Point Barrow, there weren't many walruses around. In summer, walruses tend to congregate at the southern edge of the Chukchi pack, where the ice is more broken up than ice deeper in the pack, and walruses like broken-up ice. The gaps between floes allow them to reach the sea floor to feed, while the floes themselves give them a place to rest, socialize, quarrel, and grunt. "For walruses," Francis Fay, professor of marine science at the University of Alaska at Fairbanks and a longtime observer of walruses, explained, "ice is mainly a place to haul out on, a convenient platform." The ice, though, must be fairly thick, since an adult walrus can weigh a ton or a ton and a half, and ten or 20 walruses can be crammed onto a single floe.

To climb onto floes, walruses sometimes use their long tusks, stabbing the ice with the points and leveraging themselves up. In *Farthest North,* Fridtjof Nansen described a male walrus's efforts to get up onto a piece of ice where two other males were already ensconced:

"He struck his great tusks into the edge of the ice, while he lay breathing hard, just like an exhausted swimmer. Then he raised himself high up on his tusks, and looked across the ice towards the others lying there, and then dived down again." He did the same thing with his tusks again, and this time a huge old bull "suddenly awoke to life . . . grunted menacingly, and moved about restlessly," causing the interloper to bow "his head respectfully down to the ice." Soon, however, the walrus had "pulled himself cautiously up on to the floe, so as to get a hold with his fore-paddle," and this time the old bull, "thoroughly roused," bellowed and "floundered up to the new-comer in order to dig his enormous tusks into his back." Although the new walrus was as big as the old bull and had tusks just as big, he "bowed humbly, and laid his head down upon the ice just like a slave before his sultan." Several times he went through this craven act of supplication before the other bulls would allow him to share their floe. "And it is in this friendly manner," Nansen declared, "that walruses receive their guests."

Another characteristic of ice that recommends it to many walruses: it moves. If the ice floe they are on becomes grounded, walruses generally get off it. They feed mostly on mollusks and other invertebrates on the seabed, and if they stay in one place, over the same section of bed, they soon deplete the food supply. Drifting floes of ice carry them to ever-fresh feeding grounds.

Where ice isn't broken up enough for the walruses, they sometimes break it with their heads. "There are stories of walruses rising through the ice as if it isn't even there," Fay declared. He saw them poking their heads through ice up to eight inches thick; "we could walk on it very comfortably." Nansen described the scene of a walrus butting through ice: "Every now and then we would hear some violent blows on the ice from beneath, two or three in succession, and then a great head would burst up with a crash through the ice. . . . A walrus's head," he felt inclined to explain, "is not a beautiful object as it appears above the ice. With its huge tusks, its coarse whisker bristles, and clumsy shape, there is something wild and goblin-like about it."

Walruses do their courting in midwinter, on and around ice. The females assemble on top of the ice, the males in the water, along the edge. Large bulls confront other large bulls, pointing their tusks in various directions, until one of them backs off. If neither backs off, they fight. The victorious males display for the females, "clacking and clanking," Fay said. "We don't know how they do that. We thought at first it was their teeth hitting together, but it isn't." To all these goings-on, the females pay "scant attention, yet they must be receptive," Fay remarked, "because eventually they leave the ice and join the males under water and we can guess what they are doing." In spring, males and females go their separate ways, with most of the males staying behind in the Bering Sea and the females heading north with the receding pack, swimming and hitching rides on floes, and giving birth in the Chukchi Sea. They haul out with their calves onto small ice floes—small to discourage polar bears. Polar bears like to approach their prey from the ice, but if the ice is a small floe, they are forced to approach it from the water. When they're in the water, they find it hard to defend against a mother walrus's tusks, as well as to climb up onto ice floes already crowded with walruses.

"Polar bears are in command when they're are on the ice," Fay stated. "Walruses are in command when polar bears are in the water." The bears go after the young walruses, not the adults. "[Adult] walruses are essentially tanks, impenetrable brutes," Fay pointed out. "The bears can claw and bite them with practically zero impact."

Let me recall the great white
Polar bear, . . .
It threw me down
Again and again,
Then breathless departed
And lay down to rest,
Hid by a mound on a floe.
Heedless it was, and unknowing
That I was to be its fate.
Deluding itself
That he alone was a male.

—Orpingalik, *"My Breath"*

Except when humans with harpoons or guns are around, polar bears are top dogs in the Arctic. They are the largest nonaquatic carnivores in the world, the dominant Arctic predators. Some polar bears spend occasional periods on land, but many go their whole lives without setting foot on it. Their home is the frozen ocean. They use ice as a platform for hunting seals, their favorite food. They use ice as a base on which to walk the long distances they may need to in order to find enough seals to survive on. They sleep on ice, play on ice, fight on ice, mate on ice, and often den on ice. According to Ian Stirling, research scientist for the Canadian Wildlife Service, who has studied polar bears in the wild for over 30 years, they are "entirely adapted for life on the sea ice, an integral part of the ecosystem. . . . They belong."

Anyone who has seen a polar bear, or photos of a polar bear, lounging on ice as if it were a cozy substance, sleeping with a block of ice as a pillow under its head, draped over a hummock of ice on a warmish day (looking "more like a jellyfish than the ultimate arctic carnivore," Stirling says), curled into a ball on the ice and playing with its toes, or leaning casually against the pinnacle of an iceberg, has gotten the message how truly at home on ice a polar bear is.

How a land bear turned into an ice bear isn't known for sure, but what probably happened was that 100,000 to 200,000 years ago, during the Pleistocene epoch, in a period when Arctic waters were covered with ice but the land next to them was not, a brown (grizzly) bear was wandering around looking for food when it came to an ice-bound seacoast and kept on going, out onto the ice, and there found and ate a seal, decided it was rich and satisfying fare, and stayed on for more. (In recent times people have seen brown bears on sea ice feeding on seal remains, Stirling notes; one time he encountered a large male brown bear walking on sea ice more than 35 miles from shore.)

At first, the brown bears probably scavenged on dead seals, Stirling wrote in his book *Polar Bears,* then "discovered they could catch live seals at cracks in the ice, simply by sitting still and waiting for them to come up to breathe. Bears learn quickly," he added. "Suddenly there was a new way to make a living . . . an unoccupied niche." To fit into the niche, the bears evolved quickly. Their brown coats turned white, as lighter colored bears, blending in better with ice and snow than darker ones, would have caught more seals and so had more offspring.* The bears' fat layers thickened against the cold, and their claws grew shorter and stockier, less likely to snap off as they ran across ice and climbed steep ice ridges. Their heads and necks became longer and narrower, better for poking down tight seal holes in the ice. Today, polar bears can still breed with brown bears, producing fertile offspring which are bluish brown or yellowish white as adults, but Stirling predicts that after another 50,000 or 100,000 years of separate living, they won't be able to.

The bears' behavior, too, changed when they moved onto ice. Polar bears don't hibernate in winter, as land bears do when terrestrial food runs short; they can catch seals all year long, if they hang around ice. Polar bears don't defend territories, either, as do land bears, which rely on food sources—berry bushes, spawning salmon—that show up in more or less the same places year after year. Polar bears go where the seals are, and where the seals are can change radically from one year to the next, as the sea ice changes.

A polar bear walks and walks, but it doesn't walk "aimlessly," Stirling points out. People used to think a polar bear was "part of one circumpolar population nomadically roaming at will over the arctic wastes," he explains, ". . . a cosmopolitan arctic citizen, randomly visiting any or all of the polar countries through its lifetime" (the relevant "countries" being Canada, the former Soviet Union, Svalbard, Greenland, and Alaska). People now know that although a polar bear often travels far, it doesn't travel all over. It usually has a home range, which it sticks to even where there aren't natural barriers to define the range. Scientists who

*Polar bear hairs are actually transparent, but the walls inside the hollow hairs are rough and scatter light, making the hairs appear white.

tagged bears found to their surprise that the bears came back, year after year, to the areas where they were tagged and at the season in which they were tagged, even after long sojourns on constantly moving and apparently featureless sea ice.

"They know exactly where they want to go," Stirling says. Nobody knows how they know how to get there, though. For instance, in the Beaufort Sea in spring, polar bears looking for seals are confronted with sheets of ice that extend outward from the coast for hundreds of miles, with only occasional leads in them and no landmarks past a few miles of the shoreline, or at least no landmarks "readily apparent to humans," Stirling notes. In summer, when ice in the southern part of the Beaufort melts, the bears must walk hundreds of miles north to stay with the ice, then in autumn when the sea refreezes walk hundreds of miles back to the southern part, always taking into account not only their own movements over large distances but the movements of the ice as it's propelled by wind and currents—and *still* they show up within a few miles of the spot where they were tagged several years earlier. Not only that, "bears must continuously compensate for the Beaufort Gyre in order to stay in the same place," Stirling points out, and, he adds respectfully, "somehow they do it."

To polar bears, some ice is better than other ice. Their favorite is first-year ice formed near coastlines where wind and currents regularly crack it open, and where it's thin enough for seals to break through. Most of the world's polar bears, of which there are roughly 25,000 (they're hard to count because they're spread thinly over vast areas and when viewed from airplanes appear white on white), are to be found on ice near this offshore system of leads, commonly called "the Arctic ring of life."

Although polar bears will consume a whale carcass if they find one on pack ice, and munch seaweed they pull up through holes they've made in the ice, and dine on young walruses they grab off of ice pans, mostly they eat seals, ringed and bearded. They hunt the seals in two main ways, by stalking and still-hunting.

• ***Stalking.*** In a summer stalk, a bear spots a ringed seal sunbathing on top of the ice a few hundred yards away and immediately freezes, "sometimes in mid-step," according to Stirling. It stays frozen for several minutes while it works out a plan for approaching the seal. Then it starts walking toward the seal with its head low, slowly, steadily, quietly. Any noise from its footfalls is minimized by the bear's shuffling gait and the fur around its footpads. Every 30 seconds or so, the seal raises its head to check for predators, but it may not see this white one approaching.

Some people swear that stalking polar bears cover their black noses with their white paws, or with their tongues, or with a piece of ice they hold in their paws, so that seals won't notice the spot of black against the general whiteness and be tipped off, but Stirling doubts it. In the thousands of hours that he and colleagues, including experienced Inuit hunters, have spent watching polar bears, none has reported seeing a polar bear trying to conceal its dark nose.

When the polar bear stalker is within 50 to 100 feet of the seal, it "suddenly charges, at quite incredible speed," Stirling reports, "while the horrified ringed seal attempts to get into its hole." The bear grabs the (justifiably) horrified seal with its paws and bites its head repeatedly and carries it away from the hole to gulp down the skin and blubber.

One variation on the summer stalk, according to Stirling, is "the aquatic stalk," which itself has two variations. In one, the polar bear swims under water between seal holes, breathing surreptitiously at each one with only the tip of its nose protruding above the waterline, until it gets to the one nearest the seal and submerges. After "an eternity of suspense-filled seconds" (for the watching scientist), "the water in front of the seal explodes as the bear suddenly claws its way onto the ice after its prey." Despite the suddenness, the prey usually gets away. The seal can dive through its hole in the ice and swim off faster than the bear can swim after it.

In the second variation of the aquatic stalk, a polar bear lies in a trough of summer meltwater and pushes itself through it, holding its paws out to the side instead of underneath, thereby making its approaching profile as low as possible. Stirling once saw a "hilarious" stylistic riff on this variation: an adult female "lay with the front half of her body flat in the water, but kept her rear legs straight so that her whole hind quarters towered above the ice. From a distance, it looked like the seal was being stalked by an iceberg! Curiously, this slow-moving white object did not appear to unduly alarm the seals, even when it got quite close."

• ***Still-hunting.*** In winter, when seals rarely leave the water to lie in the colder air, polar bears tend to still-hunt. A bear positions itself beside a breathing hole or lead or crack in the ice that it has determined, by sniffing, the seal used recently and therefore is likely to be using again. (Carl Kippi, a young hunter in Barrow, reports that polar bears have "a shopping list" of seal holes they check out regularly, a route they follow over the ice, sniffing each hole to find out if it is in use.) Once a polar bear finds an active hole, it lies downwind or crosswind of it with its belly and chest on the ice and its chin near the edge of the hole. Then it waits. While waiting, it lies quite still; sounds transmit well through snow-covered ice. It may lie in wait for hours, or all day, appearing to be asleep, sometimes actually sleeping.

At the sound of breath or bubbles in the hole, however, the stillness of still-hunting is shattered: with one movement the bear lunges and bites the seal and "flips it out on the ice," where it wriggles, according to Stirling, "much like a trout that has just been pulled out of a stream." A bear might sit during a still-hunt or even stand, with its front and hind feet closer to each other than normal, a stance that allows it to shift its weight onto its hind feet at the critical moment without making a warning sound. Then when the bear hears or smells a seal, Stirling writes, "it slowly stands up on its hind legs, raises its front limbs, and crashes down with its front paws" on the snow roof over the breathing hole, attempting to break the roof and crush the seal's skull all at one go.

Inuit hunters were probably inspired to do their own still-hunting by watching polar bears do it. Stirling once saw a man with the splendid name of Ipeelie Inookie approach a snowdrift on sea ice that he suspected of having a seal lair in it, "slowly like a stalking polar bear," then take "a short run and a jump. He broke through the roof of the lair with his feet, and then dove in headfirst to try and seize the pup. Then he would dangle the pup in the water on a rope in case the mother came back to check on it. Sometimes she did."

One Inuit hunter told Richard Nelson, author of *Hunters of the Northern Ice,* that he and others had seen a bear pick up a chunk or two of ice with both paws and, while standing on its hind legs, throw the ice at adult walruses, to move them away from a baby walrus they were protecting. Other people have reported seeing bears use blocks of ice to smash through icy crusts over breathing holes; to push ice or snow blocks ahead of them as blinds while they crawl toward seal holes; and to build walls of snow to hide behind while hunting. Stirling's guess is that, although polar bears are intelligent and he wouldn't put tool use beyond them, the blocks of ice sometimes found around hunting sites are more likely to have been broken off by bears in moments of "high frustration." One researcher who was watching a polar bear that had just missed catching a bearded seal reported that the bear "leaped up onto ice floes and in his fury began to toss lumps of ice about."

Usually solitary, male and female polar bears get together on the ice in springtime to mate. A male will follow the footsteps of a breeding female over the ice even when her prints are mixed with dozens of others (the female likely has cells in the soles of her feet that lay scent on the ice). He "unwaveringly" follows her track over 60 miles of ice or more. Six months after he catches up with her, she gives birth in a snow den, either on ice or on land, or occasionally even *in* land, with permafrost for a wall.

Although adult bears can weigh up to 1,500 pounds, they walk on ice so thin that humans would fall through it. Their feet are huge and "oar-like" (Stirling's word), which helps them swim, but also snowshoe-like, which helps them walk on ice by spreading their weight. On thin ice their gait is usually a toed-in shuffle that keeps the load on all four feet. Charles Brower noted that when the ice "made early" near Barrow in autumn, and great sheets of it came in from the north and scraped along the beach, he'd see half a dozen polar bears on the ice "long before it would support the weight of a man . . . the condition of which never bothered them at all. On thin ice, they spread themselves out so as to cover a large space; if very thin, they lay flat on their bellies, hauling themselves along with their claws."

Polar bears must have ice to catch seals. Stirling says there's been only one report of a polar bear catching a seal in open water, and even then the seal may have surfaced near the swimming bear because it mistook it for a piece of ice. Each summer, when the sea ice along the west coast of Hudson Bay in Canada melts, the bears that had been hunting on it go ashore, hang around, and wait for the bay to freeze again. This takes about four months, during which time they live on their stored fat. The place where the bay freezes earliest is around Cape Churchill, and in November hundreds of bears can be found in the vicinity, waiting for the ice to form.

These Hudson Bay bears, along with ones in nearby James Bay, make up the southernmost population of polar bears in the world (James Bay is roughly the same latitude as London where, incidentally, polar bears roamed at the end of the last ice age, as evidenced by the recent finding of an 11,000-year-old skull near Kew Gardens), and they may be among the first creatures to suffer from global warming. Because of higher springtime air temperatures in the region, the ice in Hudson Bay breaks up two weeks earlier now than it did a mere two decades ago, which means that female polar bears coming ashore in spring are in poorer condition after shorter stays on the ice spent hunting and eating seals. Birth rates have dropped accordingly, Stirling and colleagues found. "How fat you are translates into how many cubs you have," he says. Should the warming affect freezeup as well as breakup, delaying it by a week or two, the females waiting onshore will be even lighter when they return to the ice in the fall than they are now. Such a large drop in body weight would mean a drop in polar bear population: fewer cubs produced, fewer nursed successfully.

If someday global warming should cause sea ice in the Arctic to disappear even seasonally, polar bears, kings of the ice kingdom, creatures who left the land thousands of years ago and adapted marvelously for a life on the frozen ocean, would become extinct.

WALRUS SPORTING.

CHAPTER NINETEEN

ANIMALS II

I shall ne'er forget that great blubber-beast,
A fjord seal,
I killed from the sea ice
Early, long before dawn,
While my companions at home
Still lay like the dead,
Faint from failure and hunger,
Sleeping.

—Orpingalik, *"My Breath"*

IF THAT BROWN BEAR hadn't wandered out onto sea ice 100,000 to 200,000 years ago and been rewarded with a great meal, most Arctic seals would be very different creatures from what they are today. To understand how different they might be, you need only compare the life and times of the ringed seal of the Arctic—chief prey of the polar bear—with the Weddell of the Antarctic—prey to nothing above ice once humans stopped using some of them for dog meat.

• Weddell seals sleep soundly on the ice, sometimes for hours at a stretch, letting their heads drop as human sleepers do, often not bothering to roll over, holding a position long enough for the heat of their bodies to melt the snow and ice underneath so that when they finally vacate their spots they leave "cradles" several inches deep in the exact shape of their corpulent bodies. Ringed seals, on the other hand, take mini-mini-naps, interrupting their sleep every 30 seconds or so to scan the horizon for predators, not only polar bears but also Arctic foxes and, if they're near shore, wolves, ravens, and the occasional grizzly.

• When awake on ice, Weddell seals are imperturbable, not likely to bolt. "Fear is . . . unknown to them," Vilhjalmur Stefansson wrote, "and if you walk up to a seal and scratch him he will roll over so you can scratch him better." Ringed seals by contrast are jumpy, skittish, likely to take off at the slightest disturbance. In his book *White Bear,* geologist

Charles T. Feazel quotes "a veteran of icebreaker cruises" on the difference: "Up north, when the ship crunches through the ice, the seals slip into the water. They know that danger comes from above and that sounds transmitted through the ice signal threats. In the south, seals wait until the ship is almost on top of them before they leave the safety of the ice. In fact, I'm afraid we may have squashed a few seals who didn't get out of the way."

• Weddell seals defecate on the ice beneath them, apparently unconcerned about calling olfactive attention to themselves. Ringed seals refrain from defecating in their lairs, presumably because the smell would betray them to their predators.

• Weddell seals make a great hullabaloo on the ice, crooning, babbling, and crying. Ringed seals are "very reticent to say anything," one longtime researcher notes, thereby revealing their whereabouts. They emit "only a few very subtle barks and grunts."

• Weddell seal pups—all Antarctic seal pups—have dark coats like their parents. Ringed seal pups—most Arctic seal pups—have light coats, making them harder to spot as they lie defenseless on snow and ice. (Arctic bearded seal pups do have dark coats but with white blotches on their heads, backs, and rear flippers, which suggests to Ian Stirling that they are evolving white coats.)

• Several Weddell seals may share a breathing hole, and there may be many of those breathing holes in a small area. Ringed seals keep three or four individual breathing holes open, to have backups in case a polar bear is waiting at one, and the holes are usually more than 600 feet apart.

So, just as ringed seals helped make polar bears what they are—white-coated, long-nosed wanderers on the ice with a variety of hunting strategies—polar bears helped make ringed seals what they are—subdued, wary creatures with white-coated pups, unfouled lairs, and multiple, well-separated breathing holes.

Despite being the major bear prey—they make up over 90 percent of a typical polar bear diet—ringed seals are the most widespread and abundant of all the Arctic seals. They can be found on or near ice almost anywhere in the region, even, probably, at the North Pole. They give birth on ice, nurse their pups on ice, rest on ice (if their fitful napping can be considered restful), and molt on ice. Molting requires warmth (it would take "forever" in cold water, marine scientist Francis Fay pointed out, or at least "months"), so when spring and summer come the seals crawl out onto the ice and beaches of the Arctic and bask, nervously, in the sunlight.

Ringed seals come in two sizes, big and small. "Every Eskimo knows that," Fay said. "The range is huge. The big ones can be 260 pounds, the small ones 65 pounds." What's likely responsible for the range is ice type. Seals in shorefast ice tend to be big, those in moving pack, small. (Compared to Weddell seals, all ringed seals are small, the better to hide in snow caves.) Fay explained that since pack ice can break up at any time and "put the seals out of house and home," the seals that breed in the pack have less time to raise their young, and the young reach maturity earlier (while still small) than if they bred in fast ice. Those seals that breed in stable, shorefast ice have more time to care for their young, and the young grow a lot bigger.

Shorefast ice is in some ways the toughest ice environment of all for a mammal to live in. Seals breathe air and feed in water, and fast ice, which in winter is usually continuous and not rent open by wind or current, separates the two. To make it through winter, ringed seals must always be able to pass through the ice. At first, before the cover is complete, they breathe through leads and cracks in the ice or in gaps between floes. But as the cracks freeze over, the seals start breaking through the thin, mushy new ice from below, using their noses or the tops of their heads to do it.

Once the fast ice is about four inches thick, the seals can no longer break it with their noses or heads so they use their claws. They have claws on their front flippers especially adapted for ice; these are long, "like small bear claws," says Brendan P. Kelly, associate professor of marine biology at the University of Alaska Southeast, who's been studying ringed seals for over 20 years. The seals use the claws to keep breathing holes open once they've made them. After investigating countless breathing holes and listening to the scratching sounds ringed seals make in them, Kelly concluded that while some seals scrape the walls of the entire hole, most scrape close to the top, where a buildup of frozen splash-water narrows the opening. "You can see the rifling, like the inside of a rifle barrel," he notes. "The seal puts one flipper with its claws extended in contact with the ice, then propels itself around with the other flipper, turning its body like a corkscrew."

By midwinter, the shorefast ice could be six feet thick, and the breathing hole, subjected to daily reaming by the seal, has a characteristic form. At the surface is a dome of ice about ten inches high and an inch or so thick, produced by water sloshing out of the hole whenever a seal comes up to breathe (the seal's body is like a piston rising in the cylinder of the hole, pumping water in all directions), then freezing.

This creates "a cathedral ceiling," as Kelly calls it, into which the seal can poke its nose to breathe without being seen. At the apex of the dome is an opening the size of a half dollar through which air passes.

"I always wondered how that little hole was made," Kelly says. "It's not flared and I never saw scratch marks." He did hear, however, and several times, seals exhaling "extremely hard" as they surfaced, and he figures that what keeps the little hole in the ice open is seals blowing their warm breath on it, over and over and over.

Below the cathedral ceiling is a passageway into which the seal's whole body can fit, a lair of sorts. The Eskimos call it an *aglu.* It may be vertical or slant upward at an angle and is slightly cone-shaped, tapered toward the top. Before entering the *aglu,* a seal usually swims back and forth underneath it. "The seal doesn't charge right up," Kelly says. First it blows a train of bubbles into the passageway. "This perplexed us for the longest time," he admits. Then Kelly noticed that there were "inevitably" bits of snow and ice floating on the surface of the water and that if he himself exhaled onto the center of it, "my breath moved all that stuff to the side. It's like blowing on your soup to move the crackers toward the edges." The seal's bubble train evidently helps clear a view upward, so the seal can check for predators.

An Inuit hunter, or you or I for that matter, can tell if a seal has used an *aglu* recently, and therefore is likely to use it again, by peering down through the half-dollar-size opening in the dome. "Active holes have open water or thin ice inside, either of which looks quite black," Richard Nelson wrote in his book. "Unused holes are gray inside, showing that they are solidly frozen over."

Most female ringed seals give birth inside snow caves, which their pups may enlarge by tunneling out small side chambers for themselves. But one percent of the dens Kelly has uncovered are in ice, not snow. These are natural hollows in piled-up ice. Ice dens are safer than snow dens in one way—predators have trouble getting into them (so do scientists)—but less safe in another way. Jumbled ice piles tend to be unstable, prone to collapse. First-time mothers may be relegated to the less desirable ice dens.

To learn how the seals find their way between breathing holes, Kelly drilled several holes through the cover of a frozen pond into which he had released several seals with transmitters glued to their hair. Then he swept snow off patches of ice the same size as the holes. The seals swam toward the skylights that looked like holes as well as toward the holes.

From this, Kelly concluded they were navigating by sight. To find out if hearing played any role, he blindfolded the seals, and the seals, he observed, had trouble finding the holes. Maybe, he thought, that's because the pond is quiet as the sea is not. "Sea ice is extremely prolific with noise," he explains, "constantly creaking, groaning, popping, and hissing, from thermal stresses." He tried tapping a piece of metal against the edge of one of the holes, and the seals moved toward the tapping. "The tapping is a natural analogue of an ice-generated sound."

The seals are "hearing" differences in the structure of the ice cover while swimming beneath it, he concluded. "If we blindfold you and walk you down a hallway past the open doorway of a large, empty room, the white noise in the environment wouldn't be the same as if there were a blank wall there or some cluttered room. It's like architectural acoustics." A seal can tell if there's an ice keel or thick piece of ice somewhere and keeps a spatial memory of it, to use as a landmark when it's trying to find its way under ice to a breathing hole in the dark, or near dark, of winter.

Although ringed seals are the most dependent upon sea ice and the most adapted to it of all seals—of all *mammals*—in the Arctic, they aren't the only ice seals there. In the Bering Sea alone, there are three other species, and, according to Fay, each "occurs in greatest abundance in a special region and a special ice zone." To some mammals, he added, "the quality and quantity of ice may be as important in habitat selection as are terrain and vegetation to terrestrial mammals." By the middle of winter, the various seal species are most widely distributed in their preferred Bering Sea niches:

• ***Ribbon seals,*** which have white or yellowish white bands on chocolate-brown fur, choose to stay on thick ice floes surrounded by open water or else by very thin ice, since they don't have claws to scratch out breathing holes in solid cover.

• ***Bearded seals*** are more concerned about the ice's whereabouts than its type. Like ringed seals they have heavy claws on their foreflippers and can live in shorefast ice, but like walruses they need to be over shallow water to feed. So they are found mostly in loose, moving pack, often lying along the edge of leads (they're the ones you see in Inuit soapstone carvings, reclining like odalisques).

• ***Spotted seals*** like even more broken up ice than bearded and ribbon seals do. Without heavy claws to make breathing holes, they must breathe in open water, among small, scattered floes and brash. In spring, each family—male, female, pup—floats about on its individual ice pan, which is roughly rectangular and often has an ice hummock on it the pup can lie behind as a shield against the wind.

Harbor seals aren't ice seals—they mostly rest on land—but some associate with ice some of the time. To pup and molt, they crawl onto beaches, mudflats, and rocky reefs but also onto icebergs and sea-ice floes. All the large tidewater glaciers in the Gulf of Alaska generate icebergs, and harbor seals congregate on them. Unstable as the icebergs are, they're usually bear-free, as well as fox- and wolf-free.

Without seals, Kelly points out, Eskimos wouldn't have settled the High Arctic. "Seals provided them with the basics." " 'Meat all gone—blubber too—nothing to eat—no more light—no heat—must wait till get seal,' " the wife of a hard-luck Eskimo hunter said as she waited in her hut with two starving children for him to return with a seal. In winter and early spring, Eskimos would hunt seals at their breathing holes, by the "waiting" method. (When polar bears do it, it's called still-hunting.) Almost no Eskimos hunt this way anymore. It takes too much time, patience, and willingness to endure cold and immobility. "It's all TV stuff now," hunter Carl Kippi admits. "But I'd like to try it someday. You find the holes on cold, clear days . . ." He holds his fingertips an eighth of an inch apart. "Or you can see puffs of smoke and hear air coming from the ice, *'f-f-f-f-f.'* I look down in holes to check them, and sometimes a seal will come up in another hole and watch me do it."

In spring and summer when seals haul out on top of the ice to bask in the sun and molt and maybe get a little shut-eye, Eskimos hunted by the "crawling" method. (When polar bears do it, it's stalking.) According to Stefansson, the aim is to convince the seal that you are some "harmless animal." If you wear white clothes, hoping to blend in, the seal will think you're a bear and "dive instantly." If you pretend to be a fox, you'll be unconvincing; "they are not much larger than cats . . . and continually keep hopping around." So you "play seal." When you are within 300 yards of the real seal, you lie on your belly on the ice and wiggle "snake-fashion" toward it, or crawl "side on." When you are within 200 yards, you'll notice that the seal "becomes tense, raises his head a little higher, crawls a foot or two closer to the water to be ready to dive, and then watches you, intent and suspicious." If you manage to crawl close enough for a shot, it may have taken you as much as two hours to get there.

The hardest part of the crawling method is keeping from losing the seal once you've shot it. "The seal is lying on an incline of ice beside the hole or lead," Stefansson explained. "There are few things so slippery as wet ice, and the mere shock of instant death may start him sliding and the blood from his wound may get under him, lubricating the ice and making him slide faster." Once sliding, the seal "acquires momentum enough to take him down diagonally ten or 20 feet" into the water, in which case he'll come up diagonally through the water as well, away from the hole, "and you can't get him." As soon as Stefansson got off a shot, he'd run as fast as he could toward the seal and "slide for him like a player stealing a base in baseball."

Most hunters in the Arctic nowadays shoot seals in the water, a method known as "ice-edge sealing," which has no polar bear equivalent. Some hunters use blinds, made out of, for instance, a white bedsheet, pushing it on a sled in front of them across the ice to hide their unseal-like bodies. Kippi doesn't use blinds; he thinks people who do are just bad shots. He sneaks up on the seals by hiding behind pressure ridges. Or "I take my time."

Readers of books on the Antarctic might well ask, " 'Why all this to-do about just the right way to hunt seals?' " Stefansson wrote, and then told them. "In the Antarctic, you can secure a seal any old way. . . . Down South the seal knows no enemy. . . . The Arctic is different."

A drop of water fell on my hand.

. . . .

from hoarfrost ascended to heaven off a seal's whiskers

—Wislawa Szymborska, *"Water"*

The sky is clear and the sunlight strong. Sleek, curvaceous figures lie on the bright strand, soaking up the rays. Some are small and tan. Some are large and dark. Most are sleeping. One shifts a little, settles, dozes again. There's a mutter from somewhere, then a belch. A splash; a figure has dived into water and swum off. "Like Miami Beach," an observer says.

The splash is muted, though; the upper part of the water has two cushioning feet of slush ice in it. Along the ice edge, ranged like potted palms, are tall white bundles of splash-ice. In places where sunbathers had been lying are hollows in the precise shapes of their warm bodies, preserved by surface films of ice. Like *South* Miami Beach—very far south.

Along the coast of McMurdo Sound in Antarctica, Weddell seals gather every spring and summer in large colonies to pup, mate, and lounge on the sea ice. Their only natural enemies, killer whales and leopard seals, won't have made the long swim under ice, as the Weddells have, to reach this resort-like spot. The Weddells can hang out, snoozing and shmoozing, without fear, the dominant predator themselves, for months.

Or for as long as the beach survives. It consists of landfast ice, with a long crack in it running parallel with the shore which the tides have opened, and kept open, by lifting and lowering the ice cover. The crack gives the Weddells access to the water, for food and mates. The pregnant females start arriving in early October, swimming 20 or 30 miles underneath the ice, following leads, and emerging from the tidal crack to flop onto the ice beyond it. By the end of October, most of them have given birth, as females in other colonies along the coast have done, turning the margins of the sound into what Gerald L. Kooyman, research professor at the Scripps Institution of Oceanography, calls a "huge maternity ward."

One year in mid-November, when the pups were a couple of weeks old, I visited the Weddell colony at Hutton Cliffs near Mount Erebus, Antarctica's only active volcano, which on that fine day had wisps of steam sticking up from its peak like short, stiff white hairs. Just before I got there, a cornice of wind-packed snow had broken from the top of the glacial cliffs and landed on the ice below. It probably also landed on some seals that were loafing on the ice; two females and two pups were never seen again. "Icefalls" are a common hazard to Weddell seals, Kooyman points out. The ice the survivors were glommed on to was seven feet thick, according to Ward Testa, wildlife biologist for the Alaska Department of Fish and Game, who was doing population studies on them. The tidal crack was, I could see for myself, about a foot wide and ran irregularly across the ice, like a break in a biscuit. A few seals lay inert beside it. Others were scattered across the ice, mostly inert. One lifted a flipper and wagged it. Another opened its mouth in a huge yawn. A pup lying beside its mother adjusted itself, like any child getting comfortable in bed.

The sight was peaceful enough but oh the sound! By that time, there were males and nonbreeding females as well as mothers and pups on the ice, and all were squabbling, over access to pups, food, diving holes. The vocalizing was almost constant, and loud. Some was like human speech, not quite decipherable, as when people call out in their sleep. There were moos, baas. Sounds were coming from underneath the ice as well, long sweeping trills. Those were males, Testa said, defending their breeding territories against other males. They don't have harems, he explained, but mate with whichever females dive into or swim through their underice territories in a receptive frame of mind. They defend the boundaries of the territories by nipping and slashing at other males. When they climb out of the water to rest on the ice, they have cuts in their hind flippers and around their gonads, and the ice becomes tainted with bloody splotches. "It looks like Dieppe," says one observer.

Once mating is over, the Weddells take it easy on the ice for a while, but as summer temperatures start melting the ice, from both above and below, the areas of open water get bigger and the rest areas smaller, until the seals are jammed together on a few strips of fast ice. By then, the pups have already swum away, into areas of open water and loose ice . . . and killer whales. Testa has heard of killer whales swimming together and creating a wake that washes young seals off their ice floes.

Many adults swim away, too, but many stay behind, even into winter, when the water in the sound freezes solid. What allows a Weddell seal to live where other seals (including leopard seals) cannot are the physical adaptions it has made to ice. Its head is small and narrow for its body size, and when it's under ice it can poke its head up into small holes and cracks in the cover and get air. Its teeth are specialized for abrading ice. The upper canines and incisors are "exceptionally stout," according to Kooyman, and angle forward "to such an extent that a dentist would recommend orthodontia." Its lower jaw swings down, into the widest gape of any seal, giving the upper teeth plenty of room to scrape the ice as it sweeps its head from side to side, in 90- to 180-degree arcs. "Sometimes, when it is shaping the upper edge of an ice hole," Kooyman says, "it literally hangs by its teeth as it slashes back and forth at the ice."

Using its stout teeth, the Weddell can make, enlarge, and shape holes in the ice. Mother seals will scrape out ramps in the ice at the edges of cracks or holes so their pups won't have such a hard time climbing from water onto ice. While an adult has the swimming momentum to leap out of a steep-sided hole, a pup can get trapped there and die of starvation or exposure. Testa figures about half of the pups lost in the Hutton Cliffs colony each season die on the ice, of abandonment, starvation, or icefall, but the other half "simply disappear," probably dead in the water, unable to get out of it.

The urge to gnaw ice is apparently so compelling to a Weddell seal that one in Florida's Sea World chewed at the concrete edge of its pool and broke its teeth. Although not as hard as concrete, sea ice can take its toll on the ice-reamers, too. "The worn teeth of older animals show the price paid for countless assaults on ice holes," Kooyman wrote.

In summer, Weddells foraging under the ice probably find their breathing holes mainly by sight. There's plenty of light in the water then, pouring down through holes and cracks in the ice "like beacons," Testa says, and the bottom of the ice provides them with visual landmarks. It's thicker in some places than others, variously stained by organisms, checkered with cracks, and dotted with platelet ice (a type of anchor ice* consisting of large, flat crystals). "There are all sorts of permutations of platelet ice," Kooyman says, "from no platelet layer at all to patchy or scattered columns and mounds of platelet ice to layers that are over 35 feet thick. It makes for a pretty diverse topography. I suspect the seals are very good at picking up on ice features and using them for navigation. They live in that stuff, and must know every bit of it."

To get a look at the bottom of the ice myself, I climbed down a ladder (fighting claustrophobia) inside a metal tube the width of a manhole cover which had been set into the ice at Hutton Cliffs with a glass-covered observation chamber at the bottom from which scientists could spy on seals. I couldn't see any seals but I could hear them, chirruping as sweetly as canaries, quacking like ducks, making noises like wet kisses, knocks on wood, and rockets coming in on long trajectories. The underside of the ice was a cloudlike mix of large bumps, tiny lumps, and pits, and it was yellowish brown from all the organisms clinging to it. Projecting downward from it were clusters of what looked like clear razor blades: platelet-ice crystals. The icy ceiling had a narrow crack cutting through it, white where daylight was pouring through and, framing the crack on either side, cobalt-blue ice. At one point, a shadow passed over the crack, probably cast by a south polar skua, the rapacious Antarctic gull, which sometimes attacks the eyes of weak seal pups.

> Over plaques of virgin white—
> white no longer when they had passed!
>
> —George Allan England,
> *The Greatest Hunt in the World*

Hunting harp seals in the North Atlantic was a very different matter from hunting ringed seals. ("Harp" because the seal has a dark patch on its back resembling, depending on your Rorschachian response, a harp, wishbone, heart, crescent, horseshoe, or saddle.) Harp seals give birth on sea ice off eastern Canada at about the same time each year—end of

*"Anchor" ice is defined as submerged ice which is attached to the bottom (or something else below the surface).

February, beginning of March—and at the same time as each other, and in each other's gregarious company. By mid-March, patches of ice not more than a few miles square could be supporting a couple hundred thousand harp seals, including helpless pups with pure white fat under their pure white fur and their supple hides. These elements—near-simultaneous pupping and communal nursing—plus a nonaggressive nature (some have called it "confiding"), which inclines harp seals not to defend their pups too vigorously, as well as the highly marketable constituent parts—fat, fur, hides—made the hunting of harp seals commercially profitable for many years. Starting in the late 18th century, vessels sailed and steamed into the vast Atlantic icefields in early spring every year to harvest harp seals, first the young, then the adults.

In 1922, a young journalist from Boston, George Allan England, went out with a sealing ship to witness and describe the hunt. The ship was the *Terra Nova,* the same one that Scott took to Antarctica on his final voyage, built of greenheart and oak, "massively timbered and with iron-sheathed bows," England wrote, condemned but still in use then, its engine "only 120 h.p. or less than some racing cars, and yet pitted against the arctic ice!" Its captain was the greatest Newfoundland sealer of all time, Abram Kean, "fifty springs to the ice," a man who "knows icefields as other men know their palms" but one who treated his men like galley slaves. The men were 160 Newfoundlanders, to whom England dedicated his book, *The Greatest Hunt in the World,* "the strongest, hardiest, and bravest men I have ever known. . . . poverty-bitten, humble, heroic, cheerful, truly pious, and indomitable men who gamble with death, and who all too often lose." The villains of the hunt were the merchants who owned the ships and reaped the profits and the captains who when it came to a choice between dead seals and live men all too often chose the seals.

So important was ice to the choices Newfoundlanders had to make while sealing and fishing that they had many of their own words for it. Among those England listed were:

Sish	new and very thin ice
Ballycatters	heavy shore ice
Clumper	heaved-up formation of ice
Knot	hard ice formation
Ole man's ice	smooth ice
Glin	dazzle of the ice

H'ater fashion	shaped like a flatiron, said of ice
Rot hole	soft place in the ice
Selvage	edge of the icefield
Standin' ice	heavy ice
Stritch	said of ice "going abroad"
Swill	swell in the ice
Tickle	narrow waterway in ice
Rent	space of open water
Wedder age	outside or weather edge of the ice
Glitter	frozen mist on rig
Silver glitter	ice on rig
Conkerbills	icicles, usually on eaves

Harp seals are almost constantly on the move, traveling 5,000 round-trip miles a year, from waters of the High Arctic to waters off Newfoundland and back, with the ice "always at their heels," England wrote, "—their 'scutters' rather." In early spring when the "pack ice begins to be riven loose" in the north, they head into it, and the females look for ice to whelp on. They haul out near leads or gaps between floes since they need access to water to feed and keep up their fat stores; sometimes they make openings themselves, by diving through the thin ice of refrozen leads or sticking their heads down in the ice, and they keep the openings open not with ice claws—they don't have any—but with their bodies. Several times a day they slip through their "bobbing holes" to interrupt any ice formation. "The door of the seal's house must never be allowed to close."

When giving birth on an ice floe, a harp seal arches her body, leans onto the ice with her front flippers, and slides the newborn across the surface. Almost immediately she spins around and goes nose to nose with the pup, learning to recognize it so that when she comes back from foraging in the water she won't nurse the wrong one on the ice. An advantage of giving birth on loose, free-floating pack ice is that most land predators can't get to it. A disadvantage is that the floes are unstable, and young seals often tumble off and get crushed by the shifting ice. Or the floes break up and separate mothers from their pups. So the pups can get off the pans in a hurry, the harps have greatly compressed the weaning period. After only twelve days on the ice, fattening up on rich mother's milk, the harp pups are on their own. The mothers dive through their bobbing holes, as they have done many times before, but they never come back.

The first of the killing fields was whelping ice. "All over the ocean, then, to the northeastward of Newfoundland . . . young seals will be lying about the first of March," England wrote. "The total area covered by them amounts to thousands of square miles of icefields, anywhere from close ashore to hundreds of miles away." England often got rhapsodic at the sight of the sea ice, which ranged from "luminous gray to blinding white" and resembled "a brobdingnagian mosaic, interspersed with vast lakes and leads all wonderfully a-sparkle in the sun. Each pan was edged with a raised border of slush, making of it a vast Victoria Nyanza water-lily leaf of ice."

After several days in "hungry-lookin' ice" (no seals), the men heard bawling sounds and "knew the kill was near." What good is protective coloration, England wondered, if the pups are so noisy? "Nothing could be better hidden than [the white-coated pups] on snowy ice; and if they only had brains enough to keep their mouths shut, thousands would escape that are yearly slaughtered." (In their book, *Harps and Hoods: Ice-Breeding Seals of the Northwest Atlantic*, David Lavigne, executive director of the International Marine Mammal Association, and Kit Kovacs argue that the transparent white hairs, which absorb sunlight, are more important in keeping pups warm than in camouflaging them.) " 'Wonnerful 'unmocky ice Cap'n,' " a man on watch said, and the captain agreed. Soon the shout came from the masthead:

" '*Whitecoats!*' " The ship came to a stop against the edge of a "groaning" ice floe, and the men scrambled overboard, onto broken ice pans. Quickly they "copyed" over the pans—jumped from one to another as over rocks in a fast-flowing stream. They were agile as goats, leapt like chamois. "Sometimes they landed on hardly more than slush; but little they minded that 'loose, pummely stuff.' They scrambled up and out of it." One sealer told England that the least "little piece o' 'sish' is enough to copy on," if you're "spry."

Watching the men, England thought of war. Moving single file over the ice in their canvas jackets and spiked "skinny woppers" (boots), carrying poles and ropes and flags, shouting and "leaping, running, slipping, but ever scurrying on and on. . . . Soldiers they seem indeed. . . . To me they look like skirmishers on solid land; in New England meadows, perhaps, meadows of winter. Hills of ice, lakes, brooks, valleys; one looks for a red barn, a cow, a white farmhouse with green blinds."

Instead, on these meadows England saw brooks running red. It was the scene we have all read about. "Colour? The ice glowed with it!" "Clothing and the ice, alike, blossomed vividly." "Spots of red dotted the icescape." "Crimson trails, these, such as no otherwhere on earth exist." The ice served not only as battleground (although the male seals, "to their shame," headed straight for the water, and only half the females stayed on the ice to protect their pups) but also as butchering block and storage platform. The men stacked seal pelts on the ice, put a ship's flag on top, and left the pelts for later pickup (or piracy by other crews). Then, "fingers incarnadined," they moved to the next patch of seals. Later, on board ship, they would "salt" the pelts—that is, put finely chopped ice between the layers. " 'Salt,' they called the ice," England wrote, " 'fresh salt.' " A sealer who died on the voyage would also be salted and stowed until the end of the hunt.

The sealers' real foe was not seals but ice. "The ice-field on which the sealers had to work," Cassie Brown wrote in her book *Death on the Ice: The Great Newfoundland Sealing Disaster of 1914,* "was a treacherous, ever-shifting enemy." The ice a sealer was on could break free of the main sheet and carry him out to sea; ice could tighten around his ship so it couldn't come to pick him up and he'd be stranded; he could slip and fall between ice pans and end up under one. Although in the 1914 disaster that Brown chronicles, in which a couple of hundred men lost their lives, greed, cowardice, pride, and ineptitude played the largest part, "It was the ice that sank the ships, the ice that killed the men."

In early May the adult harp seals, relieved of their duties to nurse pups and keep bobbing holes open, drifted northward with the ice and had a "general jollification," according to England. They ate, rested on ice, and mated. The mating does sound rather jolly, as another young journalist, Pol Chantraine, described it in his book, *The Living Ice: The Story of the Seals and the Men Who Hunt Them in the Gulf of St. Lawrence.* As with walruses, female harps are on the ice, males in the water. To get the females' attention, the males "swim in circles, emitting long whistling cries and giving themselves up to all manner of acrobatic feats . . . They leap and somersault, they swim on their backs, they twirl, they bark, they dive abruptly and come flying back up through the waves at a dizzying speed, in a great flurry of flippers and foaming water." From their elevated positions on the ice, the females "contemplate this ballet, at first indifferently, then, as the males' capers become more extravagant, with greater interest. Some waddle along the edge of their lead to examine their suitors at closer range," making clucking sounds all the while and stretching out their necks. A male will mount the ice now and then and beat his flippers and "jig wildly up and down" in front of a female. If she doesn't go for this, she'll drive him back in the water by flailing at him with her flippers, even biting or scratching him "as ferociously as a tigress." If she is more taken with his act, she indulges him in his barks and jigs with whimpers and groans of her own. She then does a "strange little hypnotic ballet" herself, moving to the ice edge and stretching out her neck, after which she "retreats with quick little bounds, then draws herself up with all her strength and wiggles her chest, executing little ballerina-like pirouettes, her body arched tight as a bow, only the tip of her navel touching the ice." (*The tip of her navel!*) At that, her suitor "catapults onto the ice and wiggles about her," then "brusquely without the slightest sign of gallantry," pushes her into the icy water and, presumably, has his way with her.

After mating, they move to ice farther north, to molt. Molting sounds less than jolly, as Chantraine described it. "This time they look absolutely exhausted. . . . They sprawl by the thousands, packed tightly together [on the ice], lying in their own excrement and the thick tufts of hair they continually tear out with their claws in an effort to ease their terrible itching. They roll onto their backs and scratch their flanks with their flippers, emitting loud grunts and whimpers pitched to the level of their discomfort. Some males still carry on their necks marks of the deep wounds inflicted during the mating season. All in all, the ice-field has the atmosphere of a battlefield hospital, where the victims of gangrene slowly rot in the absense of proper hygiene."

Atlantic sealers also went after hooded seals ("hooded" because the males have tough, trunklike bags of skin on the front of their heads which blow up as big as two footballs). Great travelers, too, the hoods join up with the harps on their migration but stay several miles away, not mingling. The hoods give birth a few weeks after the harps on ice that is in even greater danger of coming apart; consequently they are in an even greater hurry to fatten up their pups and leave their floes. They have the shortest weaning period of any mammal in the world, only four days. "They give birth, feed the kid, and get off," Lavigne says.

One night, on board the *Terra Nova* during a "ravening" blizzard, England wrote a memorial verse to the sealers.

Where the old dog hood and the old harps' brood
Lie out on the raftered pack,
We tally our prey . . .

The *Terra Nova*'s tally that year was 23,157 seals, higher than any other ship's.

> *In a wild death dance we dice with chance,*
> *We Sealers of Newfoundland!*

For weeks of hard, dicey work, among the sish and the clumpers, each man earned $74.60.

CHAPTER TWENTY

ANIMALS III

In searching for eggs both he [Edward Wilson] and [Birdie] Bowers picked up rounded pieces of ice which these ridiculous creatures had been cherishing with fond hope.

—Robert Falcon Scott, *Scott's Last Expedition*

IT IS THE WORST OF PLACES, the worst of times. The Antarctic coast. The dead of winter. Temperatures down to −60°F. Wind speeds up to 80 miles an hour. Darkness around the clock. "Ground" a slab of sea ice. Yet in that place and at that time, male emperor penguins stand without shelter or food or drink or breaks for a lie-down or even much movement for two whole months, holding eggs on top of their toes. The scene is one of such extremity—thousands of hunched-over birds being lashed by blizzards—that it suggests something amiss in the rational workings of reproductive biology. *Is this hardship necessary?*

The first people to see a breeding colony of emperor penguins were men from Robert Scott's *Discovery* expedition, who looked down on one from a very high precipice at Cape Crozier in the early Antarctic spring (October) of 1902. When those men reported to their colleagues back at camp that the penguins had well-developed chicks with them, one colleague, zoologist Edward Wilson, remarked that "they must lay their eggs very early indeed." Almost ten years later, Wilson, Apsley Cherry-Garrard, and several other men from the *Terra Nova* expedition, including Scott, rowed toward Cape Crozier for another look. This time it was summer in Antarctica (mid-January). "About six feet above us on a small dirty piece of the old bay ice about ten feet square," Wilson wrote in his journal, "one living Emperor penguin chick was standing disconsolately stranded, and close by stood one faithful old Emperor parent asleep." The rest of the penguins had apparently departed already, "gone north to sea on floating bay ice." Wilson found it "curious" that with all the clean ice around, the "destitute derelicts" should have chosen to stand on the only remaining piece of bay ice, and it filthy. He figured that the penguins were waiting for the ice to carry them out to sea and wondering why it was taking so long.

"Another point most weird to see," Wilson noted, was "that on the *under* side of this very dirty piece of sea-ice, which was about two feet thick and which hung over the water as a sort of cave, we could see the legs and lower halves of dead Emperor chicks hanging through, and even in one place a dead adult." After Scott himself saw the hanging parts (which he noted included a flipper), he wrote in his journal that the birds "had evidently been frozen in above and were being washed out under the floe."

On finding the single living chick, the explorers guessed why the penguins breed when they do. The chick was "still in the down," Wilson wrote, and if its egg had been laid in summer as the eggs of other species of penguins are, it would still have been in the down—that is, without a protective coat of feathers—when winter arrived. "Thus," Cherry-Garrard concluded in his account of the visit, "the Emperor penguin is compelled to undertake all kinds of hardships because his children insist on developing so slowly."

Emperors develop slowly because they are large. They are the largest diving birds in the world, the largest of all 17 species of penguin, more than twice as large as the next largest species, king penguins. They stand almost three feet high and weigh about 90 pounds. In cold regions, having a large body mass is better than having a small one since for a given volume there's less surface area from which heat can escape and more room in which insulating fats can be stored. But growth takes time; there's a limit to how fast cells can divide. Biologists now know that it takes six months for an emperor chick to fledge and that the breeding in emperor colonies is timed so the chick is capable of feeding itself when the short summer season begins. Adults therefore have no choice but to court and lay and incubate and hatch their eggs during the long Antarctic night, under the harshest circumstances of any bird—any vertebrate—on earth.

How harsh the circumstances can be is powerfully conveyed in Cherry-Garrard's book, *The Worst Journey in the World,* which he wrote about the trip that he, Wilson, and Henry R. "Birdie" Bowers took to the rookery at Cape Crozier on foot seven months after they had rowed past it. They went to get some eggs. At that time there was a widespread belief that embryos as they grow pass sequentially through previous stages of the animal's evolutionary development, and if the penguin embryos had scales on them, as did dinosaurs from which flying birds were believed to have descended, it would demonstrate that penguins were the missing link. The three men left the expedition's quarters in June, five days after midwinter night. "It is midday but it is pitchy dark, and it is not warm," Cherry-Garrard wrote, the last time he would resort to understatement. For the next five weeks, while covering only 76 miles, the men endured "cold such as had never been experienced by human beings," temperatures as low as −77.5°F. They had to cross some of the largest pressure ridges in the Antarctic, formed by the compression of the 400-mile-long Great Ice Barrier against the rocky shoreline. They fell repeatedly into "furrows" 50 or 60 feet deep; struggled around "huge heaps of ice pressed up in every shape on every side, crevassed in every direction;" skirted "great walls of battered ice;" and blundered into a cul-de-sac formed when two ridges of glacial ice "butted on to the sea-ice," as Cherry-Garrard expressed it.

"And then we heard the Emperors calling.

"Their cries came to us from the sea-ice we could not see but which must have been a chaotic quarter of a mile away. They came echoing back from the cliffs, as we stood helpless and tantalized." Their way to the colony was blocked by a pressure ridge that had been thrown "end on" against a cliff, but Wilson discovered a "black hole, something like a fox's earth, disappearing into the bowels of the ice," and they managed to wriggle through it and on emerging stood, "three crystallized ragamuffins, above the Emperors' home."

Home turned out to be not on the thick ice of the glacial shelf but on the thin sea ice of the bay. There were only about 100 penguins on it. "Where were all the thousands of which we had heard?" they wondered. Nevertheless, the men grabbed five penguin eggs, stuck them inside their fur mitts, and "legged it back as hard as we could go" to the place where they'd left their pickling gear. The journey back was even worse than the journey out had been. At times they thought that death was inevitable, then that death was desirable: "a crevasse seemed almost a friendly gift."

As the only surviving member of the egg expedition (Wilson and Bowers died with Scott on the return trek from the South Pole), Cherry-Garrard presented three eggs (two broke) to custodians of the Natural History Museum in South Kensington, England, who were indifferent, even annoyed by the donation. "This ain't an egg shop," one said. Eventually, a professor took a look at the embryos and found the rudiments of feathers on them; the penguin was not, after all, the missing link.

Even after something was known about the emperors' embryos, little was known about the emperors. Forty years passed before anybody visited a colony of the penguins in the middle of winter again. Then in 1952, a young Frenchman, Jean Prévost, spent a full year, including "the long night of polar winter," observing 12,500 emperors in a colony at Pointe Géologie in Adélie Land, near a base of the French Polar Expedition. The colony had only recently been discovered, by sled-dog trainers out on a practice run. In the doctoral thesis Prévost later wrote, *Écologie du Manchot empéreur,* he described the penguins' breeding cycle, which is tied to the ice cycle. Both last almost ten months and begin in March. Two penguins had already made it up onto the young, flexible sea ice of March without Prévost seeing them, but he did spot the third bird coming in. It approached the other two, lowered its head, and after a few seconds lifted its head. One of the other penguins did the same, then stood facing the newcomer for several minutes before turning away. Prévost later realized that he had witnessed the first tentative search for a partner as well as the first "mutual display."

On this icy stage, he was to see many more mutual displays, since emperors perform them in all "social and family circumstances." The face-to-face display, for instance, performed not only when seeking a mate but also after laying an egg, is preceded by a "love song," Prévost reported, a sort of cackle that ends in a short or long note, emitted by one penguin with its head lowered toward the ice and its beak slightly open and its neck bent like a shepherd's crook. In one variation of that display, after the lowering of heads and bending of necks, one penguin approaches the other in a "balanced walk" and stops in front of it and, with its body arched over its feet, slowly lifts its head and points its beak toward the sky. This exposes its brood patch, the bare place on the belly of males as well as females against which an egg or newborn chick is placed before a protective flap of feathered skin is draped over it. Standing with their breasts pressed together, or with only a bit of space separating them, the two penguins are in an "ecstatic" state, or so it appeared to Prévost, with their eyelids half closed. The ecstasy doesn't last long before they simultaneously make swallowing movements and relax their bodies. Then they may do it all over again.

As for the "nuptial display," the male shows his brood patch to the female then inclines his head toward *her* belly, "as if to observe the egg,"

Prévost wrote, and touches her belly with his beak. Even at this advanced state of courtship, the female may reject him. Undeterred, he may pass his beak over her neck and invite her to stretch herself out on the ice.

A month later, the female lays a single egg, which is all the male has room for in his brood pouch. She brings her tail forward between her feet, apparently to soften the fall of the egg onto the ice, and upon seeing the egg slide downward, the male sings. No sooner has the egg touched the ice than the female maneuvers it up onto her feet, using her beak. She shows the egg to the male, inclines her head, and sings. He regards the egg and, while groaning slightly and trembling, touches it with his beak. Then he touches the female's brood pouch, looks down at his own brood pouch, becomes agitated and "insistent," pushes the female, and sometimes tries to take the egg from her by force. She widens her stance, and the egg falls from her feet onto the ice, but before he can seize it she has worked it back up onto her feet. Several minutes later she moves her feet apart again, and this time when the egg falls to the ice, the male, still trembling, rolls it between his own feet and up onto them, with his beak and "a great deal of difficulty." He and she sing. She stamps around him while he stands "passive and indifferent." They do a mutual display and regard each other's brood patches. She moves away from him, over the ice, but comes back and sings, moves away again but comes back, marches around him, then moves still farther away. He tries to follow her but this time she keeps on going. She'll be away for two months, feeding in the water, while he and the egg stay back on the ice. Already he has been without food for six weeks and he'll be without it for eight or nine more weeks, at the end of which time the egg will be hatched and she will have returned from the sea and it will be his turn to leave, having by then lost up to half his body weight.

Males with eggs aren't the only males on the ice. There are also "unemployed" males. If a male with an egg accidentally drops it, say down a hole in the ice where he can't retrieve it before it freezes, he joins the ranks of the unemployed. Males with eggs are calm and apathetic, but the unemployed are "vivacious," active, and apparently disoriented; they strike out in one direction then suddenly turn about-face and "look as if they don't know exactly what to do," Prévost wrote. One thing they *do* know is that they want eggs. At Cape Crozier, Cherry-Garrard saw males that were startled by the arrival of the humans abandon their eggs on the ice, and these were "quickly picked up by eggless Emperors who had probably been waiting a long time for the opportunity. . . . Such is the struggle for existence" that the poor birds "can only live by a glut of maternity."

So anxious were the males to sit on something, Cherry-Garrard added, "that some of those which had no eggs were sitting on ice! Several times Bill and Birdie [Wilson and Bowers] picked up eggs to find them lumps of ice, rounded and about the right size, dirty and hard." He himself watched as one bird dropped an ice egg, "and again a bird returned and tucked another [ice egg] into itself, immediately forsaking it for a real one, however, when one was offered."

While keeping eggs warm under their belly flaps, males keep themselves warm by "huddling." During particularly bad weather, according to Prévost, hundreds or thousands of males stand tightly pressed together, in a *tortue,* a cluster of birds that apparently looked to some French-speaking person like a turtle's shell (to Gerald Kooyman, it looks like a football scrum). A huddle reduces the amount of exposed body surface of each penguin to about a sixth of what it would be if the penguin stood alone. The males stand, backs to the wind, with those on the huddle's windward side peeling off after a while and working their way around until they are on the lee side of the huddle or else in the middle. The huddle thus moves over the ice, as a single body, with a spinning action, in the direction of the wind.

Almost to the day their chicks hatch, the females return from the sea. After singing and displaying, they pass the contents of their stomachs (fish "garnished with few crustaceans," Gerald Kooyman reports) into the mouths of the chicks. Then they tuck the chicks into their own brood pouches, leaving the males free to go off, at last, to eat. Within weeks, the males come back and pass the contents of *their* stomachs into the chicks' mouths, and the females then head out to sea, and so on. The parents take turns, feeding themselves, feeding their chick.

Should the female be a little late returning at hatching time and the male isn't able to wait any longer to eat, the maternal glut comes back into play. "And when at last he simply must go and eat something in the open leads near by," Cherry-Garrard wrote, "he just puts the child down on the ice, and twenty chickless Emperors rush to pick it up. And they fight over it, and so tear it that sometimes it will die. And, if it can, it will crawl into any ice-crack to escape from so much kindness, and there it will freeze."

It wasn't until 1964 that a very large emperor colony was discovered at Cape Washington (by Ian Stirling, the polar bear expert, who was then a Weddell seal expert). One reason it took so long for anyone to find that sizable a colony was that explorers and scientists go to Antarctica in spring and summer, and the colony doesn't exist in those seasons. The evidence of its existence has already gone out with the ice. Penguins that breed on land—and all species except the Emperor do—leave broken eggs, guano, dead chicks, and patches of chick down on the ground when they depart. In the emperors' case, the ground melts, and all signs of their having been around vanishes into the sea.

When Gerald Kooyman arrived at Cape Washington to study the colony 20 years later, in October (early spring), he was dazzled by the sight. Against a backdrop of the Deep Freeze Mountains, there was "a vast stain on the ice," a dark area stretching more than a mile across the horizon, made up of many thousands of penguins. As he watched over the next weeks, he noticed groups forming within the large, amorphous mass of birds, then the groups moving away from each other by several yards a day, "as if the colony were flying apart." By mid-December, groups of chicks formed a semicircle on the ice, with a line of adults running from it all the way to the distant ice edge, from which they departed on their feeding forays. The adults didn't depart right away, though. Thy milled around on the ice, in groups of 10 to 300 birds, restlessly, for hours sometimes. "It's not clear why they hesitate," Kooyman says. He figures they're just trying to assess the safety of the water, decide if predators are waiting for them in it or not. "Ice is a refuge."

Still, there are hungry chicks to feed. "Suddenly, at some unknown signal," Kooyman wrote, "the waiting birds make a single-file dash on their bellies to the edge, driving their feet into the snow like rotary pistons. Watching from the safety of thick ice I was always reminded of a fast-flowing stream tumbling over a waterfall, for in one smooth motion the front birds flop down to their bellies and toboggan along, while those behind them move up to the same spot to drop down on their bellies." The penguins gliding along on their smooth feathers and poling with their feet and flippers reminded Jean Prévost of skiers.

Emperors don't toboggan on ice, only on snow. In over a decade of observation, Kooyman has never seen any of them tobogganing on ice. On ice, they walk. "They have nails very much like crampons," says Kooyman's son Tory, his assistant one summer. "When walking on bare ice, they dig their nails in severely." Their feathers and skin hang down so far they don't have much free leg—"they've got their pants down," Gerald Kooyman says—which makes them less than agile; "They can't run or hop or take high steps." Their walking pace is slow and measured, "like a coronation march," Kooyman suggests. Going from tobogganing to walking, they use their wings and beaks to help flip themselves off their bellies and onto their feet. In fact, they often use their beaks like a third arm, or an ice pick. To climb an ice ridge, they may poke their beak into the ice and leverage themselves up.

When the adults come back from the sea, they "rocket out of the water, sometimes 10 at a time," Kooyman observes, "as if the Grim Reaper itself were in hot pursuit." Actually, the Grim Reaper *is* often in hot pursuit, "for the ice edge is a favoured haunt of leopard seals." Leopard seals patrol the water at the ice edge and snap up penguins in their powerful jaws. They may not wait for the departing penguins to hit the water but instead lunge up onto the ice and grab ones near the edge. They may even lie on the ice themselves and wait for returning penguins to plop down on a landing spot. For some reason, penguins tend to use the same, well-worn landing spots on the ice.

Once an adult penguin has returned safely, it will march "unerringly" to the group its chick is in and "honk like an antique car horn," Kooyman declares. Hearing it, the right chick among all the chicks on the ice that day dashes up to the right adult and begs for food in "a high-pitched, staccato whistle." The chicks leave the Cape Washington colony in mid-December, abruptly, within ten days of each other. One day Kooyman would see a few chicks on floes near the edge of the shorefast ice, and three days later he would see hundreds, with lines of chicks leading away from every group on the ice. Some of the chicks gather on a projecting arm of ice, then more chicks crowd in behind them, sometimes causing the first chicks to fall in the water. Penguins do not, however, as is commonly believed, shove one of their number off the ice edge as a sacrificial victim, to test the water for predators, thereby saving their own blubbery skins.

Another long-held belief about emperors that Kooyman contradicts is that the chicks need a full coat of feathers before departing so they can survive the chill of the icy water. All the chicks leaving Cape Washington still have plenty of down—a gray, soft-looking, Angora kitten fluff—on their backs. Another long-held belief is that the chicks always float away on ice floes; they have "a private yacht all to themselves," Cherry-Garrard wrote. But at Cape Washington, the chicks Kooyman has seen departing have all *swum* away, and with plenty of splash, the noise sometimes attracting the attention of leopard seals.

With 24,000 chicks a year and a total of 75,000 birds, Cape Washington is clearly a successful colony. Why? Kooyman wondered. What do penguins want in a colony anyway? He examined five other emperor colonies along the western Ross Sea coast and concluded that the most important characteristics of a site (to a penguin) are (1) stable landfast ice, (2) shelter from wind, (3) open water nearby, and (4) access to fresh snow. Fresh snow is desirable because it shields the ice from the sun's rays and keeps the ice from melting prematurely. Also, it packs around penguins' bodies during storms and helps keep cold air out. As for wind, there should be some of it to break open the ice cover close enough to the colony that the adults can feed conveniently but not so much of it that the ice under the chicks is carried out to sea. "If the chicks aren't ready to leave when the ice is," Kooyman notes, "a whole year's cohort could be lost."

Stable ice turned out to be the key component of a successful colony. In three of the six emperor colonies Kooyman studied, the areas of stable ice are small, and they are the ones with the fewest penguins in them. As the ice under the birds darkens from their guano and absorbs more sunlight and starts disintegrating, the penguins in those colonies have no other ice to move to. When he flew over one of the more successful colonies in the dark of winter and trained a military night-vision image-intensifier on it, he saw 7,000 penguins—"all males, of course"—standing on the ice, yet to his surprise they were standing "out in the middle," well away from the rocky windbreak or "any ice ridges or ice falls that could provide shelter." Evidently being on a stable plate of ice was more important to the birds than getting out of the wind, even in the extreme cold. Deeply impressed by the penguins' grit, Kooyman mused once again: "Emperors do a remarkable thing by breeding in the winter."

In answer to the burning question of how male emperors are able to stand on ice for two months straight without freezing their feet off, Kooyman explains, first of all, that they don't stand flat-footed. They rock back onto their heels, which are spongy, providing both padding and insulation. "The contact point [with the ice] is quite small, about the size of a 25-cent piece." For balance, they use the base of their tails. So the only parts of their bodies conducting heat directly to the ice are two quarter-size footpads and a half-dollar-size piece of tail.

As for why emperors breed on ice at all—probably the only birds to do so—Kooyman suggests that it may be because huddling, a requirement for males' survival in winter, is best done on a flat surface. Most of the rocky beaches of Antarctica aren't flat, but annual shorefast ice is. Emperors can go from birth to death without once touching land, the only birds in the world that's true of. (Of the 40 emperor colonies discovered so far, only two are on land, and those two are small and covered by ice and snow.) In all the time Kooyman has been studying emperor penguins, he has never seen a single one go ashore even when it was walking right next to shore. "Ice is the be-all and end-all of their lives," Tory says. "The ice is living as much as the penguins are, in its motion, its behavior, its whims, its flights of fancy" (he was an English major).

The only other penguin that breeds on the Antarctic continent outside the peninsula breeds on land. It's the Adélie, the "stereotype penguin," as one observer calls it, the "little man in evening dress." Adélies breed on rocky shores where the males scratch out hollows in frozen ground and line them with pebbles, meant to keep the eggs from being inundated by snowmelt. (The size of a colony is limited by the number of available pebbles so there's a lot of pebble pilfering and pecking of pilferers.) Unlike emperors, Adélies don't breed in the heart of winter, but they do breed in "very very *really* early spring," says Wayne Trivelpiece of Montana State University in Bozeman, who has studied Adélies for more than 25 years. In order to get from the moving pack where they spent the winter to the rocky shore where they will breed, they may have to cross as many as 60 miles of landfast ice. The males go first, walking and tobogganing single file, in groups of hundreds. "The long lines of black Adélies threading the ice," David G. Campbell wrote in his book *The Crystal Desert: Summers in Antarctica,* "are to the Antarctic what the springtime chevrons of geese are to North America or the sinuous files of wildebeest are to the African savanna."

Also unlike emperors, Adélies do run, jump, and climb over ice, and they often choose to go through rough ice. "They prefer the thicker 'white' ice and higher ice piles produced by the collisions of ice floes," Dietland Müller-Schwarze wrote in his book *The Behavior of Penguins: Adapted to Ice and Tropics.* "They stop at the edge of thin ('gray') ice, and single penguins cross these patches, running quickly. They also may rest and sleep for hours at the boundary between the two types of ice," apparently waiting for thin ice to thicken or a lead to freeze, or for wind and currents to push a couple of ice fields together so they can walk between them. "Waiting," Müller-Schwarze concluded, "is thus an important part of the Adélie penguin's strategy of migration and predator avoidance."

The predator he had in mind was the leopard seal, which likes Adélies as well as emperors. At Cape Crozier, Müller-Schwarze observed that the methods leopards use to catch penguins vary according to the condition of the ice. In October, for instance, the seals swim under the extensive cover of landfast ice, break it from below with their heads, and try to grab penguins on top. After a close call, he noted, an "escaped" penguin may remain motionless on the ice for hours. "This way he seems to be unnoticeable to the searching leopard seal under the ice."

Then in November when the sea ice off Cape Crozier is breaking up and Adélies are coming and going from their feeding areas by both swimming and walking, leopard seals hunt in the open water while the penguins "escape onto large ice floes." "Large" is key: "Birds on small floes of two to five meters [seven to 16 feet] diameter may be attacked many times from different sides of the floe," wrote Müller-Schwarze. "They always flee to the 'safe' side of the floe, but avoid the water religiously." Sometimes the leopard seal will push the penguins off a floe. Roger Tory Peterson once saw 60 to 70 Adélies "pinned down" on a small iceberg by a leopard seal that was cruising the water around it, and judging by the look of their torn flippers and bloody breasts, he guessed that two had barely made it up onto the "ragged islet of ice."

By early February, when most of the ice around the cape has melted and only a few floes are floating offshore, the leopards zero in on the departing chicks. Slow, awkward swimmers at first, the chicks don't know yet how to make evasive moves when they encounter seals. "The only pre-adaptation [they have] for the dangerous life at sea," according to Müller-Schwarze, "is their tendency to climb ice blocks stranded on the beach prior to going to sea." In a penguin version of king of the mountain, the chicks on the beach "try to reach the highest point on the ice and compete vigorously with other chicks for that spot, so that some chicks are constantly falling off, while others climb to the top." From their ship one Christmas Day, Scott and his men enjoyed the sight of young Adélies practicing their ice-mounting skills. The penguins were "quarrelling for the possession of a small pressure block [of ice] which offered only the most insecure foothold," Scott wrote. "The scrambling antics to secure the point of vantage, the ousting of the bird in possession, and the incontinent loss of balance and position as each bird reached the summit of his ambition was almost as entertaining" as had been the penguins' earlier, frantic effort to chase away a skua that threatened to eat them alive. "Truly these little creatures afford much amusement."

Once at sea, Müller-Schwarze pointed out, "the chicks still climb on any piece of ice that floats in the water in their vicinity. Often they fall off the smooth surface. If the ice floe or block is large enough, it will provide good protection against the leopard seal. But even small ice fragments floating in the sea are of advantage to the chicks, since they provide opportunities to rest and may even serve as rafts for a longer journey." Nevertheless, the potentially lifesaving urge to mount bodies of ice leads some chicks to "swim toward a leopard seal just as they do toward an ice block drifting in the water."

The greatest threat to Adélies at Cape Crozier isn't leopard seals that hang around the edge of the colony; it's *ice* that hangs around the edge of the colony *too long.* When the shorefast ice stays unbroken for a considerable distance without leads or cracks for a penguin to feed in, the female must make a longer foraging trip than usual and gain more weight in compensation, in which case she's likely to be late relieving the male on the nest, and he'll have to abandon their egg so that he can replenish his own fat stores.

"It's a very narrow time frame," Wayne Trivelpiece points out. "If a bird misses by as little as half a day, there could be reproductive failure." Although abandoned Adélie eggs aren't pounced on by adults eager to adopt them, as happens with emperors, they are pounced on by skuas which poke holes in them and suck out the embryos. Or they just freeze. Also, in heavy ice years, adults bring less food to their chicks. One year when sea ice off Cape Royds didn't break up until late January, only a quarter of the eggs in that Adélie colony resulted in fledged chicks, while in years when the ice broke up by mid-December, one-half to two-thirds of the eggs did. More than anything else, according to Trivelpiece and William Fraser, biologist at Montana State University, the timing of the breakup of fast ice determines how successful the Adélie penguins' breeding season will be.

In mid-December one year, I happened to get a look at the Adélie colony at Cape Royds from an adjoining hilltop. The nests were about two feet apart, and birds were lying on them, belly down, like overinflated, black-and-white footballs. The sea ice that stretched outward from the base of the hillside that the nests were on appeared to be solid and unbroken all the way to the horizon. I could see a few penguins far out on it, black specks on the white flats. How brave they seemed, facing that vast expanse! Undeterred, they bustled along, all business, walking quickly, hurrying to their appointment at a feeding spot who-knows-how-far away, looking pressed for time, probably truly pressed for time.

Big floes have little floes all around about 'em,
And all the yellow diatoms couldn't do without 'em.
Forty million shrimplets feed upon the latter
And they make the penguin and the seals and whales much fatter.

—Griffith Taylor, "The Protoplasmic Cycle,"
South Polar Times, Scott expedition, 1911

What the Adélies look for in the water when they find water is Antarctic krill, shrimplike creatures about two inches long which congregate in parts of the Southern Ocean in immense swarms, covering as many as 175 square miles of ocean and turning the water pink. Most marine vertebrates in the Antarctic region either eat krill (one blue whale consumes four tons of it a day) or eat creatures that eat it (those penguin-chomping leopard seals).

As for the krill, they eat algae, the primitive, mostly microscopic cells (diatoms, flagellates, etc.) that make up the plant part of plankton, the floating mass of organisms in the world's oceans. While most algae in polar seas are to be found in water that's not frozen, 5 to 10 percent are found in water that is. Nansen saw them in the Arctic, "small, one-celled lumps of viscous matter, teeming in thousands and millions, on nearly every single floe over the whole of this boundless sea, which we are apt to regard as the realm of death." Algae live on the top, bottom, and sides of ice floes. They live inside ice floes. They live not only on pack ice but on shorefast ice and icebergs. They inhabit ice in such concentrations sometimes that the ice is "dirty brown" (Nansen) or "a pale ochreous color" (Joseph Hooker, a botanist with James Clark Ross's 1840 Antarctic expedition, who was probably the first person to describe ice algae—"microscopic vegetables," he called them, which occur in "countless myriads"). Glenn Cota, research professor of oceanography at Old Dominion University, once found 246 million algae cells residing on a piece of Arctic sea ice no bigger than a milk carton.

A substance that at its warmest is about 29°F wouldn't seem to be a congenial place for organisms to settle, much less flourish, but sea ice has several things to offer microscopic vegetables. It provides a stable surface for them to hang on to. It floats, so it's closer to whatever light the sky provides in early spring (dawns, twilights) than the water is. Its surfaces are washed by, or at least sloshed by, seawater, which delivers fresh nutrients to any algae there. The most fruitful soil on a typical sea-ice floe is its underside. In constant contact with water, the underside is comparatively warm, right at the freezing point, and therefore soft, like ice cubes left too long in a drink. It's so soft it can be scraped off with a plastic spatula, says Cota, who has done that while collecting algae specimens. The algae tend to concentrate in the bottom inch of the bottom ice, he reports, usually in the bottom quarter or eighth of that bottom inch. To the creatures that eat algae, concentration is important. Krill larvae, for instance, according to Cota, can crawl onto the ice "and munch on lunch, instead of a dilute soup" in the water.

How algae adhere to the ice once they land on it isn't known, but adhere they do, vertically and upside down. As they absorb sunlight and heat up, they melt minute craters into the ice. Since they are buoyant, they melt the ice upward and dimple the undersides of the ice. "There may be a jillion craters per square yard," Cota says. Around the craters are protrusions, "like those fingers on rubber, no-slip mats or your stomach lining." On the undersides of the floes that persist into summer, algae grow into long, hanging, mucilaginous strands, like "mats of hair which move back and forth in the current," Cota says. "You see the strands when you dive under the ice. They can be over six feet long." When they slough off the ice, they float in the water, looking, Cota guesses, like tufts from a woolly mammoth.

Algae get inside ice floes when the floes are weighed down by snow or knocked askew by crashing into other floes, and seawater floods the surface and freezes and traps the algae between crystals. With each fresh flood, another layer of algae is laid down on the ice. If you sliced through some ice floes, you would see bands of entombed algae. When a floe warms up and rots, algae will colonize its sides, too, living in the water-filled holes of the candled ice.

In winter, ice algae are in a state of near hibernation, clinging to the ice in their fashion but not growing, probably living off their own preformed material. With the sun's slow return in early spring, however, they start reproducing, and the amount of algae on the ice comes to exceed the amount in the water. At a time when food is still scarce elsewhere in the frigid seas, ice algae provide a concentrated and accessible food source. "The table is set," Cota says.

Bellying up to the table are many of the creatures that make up the animal part of plankton. Adult krill, David Campbell wrote in *Crystal Desert,* "brush their thoracic baskets through the border zone of liquid and solid, mowing the ice algae in strips like a farmer cutting hay." Nematodes, which move like snakes and find their way using ice crystals as navigational aids, shoot darts through the cell walls of algae and suck out the contents; Cota once found 300,000 nematodes feeding on nine square feet of bottom ice. Amphipods, members of a family that includes the beach hoppers of warmer climes, eat almost anything, living or dead. "It's a complete ecosystem," Cota concludes, "bacteria, ciliates, rotifers, copepods, nematodes."

As the ice melts, it sows the water with algae cells. "It's like you go out and throw grass seeds on your lawn," Cota says. "It takes just a few cells to seed a population." The meltwater is mostly fresh since the ice expelled salts while forming, and the fresh water lingers for a while in a thin layer atop the denser seawater, and the newly liberated algae linger in that layer for a while, too, high in the water column, where spring sunlight can reach them and help them to reproduce, extravagantly. Within days of the melting, water that had been clear when it was covered with snow and ice is almost opaque with blooming algae. As the edge of the pack retreats northward (in the Arctic) and southward (in the Antarctic), it may be trailed by a greenish or golden brown cloud up to 150 miles long. The cloud is testimony to a recent ice cover. With plant blooms come animal blooms. "Algae are the grass that feeds the cow," Cota explains, the fuel that drives the marine food chain. The small animals that feed on the algae, either on ice or in water, are in turn fed upon by larger animals: whales, seals, seabirds, penguins.

According to Campbell, the brown pigment of some ice algae (which he thought on first sighting was paint off a ship's hull) hastens the thawing of the Antarctic pack in spring, by absorbing heat from sunlight. "The full influence of the algal pigment is unknown," he admits, "but it is clear that by subtly warming their immediate environment, these one-celled algae alter global weather patterns thousands of kilometers away. They alter the trajectories of ocean currents and the exportation of Antarctic cold to the lower latitudes. Indirectly but inexorably, these algae may affect the crops of soybeans in southern Brazil, the anchovy harvest of Peru, and the dry winds over the Sahel."

In solid freeze
The fishes lie.

—Ted Hughes,
"The Heron"

In the Antarctic autumn of 1911, Scott was checking the thickness of sea ice near the expedition's base when he found a number of fish frozen in the ice, "the larger ones about the size of a herring," he wrote, "and the smaller of a minnow." One of the larger ones was "frozen in the act of swallowing a small one. It looks as though both small and large are caught [by jelling ice] when one is chasing the other." On an earlier Arctic sledging trip, Scott had seen other fish frozen into the ice but without their heads. His guess was that seals had bitten off the heads then the rest had become incorporated into the underside of the ice cover, and as the ice sheet melted, from the top down, the decapitated fish passed through the ice sheet, from the bottom up.

Perhaps the most intriguing thing about these tableaux non vivants is that there aren't more of them. Unlike, say, octopuses, whose body fluids have a salt content similar to that of seawater and which don't freeze until the water they're in does, marine fish typically have only enough salts and other solutes in their fluids to lower the fluids' freezing points to about 30°F, which is above the freezing point of seawater. Theoretically, fish could be swimming in unfrozen seawater and still freeze. Some fish would get along fine—as long as they didn't come into contact with ice. Touching a mere mote of ice could set off a cascade of freezing in their supercooled fluid, killing them almost as fast as electrocution would. In certain fjords of Labrador which freeze over each winter, several species of fish are able to live out their whole lives in a supercooled state because they stay deep, where the pressure is so high that the freezing point of the water is reduced, and no ice forms there.

However, large portions of the polar and subpolar seas are shallow, and the water beneath the ice is just at the freezing point, and floating around in the water are many tiny, embryonic ice crystals. So why don't the fish that swim in those waters, of which there are billions, routinely bumping into ice, swallowing ice, taking in ice through their gills, resting on clumps of anchor ice, nibbling krill off the undersides of ice . . . why don't *they* freeze and die? Why aren't *their* corpses littering the frozen seas?

The answer came as a surprise even to scientists. "I stumbled on it," says Arthur L. DeVries, professor of physiology and biophysics at the University of Illinois at Urbana. As a graduate student in the 1960s, he signed up as a technician for a summer project in Antarctica ("I didn't even know where Antarctica *was,*" he claims) which involved catching fish through a hole in the ice. "In those days it took us two days to cut each hole with a chainsaw," he recalls. Some of the caught fish survived, but strangely (it seemed to DeVries) some did not. "We had trouble keeping them alive in the aquarium." The fish that did not survive turned out to be the deepwater fish. In the years that followed, he went back to Antarctica, caught more fish, took blood samples, and wrote his doctoral thesis on what he found.

What he found was that the fish that survived had antifreezes in their systems. Not the small-molecule, chemical antifreezes like those that drivers put in their cars—ethylene glycol, salts, glycerol—but large, protein ones. The fish were making the proteins in their livers and distributing them around their bodies, where the proteins lowered the freezing point of the tissues to 28°F, *below* the freezing point of seawater.

One Antarctic summer I visited a fishing hut that DeVries had set up on the ice of McMurdo Sound, something he has done almost every summer since he was a graduate student, only this time his own graduate students helped him drill a hole in the ice. At first, a Weddell seal gnawing at the edges kept the hole open, then a pipe carrying heat from a rusty, smoky, smelly, diesel-fueled stove inside the hut did. The students lowered baited fish traps into the hole, and when they brought them up they found bony deep-dwelling fish inside: naked dragonfish (*Gymnodraco acuticeps*), snailfish, Antarctic cod (*Dissostichus mawsoni*), borchs (*Pagothenia borchgrevinki*), and eelpouts (*Rhigophila dearborni*) also known as "eel crybabies."

The eelpouts live so deep in the sea that the pressure keeps ice from forming in the water around them, yet they have antifreeze in their systems. "Why do they need antifreeze if they don't come into shallow water and encounter ice?" DeVries asked himself. Other fish trapped at the same depth don't carry antifreeze. "Maybe once every five years they run into ice down there," he speculated, "and that would be enough to wipe them out." In a demonstration inside the hut, he placed an eelpout, gently, in a tankful of water that had been supercooled to 24°F and was therefore ice-free. What happened next was horrifically fascinating to watch. The eelpout thrashed in the water, as if it were trying to get off a hook. It twitched. Its mouth opened wide, its eye glazed, its head went down, its belly went up. Within minutes of being placed in the tank, it lay still. More than still: stiff. Devries picked it up like a blackjack and knocked it against the side of the tank and got a thud. It was hard-frozen. If returned to the sea, it would not recover.

"What this tells us is that in nature these fish carry ice with them," DeVries explained. He did a second, more cheerful part of the experiment. He took another eelpout out of a tankful of water that was 32°F and placed it in the tank with water that was 24°F. This time the eelpout swam "happily" around in the tank and "nothing happened." The warmer water in the first tank had melted the ice inside the fish so that when it was laid in the colder water there were no seed crystals left in its system to set off flash-freezing.

DeVries had thus shown that some deepwater fish carry ice in their systems, but "for a long time we couldn't see any," he says. He looked in parts of the fish where ice might be expected to be—blood, muscle, heart, liver, skin ("you take off its skin, and its skin will freeze")—without finding any. Finally, he discovered ice where he didn't expect it: in the spleen. "That's puzzling since the spleen is very deep in the fish and is never in contact with ice and the only way ice can get to it is through the blood and there's no ice in the blood!" He suggests that macrophages in the spleen, one of whose jobs it is to seek and destroy bacteria and worn-out red blood cells, could be seeking and destroying bits of ice as well. "Maybe," he adds, "macrophages gobble up ice crystals, subject them to salty fluids, and melt them," thus clearing the fishes' systems of icy particles.

CHAPTER TWENTY-ONE

ANIMALS IV

Reaumur, an early experimenter on cold insects, likened the recovery from freezing to resurrection.

—J. W. Kanwisher, in *Cryobiology*

FISH ARE PIKERS compared to insects when it comes to surviving freezing temperatures. Seawater doesn't usually get colder than 29°F, but air, the domain of most insects, can get much, much colder, and insects, being coldblooded, have to deal with it. Some are very good at dealing with it. Woolly bear caterpillars of the Canadian Arctic moth, *Gynaephora groenlandica,* spend eleven months of the year frozen solid in temperatures as low as −58°F yet as soon as the first leaves of the Arctic willow appear in springtime, they are out munching on them. Carabid beetles, *Pterostichus brevicornis,* of the Alaskan interior have been subjected as adults to temperatures as low as −125°F and survived.

As with plants, it's usually ice that kills insects, and not just low temperatures. Also as with plants, insects have two main strategies for ensuring that ice does not kill them: avoiding ice by supercooling and tolerating ice by freezing in the right way at the right time. Unlike plants, however, insects have other options, like flying someplace clement or crawling into holes out of the cold air. Yet migration takes energy, and holes don't provide enough shelter in regions of extreme cold.

Most insects, without doing much of anything, can supercool a few degrees. They are essentially small packets of water, and small quantities of water supercool more easily than large ones (since they contain fewer potential ice nucleators). To supercool more than a few degrees, however, insects need to do something besides be diminutive. Some make their own antifreeze proteins, like those in fish, which work the way the antifreezes do in fish but are far more potent since they have to do their stuff in colder temperatures. "A fish doesn't need a lot of punch," John

G. Duman, Gillen Professor of Biological Sciences at Notre Dame, points out.

Besides protein antifreezes, or instead of them, some insects make sugar antifreezes (including glycerol, sorbitol, mannitol, ethylene glycol, glucose, fructose, and trehalose). Sugars lower the freezing points of cells by increasing the concentration of dissolved particles. In regions where winter air temperatures run very low, concentrations of sugar antifreezes in insects can run very high. Kenneth B. Storey and Janet M. Storey, biologists at Carleton University in Ottawa, Ontario, point out that the proportion of glycerol antifreeze in the body fluid of the gall-moth caterpillar, *Epiblema scudderiana,* is nearly as high in winter—40 percent—as is the proportion of ethylene glycol in fluids added to car radiators in southern Canada is in winter—50 percent. The glycerol keeps the caterpillar from freezing down to −36°F, while the ethylene glycol keeps car radiator fluid from freezing down to −40°F. One insect, the bark beetle *Ips acuminatus,* actually produces the car antifreeze and employs it for supercooling, something Duman finds "particularly interesting" since the ethylene glycol is toxic to larger animals, like family pets, which enjoy its sweet taste and lap up pools that have leaked onto the family driveway, to their subsequent distress.

Insects' tough waxy coverings generally keep outside nucleators—snow, bacteria, dust, ice—from getting inside and setting off freezing there, but there are other nucleators inside already: bits of food, digestive microbes, blood proteins. Before winter sets in, some insects get rid of these nucleators so they can supercool. The stag beetle, *Ceruchus piceus,* stops eating in autumn, clearing its gut of bacteria and fungi, both of which are powerful ice nucleators. It also eliminates from its blood a protein that in summer, very usefully, moves lipids around the body (with its lowered metabolism, it probably doesn't need them moved around in winter). Without making any antifreezes, then, just by getting rid of internal ice nucleators, the stag beetle is able to supercool to below −4°F. By combining techniques, removing ice nucleators *and* producing antifreezes, the *Pytho* beetle of the Canadian Rockies can supercool to −76°F—a remarkable feat, Duman points out, because water freezes at −40°F even without nucleators. Thus, at a full 36 degrees below the point at which pure water is so cold it turns spontaneously into ice, some insects are able to keep their body fluids fluid.

Some insects seem to have gotten rid of internal ice nucleators permanently, for all seasons. Certain aphids in Britain apparently selected over the eons *for* proteins with surfaces that are a poor match for ice and *against* proteins that perform the same function but are a good match for ice. (They don't need to worry about food particles serving as nucleators: their food is liquid, and they suck it.)

Again as with plants, some freeze-tolerant insects don't get rid of ice nucleators but make more of them. These proteins are usually large ones, which induce ice to form early, while temperatures are still fairly high, therefore slowly, so the ice consists of small crystals outside cells and the cells have time to adjust. "Thus we have the paradoxical situation," Duman writes, "where ice nucleators, which are located in the extracellular fluid, actually function as intracellular antifreezes." Some insects even use "the ultimate nucleator, ice itself," to survive freezing, one of them by employing ice in the blood of its host to trigger early freezing in its own blood.

Frozen insects don't freeze completely. The outer limit of their survivability is usually three-quarters ice. By the time that much water in an insect's body has turned to ice, its cells have shrunk so much that the membranes are in danger of collapse. If that happens, ice has an entry into the cells, and when ice enters cells, the cells usually die and so do the insects.

Is one strategy superior to the other? Freezing does have advantages. A frozen insect loses less water by evaporation than an unfrozen one. Being frozen saves energy, by greatly lowering metabolism. Some insects that tolerate freezing can keep right on eating into autumn, since they don't have to clear their guts of nucleators. And they are free to stay around in damp places, where ice is sure to form. However, in areas that go through many freeze-thaw cycles, avoidance is probably the safer strategy, since the repeated melting and refreezing produces large, dangerous ice crystals. Some insects—switch-hitters—can avoid ice at some times and tolerate it at others.

"Name a strategy and you will find some insect somewhere doing it," declares Duman, who chose to study insects because of their adeptness at problem solving. The insects' adaptability has made them the most successful colonizers of extremely cold environments of any terrestrial animal, he points out. The "burst of insect activity in the Arctic in summer, notorious in many areas because of the abundance of biting dipterans"—dipterans being the family of winged insects that includes sandflies, horseflies, black flies, and mosquitoes—illustrates "the tremendous overwintering success of these populations."

Mosquito
Flew up singing, over the broken waters

—Ted Hughes, *"The Mosquito"*

The eggs of some mosquito species must freeze in order to hatch. Some mosquito larvae spend the winter frozen into the top layer of stagnant tundra ponds. According to Keith Miller, formerly entomologist at the Institute of Arctic Biology at the University of Alaska at Fairbanks, mosquito larvae on Alaska's North Slope stay frozen for nine months, then in spring after thawing, pupating, and hatching, they spread their wings and fly away in plaguelike hordes. "Mosquitoes, the one serious drawback of the North," Vilhjalmur Stefansson wrote, "—far more serious in the minds of all who know than winter darkness, extreme cold or violent winds." In his book *Spineless Wonders: Strange Tales from the Invertebrate World,* Richard Conniff pointed out that on the Arctic tundra when "the adult mosquitoes emerge en masse, they have about twenty minutes to mate, find a victim, get a blood meal, and lay a new batch of eggs before winter sets in again. Canadian researchers once sat still in such a swarm long enough to report that they suffered nine thousand bites a minute, a rate sufficient, at least in theory, to drain half their blood supply in two hours." Also a rate sufficient to drive people with a lesser commitment to science mad.

Sittin' on the ice till my feet got cold, sug-ar-babe . . .
Watch-in' dat craw-dad go to his hole, sug-ar-babe.

—Kentucky mountain song, from
Ballad Makin' in the Mountains of Kentucky

It's autumn in a northern forest. A wood frog lies under damp leaf litter on the forest floor. Night comes; air temperature drops; a moist spot on the frog's skin freezes. Just below the spot, liquid inside the frog starts freezing too. Ice forms in the space between the skin and muscles, between fibers of the muscles, around organs in the abdomen. "It's not a flash thing," Janet Storey points out. "It takes four to six hours." If the frog were to undergo an MRI (magnetic resonance imaging) scan during those hours—as frogs have in the lab, which is how the sequence was determined—the image would show liquid remaining in the heart and liver, then those organs growing increasingly dark as the liquid turns to ice. "You see it like a shadow passing through the animal," Storey says.

Blood starts to freeze, gets thicker and thicker, until what remains is a dense goop of blood in vessels above the heart.

Within 24 hours of that moist spot freezing, the wood frog has turned into a hard round disk. Its legs are drawn up under its body and its head is lowered over its front feet, in the water-conserving position. It is unable to move. Its heart does not beat. Its lungs do not take in air. It is stiff, not as stiff as a chicken leg in the freezer but as stiff as a crunchy semithawed chicken leg, which can be bent. If it were to be opened up at this stage, according to Storey, you would see "huge amounts of ice" inside, a large block of ice in the abdominal cavity, maybe as a solid mass, maybe as slush. You would see flat sheets of ice under the skin and between muscle groups, packed into the brain cavity, between brain and bone. The frog would be two-thirds ice.

In addition to being stiff, paralyzed, pulseless, and breathless, the frog would be severely diabetic. If a human were that diabetic, he would be dead. While a normal person has 50 to 100 milligrams of glucose per 100 milliliters of his blood, a frozen wood frog may have *4,500* milligrams of glucose per 100 milliliters of *its* blood. As with insects, the glucose lowers the cells' freezing point by increasing their concentration of solutes. It also provides energy of a sort and helps protect the cells against damage from freezing; slows metabolism (probably); and furnishes particles that serve as ice nucleators, bringing on freezing at warm temperatures and slow speeds. What triggers the huge buildup of glucose isn't short days or cold temperatures: it's ice. Within as brief a time as five minutes after the first ice crystal makes its appearance on the frog's smooth skin, the frog's liver is breaking stored glycogen down into glucose and delivering it into the bloodstream.

The disks have little trouble turning back into frogs. During Samuel Hearne's 1771 crossing of Canada's great Barren Grounds, he noted: "I have frequently seen [northern frogs] dug up with the moss, frozen as hard as ice; in which state the legs are as easily broken off as a pipe stem, without giving the least sensation to the animal; but by wrapping them up in warm skins, and exposing them to a slow fire, they soon recover life, and the mutilated animal regains its usual activity." Frozen frogs can be thawed overnight in a refrigerator, Janet Storey says, or in two or three hours at room temperature. "But it's 24 hours before they jump."

Not only wood frogs but spring peepers, gray tree frogs, and striped chorus frogs, and not only frogs but box turtles, pond sliders, garter snakes, Siberian salamanders, and many other coldblooded creatures, freeze and live. Painted turtles can do it, but only in their first year. The turtles nest as far north as Ottawa, farther north than any other land or freshwater turtles in North America. In winter, the hatchlings don't leave the nests, which are scooped-out holes in the exposed banks of rivers or lakes, only three or four inches deep. Although predators can't get to them there, the cold can. Using internal ice nucleators, the hatchlings freeze when the air temperature falls below 28°F, thaw when it rises above 32°F, freeze when it falls, thaw again, freeze, thaw. Like wood frogs, the painted turtles freeze from the outside in, and all their parts shut down eventually except their brains, which nearly shut down. The parts regain full function when they thaw, only to shut down with the next freezing. "These animals have mastered the tricks of organ cryopreservation," the Storeys conclude, "the freezing of live tissue for storage and subsequent use." After the first year, the painted turtles can no longer do these tricks, though. As adults they spend winters resting, unfrozen, on the mud at the bottom of the rivers and lakes they may have formerly nested beside, frozen.

Along the coasts of northern seas, too, many small creatures (Venus clams, for example) burrow into mud to keep from freezing in winter. When the tide goes out, a strip of seabed is left exposed to the cold air and so are they. All it takes is a short bit of burrowing, J. W. Kanwisher of the Woods Hole Oceanographic Institution noted in the book *Cryobiology,* since the cold doesn't penetrate very far down (mostly because of all the latent heat released when the top of the mud turns to ice). Other invertebrates move to deeper water in winter, which protects them from freezing, as long as the water doesn't freeze. But some seashore creatures stay out of both the deep water and the mud and tough it out in open air, where they turn largely to ice.

Go to a seacoast at low tide on a cold winter day, Kanwisher advised, and you'll find "ice coating most of the rocks and bottom. Frozen with this ice are all of the animals that normally live there. Snails are frequently visible with the foot extended, as though they had been browsing when freezing overtook them. If one is chipped out of the ice and thawed in the warmth of a hand it will usually either commence crawling or withdraw quickly into its shell. In the same way," Kanwisher went on, "a frozen mussel . . . , when thawed, may project its siphon and begin filtering a stream of seawater. These signs of normality indicate that freezing causes very little after-effect. . . . that freezing to these animals means little more than a temporary cessation of their activities."

As for ice algae, "it's surprising how much abuse they can take,"

Glenn Cota says. "You can warm cells that were frozen for months, and you'll see them move around or fluoresce. Almost all of them are intact."

> Cheerful daredevils, otters often will use the rapids of a river—the last areas to freeze and the first to thaw—to slip under the ice.
>
> —Henry David Thoreau, Journal (*Winter*)

By contrast, warm-blooded animals cannot freeze and live. The Arctic ground squirrel, the only mammal known to be able to lower its body temperature below the freezing point, will supercool to 28°F and not freeze. Most warm-blooded animals can't even lose much body heat and survive. To get through winter, they do things like grow more feathers and fluff them up (trapping air in between); grow more fur and fluff it up; add blubber; find nooks out of the wind; get small (curl up or hunch down); fly somewhere else; burrow into the earth below the frost line; decrease body metabolism and thus save energy; increase metabolism once body temperature drops below a certain point ("That's the whole point of being warmblooded," Janet Storey reminds people, "being able to regulate body temperature"); shiver (goldfinches shiver almost all winter long when perching); lay down extra brown fat in pads over shoulder blades (which act like heating pads when burned); huddle in communal nests; eat more; sleep more; tunnel into insulating snowpack and wait out the storms.

"Shrews, moles, and mice were snug and protected beneath the thick blanket of snow," David Rains Wallace wrote in his book *Idle Weeds: The Life of an Ohio Sandstone Ridge,* describing the effects that a severe winter day had on local wildlife. "The only mammals to suffer visibly from the cold were the opossums, newcomers from the South American tropics. . . . The tips of their naked ears and tails froze and turned black." The gulls and ducks you see standing flat-footed on the bare ice of a lake or pond don't get frostbitten because blood is constantly being pumped into their feet, although to keep the feet from giving up too much body heat to the ice, which doesn't need it, the blood pumped into them is cooler than blood elsewhere in the body. The birds' legs have a heat-exchange system, in which arteries and veins pass close to or touch each other so that the warm, outflowing arterial blood loses some heat to the cooler, inflowing venous blood, and blood enters the feet tepid. (Penguins' legs have the same exchange system.)

Even with such mechanisms, however, few birds besides penguins choose to winter around ice. The black guillemot of the Arctic is one of the few. Mostly white in winter, it stands on ice and rests on ice and socializes on ice. It feeds under ice and at the edges of ice. In spring it copulates on ice. It may even, perhaps, nest on ice.

"I've never actually *seen* guillemots nesting on ice but I'll bet they've tried," says biologist George Divoky, who's been studying a nesting colony of black guillemots on Cooper Island near Barrow for more than 25 years. In summer, guillemots have been spotted almost as far north as the North Pole, where there's nothing around for hundreds of miles *except* ice. Divoky has noticed the birds peering into cavities in pressure ridges, as if assessing them, then performing breeding behaviors associated with nest sites. On land they nest among rocks, "and a pressure ridge can look just like a rock talus."

Guillemots feed on cod which feed on the animals in the plankton which feed on the ice algae (guillemots also eat the plankton). Being divers and being small (only a little larger than a pigeon), they don't need more than narrow cracks or holes in the ice in order to pass to and from the underside of ice floes. Some people suspect they even live in seal holes. Alfred Marshall Bailey, respected ornithologist and author of the 1948 *Birds of Arctic Alaska,* was told by seal hunters that guillemots "live under the ice like the seals," finding air in the pockets beneath high pressure ridges. "There *are* air pockets there," Divoky allows. "I could kind of see that happening. Advantages would be lack of predation and lack of dealing with the elements, like being in an igloo. But it would be a very risky business."

Many birds that don't feed off the bottom of sea ice can nonetheless be fed by it. One day there'll be no ice around Cooper Island except for pack way off in the distance, Divoky says, then the wind will come up and drive the ice toward the island, and large pieces of ice will bash against the shore and break up. "Suddenly there'll be thousands of birds around my tent making so much noise I can't sleep," he says. "Sabine's gulls, red phalaropes, lovely white Arctic terns. When I go out, the beach is covered with billions of little amphipods, all of which have been knocked or melted off the ice, and the birds are feasting on them. It gives you an idea of the densities of these things under the ice."

Until recently, scientists didn't know where spectacled eider ducks (or "spectacular spectacled" eiders, as one biologist enjoys calling them) spend the winter. Then in 1995 two biologists with the U.S. Fish and Wildlife Service were flying over the Bering Sea 120 miles from land, "out in the middle of nowhere," surrounded by an "endless expanse of ice . . . white in every direction, as if we were flying inside a light bulb," one of them, Gregory Balogh, relates, when they saw on the general whiteness a chestnut-brown smudge. They assumed it was a walrus haul-out site. When they got closer, they "whooped victoriously," Balogh wrote. The smudge turned out to be 500 spectacled eiders packed into a single hole in the ice, the heat of their bodies keeping the hole open. A few minutes later they spotted a larger smudge, this one with 4,000 eider ducks "crammed into a tiny hole in the ice barely large enough to contain them all. Packed spectacle to tail, each eider was touching other eiders on all sides." A few minutes after that, they flew over a *really* large smudge, "a solid, seemingly congealed mass of spectacled eiders," 50,000 of them, plus one snowy owl.

In all, that day and the next, they discovered 155,000 eiders, 20 flocks of them squeezed into 30 "slivers" of open water. Except for the slivers, "the ice cover was complete for miles in all directions."

Jaws flimsy as ice
Champ at the hoar-frost

—Ted Hughes,
"The Arctic Fox"

Even black guillemots, usually on good terms with ice, would occasionally be better off having less of it around. In years when the seawater around Cooper Island stays frozen well into spring, Arctic foxes make a bridge of it. They walk ten miles from the mainland over the ice, steal the guillemots' eggs off the nests, and bury them. One year, the work of a single egg-robber meant no chicks at all for the colony. To an Arctic fox, a ten-mile walk over ice is a Sunday stroll. In the course of a winter, a fox may walk hundreds of miles across ice in its search for food. "They are great ice travelers," Francis Fay said. "As soon as the ice forms in fall, they are out on it."

"What in the world was that fox doing up here?" Fridtjof Nansen exclaimed when he spotted fresh fox tracks during his sledge trip toward the North Pole. "A warm-blooded mammal in the eighty-fifth parallel!" Nansen found it "incomprehensible what these animals live on up here," although he did notice here and there on the ice "unequivocal signs that it had not been entirely without food." Charles Brower had a chance to find out what foxes live on when he was trapping foxes on sea ice for their fur. "Here, far from land," he wrote, "they spend most of the winter subsisting on dead seals or walrus or even dead whales drifting in the pack or else following the polar bears about to eat whatever these animals leave of their kill." (Polar bears tolerate foxes as long as they keep their distance while the bears are eating.) "But along towards spring, their favorite dish is baby seals which they dig out from under the snow. Should all other food fail," Brower noted, "white foxes can always dine well on the small marine animals frozen fast to the bottom side of young ice. This may require a bit of waiting. But as soon as a pressure forms the young ice will turn over and expose everything."

To Vilhjalmur Stefansson, "the white fox is almost as much of a sea animal as the polar bear. . . . The hungry travel, and the well-fed spend long periods lying on the tops of snowdrifts, hummocks, or ice ridges, sniffing the wind." One researcher found a fox had made itself a winter den inside the carcass of a walrus.

But how, one might ask, do Arctic foxes, or any other animals for that matter, eat frozen food, all those carcasses and scraps lying around in subfreezing temperatures? Polar bears are well equipped for it, Richard Nelson notes in his book on Eskimo hunters. They "tear apart the carcasses of dead walruses along the beaches in northwest Alaska, . . . in spite of the fact that the heavy skin is frozen solid and is so tough that it cannot be butchered with a sharp axe. . . . It seems that both claws and teeth are used in the process of tearing them apart." Arctic foxes, despite what the poet said, don't have flimsy jaws either.

For many terrestrial animals, however, eating frozen food is a trial. "You can see from the tooth marks on a carcass," says William Pruitt Jr., senior scholar at the University of Manitoba in Winnipeg and an expert on northern mammals. "The cutting edge isn't clean. They are shredding the meat out." In their book *The Birds of Winter,* Kit and George Harrison tell the story of a scientist acquaintance who was in the woods one very cold December day when he happened upon a great horned owl sitting on the flank of a dead red fox, as it would on eggs it wanted to hatch. It was thawing its meat before dinner, using the heat of its live body to soften the hardened flesh of a dead one.

> If the muskrat has no longer extensive fields of weeds and grass to crawl in, what an extensive range it has under the ice of the meadows and river sides; for the water settling directly after freezing, an icy roof of indefinite extent is thus provided for it, and it passes almost its whole winter under shelter, out of the wind, and invisible to men.
>
> —Henry David Thoreau, Journal (*Winter*)

Not only muskrats but minks, otters, shrews, and beavers spend a good part of northern winters under canopies of ice. During her study of a beaver family in Bear Mountain State Park, 40 miles north of New York City, naturalist Hope Ryden learned that although all the beavers adjusted well to an ice cover once it formed, they did their best to put off the day when it did.

"One cold evening, I arrived at the pond and found most of it frozen," she related in her book *Lily Pond: Four Years with a Family of Beavers.* The only part of the pond not frozen, she noted, was a channel running from the beaver lodge to the dam, plus a small "swimming" pool in front of the lodge. She watched as the Inspector General, as she called the male, tried to keep the pool from icing over, by breaking off slabs of ice along its edges. He would do it either by pushing down on the ice with his front paws or by getting on top of it and trying to weigh it down or by "bumping" it from beneath. "One hard thud and the ice would crack," Ryden reported. After more bumps, the ice would shatter, "and the Inspector General would pop right through, like a plump show-girl bursting out of a cake."

Ryden had never heard of beavers breaking ice this way, and it raised profound questions for her, about the nature of foresight, learning, and intelligence. "Whether or not he knew it," she declared, ". . . this beaver was postponing his consignment to a subaqueous realm—an unlit, claustral world soon to be endured without the palliative of a long winter's sleep. He was also buying time in which to add more branches to his paltry food cache. Was he able to anticipate what was about to befall him and was he trying to forestall it? . . . Can beavers identify a future effect (in this case, long confinement) with an immediate phenomenon (in this case, the formation of ice)?"

Lily, the Inspector General's mate, "seemed equally determined to stave off incarceration" and joined him in breaking ice, although she was not an "underwater rammer." She preferred to do "a kind of jiggling dance until the edge upon which she perched gave way and sank beneath her." Still, over the next several nights the ice on Lily Pond thickened until finally the little family of beavers was "sealed under ice." Contemplating the ice cover, Ryden got to wondering about the kits. "How strange it must seem to them suddenly to find a lid on their world. Would the youngsters forget it was there, attempt to surface, and bump their heads?"

Three months later, all the beavers emerged through a hole in the ice, and Ryden, who had reason to believe they had exhausted their food supply and were in danger of starving, tossed birch branches on top of the ice. One of the kits captured a branch but to her surprise didn't climb out of the water to do so. He "swam underneath the ice shelf until he was exactly below the spot where the branches rested," she wrote. He hit his head against that spot, shattering the ice, "popped up through broken shards and reached for a branch, which he then pulled underwater. A few moments later I heard gnawing sounds from inside the lodge. So even baby beavers break ice!"

"When you think about it," Ryden concluded, clearly having thought about it a great deal, even to imagining what was going through the beaver's minds, "life under ice is an extraordinary adaptation."

GUILLEMOT

CHAPTER TWENTY-TWO

HUMAN I

A bitter, brittle
cold represents, as it were, a message
to the body of its final temperature

—Joseph Brodsky,
"Eclogue IV: Winter"

IN THE DEPTH of an Alaskan winter, a lone gold miner searching for his old claim broke through river ice, got wet to his knees, cursed his luck, struggled to build a fire with cold-numbed fingers, watched with despair as the tiny flame that puffed up became smothered by a falling piece of moss:

"A certain fear of death, dull and oppressive, came to him. . . . [H]e realized that it was no longer a mere matter of freezing his fingers and toes, or of losing his hands and feet, but that it was a matter of life and death with the chances against him. This threw him into a panic, and . . . [h]e ran blindly. . . . It struck him as curious that he could run at all on feet so frozen that he could not feel them when they struck the earth and took the weight of his body. . . . Then the thought came to him that the frozen portions of his body must be extending. . . . He was losing in his battle with the frost. It was creeping into his body from all sides. The thought of it drove him on, but he ran no more than a hundred feet, when he staggered and pitched headlong . . .

"He pictured the boys finding his body next day. Suddenly he found himself with them . . . [H]e came around a turn in the trail and found himself lying in the snow. It certainly was cold, was his thought. When he got back to the States he could tell the folks what real cold was."

Real cold in Jack London's story, "To Build a Fire," was −75°F. The narrator was aware of the temperature because as he looked back at the Yukon River "hidden under three feet of ice," he had "spat speculatively. There was a sharp, explosive crackle that startled him. He spat again. And again, in the air, before it could fall to the snow, the spittle crackled. He knew that at fifty below spittle crackled on the snow, but this spittle had crackled in the air." When there is no thermometer, according to Bishop Stuck, the old-timer's test of temperature is the cracking noise that breath makes at −50°F or below—no spittle necessary. Richard Byrd reported in his book *Alone* that when the air next to his Antarctic hut fell to −84°F, his breath "pinged." *Real cold.*

"Bare the hand, and in a few minutes the fingers will turn white and be frozen to the bone," Stuck wrote, describing the effects of "strong cold." He once had the somber task of burying a woodchopper who had frozen 75 miles from any settlement, in the Koyukuk region of Alaska. In the woodchopper's pocket memo book, he had noted that the air crackled when he breathed. Stuck had been out on the trail that day too and figured the temperature at the moment the man fell to the snow "never to rise again" had been "fifty-eight below zero and a wind blowing!"

Stuck considered the woodchopper's last moments. "One supposes that the actual death by freezing is painless, as it is certainly slow and gradual. But if the actual death be painless, the long conscious fight against it must be an agony; for a man of any experience must realise the peril he is in. . . . All of us who have travelled in cold weather know how uneasy and apprehensive a man becomes when the fingers grow obstinately cold and he realises that he is not succeeding in getting them warm again. It is the beginning of death by freezing."

Ice on his fingers, ice in his heart,
ice in his glassy stare

—Robert Service,
"The Ballad of Blasphemous Bill"

The woodcutter and the miner didn't really freeze to death. They died of cold and then froze. A human being's core temperature can't fall very far below normal without his vital parts—heart, brain, lungs, digestive organs—going awry and doing him in. Hearts stop beating or go into fibrillation, lungs hemorrhage and fill with blood. The lowest core temperature ever recorded in a human being who survived, according to the *Guinness Book of World Records,* was 57.5°F—25 degrees *above* the freezing point. "Miracle baby was like a block of ice after 6 hours at 40-below" went the headline in the February 24, 1994, *National Enquirer.* "ALIVE—AFTER FREEZING TO DEATH." At 2:30 one winter morning two-year-old Karlee Kosolofski of Rouleau, Saskatchewan, followed her father out of the house, without her father knowing it, as he was leaving for his job at a dairy, and the door swung shut behind her. Five and a half hours later, her mother Karrie found Karlee huddled outside the door. Karlee had no detectable pulse or breath. "She was literally like a block of ice," an emergency room doctor said. "Her legs were like something you'd find in a refrigerator freezer." Doctors hooked her up to a heart-lung machine, pumped out her blood, heated it, and pumped it back in. They detected a heartbeat when her body temperature got up to 77°F. To keep her heart beating, they gave it a couple of shocks. Five hours after she got to the hospital, she had a normal temperature. "Karlee is our little miracle child," said her grandmother.

Karlee survived because she was in what Alaskan physician William J. Mills Jr., former director of the high latitude health research program at the University of Alaska at Anchorage, calls a "metabolic icebox." An icebox is a state of diminished functioning in which a person's oxygen needs are at a minimum and which thus allows a "trace of life" to persist, sometimes for hours. It's the chilled body's last-ditch effort at self-preservation, a "mid-lethal" state; any further loss of body heat could bring death. In the cold-injury field, there's a saying, "You aren't dead until you're warm and dead." That means if you stumble across what appears to be a recently frozen corpse—stiff body, no response to cries of alarm, skin blue and cold, no apparent breath or heartbeat—you shouldn't start saying eulogies right away. Warm the corpse and keep checking for vital signs: you might just find some. If after the body is warmed it's still stiff, blue, etc., *then* you can give up.

One method of warming, common among mountaineers, is to use one person's warm body to heat someone else's cold one, but Mills doesn't recommend it. Not only does it not ensure that the patient's core will be warmed, it could cause the person doing the warming to become hypothermic as well. (A person is hypothermic when his core temperature drops 3.6°F or more.) The body-contact method should be used only if three ("3!" Mills writes in emphasis) nonhypothermic people are engaged in doing the warming.

If people don't actually die by freezing (hypothermia can kill without a single ice crystal forming), their body parts certainly do. In a sort of natural triage, extremities and appendages and protuberances—fingers, toes, hands, feet, noses, ears, chins, cheeks, nipples (breasts sometimes freeze while women are nursing babies in cold tents, according to a Lapp account), Adam's apples ("a most inconvenient thing to freeze," wrote Stuck, who once froze his), and male members ("8TH DECEMBER SATTURDAY 1804," William Clark wrote in his journal during the great western expedition, "this day being Cold . . . my Servents feet also *frosted* & his P——s a little") are often sacrificed to keep essential parts—hearts, brains, lungs, digestive organs—operating. What happens is that as soon as cold air comes into contact with a person's skin, small blood vessels near the surface of the skin constrict, thereby reducing heat loss to the outside and sending more blood to the core. "Cold hands, warm heart" is more than social small talk.

However, the body doesn't abandon the hand altogether. From time to time the constricted vessels at the surface open and vessels carrying blood away from the skin close, and blood enters the hand at up to 100 times the rate it had just been. Once the hand is rewarmed, the vessels to it close again and those leading from it open. This intermittent warming of the chilled appendage becomes more and more intermittent if the cold persists until, if the vital organs themselves are threatened with cold, it ceases altogether. In that case, the vessels at the surface stay closed, and ice starts forming in the blood-starved hand.

Ice forms first in the spaces between cells, as it does in plants and wood frogs (for, as Harold T. Meryman, editor of the book *Cryobiology,* pointed out, "regardless of the mysterious complexity of the biological matrix," whether human or plant or amphibian, "freezing represents nothing more than the removal of pure water from solution and its isolation into biologically inert foreign bodies, the ice crystals"). Also as with plants and wood frogs, most of the injury to frostbitten human beings probably comes from excessive removal of water from the cells—dehydration—rather than from any pressure exerted on the cells by ice crystals outside them, even though the crystals can grow larger than the cells themselves.

By definition, frostbite is the cooling of tissue to the point of ice-crystal formation. This happens to skin when its temperature gets down not to 32°F but to between 24°F and 20°F. "If you can avoid those temperature levels," Mills emphasizes, "your tissues cannot freeze." Degrees of frostbite range from so mild that it's essentially a cosmetic matter to so severe that it kills the body part. When ice forms only in the outermost layer of your skin, it's "frost nip." You feel a tingling or stinging or a dull ache, then numbness, with a sudden whitening of the skin. Nipping is no big deal; the tissue under the nip stays soft, resilient, and doughy. Robert Falcon Scott wrote indulgently of the nips that Henry "Birdie" Bowers, one of the men in his South Pole party, got repeatedly because he insisted on wearing a small felt hat that didn't cover his ears: "His ears were quite white . . . The patient seemed to feel nothing but intense surprise and disgust at the mere fact of possessing such unruly organs."

If deeper layers of the skin are frostbitten, the look of the skin can be horrific even when the (long-term) injury is not. "Ravaged by frostbite, everyone was getting uglier by the day," Will Steger wrote of himself and his fellow trekkers to the North Pole. "Black scraps of dead skin hung from cheeks, noses, and chins." Early in the trip, his face had "flash frozen" and his skin become "caked with mottled yellow scabs" which later hardened into "a black mask." The ugliness passed, for him and all the others. The scabs sloughed off, leaving new, reddish skin, which eventually took on a normal appearance. "In fact," Steger wrote, "one doctor told me that facial frostbite has an impact similar to a facelift. . . . On all of my Arctic journeys I've come back with a new complexion."

If frostbite extends through the skin into the nerves, muscles, tendons, and bones beneath it, tissue is probably not only ugly but dead. Little Karlee's legs were so severely frostbitten that one of them had to be amputated above the knee. It was having severely frostbitten feet that prompted Titus Oates, also in Scott's polar party, to utter one of the most famous lines in the history of exploration—"I am just going outside and may be some time"—before he stepped out into a blizzard to die of exposure rather than delay the other men on their return attempt. With swollen, "useless" feet, Oates had "become lame, couldn't keep up," Scott wrote. The very next day, Scott wrote of himself: "Amputation is the least I can hope for now. My right foot has gone, nearly all the toes—two days ago I was proud possessor of best feet."

Oates's frozen body was never found, but eight months after Scott wrote in his journal another famous line in polar lore—"Last entry—For God's sake look after our people"—a search group discovered his body and those of two of his companions inside their tent. The others were frozen in "an attitude of sleep," a member of the search group observed, but Scott had "thrown back the flaps of his sleeping-bag and opened his coat." This was probably not defiance of circumstances but "paradoxical undressing," a flush of blood rising to the surface of the whole chilled body, perhaps making Scott feel, at the very end, warm.

troops marched
—what could we do?—with frostbitten feet as white as milk.

—Donald Hall, "*1943*"

Frostbite is the soldier's curse even more than the explorer's. "Whole armies have been decimated by cold injuries overnight," Murray P. Hamlet, longtime director of the cold-injury research program for the U.S. Army, states. Cold injuries have affected the outcome of many campaigns, he added, from the skirmishes between Xenophon's Greeks and the Armenians in 400 B.C. to the American Revolutionary War in 1777–78 to the Finno-Russian winter war of 1939, the Falkland War, and the Soviet war in Afghanistan. During the Korean conflict, soldiers with cold injuries, mostly frostbite, made up 10 percent of the total wounded. When a division of U.S. Marines camped around the Chosin Reservoir were forced to retreat in −45°F cold, "they got nailed," Hamlet says. "Many died because they couldn't walk on their feet." To this day, survivors of that retreat have trouble walking: "I'm still taking care of some of those injuries."*

During World War II, 12,000 American casualties were attributable to frostbite, the most dramatic, according to Hamlet, "the waist gunners of bombers who were exposed to high-altitude cold and 200-mph wind chill." Frostbite caused 112,627 German casualties (someone was able to say with precision). On their march to Moscow, the Germans suffered 15,000 amputations in two months because of cold injuries.

Probably the first person to describe in detail what happens to human flesh when it's frostbitten, and under what circumstances it is, was Baron Dominique Larrey, chief surgeon of Napoleon's Grande Armée, which fought many cold-weather campaigns. Larrey concluded that soldiers were most likely to suffer from frostbite when "rigorous cold" was followed by thaw, and that the suffering of the soldiers was most extreme when their feet had been frostbitten at least once before. It wasn't clear to him or anybody else, though, what the best way to treat frostbite was until the time of the Korean War. During that conflict yet far from the battlefields—in Alaska—William Mills was seeing his share of frostbite victims, with a "distressing" number of amputations among them. He decided to do something about it. Alaska, he realized, was a "natural laboratory. The way of life there offered many opportunities for getting frostbite."

By comparing the various treatments Alaskans got in the field and in

*Some "cold injuries" are actually trenchfoot, which occurs in wet settings with cold but above-freezing temperatures and no ice. Early in World War I, when doctors didn't distinguish among types of cold injury, all were lumped together and called among other things "frostbite, water bite, footbite, cold bite, puttee bite, trench bite . . . chilled . . . feet . . . or merely . . . 'feet cases.' "

hospitals as well as their outcomes, Mills and fellow researchers came up with an "Alaskan system of care." The key to the system is warming of the frostbitten part *rapidly.* Baron Larrey disapproved of rapid rewarming, and the influence he exerted on frostbite treatment through his writings discouraged the practice for over 150 years. But Mills found during his studies that when tissue frozen in very cold outdoor temperatures, say −40°F, is thawed at room temperatures—that is, slowly—the meltwater in it can easily be changed back into ice since the surrounding area is still at least partially frozen. But when the tissue is thawed at *higher*-than-room temperatures—that is, quickly, within two or three minutes—the melt in the tissue does not refreeze because the area around it has already been warmed to above the freezing point.

However, rapid thawing using ovens, car exhausts, diesel generator exhaust, bonfires, and any other producers of "excessive" heat is not a good idea, as Mills points out, since all of them can burn the cold-seared flesh. "Recovery is hopeless," he concludes, "when frozen tissues are 'cooked.' " (Even Baron Larrey noted that when bivouac fires were kindled, "they were more injurious than useful.")

Under the Alaskan system of care, the best way to thaw a frostbitten part is to immerse it in a bath of water heated to between 100°F and 112°F. If possible, the water should be circulating since the frozen part will chill the warm water immediately around it. For a while, Mills used his hospital's washing machine on the warm rinse cycle.

If a warm bath can't be arranged, in extremis a blood bath might be. Indians along the Yukon and Kuskokwim rivers in Alaska, Mills pointed out, sometimes thaw their frozen extremities in the "warm gut cavity of small animals or birds (spruce hen)."

The Lapp remedy for frostbite was to rub it with snow, Johan Turi wrote in *Turi's Book of Lappland* (1931). Not any old snow, but the snow nearest the ground, *soenjas,* or granulated snow. In other chilly places besides Lappland, rubbing frostbitten flesh with snow or icy water or even ice has been common practice, probably because the thawing achieved that way is gradual and therefore less painful than quick thawing. Using cold to treat frostbite, however, is only a little less damaging than using excessive heat, Mills found. The cold prolongs chilling and allows the already melted ice crystals to refreeze and the crystals that haven't yet melted to grow.

Another Lapp remedy for frozen feet, according to Turi, is drinking the blood of a living reindeer, or else frozen blood stirred "till it is fine and smooth." Drinking fluid (it need not be blood) does give some protection against frostbite, as does food. Both stoke the metabolic furnace and provide fuel for bodies working extra hard to maintain heat in the cold.

Unlike plants and frogs, people do not have chemical ploys for keeping ice from forming in their tissues in damaging ways. "Humans for the most part show surprisingly little physiological adaptation to low temperature," Peter Marchand wrote in his book *Life in the Cold*. "Biologically, . . . we remain essentially tropical beings." What allows Australian aborigines to sleep on the ground in near-freezing temperatures and Nepalese and Andean mountain dwellers to walk barefoot and thinly clad over ice and snow, Marchand concluded, is probably psychological or cultural, not physiological, adaptation; these peoples are not fat but lean. "The highest insulation values have been recorded where they are needed least," Marchand noted, "in well-endowed Caucasians of more temperate latitudes."

Humans do have a strategy unavailable to frogs and fish: clothing. The best way for you to prevent frostbite is to keep a thin layer of warm air around you, a cocoon of your own radiating body heat, which clothes help stay intact. Wind strips away that layer, forcing your body to give off more heat locally to replenish the lost heat, then stealing that new layer and forcing your body to give up more heat to replace it, and so forth. "The warmth which active exercise stores up, the buckler of the traveller," Bishop Stuck wrote, "is borne away, and the wind falls upon him with its sword."

In 1945 Paul Siple worked out a windchill chart, meant to show the combined impact of wind and cold on human flesh. He poured water into plastic containers, passed air over them at different speeds and temperatures, and took note of how many minutes passed before the surface of the water in the containers froze. According to the chart, exposed human flesh will freeze as rapidly in +20°F air that's moving at a speed of 25 miles an hour as it will in air that's −10°F but calm. "Strong cold, though awesome," Stuck declared, "is merely a condition," while wind is "a deadly weapon . . . a purposeful, vengeful evil. . . . It pursues."*

During World War II, Vilhjalmur Stefansson wrote the *Arctic Manual*

*Some scientists consider the Siple windchill readings to be too high since they don't take into account such things as how much the subject is moving and how much sunshine he's getting.

so that men of the U.S. Army Air Corps who were downed in the northland would know how to survive until rescued. In it he noted that if you dress "properly," only your face is apt to be frostbitten, and your face is the easiest part to treat. If the weather is moderately cold, he advised, run your hand over your face every few minutes to check whether any part of it is frozen, or grimace to locate any "stiff spots." All you have to do then is "take your warm hand out of your mitten and press it to the frozen spot a few moments, until the whiteness and stiffness are gone." "In this way," Stefansson concluded, "one can walk all day facing a steady breeze at −35° or −40°F."

In the manual, Stefansson extolled the virtues of caribou-skin underwear (he got specific: "young caribou, with the hair next to the skin, made of calves anything from a few weeks to a few months old, or yearling females killed before September"). Also bearded-seal boot soles, polar bear trousers, and wolverine hood trim. In his *Guide to Wilderness Medicine,* Paul G. Gill Jr. touted Orlon, Dacron, polyester, Gore-Tex, Thinsulate, taslanized nylon, Flectalon, and Quallofil. Whether your clothing is of high-tech fabric or low-, it should be layered, since air is a great insulator, and since you will be shedding your clothes by degrees as you heat up from exertion. Even at −40° and −50°F, if there was no wind blowing, Stefansson claimed, he often ended up walking all day in his underwear.*

Some outdoorsmen consider beards a good idea in the cold; some do not. Stefansson did not. Although a beard can insulate skin on your face and act as a wind buffer, he pointed out, "the moisture of your breath congeals on it and makes for you a face mask [of ice] that is separated by an air space of a sixth or eighth of an inch from your skin"—which doesn't leave enough space for your hand to slip through and warm your frostbitten cheek or chin. Bearded men in Will Steger's North Pole expedition reported that more than a pound of ice could collect on a beard in a single day, all of which had to be picked or melted off.

Eyelashes, too, ice up. At one point during his solo stay in a hut in inland Antarctica, Richard Byrd stepped outdoors and found that "I could not see the aurora. I was blind, all right; the first thought was that my eyeballs were frozen. . . . I took my gloves off and massaged the eye

*Interestingly, George K. Swinzow of CRREL gave as his reason for not including in his paper "On Winter Warfare" accounts of the cold-season campaigns of Alexander the Great, Hannibal, and Xenophon a "sad fact of very physical significance: the soldiers of all three of these personalities did not wear pants!" He couldn't resist: "Winter warfare without pants?"

sockets gently. Little globules of ice clung to the lashes, freezing them together; when these came off, I could see again." Also eye*lids:* when former Spitfire pilot Ray Munro leapt out of an airplane over the North Pole at 10,000 feet to celebrate his 500th parachute jump and on the way down removed his goggles so he could get a good look around, his eyelids froze shut; the windchill on descent, he figured, had been −177°F.

In Stuck's experience, the nose "freezes more readily than any other portion of the body." A traveling companion of his would lay a small piece of rabbit skin over his nose whenever the weather got very cold or windy. "A little piece of rabbit skin, moistened and applied to the nose, will stay there and keep it warm and comfortable all day," Stuck concluded. "But it does not exactly enhance one's personal attractions." One day during a five-hour walk, his friend's breath condensed into two long icicles which hung from his furry nose cover, "one on each side, reaching down below the mouth," so he looked like a walrus.

Yet even body covering won't keep you warm in very cold weather if you do not move. "Stand still," Stuck warned, "and despite all clothing, all woollens, all furs, the body will gradually become numb and death stalk upon the scene. . . . For dogs and men alike, constant brisk motion is necessary."

Exert yourself, yes, but don't work up a sweat. "A sweat was a more serious matter than a chill," Will Steger insisted. Moist skin loses heat up to 20 times faster than dry skin, increasing the risk of frostbite. Clothes that are sweaty freeze too. During Nansen and Johansen's sledge trip toward the North Pole, "the damp exhalations of the body had little by little become condensed in our outer garments," Nansen wrote, and when the condensation froze, the garments "were so hard and stiff that if we had only been able to get them off they could have stood by themselves, and they crackled audibly every time we moved."

One of the most painful pieces of polar literature to read is Byrd's account in *Alone* of his trying to keep from dying of cold at the same time that he was trying to keep from dying of carbon monoxide poisoning. Something had gone wrong with his generator, and when he tried heating his hut it filled with toxic fumes. If he lit the stove he became sick and weak, yet in the extreme cold he had to have heat. He compromised by turning the stove off for two or three hours every afternoon, which meant he was both sick *and* cold. As outside air temperatures dropped into the minus 50s, 60s, and 70s, a slick, white film of ice climbed the walls inside the hut, at the rate of about an inch a day. The ceiling became encrusted with ice crystals. Water froze to the floor "as it struck." The end of the sleeping bag where he lay his head turned into a mass of ice. Before long, the room was totally coated with a "glacial film of ice. There was nothing left for it to conquer."

As the ice waxed, Byrd's resistance to the cold waned. "My flesh crawled, and my fingers beat an uncontrollable tattoo against everything they touched." He found it easy to imagine "freezing to death." It "must be a queer business," he wrote. "Sometimes you feel simply great. The numbness gives way to an utter absence of feeling. You are as lost to pain as a man under opium. But at other times, in the enfolding cold, your anguish is the anguish of a man drowning slowly in fiery chemicals."

CHAPTER TWENTY-THREE

HUMAN II

Between melting and freezing
The soul's sap quivers

—T. S. Eliot,
"Little Gidding"

THERE ARE TIMES nowadays when people *want* flesh that is cold and dead. It's something they actually pay money for. They've got tissue in or on their bodies they need or want to get rid of, and a physician induces a sort of frostbite in it, an extreme frostbite, one that comes on very fast, so fast that only aviators who bail out of airplanes at altitudes where windchill temperatures get down to −177°F would get it naturally (if jumping out of airplanes could be considered natural). There are other ways to do away with aberrant or unwelcome flesh—scalpels, lasers, radiation, ultrasound—but the deliberate freezing of living tissue to eliminate it "has a nice little niche in medicine," says Robert Berger, a dermatologist in Manhattan who regularly freezes away freckles, warts, skin cancers, and "age spots." He dabs or sprays a coolant, usually liquid nitrogen, directly on the offending places. "It leaves less of a scar than cutting or burning."

If the unwanted flesh is less accessible than skin (a prostate, for example), the physicians use metal probes to freeze it away. What they do is chill the tip of a probe to a temperature as low as −293°F, then, wielding the probe like a pencil, they touch the tip to the target tissue and hold it there for a couple of minutes. The cold draws out heat faster than the blood supply can restore it, and the tissue turns white, "white as refrigerator frost," pioneering cryosurgeon Andrew A. Gage says. Fast freezing is more deadly than slow freezing because it causes ice to form inside cells as well as outside them. The fast freezing is followed by slow thawing, which is more deadly than fast thawing because temperatures around the target spot stay close to the freezing point and allow ice crystals in the spot to fuse and grow larger. The enlarged crystals put pressure on vessel membranes, and as blood flows back through the thawed tissue, it leaks out of the damaged membranes and floods the area, producing clots, blocked vessels, swelling, a disrupted blood supply, and—devoutly to be wished—even more damage.

If the tissue is something the physician wants to be sure is good and dead, like a tumor, he may fast-freeze and slow-thaw it several times. After a week or two, the frozen tissue will slough away while tissues resistant to cold injury—nerves, blood vessels, collagen—keep on functioning. "That's what makes freezing very very very good," says Gerard Guiraudon, former professor of surgery at the State University of New York in Buffalo, who uses freezing to arrest dangerously rapid heartbeat (tachycardia) by neutralizing excitable cells. "It spares the framework."

There's little bleeding with freezing because blood vessels are shut off rather than cut off. W. Scott Melvin, assistant professor of surgery at Ohio State University, freezes liver tumors in patients whose disease is so advanced that he can't do standard surgery on them without sacrificing good liver tissue. He inserts a metal probe through an incision in the liver and, guided by ultrasound, maneuvers the chilled tip against the tumor. On the ultrasound image, frozen tissue looks black, and the border between frozen and unfrozen tissue is a bright line. As freezing proceeds, the advancing wave of ice is seen as a bright line moving outward, leaving a dark zone behind it. He can thus watch the "iceball" (the tissue freezes as a sphere around the probe tip) growing.

They shone below the ice like straws in glass.

—Dante, *"Inferno"*

During a heat wave in the summer of 1995, a 40-something white male, unconscious and bleeding from the nose and rectum, arrived at the emergency room of New York City's Bellevue Hospital. His temperature was 114°F. He was not sweating. Doctors wrapped him in a wet sheet, laid him on a stretcher with a three-inch-high ledge around it, and poured 100 pounds of crushed ice on top of him. "We were very liberal about the ice," Susi Vassallo, assistant professor of emergency medicine at Bellevue, says. "He looked like a sardine in a tin." She explains that it was important to cool him fast because as soon as a person's body temperature reaches 105.5° or 106°F, his brain cells start to die. "It's like putting your brain in a frying pan." Within an hour of his being given an "ice bath," the man's temperature fell to 102°F. Days later he walked out of the hospital, with a damaged liver and kidney but an intact brain.

The people who spend the most time with ice against their ailing bodies are probably athletes. Barton Nisonson, chief of sports medicine at Lenox Hill Hospital in New York City and physician for the New York Rangers hockey team, estimates that a third of Ranger players come regularly to hockey games wearing ice packs. Some shower with ice packs strapped to their bodies. "Ice is a superb analgesic," he explains. It numbs sensory-nerve endings and is "much safer than taking a lot of addicting drugs." If you should have an injury—"a blow to the head, a punch, a sprained ankle, an injured knee, a blunt trauma to the thigh," he lists some possibilities—you should apply ice immediately to the injured area, inside a towel or plastic bag or ice bag to avoid getting frostbitten. The cold will slow bleeding, reduce inflammation, and lower the risk of blood clots. Keep ice on the area for about 48 hours but not continuously, he advises, again because of frostbite. Ice on for 20 to 30 minutes, ice off for 30 to 60 minutes, ice on again. Some people put ice on a burn, intuiting that damage from extreme heat can be stopped by extreme cold. But frostbite is itself a burn—it destroys cell layers—and ice can deepen the injury.

Everyone knows what happened when John Wayne Bobbitt became a temporary amputee. Bobbitt's wife Lorena chopped off his penis with a 12-inch kitchen knife after he came home drunk one night (she says) and forced her to have sex (she says). Afterward, she drove away from their home in Queens, New York, in her Ford Escort and threw the severed part into a patch of tall grass in front of a 7-Eleven convenience store. Early the next day, as Bobbitt was about to go into surgery to have the open wound stitched closed—"that was all we could do," says James Sehn, urological surgeon at the Prince William Hospital in Manassas, Virginia—policemen arrived bearing the missing part in an ice-filled 7-Eleven sandwich bag (Lorena had called and told them where to look). Sehn immediately transferred the part from the ice bag to an ice bucket.

He says he doesn't know how long it would have survived without ice, but by reducing the need for oxygen "ice added to the margin of safety." One expert in replantation estimates that if such tissue were chilled by ice continuously, it might be preserved for up to 18 hours. Sehn and a plastic surgeon spent nine hours reattaching blood vessels, nerves, and urethra, and a few years later Bobbitt made a porn film, *John*

Bobbitt Uncut, in which he demonstrated quite adequately, according to Sehn, the success of the reattachment.

"A finger, arm, leg, no matter how big the amputated part is," Vassallo tells the public, "save it by putting it in a plastic bag with ice around the bag. It can stay on ice for hours, waiting to be reattached."

But not only can a person's own body parts wait on ice for hours to be reconnected; so can other people's parts. Donor organs—kidneys, hearts, livers, lungs, corneas—are nearly always in close proximity to ice from the time they are removed from one person until the time they are inserted into another. Even before the organs are removed, while they are still in the donor's body, technicians usually spread ice on them to begin the cooling process. Once the donor organs are removed, they're usually wrapped in surgical gauze and placed in a jar containing a liquid preservative, which is itself usually placed in an insulated box, usually a beer cooler, containing ice. Even as surgeons are stitching the organ into a new body, technicians are covering it with ice; every low-temperature minute counts.

"The goal is to trick the organ into thinking it never left the body," Maximilian Polyak, organ preservationist at New York Presbyterian Hospital, explains. Ice cuts down on cell swelling, which could rupture cell membranes, and "drastically" lowers the organ's rate of metabolism, reducing its need for oxygen. "For every 10° [C; 18°F] drop below normal organ temperature," Polyak says, "there's a four-fold drop in metabolic rate. By the time the temperature is down to 4° [C; 39°F], the organ's metabolism is only one-fourteenth to one-sixteenth of what it is normally." The organ is quiescent and bloodless and deteriorating yet still "very much alive." A donor kidney can be kept on ice for 48 to 72 hours, a liver for 24 to 36 hours, a heart for four to eight hours.

Although ice prolongs the dying process, it does not stop it. Breakdown is inevitable. For a major organ outside a human body, time will eventually run out.

Your statue, mine,
Perfected ice

—Reynolds Price,
"The Laws of Ice"

But must time run out? some scientists have asked themselves. What if hearts, kidneys, lungs, spleens, livers, and intestines could be preserved at temperatures below freezing instead of a few degrees above it? What if major organs could be taken out of one person's body, kept in freezers for weeks, months, years, even centuries, then inserted into another person's body? "We could eliminate time as a factor," Gregory Fahy, former head of the tissue cryopreservation section of the Naval Medical Research Institute in Bethesda, Maryland, says. For 20 years, Fahy tried to preserve organs with extreme cold. Armand Karow, president of Xytex Corp., a national frozen-sperm bank, tried for 25 years, then gave up. "I got tired of beating my head against the wall," he says. "The problems of preserving large, complex, vascularized, mature human organs are awesome." "Ice is the central issue," Kelvin Brockbank, senior vice president of Organ Recovery Systems, explains. "Ice is the ultimate enemy."

During Karow's first ten years of work he would take the heart of an adult rat—"fairly small, a little larger than a marble, the size of a thumb from the knuckle to the tip, as large as a pecan but not as large as a walnut"—and infuse it with a chemical antifreeze, then slowly lower its temperature. The rat heart would freeze solid. "There was no question about the freezing," Karow says. "It wasn't mushy." Afterward he would thaw the heart, then stimulate it, and it would contract. The lowest temperature he could freeze a rat heart to and get it to contract after thawing was −4°F; any colder and it probably had too much ice in it to recover. But −4°F isn't cold enough for storing a major organ; at that temperature things are still going on in the tissue. Besides, Karow couldn't keep a heart frozen at −4°F for more than 20 minutes and still get it to contract. Injury, he points out, is a function of time.

The next 15 years Karow spent on dog kidneys. He never was able to freeze them successfully. The main problem was uneven cooling and thawing in such a large tissue mass ("in a 60-pound dog, the size of a human fist"). "The greater the tissue volume," he explains, "the more formidable the task of pumping energy into it. Think of meat in a microwave," he adds helpfully, "how hot spots can form in it." Other cryobiologists have managed to freeze dog kidneys down to −94°F and keep them frozen for days and gotten them to sustain the life of a dog after they thawed, but the "results weren't consistently reproducible," Karow points out. "Only about two out of 100 attempts were successful, and you couldn't say which were going to be the two."

Greg Fahy used a different approach: "not dealing with ice at all." He added antifreeze to organs at such high concentrations that the freezing point in their fluids went past the point at which ice would form. As temperatures fell, molecules in the fluid moved more and more slowly, the fluid became more and more viscous, until the temperature reached about −125°F, at which point the molecules quit moving altogether, and the fluid stopped flowing. "It's the end of biological time," Fahy says. The organ had vitrified. It turned to glass. Ideally, a glassy solid should "preserve the system forever," Fahy notes, without damage to the organ. Ice rearranges the positions of molecules, usually into hexagonal shapes, but the molecules in glassy solids keep the same relationship with each other that they had in the liquid. "You get a snapshot of the liquid state," Fahy states. While a frozen kidney is opaque and white and looks like a snowball, a vitrified dog kidney looks like a normal kidney except it's embedded in a block of freestanding transparent material, "like a fly in amber."

The problem with vitrification is *de*vitrification. That doesn't mean the glassy solids turn back into liquid as temperatures rise but that they turn into ice. Ice is much more likely to form in organs as temperatures go up from the glassy state than as they fall toward it. "There's a fantastically increased chance of freezing when you want to retrieve the sample," Fahy says. Vitrification is not an all-or-nothing event, he explains. The glassy organ probably has scattered throughout it an "astronomical number of teeny tiny submicroscopic ice crystals," too small to do damage themselves but able to serve as nuclei for larger crystals to grow around as the organ warms up. The warming therefore has to be done very, very fast, according to Fahy, faster than ice crystals can grow. "You are in a race with the ice." Every time he tried to warm a glassy organ, though, the ice won the race. He wasn't able to stop large ice crystals from forming in an organ as it thawed.

Eventually, the obstacles to freezing major organs for transplantation can be overcome, Karow believes, yet he also believes there is no longer the need. Better immunosuppressants are available now for transplant patients than when he first froze a rat heart, so that tissue-matching (one reason for having a large number of stored organs to choose from) is not as critical. Also, vital organs are in such demand these days that they're transplanted as soon as they become available.

Meanwhile, freezers *are* full of smaller, simpler human replacement parts: blood cells, heart valves, stem cells, bone, blood vessels, pancreatic islets, eggs, embryos, and sperm. All are either easier to freeze than the vital organs or more tolerant of injury from freezing. Easiest to freeze are red blood cells, probably because they don't have nuclei. The first human cells to be frozen and survive, in 1949, were sperm; they proba bly freeze well because they're in suspension. Currently, an estimated two million vials are stored in U.S. sperm banks; some of the sperm has been frozen for as long as a dozen years. The freezing characteristics of sperm are idiosyncratic, though. "Some guys don't freeze, and other guys you can do anything to their samples and they'll freeze," Sue Simmons, laboratory manager at Xytex, says.

In the future, many body parts will be grown on meshes or scaffolds specifically for transplantation: skin, livers, ureters, urethras, ligaments, tendons, even breasts (they'd consist mostly of fat), maybe even fingers. "There's a whole new world of tissue engineering," Brockbank points out, "and we're going to have to have a way of storing these things. Cryopreservation is the only way we have of saving the structures." His company is designing "ice control molecules" which, like antifreeze proteins in fish, will bind to ice crystals and change the direction and speed of their growth. Although the molecules won't keep ice from forming in transplanted parts, they'll give the ice crystals a less damaging shape, "not those terrible needles."

Sparkling ice on the dead man's chest,
glittering ice in his hair

—Robert Service,
"The Ballad of Blasphemous Bill"

As of this writing, the bodies or heads or brains of about 100 people worldwide are being kept in long-term storage facilities at temperatures as low as −320°F. Another thousand people, now living, have signed up to have their bodies or heads or brains frozen when they die, or rather "deanimate," since according to those committed to "cryonics" (physical resurrection after freezing) a person who by today's medical standards is considered dead might at some later date be considered only terminally ill. The hope is that sometime in the future, say 100 years from now, a more advanced technology will permit scientists to (1) cure people of what ailed them when they passed on and (2) reverse the damage that was done by the preservation method of choice, which is freezing. There will, unavoidably, be damage from freezing.

This is how Jerry White, a Californian with AIDS, was "suspended" by freezing after his death. Equipment had been set up ahead of time in his living room, and as soon as he breathed his last and a doctor had pronounced him dead by the standards of the late 20th century, his body was placed inside a large bassinet-like vessel and covered with ice, a couple of hundred pounds of it. A medical technician activated a heart "thumper" he laid on White's chest, which compressed the heart, pushing blood through the body so the cells would be oxygenated as much as possible during cooldown. He also set up a bypass system which carried White's blood from a femoral artery through a spiral tube surrounded by ice water—"like those cooling loops on old stills," says Jim Yount, chief operating officer of the American Cryonics Society—on through an oxygenator and back into the body through a femoral vein. Then he removed the thumper. As White's blood grew cooler and more sludge-like, it was gradually replaced with a salt solution, a physiological imitation of blood. When his body had chilled to a degree or two above freezing, the equipment was shut off and, still packed in ice, the body was taken by ambulance to a "processing center." There glycerine and other cryoprotectants were added to the salt solution and the temperature of the body was lowered to *below* freezing. Next, the body went to a cryonics suspension cold-storage facility where it was placed inside a sleeping bag then a capsule, "like a big Thermos bottle," Yount says, "with liquid nitrogen at the bottom." Vapor coming off the liquid nitrogen cooled the body further, by a degree an hour, until the body reached its final suspension temperature of −320°F.

Yount explains that if the nitrogen had been used for cooling from the start instead of ice, the cooling rate might have been too great, and the outside of the body would have been frozen while the inside was still warm. "It's not snap-freezing, as you do with peas," he points out. Once White's body was frozen all the way through, it should not deteriorate, according to Yount, even after centuries in storage.

After storage what, though? How can a centuries-old frozen body ever be revived? Ralph C. Merkle, a researcher in nanotechnology in Palo Alto, California, who has a Ph.D. in electrical engineering, suggested in an article in *Cryonics* how that might be accomplished. Only the brain would be selected for full restoration, he pointed out; "secondary" tissue, like livers, could just be replaced. And only the brain's long-term memory would be restored; short-term memory would disappear, as it does routinely after traumas. If freezing injury obliterates long-term memory, "all bets are off," he wrote. Since damage from freezing would probably be "beyond the self-repair and recovery capabilities of the tissue itself," restoration would be carried out by an "external source."

The source he had in mind is an "assembler." What it assembles is molecules and atoms in the frozen brain. Freezing causes the molecules and atoms in the brain to end up in "the wrong places," Merkle wrote, and the recovery process consists of putting them "back where they belong." Since life experiences make physical changes in brain cells and synapses, everyone's memory is distinct and individual. Scientists (or whoever is willing to do the reanimating—a new profession, perhaps, the flip side of funeral directing?) would, using technology not yet available but within the realm of possibility, determine the type and location and orientation of every molecule encoded in the memory structure of the frozen brain and store the information digitally in a computer.

It would take over 100 megabytes of computer memory to store the coordinates of all the molecules in the brain of a single individual. Once computers had analyzed the coordinates and determined what changes should be made, the assembler would move the molecules, "one at a time, back to their correct locations," thereby restoring the thawed brain to a "healthy and functional state."

"It's not going to be a simple thing," Jim Yount concedes. "It's not going to be a case of stand the body up, give it a handshake and a $20 bill, and push it out the door." Nevertheless, he has himself signed up for suspension, as have a fair number of computer programmers and lovers of science fiction. A few people unable to afford suspension yet wishing to be frozen for the indefinite future have already had themselves buried in permafrost. However, as Yount points out, the temperature of permafrost is often only a few degrees below freezing. "At that temperature, the meat in your freezer doesn't do very well—and that's what we are, meat." He isn't speaking dismissively of meat: ancient Egyptians believed in the physical immortality of the body, and "we are the successors to the Egyptians."

In Greg Fahy's opinion, preserving bodies for physical resurrection "isn't totally kooky. To be fair, they might have some fighting chance of preserving something. The question is whether what they preserve is enough." Brockbank says: "All of us would *like* to see this be successful. Think of the relativity aspects for space travel; you could be in stasis while you go through space to some place very distant. But $100,000 to freeze your mother-in-law? There's no evidence to indicate that a whole organism, even a mouse, can be successfully frozen and brought back to life."

Ice, white ice, like a winding sheet

—Robert Service,
"The Ballad of Blasphemous Bill"

One person who did not sign up for whole-body preservation but who got it nevertheless was the Iceman of the Alps. One sunny day in September 1991, a German couple hiking in the Hauslabjoch area of the Tyrolean Alps at an altitude of over 10,000 feet took an unmarked path across a gently inclined snowfield, at the bottom of which was a small rocky ridge. Beyond the ridge was a gully, filled partway with ice and meltwater. As they veered left to pass around the gully, the hikers spotted "something brown sticking out of the ice," according to one of them (Helmut Simon). "Our first thought was that it was rubbish, perhaps a doll. . . ." As they drew closer they realized, according to the other (Erika Simon), "But it's a man!" A brown, bald head protruded from the ice along with a pair of shoulders and an upper back. The face, though, was hidden in slush, and the rest of the torso and the limbs were under ice. The hikers noticed a blue rubber ski clip nearby and figured the man was a mountaineer who had died on the glacier and was being melted out of it.

At a nearby lodge, they mentioned what they'd found to the caretaker, and he assumed, as they did, that it was another "glacier body." Already that year, half a dozen glacier bodies had surfaced in that part of the Alps, disgorged from the ice after as long as half a century spent inside it. The innkeeper's father, chief of mountain rescue in the area, remembered one missing person the body might belong to, a music professor from Verona who went out for a walk in the mountains one day in 1938 and never came back. As most readers know by now, the body did not belong to a music professor but to a man who was walking in the mountains 5,300 years ago and died at or near that spot, under circumstances that are still not entirely clear, even to scientists who have spent the years since the discovery studying his mummified body and possessions, which were scattered on the ice and rock around him. The scientists' conjectures about his ancient trek in the Alps as well as accounts of his modern-day treks from the glacier to a freezer in Innsbruck, Austria, where a few scientists got to see him, to a freezer in Bolzano, Italy, where many tourists now get to see him through a glass window, is told in several books, including Konrad Spindler's *The Man in the Ice: The Amazing Inside Story of the 5,000-Year-Old Body Found Trapped in a Glacier in the Alps* and Brenda Fowler's *Iceman: Uncovering the Life and Times of a Prehistoric Man Found in an Alpine Glacier.*

Ordinarily, archeologists don't bother with glacier bodies, Spindler, an archeologist and head of prehistory and protohistory at the University of Innsbruck, pointed out. They are too "recent," he explained (meanwhile admitting that he had been moved "time and again by thoughts of such a lonely, cold death"). It soon became obvious, however, that this was no ordinary glacier body. The man had a shoe on his right foot stuffed with grass or hay. Near his head lay a metal-bladed ax with a wooden handle which appeared to be old; under him was a woven grass mat and a flint dagger; and near him was a slice of birchbark with regularly spaced holes along the edge. Talk among locals was that he was a mercenary from the days of Frederick "Empty Purse," whose armies had retreated through these passes in the early 1500s.

Before anybody could figure out who he was or how old, he had to be taken out of the ice. The inn's caretaker, his father, a friend of his father's, a couple of Alpine gendarmes, a guide, a climber, and the head of the University of Innsbruck's Institute of Forensic Medicine poked, hacked, pried, drilled, picked, hammered, and chipped at the ice (and now and then at the body) with sticks, picks, a ski pole, and a jackhammer. One time they almost had the body out, but overnight it froze fast again. When they finally freed it, they carried it by helicopter to Innsbruck for determination of cause of death—something, as things turned out, they were unable to do after years of trying.

Once Spindler got a look at the ax, he declared the man in the ice to be about 4,000 years old. (Carbon dating and the finding that the blade was copper and not bronze added another 1,300 years to his age.) The Iceman was not only very old but in very good shape, considering. What's more, he had his things with him, which would give scientists a chance to find out how people lived back in the Neolithic. One of the first questions that was asked about him, though, was practical: who owned him? The spot where he was discovered was close to what had become, while he was out cold, a border between Italy and Austria. Which side was he on? Surveyors determined that he was three rope lengths into Italy. Although the Austrians had possession, he belonged to the Italians. As a compromise, the Italians agreed that the Austrians could keep him for a few years, to examine him and his belongings, then everything would come to Italy.

Another question, more urgent, was how to preserve a body that until then had been preserved in a glacier. After being out of the ice for only a day, the Iceman had thawed completely, and some people swore they saw fungus on his skin. Chemical preservation was out of the question because of contamination. So, after a brief immersion in room-temperature, late-20th-century air, the Iceman was placed inside the university's walk-in freezer, where the temperature and humidity could match those in the place where he had spent his whole prior afterlife.

Another question being asked: how could a human body stay so long in a glacier? The oldest glacier bodies found in Europe prior to this one were, judging by dates on the pieces of silver found with them, only 400 years old, according to Spindler, and they had emerged in pieces, torn asunder by the massive forces of moving ice. By contrast, the Iceman's body was in one piece, intact except for gouges it received when being chopped out of the ice. (It had lost most of its water, however, and weighed a little over 30 pounds.) Glaciologist Gernot Patzelt went to the gully where the Iceman was found, looked around the flat, rock-rimmed trench, and concluded that the main mass of the glacier would have flowed *over* ice in the trench, leaving it more or less in place. "The corpse lay frozen in an immobile mass of ice about nine feet . . . beneath the stream of the glacier," Brenda Fowler wrote of Patzelt's conclusion. The corpse could have stayed on and on, as in a protected cove, without being dragged downglacier and ripped limb from limb.

The Iceman's belongings, too, had been pretty well preserved by the ice. According to Spindler, in the few cases where ancient (Bronze Age) clothing had been found in Europe, either animal material—leather, hair, wool—survived or else plant material—linen—did, but not both. In glacial ice, both had survived and so well that "gossamer-thin" threads only six-thousandths of an inch thick still encircled the fletching of the arrow shafts. That the man was buried by nature, in ice, and not by relatives, in earth, meant also that there were more of his belongings to be preserved, not just the lone stone bead or handful of flint tools that fellow Neolithics might have laid in a grave beside him.

Over the next several years, scientists in fields including molecular archeology, paleobotany, anthropology, anatomy, and forensic radiology studied the Iceman's corpse—in tiny portions, parsimoniously doled out—and his clothes and equipment. A partial picture emerged. He was over 5 feet 2 inches tall (and probably taller before he dried out). He was in his late 40's, old for his time. His teeth were worn and his arteries clogged. He had a fungus in his lungs and the eggs of a parasitic whip-worm in his colon. A couple of his toes had been frostbitten—one bone was abnormally short, another had a hole in it. He had arthritis in one leg and hip. He might have had a stroke: X-rays of his brain showed a dark spot. Although he was bald when he came out of the ice—the upper layer of skin had been shed and hair went with it—strands found in the ice suggested his hair was wavy, dark brown, and at least three inches long.

As for his wardrobe, carefully pieced together by conservators in Germany, it included a knee-length cape of woven grasses or reeds; a fur coat or wrap so large it could cover two people; a pair of leather leggings, rather like cowboy chaps, held up by leather garters attached to a leather belt; a pouch like a fanny pack; a conical hat with fur turned to the inside; and grass-filled leather shoes. He carried with him, among other things, a longbow (unfinished, unstrung), a quiver full of arrows (most of the arrows likewise unfinished), a stave, the flint knife and old ax, and a birchbark cylinder, which probably held live embers wrapped in fresh maple leaves for starting a fire.

Six years after he was taken to a freezer in Austria, he was taken to one in Italy, inside a museum in Bolzano built especially for him (his chamber is now lined with ice tiles and his body allowed to develop a protective glaze of ice). Scientists continued to study him. From the mineral composition of his teeth they learned that he was probably born in the Italian village of Feldthurns and from the composition of his bones that he had probably lived in a valley elsewhere most of his life. From DNA in his intestine they learned that his last meals consisted of ibex meat, bread made from a primitive variety of wheat, and red deer meat, and from pollen in the intestine—probably swallowed with water he drank—that he died in the springtime.

But why did he die? One popular, unprovable theory was that he had lain down to rest on a rock, been caught in a sudden snowstorm, and died of hypothermia. Then one day in Bolzano researchers ran a CAT scan on his body and discovered something that had been indistinct in earlier imaging: a flint arrowhead, lodged in his left shoulder. The Iceman had been shot! The arrow had torn through his flesh, probably causing a great deal of pain and a fatal loss of blood. The shaft was missing so the Iceman may have pulled it out, causing more damage. "This changes everything," the museum director said. Pathologists discovered a deep cut on the Iceman's right hand, as if he had tried to defend himself against an attacker with a knife, plus small wounds elsewhere on his body. New and revised theories abounded: That the Iceman was a hunter who had strayed across a boundary into other hunters' territory. That he was a shepherd headed back to his village who was murdered by another shepherd for possession of his flock. That he was killed in a war, or by enemies who had raided his village and pursued him into the mountains, or by someone he was fleeing because of a personal feud (theft of a wife).

For all that is being learned about a very old man through ingenious scientific research, one could almost wish that at least this once cryonics would work, that the Iceman could be reanimated, brought out of suspension into full consciousness, with his freezing injuries reversed, his brain molecules restored to their rightful places, and that he could speak through that gaping mouth with its set of worn-down teeth, telling us what really befell him in the high mountains of the Hauslabjoch before the ice entombed him.

He has been in the ice too long.

—Doctor, in 1997 Chinese film *Frozen*, about a performance artist who commits suicide by "ice burial"

In other places around the world in recent years, other ice men and ice maidens have been emerging from other glaciers and permafrost and icy tombs, as if vault doors were suddenly springing open, one by one. There was the hunter in what is now British Columbia who apparently fell into a crevasse 500 years ago and was discovered as the glacier receded (goosebumps preserved on his skin) and named by local Indians "Long Ago Person Found." The teenage girl who was sacrified to the Inca gods 500 years ago atop a 20,000-foot-high volcano in Peru, was buried by ice and snow, and when the ice melted was discovered in her red-feather headdress by an anthropologist who had climbed to that altitude to view the volcano. The aristocratic young woman on the steppes of southern Siberia who was buried 2,400 years ago in a log coffin with six elaborately harnessed horses, all encased by ice when rain or melting snow flooded the tomb and froze. In 1993, Natalya Polosmak, research fellow at the Institute of Archaeology and Ethnology in Novosibirsk, led a team of archaeologists that unearthed the Siberian tomb. "When we saw the ice," she said, meaning inside the coffin, "our rapture knew no bounds." As team members poured hot water—carried in buckets from a nearby lake and heated with blowtorches—into the coffin, the ice slowly melted and revealed under a blanket of marten fur and a robe of wool and silk the tattooed body of a lady of the ancient Pazyryk sheep-herding culture, with objects of her everyday life around her, preserved by the ice, "so perfectly intact," Polosmak wrote, "that we could literally smell her last symbolic meal of mutton, unfinished on a low wooden table by her side."

CHAPTER TWENTY-FOUR

GAMES I

The painted sleighs can't get a grip

—Anna Akhmatova,
"Voronezh [O. M.]"

THERE *IS* A CHARACTERISTIC of ice that makes people rhapsodic. Not everybody: those who've flipped off front steps or smashed their cars into street signs because of it are less than enthusiastic. But for a great many people, this aspect of ice is the closest thing to a gift from Heaven. As Eric Nesterenko of the Chicago Black Hawks, who at the time had been playing professional hockey for 20 years, recalled in Studs Terkel's *Working:*

"One day last year on a cold, clear, crisp afternoon I saw this huge sheet of ice in the street. Goddamn, if I didn't drive out there and put on my skates. I took off my camel-hair coat. I was just in a sort of jacket, on my skates. And I flew. Nobody was there. I was free as a bird. I was really happy. . . .

"With the wind behind you, you're in motion, you can wheel and dive and turn, you can lay yourself into impossible angles that you never could walking or running. You lay yourself at a forty-five-degree angle, your elbows virtually touching the ice as you're in a turn. Incredible! It's beautiful! You're breaking the bounds of gravity. I have a feeling this is the innate desire of man . . . just being in pure motion."

The gift is slipperiness; pure motion and birdlike swooping are possible because ice is slippery. It's a gift because it could have been otherwise. Silver, gold, and lead are not slippery. You can't skate on cold marble or steel. That ice should be slippery may seem obvious, self-evident (slickness being so essential to its perceived character that Shakespeare included "to smooth the ice" with such examples of "ridiculous excess" as "to gild refined gold, to throw a perfume on the violet, and to add another hue unto the rainbow"). But *why* ice is slippery has not always been obvious, or evident, even to scientists. All would agree that when we encounter it ice is comparatively close to its melting point (far closer than marble and gold are to theirs), and that even when the air temperature is well below freezing there's a lubricating layer of liquid between the ice and the skate blade (sled runner, boot sole, automobile tire, whatever). The question is, how did the liquid get there?

For a long time, people thought it got there because of pressure melting. That is, the weight of a skater conveyed through the small contact area of a sharp blade puts enough pressure on the ice surface to lower its freezing point, which causes the topmost layer to melt into a film of water. For decades, this was the standard explanation taught in schools and colleges, and it's probably still being taught in some places despite the fact that physicists have long since done the numbers and determined that the most pressure a typical adult skater could exert on the ice through even a very sharp and narrow blade would lower its melting point only a fraction of a degree. To melt ice at 30°F, each skater would have to weigh over 300 pounds, at 14°F a ton and a half. "We now know the whole idea is absurd," says Samuel C. Colbeck, geophysicist at CRREL and an authority on gliding surfaces. "The ice would fracture first under the weight of the skater."

Another reason given for ice's slipperiness is that it has inherently, regardless of whether or not a skate blade or anything else is bearing down upon it, a "quasi-liquid" layer on its surface, undetectable by the naked eye, mere molecules thick, yet enough to reduce drag. English physicist Michael Faraday proposed the existence of such a layer in 1842 while investigating the adhesive properties of ice. He called it "liquid-like" because, sandwiched as it would be between bulk ice and air and in equilibrium with both at their boundaries, it wouldn't be ordinary liquid water. Its properties would be intermediate between those of water and ice. Some researchers have contended that such a layer plays the dominant role in ice's slipperiness. But as Victor Petrenko, professor of engineering at Dartmouth and an expert on the surface properties of ice, points out: "The layer is so thin that it cannot provide lubrication. To make ice slippery, you need at least several micrometers [hundreds of thousandths of an inch] of liquid, and this layer is only a few nanometers [hundreds of *millionths* of an inch] thick."

Still another explanation, put forth in the 1950s by F. P. Bowden and T. P. Hughes, is that ice is slippery because of friction. When a skater pushes his skate blade over the frozen surface of a pond, the heat generated by the rubbing together of two solids, metal and ice, raises the temperature of the top layer of ice, causing it to melt. Friction, then, gives ice its low-friction characteristic. It's the explanation most scientists embrace nowadays, that the slipperiness of ice comes overwhelmingly from friction, with only a smidgen of help, if any, from pressure melting and from some quasi-liquid layer.

Ice doesn't always give the skater a sense of flight. James D. White, associate professor of physics at Juniata College in Pennsylvania, points out that when ice is very cold, "any friction-scratching with your blade doesn't generate enough heat to melt it. You can't glide. As a one-time Minnesota skater, I can attest to that. You might as well be walking." On the other hand, when ice is warm, a skater often can't get a good glide either. During the 1992 Olympics in Albertville, France, the weather was unseasonably warm and the refrigeration system for the outdoor skating rink below par and the ice surface therefore soft, gummy, and slow, what one losing speed skater referred to as "porridge."

Even on indoor rinks, more is involved in making good skating ice than flooding a flat place with water, mimicking a good rain, and letting cold air harden it. Ice-making these days is practiced by "icemasters."

All shod with steel
We hissed along the polished ice, in games
Confederate

—William Wordsworth, *"The Prelude"*

It was an April afternoon a few years ago at Madison Square Garden in New York City. A chimpanzee in racing silks was doing cartwheels over the arena floor. A buffalo with a man on its back was stepping through a hoop of fire. A clown was juggling real pizzas. "May all your days be circus days, to the last goodbye," a ringmaster sang, and thousands of spectators filed out, taking their chicken-shaped light batons with them. "You'd never think they'd get it ready for tonight, would you?" said a security guard, looking down at the arena floor, where there wasn't a hint of the ice needed for the hockey game due to start in only three hours.

During the New York Rangers' hockey season, the ice on their rink would usually stay in place whenever there were other events (basketball, rock concerts); it was just covered over, then after the event uncovered and repaired. On this occasion the change from circus floor to hockey rink had to be made so fast that an icemaster was flown in from Canada to oversee it.

"The iceman cometh!" a workman on the Garden staff called out when Doug Moore walked in. Moore had been making ice for 40 years, currently taught advanced ice-making at a university, and claimed he could "read the ice like a golfer reads a golf green." He watched workers as they hauled off circus paraphernalia, disconnected overhead rigging from tiedown points in the floor, and pulled rubber mats and thick sheets of paper off the floor. That exposed the ice but also used up an hour and a half, half the time allotted for the changeover. The ice looked a mess to me, covered as it was with dirt, scratches, bumps, and holes (although not the large cracks I had pictured from the elephant tonnage so recently bearing down on it).

The workers started repairing the ice by getting rid of the largest pieces of dirt with snow shovels. Then they did a "dry cut," meaning the Zamboni (a motorized ice-grooming vehicle) scraped the ice surface with a blade. After that, they "burned" off the remaining dirt by hosing it with hot water and vacuuming up the dirty water. All this removed about a quarter inch of ice from a layer that was only an inch thick to begin with. "Maintaining thickness is a constant struggle," Moore pointed out. He wouldn't have time to replace the lost quarter inch before the mats and paper had to go down again for the next circus performance.

The ice had 40 large holes in it, each above a rigging tiedown point, and the workers started filling them with slush. They also packed slush into gaps next to the sideboards and tamped it down with hockey pucks. Pucks are popular tools for grooming ice, along with such other low-tech tools as spoons (to scoop out goalpost holes), turkey basters (to wet a surface), dishpans, buckets, mops, watering cans, bathtowels, broken hockey sticks (to wipe ice shards off a Zamboni blade), and, during a game, the occasional blob of referee spit, to moisten a spot of ice.

To make sure the slush hardened in time for the game, Moore turned the "chillers" up high. What the chillers did was cool the brine running through 60,000 feet of tubes imbedded in the concrete floor beneath the ice to −13°F. Because of its high salt content, the brine wouldn't solidify even at −13°. "I don't like to go that low," Moore admitted, "but I don't have any choice. I have to freeze that slush!" He didn't like to go that low because the ice already made could become too cold, therefore brittle, dry, and easily chipped and rutted by players' sharp turns and sudden stops. Later, he'd back off with the chillers and let the ice warm up. The ideal temperature for hockey ice, the temperature at which drag is lowest, he maintained, is 28°F.

For the next phase, the bathtowel was pressed into service. The Zamboni went around dragging a wet towel behind it, which created a film of water on the ice, which would freeze smooth. The towel was steaming; the water used to soak it was 160°F. Hot water, according to Moore, contains less dissolved air than cold, and "air is the greatest enemy of ice." Players help get rid of any trapped air, he said, by putting their weight on the ice, beating it up. "Ice improves, not deteriorates, with age. Brand-new ice is never as good as ice that's been skated on."

Pregame warmup was scheduled for 7:00, but at 6:55 workmen were still on the rink. One was spraying compressed carbon dioxide from a can at the slush along the boards, giving it a last-minute cold blast to harden it. Two men were gently patting spots on the ice with hockey pucks, to level them. A player from the Pittsburgh Penguins, in full black padded warrior regalia, appeared at an entrance and took a look at the men. "You guys going to be ready on time?" he asked doubtfully. They were, just. Penguins captain Mario Lemieux got five goals during the game, a hat trick plus two; even Ranger fans cheered. A Zamboni driver could be heard to mutter, though, "All that work and the home team loses." The game was short, which was good. "Fast is a tribute to the ice," Moore said. Pucks didn't bounce, passes were crisp. Bad ice usually means more than bad hockey. It means injuries. "Bodies a fraction of a second out of kilter are that much closer to a pull, a hook, or an elbow," a Rangers coach pointed out. If skate blades get stuck in ruts or are slowed by soft ice and a player tries to change direction, his body makes the turn before the blades do, and the knee takes the strain.

Hockey ice is not the same as speed-skating ice, and speed-skating ice is not the same as figure-skating ice or curling ice or ice on bobsled tracks, all of which are different from each other. At the Pettit National Ice Center in Milwaukee, ice on the speed-skating oval is kept at 20°F. Speed skaters need the harder ice to make fast times; the brittleness of the ice doesn't worry them since they don't make the sharp turns that hockey players do. Their skate blades are also longer, thinner, and flatter than hockey players' blades and don't cut into the ice as much; "all you'll see are fine little cuts in the ice even after 10 or 15 pairs of skaters have gone around the track," says Jim Gulczynski, Pettit's director of operations. Meanwhile, the air over the oval is kept warm, 62° to 64°F, to enhance any microscopic film of water on top. Warm air plus cold ice add up to great gliding. Within weeks of the rink's opening in 1992, skaters at Pettit had broken eight national speed records.

As for figure skaters, the ice at Pettit is kept warmer, 26°F, therefore softer, cushioning their landings after big jumps and allowing them to hold their edges. It's also thicker than speed skaters' ice, 1 1⁄4 inches instead of 3⁄4 of an inch, so when they dig into the ice to get purchase they won't hit cement. As for ordinary weekend recreational skaters of the "slip and slap" variety, as Gulczynski calls them, the ice at Pettit is kept at the same forgiving temperature as for figure skaters, 26°F.

> Another middle-class innovation over which Goethe presided . . . and which became a popular fixture for the court during the winter months, was skating, on a specially flooded meadow near the walls. . . . The ice became an open-air ballroom with a wind-band and masks, and at night torches, braziers, and fireworks; one of the pages recalls being dressed as a demon and skating through the falling snow with a Roman candle attached to his horns.
>
> —Nicholas Boyle, *Goethe: The Poet and the Age*

Skating of the slip-and-slap variety has had periods of astounding popularity. In New York City in the late 19th century, as many as 7,000 people would show up to skate on a lake in Central Park. An article in the *Daily News* of December 1860 ventured, "If it had been possible to set the ice up edgewise and skate on both sides of it, every square yard would have its steel-shod tenant."

The first covered rink for recreational skating was probably one built in Quebec City in 1852. It's not known now what the quality of the gliding surface was, but oh the social lubrication! Artificial rinks "revolutionized the social life of the colony," Edward Cavell and Dennis Reid wrote in *When Winter Was King.* People held large costumed balls on the ice with military bands playing waltzes. Ladies and gentlemen would sneak off "at peep of day" to ice rinks to practice skating on the sly, according to an 1862 edition of the *Cornhill Magazine* quoted by Cavell and Reid, "hoping that wriggles and contortions which precede the acquirement of that ease and apparent absence of force, which mark a finished skater, may escape the observation of more experienced connoisseurs."

It's this apparent absence of force that John Cheever celebrates in his book *Oh What a Paradise It Seems,* in which a man named Sears goes ice skating on a frozen pond near his daughter's house in Connecticut. "The pleasure of fleetness seemed, as she had said, divine," Cheever wrote. "Swinging down a long stretch of black ice gave Sears a sense of homecoming. . . . At long last, at the end of a cold, long journey, he was returning to a place where his name was known and loved and lamps burned in the rooms and fires in the hearth. It seemed to Sears that all the skaters moved over the ice with the happy conviction that they were on their way home." Most of them, like him, didn't do anything fancy but "simply went up and down, up and down, completely absorbed in the illusion that fleetness and grace were in their possession and had only to be revealed."

It is Arctic weather under your feet

—Mary Mapes Dodge,
Hans Brinker; or, The Silver Skates

As fleet as ice skating is, some people have tried to make it fleeter. They have stepped out onto frozen water not only wearing skates on their feet but carrying in their hands sheets, sails, kites, or umbrellas, hoping to get the wind to propel them faster than their muscles alone could. "Skate sailing next to ski jumping is the most bird like of winter sports," Elon Jessup proclaimed in his 1926 book *Snow and Ice Sports: A Winter Manual,* "bird like" because it gives you the "sense of flying." You are "a human sailing craft unto yourself," Jessup wrote, switching from simile to metaphor. "Your body is the mast, your feet the hull, and the iceskates are keel and steering gear . . . [You are going] 30, sometimes 40 mph, which is probably the greatest speed people have made on two feet."

Even in 1926, humans could reach speeds on ice greater than 40 miles per hour by being on a sailing craft rather than being one. The fastest nonmotorized conveyance in the world was then and still is the iceboat, the iceboaters say, and for a period of time in the late 19th and early 20th centuries the iceboat was the fastest of all conveyances *including* motorized ones: trains, boats, automobiles, and airplanes. In 1907, when the fastest that airplanes could do was 40 mph, an iceboat could do 144.

When you first see iceboats, they may look to you like sailboats, particularly if they're far out on a frozen lake in a regatta. But after watching them for a while you'd never mistake them for sailboats. They move on "hard water," as iceboaters call ice, in a horizontal, flat plane, not bobbing up and down, as summer sailboats do on "soft water." They zip across the surface as if motorized. Also, their sails tend to be trimmed tighter than sailboat sails. "To the amazement of the uninitiated, [iceboats] are always going to windward," Raymond A. Ruge, former commodore of the Hudson River Ice Yacht Club, noted, "since their lack of resistance [to the ice surface] results in speeds that pull the apparent wind around toward the bow and thus require tight sheeting almost all the time."

Iceboats did start out as sailboats, however, probably in Holland. You can see them in 17th-century Dutch engravings, big and square-sailed, with wooden frames under their hulls and, presumably, runners under the frames. Some were used for ferrying supplies across frozen canals, but others were strictly for joyriding. In *Hans Brinker,* a boy named Ben stands watching iceboats ply the frozen Haarlem Meer on a winter day when "every ice-boat in the country seemed afloat or rather aslide." The boats "were of all sizes and kinds," Mary Mapes Dodge wrote, "from small, rough affairs managed by a boy, to large and beautiful ones filled with gay pleasure parties." The deck of one boat was "filled with children muffled up to their chins" singing a chorus in honor of Saint Nicholas:

> *Friend of sailors, and of children!*
> *Double claim have we,*
> *As in youthful joy we're sailing,*
> *O'er a frozen sea!*

Ben hitches a ride with his buddies on a "shabby" iceboat which nevertheless gives them "a grand sail, or ride; . . . perhaps 'fly' would be the best word; for the boys felt very much as Sinbad did when, tied to the roc's leg, he darted through the clouds."

The pleasures of reaching flying speeds without actually flying ("B-r-r-oooo! How fast we're going," Ben shouts. "This is glorious!") should make iceboating one of the most popular sports in the world. It is not. "Conditions are almost impossible to get right," one longtime iceboater explains. By "conditions" he means mainly the ice. "We spend most of our time waiting and wishing for ice. Plus sharpening our runners." To go iceboating, you need not only a suitable body of water—lake, river, bay, or estuary with room for a "long-distance slide"—but air temperatures cold enough to freeze the water thick enough to support boat and boaters (the heaviest iceboats need nearly a foot of ice underneath) yet not so cold that all precipitation falls as snow and the snow that has already fallen doesn't melt. "Snow is *the* enemy," iceboaters point out. When ice has two or three inches of snow on top of it, most iceboats bog down.

As for wind, there should be little of it early in winter so the water won't be too agitated to freeze yet once the water has frozen there *should* be wind, to fill the sail. Even then, the wind shouldn't blow too hard or in too "puffy" a fashion since it could cause the speedy, unstable boats to tip over. "An ideal day of iceboating," one writer concluded, "requires a sort of harmonic convergence of the elements." Places in the world where the elements converge often enough for an iceboating tradition to develop include the "melt belt" of North America—parts of New England south of the snow belt and parts of the Midwest and Canada north of it—Poland, northern France, Germany, Belgium, Denmark, Sweden, Estonia, and Russia from the Baltic Sea to Leningrad.

"Iceboaters are always looking for the perfect ice, like surfers for the perfect wave," says Peter Zendt, a third-grade teacher from Bay Shore, New York, who owns two Scooters (boats that float on water as well as glide on ice), one of them almost 100 years old—but they'll take what they can get. "On Thursday nights the phone rings off the hook," Zendt says. He'll call someone in, say, upstate New York and ask about conditions there, and that guy will tell him that he talked to somebody in Boston who passed on the word from somebody in southern New Hampshire that a certain river in Maine had ice last weekend and might still have it this weekend . . . There are conference calls, hotlines, newsletters, e-mails. One February day the newsletter of the Lake Ronkonkoma, New York, iceboating club announced that "the only sailable ice in the Western World" was on Lake Ronkonkoma. That weekend, squeezed onto its two-square-mile surface, the minimum size for iceboating, were about 30 iceboats, including a couple of Hudson River stern-steerers, the oldest and largest of all iceboats. In the 1880s, wealthy men living in mansions along the Hudson River, descendants of Dutch settlers, began staying home in winter and racing iceboats on the frozen river instead of going to the Caribbean. (Iceboating evidently involved enough discomfort and hardship that the sport was considered manly.) Hudson River stern-steerers have wooden frames instead of hulls, up to 1,000 square feet of sail, and a basket-like cockpit in which the skipper and passenger lie flat on their stomachs.

Also at Ronkonkoma that day were several Skeeters, the fastest of the fastest nonmotorized conveyances in the world. Developed in the Midwest in the 1930s, Skeeters are long, skinny, light, and limber, so they look like big waterbugs, and are steered from a blade under the bow instead of the stern. There were several DNs on the lake too; DNs are the VWs of iceboats. "DN" stands for *Detroit News,* the newspaper that underwrote their design in the late 1930s so ordinary folks could build their own iceboats from kits. Today there are more DNs in the world than any other iceboats, over 3,000, and every other year there's a DN competition in Lappland because it gets the first black ice in Europe, in October.

At Ronkonkoma, I got a ride in a Yankee, a type of Skeeter, with a man who'd been iceboating since he was ten. We sat leaning back with our legs under a tiny deck. No sooner had the skipper pulled the sail tight than we took off, as abruptly as if he'd floored an accelerator. "It's like a catapult shot if you catch the wind right," one boater says. Close to the ice, we could see the mottled surface as it flashed by, hear the scraping of the runners on the crusty top, feel in our bodies every bump of ice and hardened snowmobile track we crossed. We were zigzagging around, tacking this way and that, with the cold air beating against our faces. "B-r-r-oooo!" indeed.

Ronkonkoma's club secretary had a Yankee, too, and four weeks earlier it had gone through the ice on a lake in Connecticut. He still had the jitters. The iceboat was going 60 when it hit a ten-foot wide crack and "torpedoed" into the water then slid under the ice on the far side. If the headstay hadn't taken the hit of the ice edge first, transmitting such force to the mast that it "exploded" in 26 places, the secretary said, the ice would have decapitated him.

"In an age enamored of extreme sports, why isn't there more interest in iceboating?" Ron Sherry, champion iceboat racer from Detroit, wonders. "Hey, we've got collisions, we've got carnage, we've got broken equipment, we've got 'swimming' [what falling through ice is called], we've got speed. Coming around the marks at 95 miles an hour in a 45-pound fuselage is scary stuff."

> Remember, water is only ice that has temporarily melted.
>
> —Iceboaters' saying

In addition to iceboaters, Ronkonkoma's boatable ice had on it that weekend ice skaters, ice kiters, ice surfers, and ice sledders. A woman wearing skates was pulling behind her a padded car seat with a baby buckled into it. (In Hans Brinker's day, Mary Mapes Dodge wrote, "the sleepiest of servants" would push "some gorgeous old dowager, or rich burgomaster's lady" over the ice in a pushchair "mounted upon shining runners [and] heavy with footstoves and cushions.") A teenage boy was zooming about on an "ice sled" he'd made out of skateboard trucks and a sail. According to Bert Rufenach, Ontario denturist and champion ice surfer, "ice surfing is the fastest non-motorized, non-gravity-based, standup sport in the world." ("Standup" because iceboaters sit or lie.) He designed and built his own *Ice Hawk,* which has three "machete-like" blades and can go 25 miles an hour in a 5-mile-an-hour wind (and 60 in a 35-mile-an-hour wind). The best surfing ice, according to Rufenach, is so new and clear it's dark as water:

"I've sailed on black ice on Lake Ontario at night with the city lights three or four miles away and it was like being suspended in time. The ice is so smooth and slick you don't feel anything under your feet. Sometimes when it's almost unsafe, you can hear a high-pitched tinky sound of the ice cracking. I remember going 30 or 40 miles an hour and it was like I kept tapping a crystal glass as I went, *tink tink tink,* very beautiful."

Another qualifier to Rufenach's superlative besides standup was "non-gravity-based." Tilt an icy surface, and you've got a different way of having fun on ice. At Montmorency Falls in Quebec, a giant hill of ice forms every winter from mist coming off a waterfall and settling in front of the falls and freezing there, layer by thin layer; Victorians would travel to see it, as they would other natural monuments like Niagara Falls, and some would slide down the hill's steep slopes, innocently hoo-hooing, no doubt. For people too timid to slide down the hill themselves, according to Cavell and Reid, "a local lad could be hired to demonstrate the phenomenal speeds achieved by a toboggan launched from the top."

Out of such merry plunges on ice and snow grew the competitive, sophisticated, modern "sliding sports" of bobsled, luge, and skeleton. The sports are more demanding than they might appear to people watching the Olympics on TV, and so is making the ice for them. Ten groomers work 12 hours a day for two weeks to prepare the ice for the bobsled-luge-skeleton track at the Utah Olympic Park in Park City, Utah. The groomers groom by hand, Craig Lehto, director of the track, explains, using "quite archaic" methods. Each is responsible for his own 25- or 30-yard section of track, and he walks up it at night when the air is coldest, spraying a fine mist of water over the concrete base. The concrete is chilled by refrigerated ammonia which runs through pipes winding through every square inch of it, so even when outside air is 50°F or 60°F, the mist freezes within 20 seconds of touching down.

Once the groomer reaches the top of his section, he waits a minute and a half, has a smoke or a yawn, then turns and walks down, spraying as he goes. If the air temperature is cold, he walks quickly, if it's very warm, slowly. All night for eight to ten nights he walks up and down spraying his section, completing roughly 1,200 round trips a night and laying down 29,400 ultrathin ice layers, which together add up to two to three inches of solid ice. By keeping the layers thin, he can make the ice match, curve for curve, the preformed contour of the concrete walls.

However, at this stage the ice is "pebbly" because it flash-froze from water droplets. The groomers scrape away the pebbles by hand with blades, like big razor blades, then go after the "bumps," or unwelcome high spots. One way of detecting bumps, according to Tracy Seitz, who's in charge of ice for the track, is watching the first luges come down and seeing if the riders are compressed onto their sleds at any point, since that would reveal the presence of a low spot. Another way is noticing if the riders' feet pop up.

"It's very, very hard to get perfect ice," Seitz points out. "Ice is almost alive. It expands fast and contracts fast. Its hardness and resiliency are constantly changing." Once the ice is made, the groomers still have to keep fixing it up every day the track is in use. "We've got a . . . I don't know the word, *vulgar* and simple and brutal way of repairing the ice," Lehto says. The groomers slap slush "very very hard" onto rutted sections, pack it down with the backs of shovels and trowels, and level it with blades, by hand. "It's difficult, laborious, tedious, monotonous work and takes years of experience to do well." Yet it must be done well. "Every little nick or bump in the ice can take hundreths of a second off a finish time," Lehto explains and concludes, "Luge and bobsled performances are more dependent on the people who prepare the surface for the athletes to compete on than any other sports."

As independent as a hog on ice

—Folk saying

Still more fun can be had on ice by imparting fleetness to a thing, sending it, unaccompanied, across ice. Almost five centuries ago in Scotland, people were probably flinging smooth stones out over the ice of their wee lochs, enjoying the sight of them sliding a long way out, then after a while not just throwing the stones out but aiming them at some spot on the ice, and so inventing the sport of curling. (A few people argue that it was the Dutch who invented curling, citing as evidence the miniature figures in Pieter Breughel's mid-16th-century painting *Hunters in the Snow* who are engaged in a game on ice that looks like curling, but more likely they were playing with frozen clods of earth.) Today's curling stone is a flattened sphere with a handle on top, which looks sort of like a teakettle. It weighs up to 44 pounds and is made of granite so fine-grained there's little space for moisture to insinuate itself and freeze and expand and crack the rock.

A 44-pound rock isn't something you'd fling onto the ice like a skipping stone, however. Instead, curlers hold it by the handle and slide it over the ice with an underhanded motion that reminds some people of a bowling shot, comprising as it does a backswing, swing, and follow-through. On release, the handle is sometimes rotated so that the rock spins gently over the ice and describes an arc, rather like a hook in bowling: the curl. Like first-stage bobsled and luge ice, curling ice starts out pebbly, but it is kept pebbly, not shaved flat. "You can't throw a stone on flat ice," explains Raymond ("Bud") Somerville, former county clerk in Superior, Wisconsin, who was five times U.S. curling champion (about whom a sportswriter once wrote that he was "to curling what Babe Ruth was to baseball, Gretzky to hockey . . . a giant among curlers"). As slippery as flat ice is, it produces too much drag for a rock to slide across; pebbles give the rock fewer ice points to ride upon.

The pebbles are made shortly before game time by a person who walks back and forth over the curling "sheets" (ice equivalents of bowling alleys) spraying extremely fine droplets of water on the existing ice, where they freeze. The colder the water, the higher the pebbles. High pebbles are undesirable because the rock will bust their tops off as it moves and be slowed down. To make the ideal low, rounded pebbles, ice-makers spray their sheets with water heated to about 160°F.

The curling shot is something lovely to behold if done well: graceful, with the front knee deeply bent during follow-through and back leg stretched out behind, the player often continuing to slide a fair distance down the ice with his arm still reaching out and his eyes directed toward the rock as it makes its slow, rotating way toward the target ("house") at the other end.

It's also a pleasure to watch curlers get around on the ice, as when they switch sides after an "end" (inning), since they have their special way of doing it. They wear mismatched shoes, with the sole of one smooth—Teflon, ceramic brick, or stainless steel—and the sole of the other rougher—crepe usually. The former is the "slider," the latter the "gripper." A curler moves by keeping the slider flat on the ice while making pedal motions with the gripper, as he would if the slider shoe were a scooter.

To anybody watching it for the first time, though, the most arresting thing about curling isn't the graceful toss or the goofy walk: it's the sweeping. That's something you don't see elsewhere in the world of sports. To curlers, brooms are equipment. After a player releases a rock, two members of his own "rink" (team) will usually walk or run down the sheet and sweep the ice ahead of the rock as it goes. There's nothing leisurely about this activity. The curlers sweep vigorously, madly even, leaning over and giving the ice a good scrub, like housewives on a tear. They sweep to help the rock slide farther or straighter than it otherwise would have done. Sweeping can add as much as 20 feet to a glide. A rock that seems within an inch of stopping can be coaxed another foot or so over the ice with diligent sweeping.

The sweeping probably gets its effect from cleaning and polishing the ice ahead of the rock as well as warming it. Debris, including busted pebble heads, slows a rock and makes it curl. "Dirt is the enemy of curling ice," says Steve Smith, former icemaster at the Ardsley Curling Club in New York. "A single hair can cause a rock to jump 45 degrees. I get accused all the time of dropping my beard hairs on the ice." Nevertheless, he believes sweeping gets its main effect by "pre-heating" the ice surface through friction, enhancing the liquid film on top.

One way players judge the changing condition of the ice is by keeping track of its "speed." "It's 19-second ice," players will say, or "It's 24-second ice," referring to the time it takes for a rock to glide from one "hog line"* to the center of the other house or to the place where it stops moving. Although you wouldn't think so, 24-second ice is faster than 19-second ice. That's because when ice is fast, a player needs to throw his rock with less force than when ice is slow, in order to reach the target, and the rock spends more time en route. "Ideal," according to Somerville, "is 23-second to 24-second ice." Another way to judge the ice is by watching the curl. If the curl has a high degree of arc, the stone is grabbing the ice and the ice is considered "swingy." "That can happen later in the game when the pebble gets worn down," Somerville says. He likes ice "somewhere between swingy and straight."

*"Hog" refers to a curling stone that lags, or doesn't go far enough down the sheet because a player didn't throw it hard enough. (A stone must go over the hog line to be in play.) It was first called hog in the early years of the game, Charles Earle Funk surmised in his book *A Hog on Ice and Other Curious Expressions,* because of "its unwieldiness and its inertness, becoming partly frozen into the ice," just sitting there interfering hog-like with later play. Maybe for that reason, people started using "hog on ice" to mean a person of "cocky independence, supreme confidence," despite the fact that a real, live hog would be hapless on ice, unable to keep its footing, needing to be dragged off the ice, not at all independent.

The ice of life is slippery.

—George Bernard Shaw,
Fanny's First Play: Induction

For one person at least, the slipperiness of ice is not just an unusual property that allows a variety of games to be played on an otherwise solid substance but something more . . . well, slippery, illusive, deceptive:

When writer Stephen King was asked not long ago by a reporter who admitted that he had until that moment been too sheepish to confront the author with the question, "What *about* ghosts, really?" King reluctantly replied:

"I guess the closest I can get is what Mike Noonan [the main character in his novel *Bag of Bones*] thinks when he's walking along the road at dusk and he says it looks like there are faces in the leaves and it feels like reality is thin and that the world is something that we skate on—it's like ice and we're always turning toward home. Just the act of turning . . . to me, when you skate, when you turn, there's this sort of giddy feeling of being in control and out of control at the same time, this real swoop like the corner on an amusement-park ride. Except that's clattery and noisy and defeats the whole sense of wonder, whereas ice skating is this quiet thing. I think the metaphor serves because reality is slippery. People slide on it and break their bones all the time. And if the ice is thin you can fall through."

CHAPTER TWENTY-FIVE

GAMES II

> We had our own New Year's celebration, a football game on the ice. The ball was made of seal-gut.
>
> —Robert Bartlett,
> captain of the *Karluk*, when it was
> trapped in sea ice and drifting toward its doom

PEOPLE SOMETIMES HAVE FUN on ice in spite of its slipperiness, not because of it. Simply because it is there: miles of playing fields and promenades and tracks, hundreds of feet of climbing wall. Polar explorers often left their ships, even occasionally in dire circumstances, to sport on the sea ice. When the screw-yacht *Fox* was hemmed in by ice in Melville Bay in August 1857 and Captain Leopold M'Clintock observed that "this detention has become most painful," he also observed that the "men enjoy a game of rounders on the ice each evening." After Fridtjof Nansen left the ice-locked *Fram* to make a try for the North Pole, the ship's crew celebrated Norway's national day by holding a festival on the ice. "The programme offered a rich variety of entertainments," Captain Otto Sverdrup wrote. "There was rope-dancing, gymnastics, [and] shooting at running hares."

Before Nansen set out, he and his shipmates had been taking regular strolls on the ice, undeterred by the cold, Nansen claimed, "though sometimes one or another of us does not take quite so long a walk as usual when a strong wind is blowing." When commander Sherard Osborn was en route to Baffin Bay in 1850, he had the bark *Pioneer* drawn up alongside a floe where the ice was "hard, affording good exer cise for pedestrians," he noted. "To novices, . . . the idea of walking about on the frozen surface of the sea was not a little charming. In all directions groups of three and four persons were seen trudging about."

In parts of the world where people lived more settled lives near bodies of water that froze in winter, they were long accustomed to strolling on ice. "Once the St. Lawrence had frozen solid, . . . it became a grand boulevard for the inhabitants of the city [Quebec]," Cavell and Reid wrote in their book *When Winter Was King,* "soldiers and merchants with their families, habitants and Indians, all mingling freely in the weak winter sunlight." With its topping of ice, the river became "a bridge, a playground, a crystalline freeway connecting cities and neighbours."

The crystalline freeway probably had snow on it most of the time, therefore the slipperiness of the ice didn't upend the promenaders or make them walk like ducks. Yet there are activities some people engage in on ice that require more than a snow cover to counteract ice's slipperiness. Take dirt biking. Buzz Arndt, owner of Buzz's Barber Shop in Brooklyn Park, Minnesota, started riding his motorcycle on frozen Minnesota lakes in the early 1970s. But every time he got going more than a few miles an hour, his tires would spin out on the ice. He tried belting snowmobilers' cat's claws onto the tires, and although they helped him hold a straight course, he'd still spin out in turns. So would his friends.

"We were sticking all kinds of things into our tires to get traction," he remembers, including sheet-metal screws. The screws' rough protruding heads gripped the ice well enough, but they were so short they pulled right out of the rubber. Longer screws punctured the inner tubes. Also, the slots on the screw heads kept getting plugged up with ice. The bikers experimented with other screw heads, ones with angle-sided slots, star-shaped slots, and V-shaped slots. The Vs "cleaned best," Arndt says. Some of the tires split because there was not enough rubber between the Vs so they tried skinnier Vs. At last they had the "perfect" tire studs, and factories started making them exclusively for racing motorcycles on ice—"they aren't good for anything else," Arndt says—and christened them "Cold Cutters."

Now, in late fall or early winter, Arndt can be found at home with a drill bit inserting 1,250 Cold Cutters into his motorcycle's back tire, one by one, and 890 Cold Cutters into its front tire. Each screw sticks up $\frac{3}{16}$th of an inch above the rubber's surface and lies at a slight backward angle to it so the head will be less likely to become clogged by ice. With his tires thus all studded up, Arndt can ride his bike 60, 70, 80 miles an hour across the surface of a frozen lake, even accelerating in the turns, successfully defeating the slippery property of ice. "What a rush!" he says.

> My attachment to the world has been reduced to a few thin points of steel sunk half an inch into a giant popsicle.
>
> —Jon Krakauer, "Straight Up Ice"

The people who truly take their pleasures on ice because it is there—following in the footsteps of Sir George Mallory, who was the first person to use that phrase in relation to Mount Everest, where he perished in 1924 on his third try for the summit—are the ice climbers. Like ice bikers, ice climbers need something to help them get a grip. Nowadays they are able to ascend very steep, even vertical, even past vertical ice by essentially nailing themselves to it, kicking steel points into it with their booted feet, driving steel picks into it with their gloved hands. "It would be nice if man could climb on ice as freely as the ape climbs trees and rock," pioneer ice climber Yvon Chouinard wrote in his book *Climbing Ice*. "Unfortunately, ice is not a natural medium for man. To perform on it, he needs his tools."

Needing tools is the main way that ice climbing differs from rock climbing. While the rock climber can grasp his climbing medium with bare hands and nearly bare (thin-slippered) feet, the ice climber grasps his medium mostly with hardware: crampons and ice picks. Being at one remove from the medium thus induces in the ice climber a "high degree of uncertainty," according to Mark Moran, geophysicist at CRREL, for whom ice climbing is, he confesses, "a vice, considering that I have a 2 1/2-year-old son." The variability of ice contributes to the uncertainty. "With ice, you can never really know how well you are fixed," Moran says. "You can never tell how much movement your point placements will tolerate."

While rock doesn't change much week to week or even year to year, ice can change in days, or fractions of days. "A ribbon of hardened water may exist only in the morning hours after a hard freeze," Jeff Lowe wrote in his book *Ice World: Techniques and Experiences of Modern Ice Climbing;* "by late afternoon, the only remains might be a wet streak on the rock." Andrew Embick noted in *his* book, *Blue Ice and Black Gold: A Climbers Guide to the Frozen Waterfalls of Valdez, Alaska,* that while ice in a certain formation on a 15°F day with the wind blowing could be like "concrete, porcelain, ceramic," on a warmer day it could be like Styrofoam. "The medium [ice and snow] is a plastic material," Chouinard concluded, "the most variable of all plastic media."

Another way that ice climbing differs from rock climbing is the sometimes violent bashing that goes on. "Think about it," Mark Moran says. "You are taking a big hammer with a long spike and physically striking the medium; there's an explosion of crystals; pieces tumble down the side of a mountain. It's dramatic." Where ice is concerned, though, the willful destruction of a natural object isn't considered environmentally objectionable, as is often the case with rock. "Ice heals itself," Moran notes. "The holes I made Saturday [on a frozen waterfall] are gone today." "Ice is a renewable resource," Yvon Chouinard wrote; while climbing on rock "in poor style can ruin the route for future parties, . . . no one should care what you do with [ice]. Even a line of bucket steps will be melted smooth in a day or two."

There are two main kinds of climbing ice, alpine ice and water ice. Alpine ice starts out as precipitation, usually snow, and changes into glacial ice. People have been climbing alpine ice for centuries, particularly people in the Alps. Sixteenth-century shepherds crossed high, icy passes with nail-studded horseshoes on their feet. Nineteenth-century gentlemen climbed glaciers in hobnailed boots, often on steps their hired guides chopped out for them. Even with steps, however, climbers couldn't go up slopes that were more than gently inclined. Then, just after the turn of the last century, a climber introduced ten-point crampons, which when affixed to boot soles and pressed all at once into the ice or snow allowed people to climb somewhat steeper slopes, flat-footed. Still, most people couldn't make it up glaciers with inclines greater than 50 degrees until the early 1930s when some climbers added a pair of spikes to the front of their boots, forward-pointing spikes with which they could "tiptoe" up mountainsides.

Several decades later, Chouinard and a fellow climber developed ice picks whose curves matched the arc of the swing required to plant them in the ice, which increased chances that the points would stay in the ice instead of, as had been happening with straight-headed picks, popping out and taking big divots of ice with them. With these and other new tools, people could climb not only steeper alpine ice but water ice. Water ice is frozen from the liquid and is usually even steeper as well as harder and more brittle than alpine ice. Most of it that's climbable is in the form of frozen waterfalls. In *Blue Ice and Black Gold,* Embick acknowledges that waterfalls are beautiful in summer, but he points out that in winter they are not only beautiful but their beauty can be "viewed more closely and experienced more directly" by people willing to climb them. He himself has climbed hundreds and in the process seen such icy delights as—he listed some—"diaphanous sheets, glittering hanging icicles, massive emerald and white pillars, delicate lacework, flashing sunlit ribbons, dark caves, evanescent verglas, huge pedestals and buttresses as well as wild scoops, obscure cauliflower and artichoke shapes, [and] glistening slabs."

Jeff Lowe and Mike Weis, the first people to ice-climb Colorado's

Bridalveil Falls, saw it the first time in summer, unfrozen. "As you look up at the roaring column, face dripping with spray," Lowe wrote in *Ice World,* "it is difficult to imagine any force formidable enough to still such awesome power." Cold was the stilling force, gradually applied. "In late autumn, the edges of the main fall sprout delicate white wings of ice," Lowe observed, chronicling the changes. "By December the whole cascade has hardened into a brittle pillar, narrow and spotty, with many cauliflowers and chandeliered sections." By January 1, 1975, when he and Weis free-climbed the fall, it had become a "beautiful, convoluted shaft of ice" 400 feet high with "brittle bulges, insubstantial pillars, and numerous overhangs."

Some ice routes offer climbers both alpine ice and water ice. Some offer both ice and water. In 1977, when Henry Barber and Rob Taylor made the first climb of the world's highest freefalling waterfall, Norway's Vettifossen, 1,000 feet top to bottom, the water kept on flowing "full-tilt boogie" behind the ice, Barber reports; "all we were hearing was water." Many ice routes offer ice and rock, either both of them at once or one right after another. Such "mixed climbing" is particularly popular in Scotland, which has a lot of iced-up rock, crags where wet mists have laid down a glaze of ice. The great climber Alex Lowe even made an ascent of an iceberg once, off the coast of Antarctica. He was on a sailing junket, saw some large bergs floating around, got out his climbing gear, which "I never expected to use at sea," and scaled a berg on the spot. "Pretty much like climbing anything," he pronounced.

another world . . .
. . . inverted
cumulus and azure

—Heather McHugh, *"Seal"*

Across ice, up the side of ice—how about under ice? Sure enough, some people have made a sport out of ice diving. It's something scientists have been doing for years, out of necessity, so that they can observe their study objects—penguins, seals, algae, sea squirts—in ice-covered waters. When I first heard about it, it had all the claustrophobic appeal of a sailor's corpse being slid through a hole in the ice for permanent entombment at the bottom of the sea. Yet people I've talked to about diving in the polar regions don't speak of feeling trapped beneath the ice but of feeling free and calm, with all the space, light, and beauty down there. "I was swimming down the nave of a vast cathedral," Vince

Rhodes wrote in *Sport Diver* magazine after his first excursion under Arctic ice, "the walls of the channel glowing as though made of stained glass." Andrew Driver, one of the leaders of the dive trip Rhodes was on, considers below-ice vistas to be "space-like, ethereal."

In the Arctic, divers choose where to dive by looking not at the underside of the ice but at the upper side. "What you see above is what you'll see below," Driver says. "If it's rough on top, it'll be rough below." Rough is good. "The worse the walking on the surface," R. Todd Smith wrote in *Beneath a Crystal Ceiling: The Complete Guide to Ice Diving,* "the more beautiful the formations underneath." Formations may include jagged keels, rounded keels, keels with chambers and tunnels eroded into them, all of which divers can explore. "The joy of Arctic diving is the contour," Rhodes concluded. He swam through one "yawning" ice cave that had many side passages where the light changed gradually "from white to teal to royal blue."

Scientists, too, have their moments of aesthetic pleasure in the under-ice world. In early spring in McMurdo Sound, says James Stewart of the Scripps Institution of Oceanography in LaJolla, Calif., who was in charge of U.S. research diving in Antarctica for 30 years, the water below the ice is as "clear as air. It is essentially distilled. You can see right through it." There's little plankton in the water after six months of darkness, and there's no sediment either since the ice cover stops waves from stirring up the seabed. "When someone under the ice shines a flashlight," Stewart relates, "you see the flashlight in his hand and the spot where the light hits something, but you don't physically see the beam. There's nothing in the water to backscatter light. It's like you put a layer of ice seven feet thick across the top of the Grand Canyon and stepped through it. *That's* how you feel under Antarctic ice. As if you are hanging in space."

With all that great visibility, what is there to see under Antarctic ice? In the course of a single dive under the ice he made in McMurdo Sound, Jim Mastro, for five years U.S. scientific diving coordinator in Antarctica, came across, among other things, "a dazzling array of sponges: long, thin yellow fingers, giant white vases, pink staghorns; bright yellow cacti; deep, green globes; and bright red clusters." Lavender hydroids and orange sea anemones and "round, bright yellow-green mollusks." Mastro was struck by how, "just a few meters beneath the ice, we are discovering a richness and diversity of animals to rival some tropical environments."

My first ice-canoe team was composed of veteran parachutists, people who were willing to throw themselves out of a plane for a little amusement. That's the kind of attitude you need.

—Yves Gilbert, in "Trial by Ice" by Michel Beaudry

One of the strangest things you can do with ice in the name of recreation or physical challenge or perverse satisfaction is ice canoeing. *Canoeing?* Even those who do it know it sounds bizarre. Every winter Quebec City holds a carnival whose highlight is a canoe race across the Saint Lawrence River, which always has a lot of ice in it at that time of year. Still, 130 to 150 men and women in tights and crampons and *tuques* (wool hats) climb of their own free will into 28-foot-long canoes, five to a canoe, and paddle, row, run, and "scoot" over ice and through water to reach the opposite shore and come back faster than other canoers. They get wet, they get sore, they get tired, they get disoriented, and occasionally they get lost. "You have to be a little crazy to do it," one of them says proudly.

Ice canoeing has nothing to do with iceboating. The ice that ice boaters want is flat, smooth, and snow-free. The ice that ice canoers get is lumpy, cracked, slushy, hard, jagged, smooth, waterlogged, thick, thin, piled-up, dirty, snow-encrusted, and/or soupy. "The ice is always different," Yves Gilbert, engineer and veteran ice canoer, tells me. "One day it's very flat, the next day very rough." Currents, winds, tides, and icebreakers trying to keep open a channel for ships all the way to Montreal tear up the ice, but pieces of ice often come together and form a cover, or part of a cover, at Quebec City.

Ice canoeing isn't a made-up sport, either, like one of those TV gladiator events, wrestling a backhoe while wearing a cape. In earlier times, French settlers on the south side of the river who wanted to cross to the north side, to fetch supplies or see a doctor early in winter before the river froze and formed natural ice bridges, would make the hazardous crossing through water and ice in wooden canoes, usually hewn out of single logs. Then ferryboats appeared and metal bridges and airplanes, and by 1953 when organizers of the winter carnival decided to stage a race in the traditional ice canoes, only a few people were still using them, mostly people living on small isolated islands in the river. Nearly all of the early races were won by crews from those islands, in canoes from those islands, which although not hewn from single logs were also wooden, with metal plating under the keel to protect it from the abrading ice. The crew that won most often came from a single island—Ile-aux-*Canots,* as it happens—with a single family living on it. The island was small but the family was not; the winning five-man team was made up entirely of brothers.

Nowadays, almost all ice canoes are made of fiberglass; they are longer, lighter, tougher, and easier to push over ice and propel through water than wooden ones. Their hulls tend to have a shape in cross-section somewhere between a U and a V. The U shape (flat bottom) is faster on ice but the V shape (pointed bottom) is faster in water. The U is more stable on ice and rides up on it like an icebreaker, but the V can pass through narrow gaps in the ice cover that the U might get wedged into. "It's hard to reconcile these things," an ice canoer from Alberta admits. "Designing for both ice and water is always a compromise." In technique as well, racers have to reconcile the demands of ice and water. Only the bowman carries a paddle. He uses it when the canoe rides in openings in the ice cover too narrow for oars. The rest of the crewmen carry oars and use them in open water, as well as in water that's not so open. Each oar blade has a metal spike on the tip, meant to snag ice and give the oar stroke purchase.

On water, the bowman gives commands—*"Glace avant!"*—but on ice it's the ice captain, one of two canoers in the front seat, who does. *"Gauche!" "Droite!" "Cap!"* (Keep going in the same direction.) *"Lentement!"* The ice captain also decides, based on his assessment of the ice, whether the team should run or row or scoot through or over the ice. To scoot—*trottiner*—racers keep one leg inside the canoe, knee bent and on the seat, and the other leg outside, hanging free with the foot touching the ice, propelling the canoe as if it were a scooter. Over their boots the racers wear waterproof coverings fitted with crampons which they make themselves, out of plastic cold-resistant to −25°F. One young racer explains how to make them, as if he is sharing a recipe. "Heat the plastic for ten minutes in a 250-degree oven. Cut it into an oval-shaped piece and mold the piece around your boots. Pound Ski-Doo nails in the bottom." Allow to cool. Tie on.

Race day at Quebec City's 1993 carnival was Valentine's Day. Seventeen canoes assembled in the harbor to await the starting gun. Sponsors included a car dealer, a bagel maker, a commuter airline, a hotel, a city, and a deodorant. Outside the harbor, the river looked like scrambled egg whites. An icebreaker had evidently been clearing an ice jam several miles upstream and was churning up a lot of *bourguignon* (brash), which had floated down and covered most of the surface at Quebec City. As the *bourguignon* flowed past the bank on which I was standing, enjoying the panoramic view, it made a soft, brushing sound.

The gun sounded, and the canoes headed into a mix of ice and water. I was surprised to see that within minutes they were widely scattered, going separate ways. Some of the canoers were rowing but some were scrambling on foot over mounds of ice and dragging their canoes with them. From where we stood, it was hard to figure out how they were making their decisions, why one crew's path was so different from another's. "They have to judge on the basis of what they are seeing," Yves Gilbert had explained earlier, and what they are seeing in one place can be quite different from what others nearby are seeing. Piles of ice shorten the views. "Sometimes all we can see is the first ice," Gilbert said. "We have to *imagine* there will be some water or a flat area ahead." Typically, ice canoers make course corrections every 50 to 80 feet because that's how far, typically, they see ahead.

The worst ice for ice canoers, according to Gilbert, is frazil; the crystals can clog the river to a depth of many feet. "It's very hard to row through and you can't walk on it. It sticks like glue to the boat. We spend a lot of work to go a little." When frazil mixes with *bourguignon,* you get "a real soup."

To us the canoes appeared to be meandering and the crews to be out of them more often than in them. When the ice is very uneven, canoers step clear out of the boat and run up and down the hummocks pulling the canoe. On flatter ice, they'll stay half in the boat and scoot it along, moving it ahead with that pedaling foot. When they come to water, the racers often stay in scooting mode if the water is, say, a 50-foot open stretch between two floes and they don't want to bother pulling their legs back in the boat and turning around to face the other direction in order to row only to have to turn their bodies around again and stick their legs out a few minutes later when they reach ice again; they just keep on scooting, through water, using their dangling legs as paddles.

While the canoe's weight is spread on ice, the canoer's is not, and sometimes a racer falls through. Whenever he steps out of the canoe, he is supposed to keep both hands on the gunwales so if the ice gives way the current under the ice won't sweep him away. So long as he hangs on and his teammates keep the boat moving, he'll make it back onto ice when the boat does. Ice canoeing is safe, Gilbert insisted. "We've had very few accidents. It's not like skiing."

The winner, for the fourth year in a row, was the team sponsored by the hotel Chateau Frontenac. Ile-aux-Grues, Gilbert's team, came in second, a minute behind. Both teams made it across the river and back in less than an hour. The ice was "easy," Gilbert reported. "Slushy," the captain of an all-woman team added. Generally, the more ice there is, the tougher the race. The year before, the river had been 95 percent covered with ice which winds had pushed against the north shore into ridges up to eight feet high. "Nobody used the oar," Gilbert said. The teams spent most of their time hauling their 250-pound canoes up and down ice ridges. Only two teams finished within the two-hour limit. Neither was Ile-aux-Grues. "The big adversary is not the other canoes," Gilbert confided. "It's the river. Each year we win or lose the race against the river. If we win that, *then* we look for other canoes."

CHAPTER TWENTY-SIX

USES I

Ice that merely performs the office of a burning glass does not do its duty.

—William Scoresby Jr.,
Journal of a Voyage to the Northern Whale-Fishery

WHILE ON A WHALING TRIP to Greenland in 1822, Captain William Scoresby Jr. made optical lenses out of ice. He chopped pieces from an iceberg with an ax, scraped them into convex shapes with a knife, and polished them with the heat of his hand held inside a woolen glove. Using the most transparent of these chunks, he was able to focus the sun's rays and "produce a considerable intensity of heat," he wrote. "I have frequently burnt wood, fired gunpowder, melted lead, and lit the sailors' pipes, to their great astonishment. . . . Their astonishment was increased," he added, "on observing that the ice remained firm and pellucid, whilst the solar rays emerging therefrom were so hot, that the hand could not be kept longer in the focus than for the space of a few seconds."

Making lenses of ice was a curiosity, a shipboard diversion (the other sailors "flocked around me for the satisfaction of smoking a pipe ignited by such extraordinary means"); the lenses didn't have a whole lot of practical value. Yet on his voyages in the Arctic, ice sometimes did serious duty for Scoresby. It gave him solid structures to hitch his ships to during storms and a source of fresh water in the midst of salty seas. Although ice was often a curse to him, obstructing and delaying his ships' progress, it also served him.

Other people in other times, the vast majority of whom never got within ice-blink distance of a polar sea, have found their own ways to make ice work for them. The chief characteristic of ice that makes it useful to human beings is its comparative cold. Ever since Mesopotamians began selling ice in settlements along the Euphrates 4,000 years ago, people have used ice to chill things: their drinks, their food, their fevered bodies. Particularly their drinks . . .

The summer palaces on slopes, the terraces,
And the silken girls bringing sherbet.

—T. S. Eliot, *"Journey of the Magi"*

Ancient Egyptians and Athenians as well as Mesopotamians would throw ice or snow directly into their wine and water to cool them. The Roman emperor Nero reputedly had snow placed *around* a goblet or flask so the snow would cool the water inside yet not taint it. In 17th-century Spain and Italy, ice or snow was sometimes slipped into glass pockets in wine vessels so the wine could be chilled without being diluted. "For warm wine I cannot swallow!" an Italian poet declared, and evidently he didn't have to.

In the days of the Turkish empire, snow and ice were thrown into fruit beverages, too, to chill them, a practice that spread to Italy. Eventually, fruit beverages were not only chilled by ice but frozen *into* ice, or rather "ices," as British food writer Elizabeth David pointed out in her richly detailed book, *Harvest of the Cold Months: The Social History of Ice and Ices.* Ice is frozen water, pure and simple; ices are frozen desserts or refreshments, often complex. Figuring out how to make good ices over the centuries, however, must not have been easy. According to David, one 18th-century French confectioner stated in his book on *glaces* that "the early ices were composed mostly of water with a little fruit and powdered sugar and . . . were nothing more than pieces of ice as hard as concrete." The trouble probably was that they didn't contain enough sugar. Confectioners of the day didn't fully appreciate the antifreeze properties of sugar, the way it slows freezing and keeps ice crystals from growing large. Later in the evolution of ices the opposite problem arose: so much sugar was added to dishes that freezing was suppressed and concoctions wouldn't harden, and people ate—and apparently enjoyed, according to David—"sweet snowy mush."

What ultimately made it possible for confectioners to freeze and not merely chill sweet liquids was the discovery, probably made in Naples in the 17th century, that salt added to ice or snow lowered its freezing point, causing the ice to melt and the heat required for the melting to be drawn out of any sweetened liquid in an adjacent container.

Once the makers of ices got the hang of things, they were limited only by their imagination and the ingredients available to them. Wealthy Neapolitans of the late 17th century, David noted, partook of "sunrise sherbet," a mix of frozen milk and pulverized cinnamon bark thickened with minced candied pumpkin. English noblemen of the late 18th century enjoyed tea cream, champagne water-ice, and plum-pudding ice. Habitués of late-18th-century Parisian cafes savored *fromages glacés à la crème,* or custard ices shaped to look like wedges of Parmesan cheese.

To this day you can get ices that masquerade as something else. I saw a photo not too long ago in *National Geographic* of a Sicilian waiter serving ices that looked just like squid, pasta, mussels, and raisin buns to diners in need of that particular perversity. You can also get almost every other kind of ices imaginable, from simple shaved ice with fruit syrup dolloped on top—often served from street carts—to sorbets and gelatos and frozen yogurts and parfaits and granitas (mix of pureed fruit, water, and sugar which once it has started to freeze is repeatedly raked with a fork to produce a coarse, grainy texture); mousses; bombes (light ice cream surrounded by a thick rind of ice); Baked Alaskas (ice cream surrounded by hot meringue); and assorted Smoothies, Icees, and Freezees.

You can also get ice cream in almost any conceivable (and inconceivable) flavor; Baskin-Robbins alone has come out with more than 700, including Bubble Gum and Pralines 'n Cream, the only flavor of theirs that has outsold vanilla. Enterprising restaurant chefs have created signature flavors for their dessert ices: oatmeal, sweet potato, lavender, roasted garlic, mint-risotto, purple corn, curry, beet-yogurt-swirl. Ice milks, being lighter than ice creams, can be savory as well as sweet, former *New York Times* food editor Molly O'Neill observes; she highly recommends basil-mint-and-hot-chili ice milk. In his book *Make Your Own Ice Pops,* Mathew Tekulsky offers recipes for homemade Popsicles, including the adults-only Pear–Red Wine and Bloody Mary Ice Pops. Tekulsky credits the birth of frozen treats-on-a-stick to the time in 1923 when "inventor Frank Epperson left a glass of lemonade (with a spoon in it by mistake) on a windowsill . . . and the lemonade froze to the spoon during the night."

Before you roll your eyes at the preposterous reach of some of these concoctions, you might bear in mind that in 19th-century New York, at Delmonico's restaurant, diners could order pumpernickel rye and asparagus ice creams.

> Maine ice for the juleps of Charleston, northern January cooling Jamaica's rum.
>
> —Donald Hall, *Seasons at Eagle Pond*

But where, a reader might ask, in all the centuries before people had freezers, did they get the ice they needed for chilling their wine and making their ices? For a very long time, they got it from hills and mountains, in the form of natural snow and ice, which they brought down in carts or on the backs of animals. The system had drawbacks, including inconvenience, theft, and dirt. At some point, to get a better product, a few Italians set up "ice reservoirs" or "ice lakes." Englishmen got ice from their estate ponds (during cold spells) or from Scottish lochs. Inhabitants of Russia, Scandinavia, Canada, and the northern United States took ice from nearby lakes, ponds, and rivers that froze over naturally each winter.

But this left a lot of people in warm places without ice (and in many cases not even wanting ice), a situation that Bostonian Frederick Tudor decided to change, starting around 1805. "In the manner of a drug pusher," Gavin Weightman wrote in his book *The Frozen Water Trade,* Tudor "set out to create a taste for cold that would get customers hooked." Through an almost obsessive determination, Tudor eventually succeeded, although it took a couple of decades, during which he endured skepticism (a "slippery speculation," one paper wrote of selling ice), outright ridicule, bankruptcy, and even jail time. Tudor's first shipment of ice, cut from a pond on his father's farm near Boston, went to Martinique in the West Indies in 1806. After that, he sent ice to Cuba, Jamaica, Brazil, the American South, and India, among other overheated places. Once Tudor showed it could be done, other people started doing it. For years ice was America's second-biggest crop by weight to be transported by trains and ships (the first was cotton). The ice went to Calcutta and Madras, Hong Kong, Rio de Janeiro, and Yokohama; to Charleston, Savannah, New Orleans, Mobile, and Key West; even to Washington, Baltimore, Norfolk, Philadelphia, and New York. More than three decades after Tudor first sent ice to Martinique, a rival company sent New England ice to England. It made a big splash there, too. The London office kept a foot-thick block of it in the window so passersby could see how clean and clear American pond ice was; the block was replaced every day so it seemed never to melt. People would walk in off the street and touch the ice to make sure it wasn't glass. The ice came from spring-fed Lake Wenham near Boston, and by association "Lake Wenham" came to be synonomous with quality. "Everybody has the same thing in London," William Makepeace Thackeray wrote in his 1848 story, "A Little Dinner at Timmins's." "You see the same coats, the same dinners, the same boiled fowl and mutton, the same cutlets, fish, and cucumbers, the same lumps of Wenham Lake Ice, &c."

> He . . . unroofs the house of fishes, and carts off their very element and air, . . . like corded wood, . . . to wintry cellars, to underlie the summer there.
>
> —Henry David Thoreau, *Walden*

To obtain the enormous amounts of ice needed for the ice trade, new tools and techniques for harvesting it were developed, chiefly by Tudor's foreman, Nathaniel Wyeth, after which a typical commercial ice harvest went like this:

On a large pond in New England—let's call it Pellucid—ice has been growing since late December. It is now mid-January, and the air temperature is 10°F. The workers—farmers and other locals glad to get paying jobs in winter—drill holes in the ice to see if it's thick enough to cut. If it isn't, they may "sink" the pond. Using narrow-ended chisels, they punch holes in the ice so water rises through the openings, floods the surface, and freezes, adding an inch or two of ice to the total.

By mid-January, the ice on Pellucid is 15 inches thick, and the harvest begins. Over the next several weeks, the men will work ten hours a day seven days a week taking ice from Pellucid even as thousands of other men are doing the same on other lakes, ponds, and rivers around New England. The first thing they all do is scrape snow and dirt off the surface of the ice. Horses in spiked shoes pull a scraper behind them, followed by "shine boys." The job of the shine boys is to shovel up the horse droppings and the ice underneath the droppings that has been tainted by them as well as by splashed urine. The boys pile the scrapings on a sled and spray the ice left behind with formaldehyde, which gives the ice a telltale shine and the boys their name.

Next, the men mark the boundaries of the ice field, which at Pellucid is to be 600 square feet. Using a single-bladed plow, they carve a three-inch-deep groove along the perimeter, then take a two-bladed plow—Wyeth's revolutionary invention, which is like a sled only with blades instead of runners—and set one blade in that groove, so when the plow is pulled along, a second groove is cut parallel to it 44 inches away. Then they set the outside blade in the second groove and cut a third one 44 inches away, and so on; they score the ice row after row until the whole field is striped. They do it again at right angles to the stripes until the whole field is checkered.

Using very long saws, they cut through the bottom few inches of the ice cover to make "floats," or sections of the field consisting of many squares. Then, using pike poles three times taller than they are, they propel the floats through a channel they've opened up in the ice, breaking the floats up on the way to shore into smaller and smaller ones with fewer squares, using long-handled splitting bars. (There's a whole armamentarium of tools made especially for harvesting ice, including ice hooks, needle bars, breaking spades, hooking chisels, ice hatchets, calking bars, fork bars, and feeding forks.) Once opened, the channel isn't allowed to close. Men sweep ice cakes back and forth through it to keep other ice—a skim—from forming. "Sometimes the men worked by moonlight, if the boss felt there would be a change in the weather," Dewey D. Hill, owner of an ice company in Utica, New York, and Elliott R. Hughes wrote in their book, *Ice Harvesting in Early America*. "It was cheaper to do this than to try to break ice out in the morning after a cold night."

A person might think that with so much frozen water being so definitely removed in this manner from American lakes, ponds, and rivers year after year (nothing more definitive than being shipped to India) that the bodies of water might be affected, their levels lowered the rest of the year, their volumes reduced. "That's one of the questions people always ask us," says Bruce Nelson of the Wee-Kut Ice Company of New London, Minnesota, which harvests ice to this day, only with circular saws instead of plows and tractors instead of horses. Nelson passed the question on to a hydrologist at Minnesota's Department of Natural Resources who told him that the amount of water Wee-Kut removes from a 400-square-foot ice field on a local lake each year is less than the total amount of water that evaporates from the surface of the very same lake during a single hot summer day. Much less, the hydrologist said.

> It may be that he lays up no treasures in this world which will cool his drink in the next.
>
> —Henry David Thoreau, *Walden*

Once they had gotten ice from somewhere, where did people in earlier times store it so it wouldn't melt unduly before they needed it, usually in summer? The answer is (1) in pits, caves, or trenches underground; (2) in cavities mostly underground but with roofs or entry structures above ground; or (3) entirely above ground, in insulated houses. In her book, *Ice and Icehouses Through the Ages,* Monica Ellis describes "the cheapest," which were ice stacks. To make one, ice and snow were piled onto a high, dry spot and "rammed" down; doused with water so the pieces would freeze together into a steep cone; and covered with wheat straw, then wood faggots, then three feet of thatch. Surrounded by a ditch for meltwater, Ellis wrote, it could last for "many months."

Well-off Italians kept their frozen stores in ice vaults, chambers that one British visitor of the 17th century described as "sunk in the most solitary and cool'd places, commonly at the foot of some mountain." Persians made their ice in pools and kept the ice in pits, which were often covered with giant mud-brick domes. Chinese fishermen, before setting out in their boats, would pick up ice for preserving their catch from one of the small steep-roofed icehouses that stood along the coastlines. Landowners of late-18th- and 19th-century England built icehouses to go with their mansions; Ellis estimates that there were once 3,000 of these domestic icehouses, which tended to be massive double-walled brick cellars shaped like eggs. On American farms, according to Eric Sloane writing in the book *An Age of Barns,* icehouses were often the fanciest buildings in the yard. While most outbuildings—the wagon shed, sheep barn, cider house—had a traditional shape, the icehouse did not, thus "the farmer could express himself freely." Usually this icehouse was above ground and made of wood, often painted white to reflect sunlight. Between its outer and inner walls, according to Hill and Hughes, the farmer would stuff "whatever he had on hand": tanners bark, charcoal, sawdust, hay, straw, shavings.

All ice-storage places, no matter what their style or period, had to have some insulating material to protect the ice from warm air and ground. Still, some ice in storehouses was always lost to melting and evaporation. In even the best icehouses, shrinkage was 10 to 30 percent. That meant, though, that 70 to 90 percent of the ice stayed around, sometimes for years.

Most people in cities didn't have their own icehouses. They had ice*boxes* and relied on icemen to stop by and fill them. According to Hill and Hughes, these iceboxes were often little more than "wooden boxes with a cover," and many owners wrapped newspapers or rags around the ice inside to keep it from melting. Hill's company informed them that "this was false economy as the ice had to melt to cool." In order to be chilled, lemonade (or milk, wine, whatever) must give up some of its heat to the ice, and the heat taken up by the ice causes it to melt. As Utica Ice advised its customers, an ice block that melts in a home icebox is merely doing its duty. Shrinkage is not only unavoidable but desirable.

As the winter was so mild, ice is scarce.

—Ludwig von Beethoven,
[in a letter from Vienna, 1794]

In his plan to get people hooked on ice, Tudor succeeded almost too well. In parts of the world where the trade flourished, having ice became less a privilege than a right. In early July 1837, the ice ship to Calcutta was overdue, and one Emily Eden complained in a letter to a friend in England: "In the absence of ice, great dinners are so bad. Everything flops about in the dishes." In 1850, the ice "failed" in Bombay—the ice ship was late—and the distressed, ice-loving populace accused Tudor, the "Ice King," of laying down his scepter.

Also in 1850, the ice ship to Apalachicola, Florida, a steamy, cotton-exporting city on the Gulf of Mexico, was late. The French consul there was forced to face the terrible prospect of hosting his Bastille Day banquet with warm wine. Unbeknownst to him, however, a local doctor, convinced that cold instead of heat should be used to treat patients with malaria and yellow fever, had built a machine that could make ice. The doctor, John Gorrie, had mentioned this to a friend, Alvan Chapman, a pharmacist. Years later, Miss Winifred Kimball of Apalachicola recalled what Chapman had told her of that conversation when she was a child and he was an old man sitting in her family's house sipping a mint julep:

"John Gorrie came to me one day in the last of June 1850, and he seemed unduly excited. . . . 'Well, have you found a way to freeze all your patients?' 'Not exactly,' he answered with a glimmering of that infrequent smile, while he paused. 'I have made Ice.' This he said quietly." Gorrie took Chapman to see his ice-making apparatus, then the two of them let the French consul in on the secret. Chapman suggested that Gorrie make ice for the Bastille Day banquet, and the consul came up with a great way to introduce it. He offered a public wager, at a dollar a bet, that the wine at his banquet would be iced.

As the bets rolled in, by Miss Kimball's account, "Gorrie and Chapman were keeping that queer little ice machine busy." Bastille Day arrived but the ice ship did not, and the first toasts of the banquet were in warm red wine. It appeared to all who drank it that the consul had lost his wager. Then Chapman rose and announced that "an 'American has produced the ice which will cool the champagne.' There was a pause. . . . [Then] four waiters entered, each carrying a silver salver and upon each salver there rested a cube of ice." (Actually a brick of ice, judging by its dimensions, ten by eight inches.) Gorrie appeared triumphant, Chapman told the young Miss Kimball. " 'We were all quite sure that a new industry had started.' "

Gorrie was not the first to make ice artificially. It had long been done using water-filled molds with salted snow or ice packed around them. But Gorrie and others who were attempting around the same time to produce ice in quantities large enough for commercial use were taking advantage of other physical phenomena that lower the temperature of liquids, which were brought to their attention by scientific experiments. One was evaporation (think of the cooling effect of sweat); another was the expansion of gases (think of rising air cooling as it spreads out). In both cases—liquid becoming a vapor and vapor expanding into a larger space—the molecules speed up, and the heat fueling the faster movement is absorbed from the surroundings. If water is part of the surroundings and the amount of heat removed from the water is sufficiently great, the water will freeze.

Gorrie's machine worked by vapor spreading into a larger space. A piston would compress ordinary air in a vessel to an eighth or tenth of its normal volume, then release the pressure, allowing the air to expand. The chilling effect of the expansion would pull heat from brine around the vessel, and the chilled brine would in turn pull heat from the water *inside* the vessel. That water would freeze from the bottom up, in films, to form bricks, the likes of which appeared on the consul's silver salvers.

To extract the ice bricks from his machine, Gorrie spread a thin coating of oil or grease on the inside of the vessel with a sponge. Miss Kimball considered it odd that Gorrie used grease instead of heat for this purpose. The reason he did so, she concluded, was probably his single-mindedness: "He would have seemed to have centered his mind upon freezing water until the usefulness of the melting side of it escaped him."

Despite Gorrie's triumph at the Bastille Day banquet, it did not start a new industry. That year he got a British patent for his ice machine and the following year an American patent, but he failed to get financial backing for the ice-making plant he hoped to set up and a few years later died, a disappointed man. Americans may have had a huge appetite for ice, but most of them weren't ready yet to have it come out of a machine instead of a hole in the ground. Some thought that making ice yourself was the devil's work.

Nevertheless, other inventors were soon developing other ice machines, using instead of plain air volatile gases such as ether and ammonia which expanded faster than air and therefore cooled better. James

Harrison, a Scot living in Australia, constructed one of the first practical "can-type" ice-making machines, with sulfuric ether as a refrigerant. When Harrison introduced his machine at the 1862 International Exhibition in London, however, the public didn't recognize it as a practical thing. Of all the "mechanical wonders" on view at the exhibition, the *Illustrated London News* reported, the ice-making machine created the greatest sensation. Both the "learned" and the "ignorant" in the crowd were "surprised to see miniature icebergs rise up before their eyes, the result of the labours of a powerful steam-engine and a quantity of very hot-looking apparatus." Many onlookers concluded that the whole thing was a magic trick, with ice blocks being handed out to the machine's operator by a man hidden inside the box.

Although some of the new ice-making machines leaked gas or exploded, over the years they became more reliable and efficient, and as populations grew and rivers and lakes became more polluted, machine ice began to make inroads on pond ice. In 1886 natural-ice sales were 25 millions tons, higher than they had ever been, but also higher than they ever would be again. In the northern U.S. the mild winter of 1890–91 produced an "ice famine," and ice-manufacturing plants sprang up to fill the need. While in 1889 there had been about 200 ice plants in the country, only 20 years later there were over 2,000.

Those who harvested ice and those who made it competed fiercely for customers; they "co-existed in a state of belligerency," one observer noted. The adherents of natural ice claimed that natural ice was harder and colder and lasted longer than machine ice, as well as less brittle and likely to shatter. The proponents of artificial ice (which was of course not fake but genuine ice) maintained that it was cleaner than natural ice since it didn't have sawdust or moldy hay clinging to it or something the shine boys had missed. Occasionally disputes went beyond the verbal; at one meeting called to work out their differences, a machine-ice man reportedly took a swing at a natural-ice man, and the latter grabbed the arm of the former and bit off his thumb.

Eventually, both ice industries lost out to mechanical refrigerators, which people kept on calling "iceboxes." The home refrigerators not only kept food cold but produced their own ice; the first ice-cube tray was probably a muffin baking tin, stuck among the coils of a refrigerator's evaporator. In one ad, a housewife standing in front of her fridge informs a woman visitor, "Time was when you could get a line on a family's social status by the make of their car . . . Nowadays you can tell whether the so and sos are worth knowing by the number of ice cubes per minute their ice box is capable of throwing off."

glass, air, ice, light,
and winter cold.

—Donald Hall,
"*Waiting on the Corners*"

Gorrie didn't set out to make ice but to chill air. Air that was cooled in his machine was meant to be pumped though ducts into sickrooms, not to stay inside the machine and chill brine. He probably decided to make ice because he figured the market for it would be better than for room coolers. Before then, ice had only occasionally been used to cool rooms, for instance during London's 1844 summer social season when ballrooms were, according to Elizabeth David, "stifling" and there developed a "craze for air-conditioning by means of massed blocks and pillars of ice . . . in great pans in which water lilies had been placed."

Around the turn of the last century, ice was used for a short time in the United States for "comfort cooling" in public places. At the high school graduation ceremonies in Scranton, Pennsylvania, in June of 1901, the air temperature outside the school auditorium was 90°F while inside, where 1,400 guests and students and teachers were packed into a space meant for 900, it was 76°F. The 14-degree difference was achieved with ice. According to the 1903 *Transactions of the American Society of Heating and Ventilating Engineers,* ice blocks were placed on slatted wooden shelves in front of an air inlet, and a fan blew air around and between the blocks and on into the auditorium, where it sank over the audience and exited through registers in the aisles. That evening, three million cubic feet of air blew across 13,000 pounds of ice. Some members of the society were concerned about the expense of purchasing so much ice. "It is a good deal easier to design an apparatus that will cool a room," one of them pronounced, "than it is to design a school board that will pay the bill."

In theaters, cooling by ice was sometimes considered worth the expense; people went to places of amusement to get into the cold. Keith's Theatre in Philadelphia kept ice under the floor—"You can look down the floor register and see the ice in there," a patron remarked—with natural drafts circulating the ice-chilled air. Keith's used two tons of ice a day, one ton per performance.

For decades, railroad passenger cars were air-conditioned by ice. Icing stations were set up along rail lines at regular intervals, and local men and boys would load ice blocks onto racks under the cars or into boxes on the sides. "The job is a wet one," a young member of an "ice gang" reported in the summer of 1946, "as a certain amount of water lies in the bottom of the ice box and sloshes out as soon as the door is opened." Dining cars were iced through the roof, which "requires a special technique," he explained, "much the same as riveters use on construction work when red-hot rivets are tossed through the air from a man on the ground to a catcher high up on the girders. Only in this case we toss hunks of ice weighing five or six pounds. The ice is caught just below the peak of trajectory by the man on the roof of the car."

Airlines occasionally used ice to cool cabin air before passengers came aboard. And automobile drivers, when they passed through hot country, would sometimes set tubs of ice on the floor or canisters of ice in the windows or on the dashboard and let outside air pass over them. One problem was that the effect was proportional to the speed of the airflow; slow the car down and the car heated up.

One ingenious inventor even made a "ventilating rocking chair" that relied on ice. Chilled air would puff toward the face of the person sitting in the rocker from a nozzle attached to a box in which the ice was stored, the airflow presumably activated by the to-and-fro motion. "With the addition of, say two cents of ice per day," promised an ad in a Philadelphia newspaper in 1859, "the luxury of pure air may be fully enjoyed within doors, and the heat of the summer, or the vitiated atmosphere of a close apartment defied."

CHAPTER TWENTY-SEVEN

USES II

a core of sugar.
. . . a transitory Icehenge

—Robert Finch,
"Train Window"

ONCE PEOPLE COULD MAKE their own ice whenever they wanted it, they no longer needed icehouses. Today there are few icehouses left in the old hotspots of ice use, and few of those have ice in them. Iceboxes have gone the way of icehouses, and so too have icemen. Yet here and there you can still find places where people use both. One warmish May day, I went out with two icemen while they delivered ice to Amish families on farms in northeast Ohio. These Amish were ones who eschewed not only electric refrigerators but gas ones, out of the same respect for their forefathers' beliefs that kept them riding around in horse-drawn buggies and wearing hooks and eyes. Still, they needed to preserve the food they grew, shot, butchered, and milked, so they had iceboxes, and it was these men's job to fill them.

The plant that made the ice the icemen delivered was the Millersburg Ice Company in Millersburg, Ohio, owned by brothers Lew and Phillip Ritchey, whose father bought the plant in 1936. The Ritcheys make up to 25 tons of block ice a day in pretty much the same way their father did, in steel cans. Each can is four feet deep, 11 × 22 inches wide at the top, and an inch narrower at the bottom so the ice won't get jammed in. There are 12 cans to a grid and 28 grids, for a total of 336 cans. Men fill the cans with city water, slightly softened, then lower each grid of water-filled cans by crane into a tank of brine water chilled to 12°F. Into the center of each can, they insert a tube with air bubbles coming out of it. "The bubbles churn the water in the middle and slow the freezing," Lew Ritchey explains. "You could put a propeller in there and it would do the same thing." Even as water at the margins is giving up heat to the brine and freezing, the bubbled water in the middle stays liquid, and minerals expelled by the ice at the margins collect there. If all the water in the can were left to freeze unchurned, the minerals would be dispersed throughout the ice and the ice would appear white. "Customers don't like white ice," Ritchey says. "It probably cools just as much, but it doesn't look that nice."

It takes about 24 hours for the margins to freeze, after which the liquid in the center is pumped out and the hollow filled with more water. The replacement water has some minerals in it but not in the concentrations of the old core, and it is allowed to freeze undisturbed. Twenty-four hours later ("any faster," Ritchey says, "and you'd get brittle ice"), one man "pulls" the ice. Pulling consists of lifting a grid out of the chilled brine with a crane, lowering it into a trough of warm water, lifting it out of the trough three minutes later, by which time the bonds between ice and metal should have loosened (what Gorrie accomplished with grease), then tilting the grid so that blocks of ice, each about the size of a three-drawer file cabinet and weighing 300 pounds, go thumping out of their individual cans onto a wooden runway. All of them have ghostly whitish cores, I notice, but customers must not mind those. The blocks are moved by conveyor belt into a cold room with a floor of white oak. "White oak doesn't rot like red oak," Ritchey explains. In the cold room shallow cuts are made on the surface of each block so that during delivery an iceman can split the block apart with an ice pick into smaller ones weighing 25, 50, 75, or 100 pounds; Amish families don't generally want the whole 300-pound block, unless there's a wedding.

At 6:30 a.m., the icemen come to the Millersburg plant. They are Judson Miller and Marion Weaver, sons of Mennonites. They buy five tons of ice, or 35 blocks, and stack them in the back of the truck. They drive off, into gently rolling countryside, almost all of it owned by Amish. The farms look prosperous and well-tended, with cows in the barns, goats in the gulleys, spotted hogs in the mud. Apple trees are white with blossom. Black buggies stand, unhitched, in driveways. Pale gray dresses swing from clotheslines. Squirrel tails hang from outhouse roofs. A woman in an apron and cap is spading a garden. A boy in suspenders and soup-bowl haircut is running across a lawn shouting, *"De Eismann ist do!"*

In 100 miles, the icemen make 70 stops, bringing ice to the Schlaback, Hershberger, Kaufman, Stutzman, and Borntrager farms, where women named Edna, Anna, Ada, Irma, Naoma, and Louella lead the way to iceboxes in kitchens or on porches or in basements next to old Maytag ringer washers. Many of the iceboxes are converted freezers, with a rack in the middle for ice. Few families have iceboxes originally built for ice. At the icemen's next-to-last stop, a woman tells them she'll need an extra 300 pounds of ice on Monday; there is to be a wedding.

In the spring, according to Weaver, 100 pounds of ice usually lasts a family a week, in the summer three days. For him and Miller, the season starts on May Day and ends on Halloween. On this May day, the icemen don't get back to the Millersburg plant until 6:30 p.m., twelve hours after they left it. "We have to drive farther and farther to deliver the ice," Weaver remarks, a little wearily. "In the last six years we've lost a lot of customers. Traditions are changing in the Amish church. Some communities are switching from ice to kerosene or gas refrigerators. The ice age is ending."

The ice age may be ending for block ice but definitely not for all manufactured ice. "There's probably more ice being made and sold in the U.S. today than ever," Lew Ritchey claims. The ice that people are buying more of is "bag ice," also known as "packaged ice" or "cube" ice, although the ice doesn't really come in cubes. "They're cube-*size* ice," Ritchey explains. "We write it on our label." (Immoderately, another ice plant calls its cube ice "frozen diamonds.") Millersburg sells its cube ice to groceries, convenience stores, and filling stations, which sell it to picnickers, campers, and boaters. "People used to go in the backyard for a picnic," Ritchey notes. "Now they're mobile and affluent and travel and have barbecues, and they want their beer *cold*."

Bag ice is made in a different way from block ice. At Millersburg, water is dribbled evenly and continuously down the faces of 40 vertical plates, each two and a half feet high, that have been cooled to 0°F and stand close together in rows, "like cards in a deck," Ritchey says. After 45 minutes, each plate has a layer of ice on it an inch thick. Harvesting it consists of pushing stainless-steel knobs against the ice layer, breaking it up into cube-size pieces that land on a screen through which the "snow," or tiny bits, falls. If the so-called snow were to get bagged along with the so-called cubes, it could act as a mortar and bind the cubes together into an unwieldy clump.

When a small but fierce volcano vexes me sore inside,
And my throat and mouth are furred with a fur that seemeth a buffalo hide,—
How gracious those dews of solace that over my senses fall
At the clink of the ice in the pitcher the boy brings up the hall!

—Eugene Field, *"The Clink of the Ice"*

Besides all the ice cubes and cube equivalents being made each day by commercial ice plants like Millersburg, countless others are being made by restaurants, schools, bars, hotels, grocery stores, stadiums, fishing boats, amusement parks, hospitals, and private homes. Most of these cubes aren't cubic, either. They may be shaped like pillows or half aspirins, airplane wing foils, class rings, or large, pitted olives. A common restaurant cube I've seen is a faintly lopsided cube shape with a dent on top, into which a fingertip might fit. The most common cube made in homes, outside a tray, is the crescent, which refrigerators create by spraying tap water into crescent-shape molds then upending the molds and letting the crescents fall into a bin. "Terrible ice," says Larry Lozar, director of sales and marketing for the largest maker of ice-makers in the world, who might not be considered impartial in the matter. He explains that dissolved minerals and air have no place else to go and therefore stay in the crescents—as they do in cubes in an ice-cube tray—and end up looking partly cloudy—as cubes do in an ice-cube tray.

What his company makes are "gourmet cubes," shaped like tiny flowerpots, to minimize clumping. Their chief selling point, though, is their clarity. To make them, ordinary tap water is sprayed in two fine mists upward into an upside-down evaporator that is cooled to precisely 32°F, and since minerals tend to freeze at temperatures below 32°F, only the pure water freezes, layer by clear layer, while the minerals drain away. Many people passing Lazar's booth at a recent kitchen-and-bath show in Chicago thought the cubes on display were plastic. In Frederick Tudor's day, the ultimate compliment to ice was mistaking it for glass.

> I can't bear iced drinks . . . the iceberg, you know. Perhaps some champagne, though.
>
> —Millvina Dean; at eight months, the youngest survivor on the *Titanic* (her father did not survive)

It was an American engineer who introduced to the world the rubber ice-cube tray, which could be twisted to oust ice cubes instead of, as with a metal tray, having to have warm water run over it to loosen the grip of the ice. That was in the late 1920s, and the engineer, according to *Heat and Cold,* "got the idea while duck hunting, after noticing that ice frozen on his rubber boots easily broke off when the boots flexed as he walked." Whether ice-cube trays are of rubber, metal, or plastic, water in the individual compartments freezes from the outside in and the top down, trapping air bubbles and impurities in the center, giving the cube a cloudy eye. An ice cube also melts from the outside in so its center is the last part to go, which is why it's good manners to throw away what's left of a guest's ice cubes when you're fixing her a new drink, replacing the mineral-laden ice remnants with fresh cubes. ("I know the ice in your drink is senile," Anne Sexton wrote in a particularly sour mood.)

Ice cubes keep on cooling drinks until they are completely gone. The heat removed from the liquid goes toward melting the cubes (causing the water molecules at their surface to vibrate so fast that they break free of the ice lattice and float away as icy liquid), not toward raising the cubes' temperature. Down to the last sliver, the cubes stay at 32°F. They can't get warmer than that without melting away, and being in liquid ensures that they won't get colder than that, either.

Like icebergs, ice cubes float with most of their mass under water, seven- or eight- or nine-tenths of their mass depending on their density. Scale does not apply. You could very well say when discussing some political or social scandal, "Oh that's just the tip of the ice cube!"

> Snow preserveth fish from corruption
> as also flesh from putrefaction
>
> —Galen,
> Greek physician, 2nd century A.D.

There's another kind of ice you see a lot in stores besides cubes: "flaked" ice, which resembles pieces of broken glass. At a fish market in my neighborhood, flaked ice undergirds and surrounds and occasionally rests on the eyes and fins of 85 different species of fish and shellfish. The manager there explains that regular ice cubes are too hard and will dent fish, while crushed ice compacts too much, like wet snow. The store makes a ton of flaked ice every day, in a machine that sprays water onto a rotating drum, where it hardens into a layer as thin as a skim of lake ice and, as the drum rotates, breaks apart and falls in pieces, like shattered glass.

Long before the fish get to the store, though—in fact, from the time they die—they are on ice. Most large fishing boats these days have their own ice-making machines on board, to preserve the catch while the boat is at sea. The machine on the *Andrea Gail,* the swordfish boat whose terrible fate Sebastian Junger conjured in his book *The Perfect Storm,* was capable of producing three tons of ice a day. "Commercial fishing simply wouldn't be possible without ice," Junger wrote. "Without diesel engines, maybe; without loran, weather faxes, or hydraulic winches; but not without ice. There is simply no other way to get fresh fish to market."

My neighborhood fish market's manager tells customers how to home-freeze the fish they've just caused to be removed from their beds of flaked ice: place the fish in a container, pour water over it, set the container in the freezer, wait for the water to freeze, take the container out of the freezer then the block of ice with its embedded fish out of the container, wrap the block of ice in freezer paper, and set it back in the freezer.

The ice, Don Schlimme, professor emeritus in the department of nutrition and food science at the University of Maryland, points out, acts as a "pretty good barrier" to oxygen, and fish, being very high in unsaturated fatty acids, are particularly susceptible to spoiling when exposed to oxygen, "even worse than pork." Some commercial outfits go so far as to put an ice glaze on the frozen fish they sell. The main point of the ice barrier, according to David S. Reid, professor of food science and technology at the University of California at Davis, is to have as little "vapor space" in your frozen food as possible. That way, there'll be not only little oxygen reaching the food but little moisture leaving it. Reid explains that water migrates to the coldest part of the food, and when the coldest part is the surface (right next to cold freezer air), ice crystals will form on the surface, and moisture will evaporate off the crystals if there's space for it to evaporate into.

"The more vapor space there is," Reid concludes, "the more the water molecules can come out of the food, and once they're out, it's hard to get them back in." What happens with frozen food is that, as freezer temperatures cycle and freezer doors open and close, the outside of the frozen food warms in relation to the chilled interior, but the water that migrated to the surface when it was the coldest part and turned to ice there doesn't migrate back to the interior. The crystals stay on the surface or else evaporate and collect as frost on other cold surfaces in the freezer, usually the coils or the inside of the food packages. An eight-ounce package of frozen peas that has a small hole in the cover could turn into a two-ounce package by the time you take it out of the freezer. A chicken you didn't wrap tightly could end up with dry, discolored skin: "freezer burn."

Reid warns, "Freezing must not be thought of as a magic technology which prevents all change." Although freezing reduces the speed at which molecules move and thus the frequency of their contacts, it doesn't keep them from moving altogether. Chemical reactions continue, which is why people are advised to blanch many of the vegetables and fruits they plan to freeze. Still, lowering freezer temperatures can prolong shelf life. At 0°F, green beans can last a year, at −8°F at least six years. "In Japan, where they're great aficionados of fish," Schlimme notes, "they insist that the fish be kept colder than −14°F. Fish gets bad more slowly there."

There's a story that when Clarence Birdseye was on a fur-collecting trip to Labrador in 1916 he noticed inhabitants leaving fish and game right on the ground, outdoors, where it froze quickly in air temperatures that ran as low as −50°F. Weeks or months later, when they needed the meat or fish, they told him, they would bring it inside and thaw and cook it, and it would taste just fine. That inspired him to try fast-freezing food commercially when he got back to the United States. In those days, commercially frozen food was frozen slowly; the meat or fish would simply be placed in a room with cold, calm air and left there until it hardened. Birdseye and others who were experimenting with fast-freezing around that time realized that it not only shortened the time during which bacteria and molds could spoil the food but produced smaller, less damaging ice crystals than slow freezing did.

Early fast-freezing systems included the "diving-bell," the "floating pan," the "fog freezer" (all of which involved immersing the food in an edible refrigerant), and—how can one not applaud?—the "National Continuous Individual Berry Freezer." With it, single berries, accompanied by cooled sugar solutions, would drop down a hollow shaft through a series of perforated plates and inverted cones until, on exiting the last cone and falling into a hopper, each little berry was frozen solid.

Today, most foods are fast-frozen by air blast, most often in a tunnel. A mesh belt carries the food through an insulated tunnel in one direction while cold air blows through the tunnel in the other direction at speeds up to 3,500 feet a minute. In some cases the belt spirals and the air blasts down at the food from the top of the spiral, or else upward from the bottom, or both upward and downward from the middle. In some cases, the air isn't blasted past the food "like a gale," Reid says, but directed toward it at right angles, through perforated plates, as from a showerhead, with the tiny jets of air creating turbulence over the surface of the food and breaking up the static film of air, increasing the wind-chill effect. By pushing down on the food, the air jets also keep pepperoni toppings on pizzas from zooming off, as they had been doing, Reid says, "like flying saucers."

> Milk can be frozen into bricks and handled like bricks. . . .
> Meat can be . . . handled like chunks of wood. . . .
> baked beans . . . [frozen] in separate kernels . . .
> like a bag of peanuts.
>
> —Vilhjalmur Stefansson, *Arctic Manual*

Almost anything can be frozen and, it seems, has been; for all I know there are frozen bean kernels for sale somewhere. All foods contain some water: an uncooked chicken is 74 percent water, prime grade beef 45 percent water, utility grade beef 62 percent water, North Atlantic halibut 76.5 percent water. A pea is 78 percent water and a potato 80 percent water. Of the billions of pounds of food frozen commercially in the United States each year, potatoes are the food most often frozen, followed by peas, fish fingers, chicken, and orange juice. In an article in the *New York Times* entitled "Freezers: The Rewards of Using Them Wisely," I was interested to see on the list of things a person might consider freezing at home such items as raw eggs (each in its individual ice-cube-tray compartment); leftover pancakes; olives in their own liquor; brown sugar ("won't harden"); and cream ("whips very quickly if it contains a few ice crystals").

When I asked chef Jacques Pépin if there are any foods that are better when frozen than fresh, he shot right back, "Little baby peas." The sugar in the peas, he explained, turns into starch in two or three days, and unless you buy and cook them right away, they lose their sweetness. "Frozen, you're getting them practically on the vine."

> Chewing frozen raw bear meat is like eating raw oysters; half-frozen it has . . . the consistency of hard ice cream.
>
> Vilhjalmur Stefansson, *Arctic Manual*

Almost everything can be frozen, but not everything should be. With a certain free-verse flair, the aforementioned *Times* article named several foods that don't do well with home freezing:

> Hard-cooked eggs become tough . . .
> Boiled white potatoes become mealy.
> Boiled frosting turns sticky.
> Most gelatin dishes weep.

The reason boiled frosting turns sticky is that the coldest part of the frosting is at the bottom, and water migrates there and joins other ice crystals, and together they push the sugar solution toward the top. The reason hard-cooked eggs become tough, and boiled white potatoes mealy, and gelatin dishes weepy is that freezing reduces the amount of liquid water in these foods, which allows the molecules of other substances—the yolk lipids, the starches, the gelatin chains—to come closer together, and once they are together they are loath to part, or to let the water return to the places where it had been.

"In what appears to be a very simple process," David Reid concludes about the freezing of food, "things get very complex." Put a strawberry—or a tomato or a cucumber for that matter—in a freezer and it'll be mush when you thaw it out. A strawberry doesn't have much internal structure to begin with; it relies on water inside its cells to plump it up. When a strawberry freezes and ice crystals damage its cell membranes, it loses even that minimal structure. According to Reid, the strawberry is "the holy grail of frozen-food research. (Also lettuce.)"

Food researchers are working too on the monumental problem of how to keep frost crust off ice cream. As Reid notes, ice cream is one of the few foods you choose to eat frozen, and "crystal size *is* the eating experience." Well-made ice cream has minuscule ice crystals blended throughout, but as it sits around in a typical freezer in fluctuating temperatures, the crystals change size, like snowflakes in a metamorphizing snowpack. "Lilliputian physics" are involved, John Bedbrook of DNA Plant Technology Corporation points out. "It's a dynamic situation." Small crystals keep getting smaller, large crystals larger. The large crystals, having expelled sugars and other substances that impart flavor, are crunchy and watery, which cancels some of the luscious mouth feel and creamy texture and taste you bought the ice cream for in the first place.

> I suppose these ginks who argue that . . . because the rich man gets ice in the summer and the poor man gets it in the winter things are breaking even for both. Maybe so, but I'll swear I can't see it that way.
>
> —William B. "Bat" Masterson, reporter for the *New York Morning Telegraph*, only minutes before he died, October 1921

It seems churlish to make distinctions about kinds of ice and degrees of clarity and superiorities of cube shapes when having ice at all in hot weather can be considered a wondrous thing. In Gabriel García

Márquez's book *One Hundred Years of Solitude,* José Arcadio Buendía took his children one day to a gypsy fair in the village of Macondo, paid thirty reales, and entered a tent where a giant with a nose ring was guarding a pirate chest. "When it was opened by the giant," García Márquez wrote, "the chest gave off a glacial exhalation. Inside there was only an enormous, transparent block with infinite internal needles in which the light of the sunset was broken up into colored stars. Disconcerted, knowing that the children were waiting for an immediate explanation, José Arcadio Buendía ventured a murmur:

" 'It's the largest diamond in the world.'

" 'No,' the gypsy countered. 'It's ice.'

"José Arcadio Buendía, without understanding, stretched out his hand toward the cake, but the giant moved it away. 'Five reales more to touch it,' he said. José Arcadio Buendía paid them and put his hand on the ice and held it there for several minutes as his heart filled with fear and jubilation at the contact with mystery. Without knowing what to say, he paid ten reales more so that his sons could have that prodigious experience. Little José Arcadio refused to touch it. Aureliano, on the other hand, took a step forward and put his hand on it, withdrawing it immediately. 'It's boiling,' he exclaimed, startled. But his father paid no attention to him. . . . He paid another five reales and with his hand on the cake, as if giving testimony on the holy scriptures, he exclaimed:

" 'This is the great invention of our time.' "

CHAPTER TWENTY-EIGHT

USES III

The sonic qualities of ice turned out to be . . . fantastic.

—Terje Isungset,
composer of *"Iceman Is,"* played on
instruments carved from ice

ALTHOUGH COLD is the characteristic of ice most obviously of use to humans, ice has other traits that can make it productive on their behalf: transparency, floatability, slipperiness, purity, adhesiveness, hardness, availability. Above Lew Ritchey's desk at the Millersburg Ice Company is a photograph of an eagle with soft-edged feathers and a rounded beak; apparently, it had started to melt. Ronald Reagan's cook, Ritchey explained, had carved the eagle out of a block of Millersburg ice when the president came to dedicate a nearby library and stayed on to dinner. For this ceremonial purpose, ice's ability to impart cold wasn't important. The eagle wasn't there to chill food or the president. What made the ice useful on that occasion was its hardness, plus its transparency.

The eagle had sat on a banquet table, which is where most ice sculptures sit. Most ice carvers are, or were, cooks. Ice sculptures are associated primarily with communal eating. Joseph Amendola, former consultant to the Culinary Institute of America, noted in his book *Ice Carving Made Easy* that in the Middle Ages when dining became a form of entertainment and a means of displaying wealth, cooks began making complex, artistic table decorations out of edibles, including sugar, marzipan, pastry dough, and ice. Ice was prized for its "strength (comparable to that of concrete)," Amendola wrote, as well as for its "light-reflecting translucency." Then in the late 1800s when Russian service came into general vogue and banquet courses were plated in the kitchen, leaving the center of the table (where the serving dishes had been) in need of filler and diners in need of something diverting to look at and talk about between courses, the impulse to decorate tables with ice grew even greater. Sometimes ice was incorporated into the dishes themselves. At the Savoy in London in 1892, Amendola pointed out, master chef Auguste Escoffier created *pêches Melba,* "an ice swan that cradled on its back a scoop of ice cream topped with poached peaches and spun sugar," named for opera singer Nellie Melba, Elsa in Wagner's *Lohengrin,* in which a swan turns into a man. Swans are still *the* favorite subject of ice sculpture, not because of Melba but because they are inherently elegant and otherworldly, therefore suited to the elegant, otherworldly look of ice.

Over the years, homemakers who wanted to make their own ice carvings would consult "lesson" books. The early-20th-century *Artistic Sugar Work and Ice Sculpting* suggested making an ice vase embellished with pink caramel roses, and the 1947 *Fancy Ice Carving in Thirty Lessons* recommended that carvers make ice gondolas to serve appetizers in and ice sailboats to present fish upon. Amendola's own how-to book instructs readers on ways to carve not only ice bowls and plates (fluted, scored, beveled) but tigers, ducks, bears, dogs, squirrels, rabbits, peacocks, parrots, clam shells, swordfish, and swans (floating, rampant, kissing).

Typically, an ice block made especially for carving weighs 300 pounds and measures 40 by 20 by 10 inches, but it can be smaller or much larger. Whatever its size, it should be as transparent as an ice machine (or pond) can make it. A block with a ghostly core may be acceptable for an icebox but not for an ice sculpture. According to Ritchey, opaque ice takes chiseling about as well as transparent ice, but it doesn't do the same things with light. Many professional carvers have their own ice-making machines; one kind freezes the water from the bottom up instead of the top down so that air bubbles rise and escape from the surface and minerals collect in a thin layer on top, which the carver can just slice off, like a rind.

Transforming a prism of ice with sharp angles into a swan with feathers takes tools: chain saws (for rough shaping), chisels (for details), die grinders (details), heat guns (shaping, smoothing), rakelike chippers (shaving, scoring), and steam hoses (melting, dimpling). I once watched an ice-carving competition at a restaurateurs' convention, and it was a noisy affair. A dozen or so men—all men—were laboring away at enormous ice blocks, hacking, sawing, chiseling, and grinding as fast as they could (the competition was timed). It was odd seeing something so ethereal-looking attacked so aggressively.

There are ice-carving competitions at winter festivals, too. Ones in Sapporo, Japan, and Harbin, China, feature elaborate ice and snow sculptures, from writhing dragons and small cities to cats so big that people can slide down the tongue. Ice carving is an old tradition in northern Asia, Douglas Yates pointed out in the booklet *Alaska Ice Postcard Collection: New Ice Carving from Fairbanks, Alaska*. "Candle lanterns of delicate filigree . . . or animals cut in ice enlivened village lanes in countless settlements." At Fairbanks's own 1990 Winter Carnival, a Chinese ice-carving team created a laughing Buddha, a Japanese team produced a phoenix rising, and an Alaskan team carved a 16-dog sled with musher.

One man in New York City has been given credit for making ice sculpture cool. He has carved ice sculptures for fashion shoots, art gallery openings, Wall Street broker-recruiting parties, and music awards shows. He is a brooding, intense, young, wired, chain-smoking former chef named Joseph O'Donoghue, or "Joe Ice." The first sculpture he ever made was a horse's head, on a night when his parents were coming to dinner at the restaurant where he cooked and the regular sculptor wasn't around. "I didn't know how to make ice smooth so it was a Rodin-type head," he said, meaning lumpy. Six weeks later he sold his first ice sculpture, of Mickey and Minnie Mouse holding hands, and over the next six years, 4,000 more.

One afternoon in his Brooklyn studio, I watched him sculpt a seabird rising from waves for a hospital's anniversary dinner. An hour earlier he had taken a block of ice out of the freezer and left it on a table, to temper. He had made the ice himself in one of the machines that freeze water from the bottom up, and it was breathtakingly clear, lusciously limpid, with only a couple of tiny milkweed-like tufts of white showing where the points of the ice tongs had touched it. The points entered the ice only an eighth of an inch, at most a quarter of an inch, which he found amazing since with that minimal penetration he could lift a 300-pound block. I couldn't see a single air bubble in the entire block, and if there were one, he told me, I would definitely see it. "Even a tiny bubble can create a big light."

The first thing Joe Ice did was saw a layer off the bottom of the block, to be sure the sculpture sat flat. "The bird's bill could spear your guts if it fell off the table," he pointed out. Then, holding a lit cigarette between his lips, listening to rock music so loud that even the chain saw couldn't drown it out, and standing wide-legged in leather chaps like a gunfighter, he began sawing large pieces out of the ice block, getting rid of everything that wasn't the bird. After a while he switched to an electric drill, which sent ice chips flying six feet in the air, like little whale spouts. He explained that he was designing as he went along. He wanted there to be flow in the sculpture, movement. The waves should look as if they are rising, the bird as if it is taking off.

He always keeps in mind the fact that ice was water. "If water was going to animate," he asks himself, "what would it animate *to?* If you splashed water and it stopped . . . how much can a splash actually do?" He also keeps in mind the fact that ice *will* be water. He tries to carve the sculpture so it will melt gracefully. "I control the dripping. You don't want to make an angel that is going to get uglier. You don't want it to get to the point where it looks like a pig." He tries to do streamlined versions of everything. "I don't carve little cheeks and things." The finished seabird did appear to be taking off, the waves to be rising. So that the bird would be shiny, he ran a blowtorch flame over its surface, which retreated almost imperceptibly, like a candle growing shorter.

But thou didst hew the floods
And make thy marble of the glassy wave.

—William Cowper, *"The Task"*

The glassy wave was on the Neva River of Russia in the winter of 1739–40, one of the coldest winters in that long stretch of cold winters, the Little Ice Age. Anna Ivanovna, the fat, malicious, tyrannical empress of Russia and niece of Peter the Great, ordered a palace built out of ice from the river Neva (instead of wood from the forest or rock from the quarry) to distract her subjects from the terrible cold of that winter as well as from a series of executions she had had carried out for minor offenses. The palace was to be a setting for a wedding, or rather a wedding *night*—not a romantic wedding night, either, but a nightmarish one. One of her princes had converted to Roman Catholicism, which so displeased her that she was forcing him to marry a servant of hers so ugly that the woman was known by the name of a dish of roast pork and onion sauce. He and his bride were to spend their first night together in the ice palace, upon a bed of ice.

They rode to the palace inside a cage mounted on the back of an elephant, followed by other couples on camels and horses and sleds pulled by wolves and pigs. Once there, they were taken to a bedchamber, stripped, and left for the night, with guards outside the door to make sure they didn't get out. The bedchamber had walls, ceiling, and floor made of ice. In it were a "large, elaborately curtained bed, furnished with a feather mattress, a quilt, and two pillows, a nightcap laid out on each—everything skillfully carved in ice," as Mina Curtiss wrote in her book *A Forgotten Empress.* Beside the bed was a small table—of ice—and a low stool—of ice—with two pairs of slippers—of ice—and a dressing table—of ice—laden with boxes, bottles, and candlesticks—all of ice—over which hung a prettily carved mirror—of ice.

It's known that the prince and the pork dish survived their wedding night, but it's not known what their reaction was to the little ice nightcaps and slippers so artfully laid out for them. By one account, they ran around all night and slapped each other to stay warm. The empress's little joke, according to Fred Anderes and Ann Agranoff, authors of the book *Ice Palaces,* was the first "well-documented" ice palace ever built. Before that, Russian villagers had built small ice structures, but "little is known of them." Although it would seem difficult to top that first palace, the *peak* era of ice palaces came a century and a half later when North American cities—Montreal, Saint Paul, Quebec City, Ottawa, and Leadville, Colorado—built ice palaces as centerpieces for their winter carnivals. These ice palaces tended to be large and complex, with turrets and towers, arches and outworks, sometimes parapets or barbicans, palisades, flying buttresses, stockades, and sally ports, all carved out of ice.

Montreal kicked things off in 1883 by constructing a simple castle, meant mostly to test the building characteristics of ice. The architect who'd been in charge of much of the cut-stone work for the Canadian Parliament buildings was in charge of the cut-ice work for the ice palace, which came from the Saint Lawrence River. He also designed the structure, which featured a 90-foot steeple and a roof of green boughs covered with sheets of icicles (almost all later palaces would be roofless). Ice's building characteristics must have proved themselves because the following year Montreal's palace included a main tower and towers at all four corners, plus curtain walls, porches, and an apse. Besides the palaces themselves, carnivals sometimes had smaller, freestanding ice structures such as grottos, temples, fountains, windmills, chime towers, mazes. Montreal's 1885 carnival included a seven-story Tower of Babel with a snowshoer on top carved in ice holding an electrically lit torch of ice. For *its* 1894 carnival, Quebec City built a walk-in ice beer bottle.

The ice palaces were meant to be seen inside as well as out. "The full beauty of St. Paul's 1886 palace is not experienced until its gleaming interior is seen," a carnival publication proclaimed. "It is divided into great apartments, spacious halls, vast corridors and wide chambers, all with walls transparent and glistening." In its 1887 palace, *West* Saint Paul presented festival-goers with not only a skating rink but a retiring room, a gentlemen's coatroom, an oyster room, a hot-lunch room, a toboggan slide, and a bandstand. The inside of Leadville's 1896 palace contained both an unheated ice rink and a heated ballroom, separated by glass panels through which skaters and dancers could watch each other moving to the music of the Fort Dodge Cowboy Band playing on a balcony between them.

At night, newfangled electric lights lit the palaces. "It is not until evening that one sees to perfection the enchanted castle of gleaming sea-green united ice," a journalist wrote of Montreal's 1885 castle, "when with . . . the electric illuminations glittering through its towers and turrets, it seems much more like the marvelous imaginings of some opium sated dream than a real tangible thing."

At the end of the carnivals, the ice palaces were "stormed." Hundreds of men would attack them with rockets, bombs, guns, Roman candles, and mortars while other men defended them, lobbing fireworks down onto the invaders. The attackers won, every time but one, demolishing the palace—although sometimes the weather (unseasonable warmth or rain) did the work of demolition.

In the 20th century, North Americans built ice palaces only sporadically. Saint Paul created one in 1937 meant to cheer people up during the Depression. In the 1960s, Quebec City erected several modernistic palaces designed by professional architects. By 1979, however, the Quebec carnival organizers had decided that ice was too expensive to build with and that henceforth their palaces would be made of cast snow, with lines traced on the snow to suggest ice masonry. "No longer, alas, do they possess the very special transparency, solidity, and sheen of block ice," Anderes and Agranoff wrote of snow palaces, ending their book on an elegaic note. "The great age of ice palace building is, alas, over. No true ice palaces (as opposed to palaces of snow and sculptures of ice) have been built since 1978."

Without their knowing it, however, a village in northern New York State had kept on building ice palaces every year to go with its winter carnival. Its palaces were not nearly as elaborate as the ones Saint Paul and Montreal built at their peak, but they were palaces nevertheless, with towers and throne rooms and now and then a dungeon. One year I went to that village, Saranac Lake, as the ice palace was being built, to see how it was done. The palace was erected right in town, next to Flower Lake, where the ice came from. By the time I arrived, it was half built and already looked like something in an "opium-sated dream." Against the white of the snow-covered ground and the black of the pines and shrubs and leafless trees and utility poles and street, bam! there it was, the beautiful aquamarine blue of an upended swimming pool. A person could sit in the Burger King across the street from the palace with a Coke (or other opiate) in hand and watch it rise, block by block. Or she could walk across and shove a few blocks around herself. All the builders were volunteers, including the mayor, William Madden III, who designed the palace, an Irish castle with five towers.

As some volunteers were laying down ice blocks, others were out harvesting more. Each block—about the size of a hay bale—was set on the ground behind the palace, in a row with other blocks. One by one they were picked up by a crane and lifted up to a volunteer straddling the palace's back wall, which was then about 18 feet high. The man guided each block onto a spot he had just slathered with slush from a bucket, then he packed more slush in the space between it and the last block, like any good stonemason. "Slush freezes hard as mortar," the mayor pointed out.

When they reached the top of the walls, the volunteers built an overhang, by turning blocks sideways. Then they made merlons (those upright rectangles that people hide behind when they're shooting missiles through the crenels), by sawing blocks in half. Along the back wall, they made an ice slide and sprayed it with water so the surface would be fast. Then they built ice thrones inside for the festival's king and queen to sit on. Their last duty was to install ice carvings over the entrance gate, a pair of "swaneagles," the two ice-carving favorites combined in one noble bird. The finished Irish castle was spectacular, with more sunlight passing through it in the upper part than in the lower, so it seemed both buoyant and massive. At night colored spotlights would be trained on it, but in the daytime it glowed its own green-tinged, gorgeous blue.

The backhoe operator looked it over. "It's like working in a dream," he mused. "Nine days ago this structure didn't exist. Thirty or 40 days from now it won't exist. It's out of the lake, back into the lake. Ice is a very inexpensive medium; we use it over and over and over again."

It was a miracle of rare device,
A sunny pleasure-dome with caves of ice

—Samuel Taylor Coleridge,
"Kubla Khan"

People who'd like to spend more than a few hours in an icy structure should consider checking into an ice hotel. There's one in Kiruna, Sweden, 125 miles north of the Arctic Circle, the realized dream of an environmental engineer who built backyard ice forts as a kid. Like ice palaces, the ice hotel is built anew each year. The walls are of pressed snow but the pillars and ribs supporting the vaulted ceilings are of ice, and so are the windowpanes and furnishings. There's an ice bar that serves blue vodka cocktails and an ice chapel where couples, dressed in insulated suits, can get married. Inside air is kept at 23°F (hair dryers are forbidden). Not long after the hotel opened, Bob Spitz, an American who swears he hates cold, visited on behalf of a travel magazine and reported finding in his "cavernous" bedroom a bed resting on 450 ice bricks, two Adirondack chairs and matching sofa and coffee table, "all ice of course," plus an ice-brick fireplace in which there burned a "Beverly Hills–style electric flame." A carver was still at work on the chandelier which when finished would have 168 individually cut, tear-shaped pieces of ice hanging on a nylon cord above a mirrored ball, which was itself a "beveled polyhedron of ice."

The ice hotel is really an igloo; igloos are made of snow.* In his book *Confessions of an Igloo Dweller,* James Houston, who lived in the eastern Canadian Arctic at a time when many Eskimos lived in domed snow houses, pointed out that he had seen only one ice igloo in his life, "and it was used not to live in, but as a cache for meat." When Houston was asked by a radio talk-show host in New York City about all those "ice igloos" up north, he gently set the host straight. "Ice is terribly cold," he explained, "and so is the ground." It is the air in snow, even hard-packed snow, that makes it a good insulating material for people's houses.

However, builders of igloos did sometimes use ice—for windowpanes. As Stefansson pointed out in his manual, ice windowpanes are installed only in midwinter, when temperatures are below freezing and the ice not likely to melt. Usually there's only one ice window per igloo,

*In the Eskimo language, Vilhjalmur Stefansson explained in his *Arctic Manual,* igloo (or *iglu*) means (or meant in 1944) any shelter for man or beast, so that a railway station was an igloo and so was a cathedral and a cow barn and the Empire State Building. Even when the word referred to a dwelling, an Eskimo could mean by it a wood-and-earth or a stone-and-earth or a bone-and-earth structure, if that's what the Eskimos in his area lived in instead of the domed snow house we think of as an igloo. In the century before Stefansson wrote his book, by his estimate more than half of the Eskimos in regions such as northern Alaska, northeastern Siberia, and Greenland had never heard of domed snow houses. Or if they had heard of them they didn't believe what they were hearing because it seemed "unreal." Only 15 to 20 percent of Eskimos lived in snow houses, and only during the cold part of the winter.

set between the entrance and an air vent "called the nostril," according to Houston, "which was stuffed with a mitten at night." He advised that the proper ice for a window is a slab four to six inches thick taken from a lake; "salty sea ice is too murky."

In the course of telling a story about an igloo and Tukik, the moon—one of the Eskimos' few male gods—Houston managed to explain how to install an ice window. The story was that a man built a new igloo for his strong-minded wife after she complained that the igloo they were living in was too "thin, icy, and sooty with lamp smoke." After he completed the new snow house, the man "went with his wife's brother to the fish lake where they dug out from beneath the snow a smooth, clear piece of freshwater ice that they had cut and stored when the ice was new and unclouded and just about the thickness of his wrist." The men then cut a square hole in the new igloo's wall above the door, angling the cut inward so the piece of ice would fit snugly, after which they set the ice in the hole and "glued the window edges tight with fine-cut snow." Using a mitten, the wife's brother cleared the ice pane "until they could see a pale shaft of moonlight illuminating the center of the wide, snow sleeping platform."

A day or so later, Houston was hunting on sea ice with a couple of men when the older one remarked, "Tukik, the moon, is big tonight. He's on the move hunting for ice windows to look through." The old hunter told Houston that the women of his village had told the men of his village that when "Tukik comes and shines down through the window at them he can make them pregnant." He asked other, younger hunters about this, and they confirmed what the old man had said. " 'We're never there when it happens,' " the young ones told him. " 'But it must be so.' And they lay back and drank their tea, contented with the powers of the moon."

> *We are afloat*
> *On our dreams as on a barge made of ice.*
>
> —John Ashbery, *"Voice of the Poem"*

One of the weirdest things to be almost built out of ice was an aircraft carrier. It was late 1942. Allied aircraft lacked the range to escort merchant ships across the North Atlantic, and German U-boats were sinking ships many times faster than shipyards could build them. Some sort of mid-Atlantic landing field was called for, but materials for building conventional aircraft carriers were in short supply. So when Geoffrey

Pyke, director of programs for Lord Mountbatten, suggested making aircraft carriers out of ice, Mountbatten and Churchill didn't laugh; they listened. Pyke was a man about whom Canadian scientist Dean C. J. Mackenzie, who was later involved in the ice-carrier project, wrote in his diary: "He never does anything because other people do it. He lands in this country without any gloves and with only a light raincoat to embark upon a trip into the Arctic weather. When we were in the mountains he went skiing in the most grotesque outfit—a little hat on the top of his head with his muffler and raincoat and a pair of spats. . . . All together he is a most unusual type and most people think he is absolutely mad. He is not mad. He thinks in a most unorthodox way. . . . He has moments of what amounts to intellectual intoxication when he is seized with his ideas."

Pyke became seized with the idea of "ice warfare" and chose to share his intoxication with Lord Mountbatten in a memorandum 232 pages long. Pyke argued that ice would be an "ideal" construction material for a floating air base because, among other things, a vessel made of it couldn't be blown up and—"its greatest attraction"—couldn't be sunk. After reading a one-page summary of Pyke's "bergship" plan (code name "Habbakuk") Churchill wrote a note to his chiefs of staff in which he too showed signs of intoxication. "I attach the greatest importance to the prompt examination of these ideas," he declared. "The advantages of a floating island or islands, even if used only as refuelling depots for aircraft, are so dazzling that they do not at the moment need to be discussed. There would be no difficulty in finding a place to put such a 'stepping-stone' in any of the plans of war now under consideration."

British and Canadian scientists who were instructed to evaluate the plan—some of them mighty skeptical—quickly realized that ice from an "icefield"—that is, naturally occurring sea ice—wouldn't work as a construction material since most of it was too thin, say two or three feet thick, to withstand waves in the North Atlantic. Ice would have to be *made.* But first, scientists needed to know more about the physical properties of ice than they did at the time. Researchers duly ran tests on ice, large-scale and small-, with disappointing results. Some beams of ice broke under stresses of less than 70 pounds per square inch (pine beams took stresses of more than 11,000 pounds per square inch). Pure ice, they concluded, would be too brittle to build with.

Then they learned of work carried out by two chemists at the Cold Research Laboratories of the Polytechnic Institute of Brooklyn, under Pyke's direction, in which a small amount of wood shavings or pulp was mixed into water, and when that water froze the ice it produced was very strong. The chemists were able to balance the weight of a "medium size motor car" on a one-inch column of the stuff. Evidently tiny fibers embedded between ice crystals kept surface cracks from deepening into total fractures, eliminating brittleness as a factor. The chemists named the ice "pykrete," for Pyke + concrete.

In June of 1943, to test a system for keeping ice in a vessel frozen, the Canadians had built a model bergship and launched it on Patricia Lake in the Canadian Rockies. Made from pure ice, it was 60 feet long, covered in wood, weighed 1,000 tons, carried a one-horsepower engine, and had pipes in it that circulated cold air. It survived a hot summer, not deteriorating and not attracting attention despite the fact that, according to Lorne Gold of Canada's National Research Council and author of the book *The Canadian Habbakuk Project,* "it looked like a floating chicken coop out in the water."

Engineers designed a full-size bergship which would use pykrete, be 2,000 feet long and 300 feet wide, weigh two million tons, have a displacement 26 times that of the *Queen Elizabeth* and room aboard for 200 Spitfires plus 1,500 crewmen. The hull would be a square, hollow beam of pykrete with 35-foot-thick walls covered by cork (to limit melting) and timber (to limit creep; also to guard against "waves and enemy action," Gold wrote). Cold air would be circulated through cardboard tubes frozen into the hull so the pykrete would stay frozen, even in the tropics. "Interior voids," according to Gold, "that were not required for buoyancy, machinery or space for cargo and living quarters would be filled with ice."

Building a single such bergship would take 1.7 million tons of pykrete. Making that much, or several times that much for several bergships, presented the engineers with a major challenge. After a while, it began to look as if the Habbakuks, as Pyke referred to them, wouldn't be ready for use in the war with Germany but might be ready for the war in the Pacific. Churchill decided the Americans needed to become involved, so when the Allied chiefs of staff met in Quebec City in August of 1943, he gave a demonstration of the powers of pykrete to President Roosevelt, in his hotel room. " 'Wait till you see this!' the Prime Minister said excitedly," David Lampe recounted in his book *Pyke: The Unknown Genius.* A small block of regular ice was placed in a large silver punch bowl and hot water from a silver pitcher poured over it. " 'See!' smiled Churchill after a few minutes. 'The ice melted!' Roosevelt nodded." A small block of pykrete was placed in the other punch bowl and doused with hot water. "Churchill waited several minutes, then sat back in his chair and said quite happily, 'Hasn't melted at all!' "

The Americans never got involved (Roosevelt's science adviser reportedly told him, "It is the bunk"). Before long, Allied aircraft increased their range and islands became available as stepping-stones, and there was no longer a need for giant floating islands of ice. No bergships (except for the one that drifted about during a long, hot summer on Lake Patricia) was ever built. The whole idea could be considered an oddity of war. "Yet, looking back," biochemist Max Perutz wrote after the war was over, "it is easy to understand how this daring venture came to fascinate men's minds and was welcomed as a possible solution for one of the most difficult military problems facing the Allies. . . . In fact, I think that had not the course of the war and the state of our armaments changed, the bergship could have been constructed."

As for Pyke, he never took out a patent on pykrete. It melted away into history, if not into Roosevelt's bowl. Since the end of World War II, a few structures—runways, temporary docks, mines—have been built with fiber-fortified ice, but very few. For one thing, it takes money to get the pulp to where the ice is. For another, according to George Blaisdell, research civil engineer at CRREL, there really isn't the need, at least where airfields are concerned. "Ice in its natural state is strong enough to support any aircraft," he states. "You don't have to mix it with pulp." The heaviest airplanes in the U.S. military, C-5 Galaxies, routinely land on runways near the U.S. Antarctic research station in McMurdo Sound where, with cargo and passengers aboard, they put a load on the ice of up to 840,000 pounds.

Off in the distance the wind—all ice and feeling—
Invents a tree and a harp, and begins to play.

—Mark Strand, *"Precious Little"*

People have come up with plenty of other uses for ice, including as a camera lens, as a skin toner during a facial, for preserving bodies and drying out waterlogged books, to clean garbage disposals, keep Christmas trees fresh and hamburgers juicy, create musical intruments and ice lanterns, make ice wine and ice beer, even strengthen soil too weak to tunnel through. For Boston's multibillion-dollar Big Dig project, undertaken to connect the Massachusetts Turnpike with Logan Airport, engineers deliberately froze ground that was too soft, wet, and unstable for them to drill into. They pumped a brine chilled to −22°F through 1,600 steel pipes they'd inserted into the soil and kept pumping for four months, at the end of which time the ground was frozen to a depth of 50 feet. "It had a feeling comparable to permafrost," Philip M. Rice, a resident engineer for the project manager, reports. "Artificial ground freezing was a means of last resort. There was no other good way of stabilizing that soil."

Dartmouth physicist Victor Petrenko has even made solar cells and transistors out of ice, as a way of studying ice's properties. Like silicon, ice is a semiconductor, except that protons carry the charge instead of electrons. Using an ice transistor, he could actually build a small computer and get it to work if he wanted to, but "that's not a goal," he says. For one thing, it would have to be kept in a refrigerator.

Probably the grandest use of ice, and surely the one that requires the most ice to execute, is as a very large telescope for finding a very small thing—a neutrino. A neutrino is a subatomic, chargeless particle whose most distinctive characteristic is that it has almost zero mass. In the time it takes you to blink, trillions of neutrinos will have passed through your body without affecting it or being affected by it. "Neutrinos are unsocial," Robert M. Morse, senior scientist at the University of Wisconsin at Madison, concludes. "They don't talk to anything in nature."

It is this very unsociability that makes neutrinos of interest to Morse and other astrophysicists. Neutrinos are produced in distant galaxies by objects the scientists would like to know more about. The objects aren't ordinary stars but old ones that have supernovaed and turned into other objects: pulsars, exploding white stars, black holes, colliding neutron stars. For their small size, according to Morse, they put out an enormous amount of energy, "like a little kid with a big opera voice." In addition to neutrinos, the objects emit light rays, gamma rays, X-rays, and infrared rays, forms of energy that *are* sociable and do interact with matter, including background material left over from the Big Bang. On their way to Earth, they smash into matter which blocks or reprocesses their signals, while neutrinos keep right on going and arrive here unscathed, the same as when they were born billions of years ago and billions of light-years away.

Morse compares one of the high-energy objects that scientists want to know more about to an old potbellied stove (he's used to speaking to schoolchildren about these matters). "We know the stove is warm," he says, "but we don't know what's inside producing the heat." Trying to get information about the objects from other forms of radiation is like touching the outside of the stove; trying to get information from neutrinos is like "opening the stove door and peering inside."

The bad news is that before scientists can use the handy neutrinos to get information, they have to find them. "How do you detect the undetectable?" Morse asks, not entirely rhetorically since scientists have worked out a plan for indirect detection. It happens that very very very rarely a neutrino does collide with an atom of ordinary matter, and the impact of that collision produces a muon (like an electron but 270 times heavier) plus other subatomic debris. The neutrino stops in its track like a cue ball that just hit a billiard ball head-on while the muon, like the billiard ball, heads off in the direction the neutrino had been going. The track of the muon extends the former track of the neutrino. As it travels along, the muon trails behind it a wave of faint blue light, "like the shock wave of a boat going through water," says Morse, "or a plane hitting the sound barrier." Known as Cherenkov radiation, this blue light is what astronomers are after. When they find it, they should be able to figure out, by calculating backward, what path the neutrino had been on before the collision occurred and therefore what part of the sky it came from. If they find "hot spots," parts of the sky where clusters of neutrinos came from, they can deduce the presence of a high-energy object there. They can also begin to tease out the personality of the object, by noting its energy spectrum and flux.

Ordinary optical telescopes won't pick up the blue light, so neutrino hunters have created other kinds of telescopes, all made up of some clear medium and lots of it. A clear medium is needed because any blue light would probably have to travel a long way to reach a detector, and lots of the medium is needed to increase the likelihood that there'll be a collision within it. Some researchers have chosen as their clear medium ordinary water in lakes, oceans, or large underground tanks, and some have chosen heavy water, but Morse and fellow Wisconsin astrophysicist Francis Halzen decided to try ice, at the South Pole.

Not only is there plenty of ice at the pole—it's almost two miles thick there—but the ice is almost pure. "If you're going to use ice or any other solid material, there couldn't be a better place on Earth than the South Pole," Buford Price, professor of physics at the University of California at Berkeley, says. "It's extremely clean." Also, the same part of the sky can be observed continuously from the pole. "The Earth doesn't turn for us," Morse says. Over three austral summers, in a project called AMANDA (for Antarctic Muon and Neutrino Detector Array), an international team of investigators, including Morse and Price, melted holes in the polar ice cap, 19 of them, each 2 feet wide, the deepest 2,400 feet deep, or as far below the surface, Morse points out, as one and a half Eiffel Towers are high. They lowered into each hole a cable with photodetector bulbs strung along its length, "like Christmas-tree lights," according to Morse, 750 of them, each bigger than a basketball and sensitive enough to register the hit of a single photon.

The bulbs act like regularly spaced buoys, Morse says, returning to the boat analogy, "rocking and jiggling" whenever bow waves (the blue light) reaches them from passing boats (the muons).* In the five years since the detectors were locked into the ice, they've registered hundreds of hits from the blue light of muons—but not the high-energy muons the astronomers are interested in, only the lower-energy muons produced when cosmic rays from the sun strike our atmosphere. "Those things rain down on us by the billions," Morse says. AMANDA tried to filter out most of these ordinary, "atmospheric" muons by hanging the detectors pointing downward. "We look through our feet," Morse says. Most of the low-energy muons must pass all the way through the Earth to reach the detectors, and during the journey most of those get waylaid, while the neutrinos, and the high-energy muons they produce in a collision, do not.

Still, no high-energy muons have yet registered on the South Pole detectors, and AMANDA participants have decided they need a bigger telescope, involving still more ice. "It's impossible to build too big a neutrino detector," Morse maintains. The new ice telescope will contain almost 5,000 photo tubes (each with a computer chip connected by Internet to computers in the university offices back home) frozen into 80 holes drilled in one cubic kilometer—a quarter of a cubic mile—of polar ice. It will (of course) be called IceCube and will be built at a bargain price, as the astrophysicists figure it: 25 cents per ton of ice.

> Here's a quick-freezing icy sweatshirt for you, Stretch! This'll cramp your style real good!
>
> —The Iceman, *Marvel Comics*

There's one person who can put ice to almost any use. He's Bobby ("The Iceman") Drake, one of the original X-Men in the comic-book

*The reason a muon creates a shock wave is that it travels faster through ice than light does; for the record, light travels three-quarters as fast through ice as it does through a vacuum.

series about alienated teenage human mutants with superhuman powers which they use to fight evil forces and get ordinary non-mutants out of jams. The Iceman creates objects and structures out of ice at will. He makes ice lassos that bind the chests of enemies like iron. He flings slick patinas of ice onto the ground for his pursuers to slip on. He freezes millions of tons of raging river about to engulf fellow X-Men. He coats the inside of a crumbling tunnel with ice to fortify it; immobilizes a troublemaker by freezing the flow of blood to her brain; and tosses into the air swaths of water that flash-freeze into ice bridges he can run across and ice scaffolds he can spy from. When someone shoots at him, he throws up an instant shield of ice for bullets to bounce off of. If he wants to bash in a perpetrator's head, he is in immediate possession of a spiked club of ice. When he spots an incoming missile, he transforms it into a giant Popsicle. And when the evildoers are captured, he builds an ice jail to hold them.

Despite the awesomeness of the Iceman's "ice powers," however, all too often his enemies are able to break open their ice bonds, or melt the ice barrier he erected, or shatter the block of ice he froze their bare feet into. "No Dice, Ice!" a woman taunts him. The Iceman has cracks in his ice armor. He is self-doubting, maybe a little wimpy. He has his weaknesses, like his weapon of choice: ice.

CHAPTER TWENTY-NINE

OTHER FORMS OF ICE

hot ice and wondrous strange snow.
How shall we find the concord of this discord?

—William Shakespeare,
A Midsummer Night's Dream

AS IF DIAMOND DUST, needles of lake ice, discs of frazil, whorls of hoarfrost, and fibers of ground ice were not enough, scientists have produced in their laboratories nearly a dozen other forms of ice, all new to Earth so far as they know, ice that does not float in water, whose crystal structures can be cubic or rhombic or monoclinic instead of hexagonal, or if they are hexagonal the hexagons are compressed, cramped, warped. Ordinary ice—those disks, whorls, etc., the subject of this book until now—is Ice I. To make Ices II through XII, scientists take water, either as liquid or as ordinary ice or as another of the new forms of ice, and subject it to high pressures and low temperatures until, to relieve the strain, the molecules abruptly rearrange themselves into new patterns that allow them to squeeze into smaller spaces. Producing new forms of ice is essentially a process of removing some of the open space at the heart of ice crystals, in one way or another, and cramming more water molecules in.

What makes the cramming possible is the flexibility of the hydrogen bonds between the molecules. Under great pressure, usually achieved in the lab by putting a sample between two anvil-shaped diamonds and squeezing it, the hydrogen bonds bend or twist or shorten or lengthen or otherwise distort, allowing the water molecules to take up new positions in the crystal lattice. "The bonds are very adaptable," Peter Kusalik, a liquid-state theorist and associate professor of chemistry at Dalhousie University in Nova Scotia, points out. "They can be changed very slightly in a great variety of ways." The bonds are also very robust. For despite all the skewing and folding and twisting they go through to produce the new ices, the bonds stay fairly normal. They bend more often than they break, and they keep the same number of close neighbors (four). They just move the next-closest neighbors, and the next-closest, closer.

In Kurt Vonnegut's novel *Cat's Cradle,* a brilliant scientist but socially stupid human being invents a new form of ice, which he calls *ice-nine.* Ice-nine turns out to be a weapon more powerful than the hydrogen bomb, which he also helped invent. He realizes with what seems insufficient horror that if ice-nine were to be released in an uncontrolled way from its thermal container, it could wipe out all life on Earth. He didn't start out to make a weapon but to harden mud. A marine general had complained that his troops were tired of getting stuck in mud during combat, so the scientist decided to freeze the mud and thereby firm it up. Inspired by an odd pile of cannonballs he spotted on a courthouse lawn, he concocted a new stacking pattern for the ice molecules, one that allowed the ice to stay frozen at room temperature—in fact, at temperatures as high as 114.4°F. Since mud doesn't often get that hot even with marines fighting on it, any of the mud frozen into ice would stay frozen. A mere grain of ice-nine would do the job: acting as a seed or nucleator, it would "teach" any water molecules it came into contact with to form ice crystals according to the same odd cannonball stacking pattern, and those crystals would in turn teach the water molecules *they* touched to form the new pattern, setting off a chain reaction of freezing. A single marine could carry enough ice-nine under the nail of his little finger "to free an armored division bogged down in the Everglades," the head of the inventor's research lab told the narrator; "infinite expanses of muck, marsh, swamp, creeks, pools, quicksand, and mire" would become "as solid as this desk."

Freezing wouldn't stop with the mire, however. Streams flowing through the swamps and marshes would be seeded by ice-nine, and so would the rivers and lakes that fed the streams, and the springs that fed the rivers and lakes, and the oceans fed by the rivers, and the rain that fed them all. When rain fell, "it would freeze into hard little hobnails of ice-nine," the novel's narrator realized, "—and that would be the end of the world!"

The beginning of the end came when the ailing dictator of a Caribbean island nation swallowed a sliver of ice-nine. "Marble! Iron!" the dictator's doctor exclaimed. "I have never seen such a rigid corpse before. Strike it anywhere and you get a note like a marimba!" Its blood transformed into ice-nine, the corpse plunged through a hole in the castle wall and slid into the sea, which as far as ice-nine was concerned was one big bowl of supercooled liquid, primed for nucleation. Thus spake the narrator:

"There was a sound like that of the gentle closing of a portal as big as the sky, the great door of heaven being closed softly. It was a grand AH-WHOOM. I opened my eyes—and all the sea was ice-nine. The moist green earth was a blue-white pearl." The narrator survived (by hiding in a torture chamber) and so did the inventor's children (by hiding in a dungeon), plus some ants. The ants formed "tight balls around grains of ice-nine. They would generate enough heat at the center to kill half their number and produce one bead of dew. The dew was drinkable. The corpses were edible."

Not one of the new forms of ice made by real-life scientists in real-world labs on Earth could cause even a modest AH-WHOOM. All of them can exist only under high pressures or low temperatures or both. Once they are removed from pressure vessels and freezers, they not only can't teach other water molecules to stack according to their novel patterns, they lose those patterns themselves.

"Put a sample of any of them on a table and it will explode," says Barclay Kamb of the California Institute of Technology, who a couple of decades ago worked out the crystal structure of several of the high-pressure ices. "Like a book opening," he says. He uses his hands to show an unfolding. "A flowering. It exfoliates and decrepitates, turns to powder." The clear ice becomes opaque, from all the expansion cracks in it. "A fuzzy white mass springs out," Kamb says. "It looks kind of like hoarfrost, with a lot of feathery blades." To Dennis D. Klug, research scientist at the National Research Council of Canada and a longtime investigator of high-pressure ices, it acts like popcorn blowing up. "It's really quite cute," he says.

The real-life ice-nine, called Ice IX, is "not as interesting as Vonnegut's," Klug admits, sounding apologetic on behalf of nonfiction. Discovered by Edward Whalley, Klug's mentor and colleague at the National Research Council, the benign IX is made by radically cooling Ice III. Ice III's crystal is four-sided—"you can't get a six-sided snowflake out of it," Klug says—and so is Ice IX's except that it's "ordered." Ordered means that all the molecules in a crystal point in the same direction. "They line up like soldiers," Klug says. Ordinary ice is "disordered," which means, according to the so-called ice rules, that each water molecule can point in a different direction. The molecules in a disordered form of ice have six possible orientations; those in ordered ice only one.

One of the ices whose structure Kamb determined using X-ray and neutron diffraction is one of the densest, Ice VI. It has crystals nearly 1 1/2 times as dense as those in ordinary ice, a feat it accomplishes by "doubling," Kamb explains. "Two entities co-exist in the same space." A pair of separate lattices interpenetrate, with the molecules of one lattice occupying the cavities of the other but without the lattices touching each other or having bonds between them. Kamb calls VI a "self-clathrate" (a clathrate being a crystal whose lattice holds a "guest" molecule, of for example methane, but in this case the guest molecule of the ice is ice).

Hermann Engelhardt, Kamb's colleague at the California Institute of Technology, who has also worked on the high-pressure ices, describes Ice VI as "very beautiful." (Dennis Klug insists: "All these crystal forms are beautiful.") You can make Ice VI by putting pressure on a sample of Ice I, and that gives you, first, Ice II (an ordered form of Ice I, discovered in 1900), and then, as you increase pressure, Ice V, and finally, as you increase pressure and *raise* temperature, Ice VI. If you continue adding pressure to Ice VI and raise its temperature above freezing, you get Ice VII. Like Ice VI, Ice VII consists of two separate but equal, interpenetrating frameworks, but they are cubic. Now if you *cool* Ice VII, you get Ice VIII, which like VII is a pair of coexisting cubic crystals but *ordered* ones this time. And so it goes, as the narrator of another Vonnegut novel, *Slaughterhouse-Five,* would say. Each new form of ice, and variations within the form, is produced by a different combination of pressure and temperature.

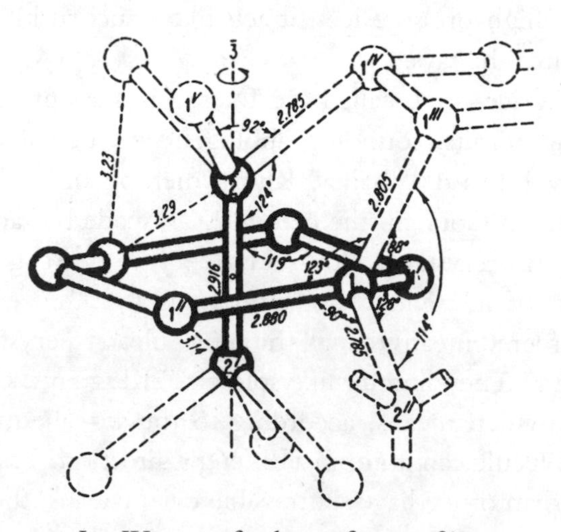

Ice IV: one of a dozen forms of ice

Although all the supernumerary ices are less stable than ordinary ice, some are less stable than others. Ice IV is particularly unstable. "It is a mystery why it can be made," says Engelhardt, who was working on Ices V and VI when "suddenly a new form appeared." For a long time, he says, he was the only one who could make it—"you have to cool it down in a certain way"—and even then, he could make it only once in every 100 tries. "It always has to compete with neighboring phases—VI, III, and V. It's always in their region of stability. It doesn't have one of its own." Kamb, whose California license plates read "ICE IV," refers to IV as "peculiar" (plus beautiful). The way its open space is reduced is by having a hydrogen bond pass through a ring of six water molecules. That makes for a particularly crowded structure, which is probably why it is so unstable.

Not long ago, several teams of physicists from Japan, Europe, and the United States took Ice VII and VIII and subjected them to 1.5 *million* times normal atmospheric pressure—and got Ice X. At those exceedingly high pressures, "the water molecules disappear," as Dartmonth's Victor Petrenko points out. A proton from one of two neighboring water molecules moves onto the middle of the hydrogen bond between them, and you can no longer tell which molecule the proton belongs to. It belongs to both molecules, or neither. The ice is just a mixture of oxygen and hydrogen atoms in long chains. "It's still ice, but it has quite different properties."

The main reason scientists create these strange, new forms of water ice is so that they can learn more about the hydrogen bond, by observing ways it can be perturbed. "Ice is not just an esoteric subject out in the corner somewhere!" Klug insists. The hydrogen bond is one of the most important interactions in all of nature, a key element in biological, geological, and chemical systems, and, as Engelhardt points out, "the greatest variety of hydrogen bonds are in ice."

Another reason for creating forms of ice not native to this planet is that they might well be native to other planets. Ice II is the ice form that requires the least pressure to make, yet there's not enough pressure anywhere on Earth, even under the heaviest glaciers, to produce it naturally. Elsewhere in space, though, in the larger icy planets and moons of our solar system, there are undoubtedly regions of extremely high pressure with high-pressure ices in them. "A slice through a Jovian planet," Igor Svishchev of Dalhousie University has suggested, "might reveal a nearly complete phase diagram of water ice."

One type of ice believed to be bountiful in space is Ice Ic, the "c" standing for cubic. Ice Ic is the same as ordinary ice, which is officially known as Ice I*h,* the "h" standing for hexagonal, except that its crystal structure is a cube. Ic has the same size and density as Ih, and it too floats on water. It forms when Ih is chilled below −135°F, a common enough temperature in space. Whalley once suggested that there might be cubic ice on Earth, in the very cold, upper reaches of our atmosphere, but nobody has been able to prove it.

There's yet another form of ice, which doesn't rate a number or even a letter because it doesn't have crystals. If the dictionary definition of ice, "a crystalline form of water," were adhered to, it might better be called "solid water" than ice. It is known as "amorphous" ice. It forms when water cools so fast that the molecules can't organize themselves into lattices and therefore stay in the random arrangements of liquid water. It's thought that in the body of a comet, where water vapor condenses on a very cold surface in a vacuum, the ice could be amorphous.

Recently, a European team found Ice XI, which is an ordered form of Ice I, and another European group discovered Ice XII, which has the bent bonds of Ice V but bent even farther. Already ice has been shown to have a greater variety of forms than any other solid, and scientists aren't through identifying new ones. In the future, there is sure to be an Ice XIII, an Ice XIV, an Ice XV, and so it goes. "It's an exercise that will continue," Klug says. Nobody expects that any of the new forms will have the earth-destroying capabilities of ice-nine. As Klug points out, "Nature has had a long time to make ice-nine and never succeeded."*

Although each of the other forms of ice has its special characteristics, "perhaps the most intriguing form is Ice I itself," Kusalik says. "It's exceptional because at the pressure at which it exists it's less dense than its liquid. If it weren't—if it didn't float—the Earth would be a very different place. Oceans would fill with ice from the bottom up, and marine life would die." A catastrophe of a kind not unlike the one produced by the ice-nine–riddled corpse of a Caribbean dictator slipping into the great bowl of the sea: much of the world as blue-white pearl.

*Some scientists suggest that mad cow disease is spread by an ice-nine–like process, in which abnormal proteins in the brain called prions act as seed crystals, teaching normal proteins as they form to fold in abnormal ways, and these "infect" other proteins, which build up in the brain and kill cells, leaving it full of holes.

CHAPTER THIRTY

ATMOSPHERE I

Freeze, freeze, thou bitter sky

—William Shakespeare,
As You Like It

IT STARTED RAINING in Batesville, Mississippi, on a Wednesday night in February 1994, kept raining all day Thursday and into Friday. As soon as the first rain hit the ground—the lawns, streets, pines, awnings, pecan trees, stop signs, Lenten roses—it froze, and the rain that fell on top of that froze, too, and the rain that followed froze on top of what was already there, until by Friday so much ice had built up, 12 inches of it in places, that tree limbs broke under the weight and power cables, their circumferences fattened to three times normal, drooped so low they pulled down the poles that had been holding them up.

"It sounded like guns going off," Rita Herron remembers. She and her husband, Harold, owned a dairy farm in Batesville with a herd of 50 cows. "Limbs were snapping off the trees, cracking and popping all around," she says. "You felt like you were under siege." When several of their cows took shelter in a thicket of oak and cedar trees, as they often did during storms, branches laden with ice broke off and crushed seven of them to death. When power lines went down, the farm was left without electricity for 16 days. The lights and phone and water pump didn't work. "To flush the commode, we caught water in five-gallon buckets from ice melting off the roof," Rita Herron says. "We washed our hands and faces in a teacup."

The worst was that the milking machines didn't work. The Herrons were forced to milk the surviving cows by hand. Not for the milk—they couldn't get it to market anyway—but for the cows. If the cows weren't milked twice a day, they could sicken or die. With machines, it had taken an hour and a half to milk 50 cows. Without machines, it took Rita and Harold Herron and their son and daughter and their daughter's boyfriend and Harold's 78-year-old father, working nonstop every day from before sunup to after dark to milk 43 cows, and many of those got sick anyway. Half had to be sold, at slaughter prices. "We still owed on them," Rita Herron says. "I was wringing my hands. 'Lord, what is your plan for us?' "

The utility company called it "the worst ice storm of the century." More temperately, a meteorologist said it was "the most widespread and worst ice storm in the Southeast in 20 or 30 years." Jane Pund, whose pig farm near Batesville was without electrical power for 23 days, had her own way of judging how bad the storm was: "My father-in-law never saw such ice in his life and he's 82." The utility company had "tracked the movement of the ice on a big map," it reported in its newsletter, "like a hurricane tracking map." The ice storm cut through four states, it noted, "first struck in southeast Texas, . . . then moved north across Louisiana, freeze-wrapping Arcadia, Grambling, and Minden. It swept into Arkansas [and] finally inflicted its worst damage on Mississippi, particularly to Coahoma and Washington counties."

Roads were blocked by downed trees; drivers carried chain saws so they could cut their way home. In Mississippi alone, catfish, pecan, and timber industries suffered $1 billion in losses. Citizens tied ribbons—white, of course—around trees to show their appreciation for crewmen who worked around the clock to restore power. Twenty-one thousand poles had been pulled down, 2,600 miles of power lines needed to be replaced. There were "inconveniences," too, the utility company pointed out: no power for pumping gas, no power for operating ATMs, and "the National Guard began delivering, of all things, ice."

the town is frozen solid, leaden with ice.

—Anna Akhmatova, *"Voronezh [O.M.]"*

Four years later and a thousand miles north, "ice rain," as one official described it, fell for six days on large parts of Quebec, Ontario, Maine, Vermont, New Hampshire, New Brunswick, and northern New York. In Canada, 1,000 electrical transmission towers, 200-foot-high steel pylons, collapsed under the ice load "like exhausted marathon runners," according to the *Canadian Geographic,* or "almost like they'd melted or been hit by a huge fist," according to the *Power Workers' News.* More than two million people were without power, for as long as a month. "Montreal on the Verge of Chaos, in the Dark," went a headline in the *Montreal Gazette.* "Thousands forced into shelters. The army and police patrolling the streets." Soldiers wore combat helmets to protect their heads from "ice bombs" falling off trees and buildings. One Montreal policeman was sitting in the front seat of his squad car when a 50-pound chunk of ice crashed through the roof and landed in the back seat.

Trees that grow in the north are adapated to handle some ice, Tim Fleury, a forester with the University of New Hampshire's extension service, pointed out, a half inch or so of glaze, but this ice storm laid up to three inches of glaze on some trees, and three inches of ice on a 50-foot-tall maple can add up to *eight tons* of extra weight. Conifers did better than most because they tend to shed ice. "Folded against the trunk by the weight of ice and snow, conifer boughs act like a collapsing umbrella," Bernd Heinrich wrote in his book *The Trees in My Forest*. One editorial suggested that if the transmission towers had been designed more like conifers, with their center of gravity low, instead of, as they were, like the deciduous "slender, full-headed red maples, thousands of which were weighed down and then ripped apart in the great ice storm of 1998," the towers might not have collapsed as they did.

For all the trouble that ice storms cause—the one in 1998 was described afterward as "the most damaging natural disaster in Canadian history"—they do offer something in compensation. Even Jane Pund on her pig farm in Mississippi admitted after the 1994 ice storm in the U.S. southeast, "It was a little pretty." For Mark Twain, writing in his book *Following the Equator* (1897), an ice storm was beyond pretty:

"In America the ice-storm is an event. . . . Usually its enchantments are wrought in the silence and the darkness of the night. A fine drizzling rain falls hour after hour upon the naked twigs and branches of the trees, and as it falls it freezes. In time the trunk and every branch and twig are encased in hard, pure ice; so that the tree looks like a skeleton tree made all of glass—glass that is crystal-clear. All along the underside of every branch and twig is a comb of little icicles—the frozen drip. . . .

"The dawn breaks and spreads, the news of the storm goes about the house, and the little and the big, in wraps and blankets, flock to the windows and press together there, and gaze intently out upon the great white ghost in the grounds, and nobody says a word, nobody stirs. All are waiting; they know what is coming . . . The sun climbs higher, and still higher, flooding the tree from its loftiest spread of branches to its lowest, turning it to a glory of white fire; then in a moment, without warning, comes the great miracle, the supreme miracle, the miracle without its fellow in the Earth; a gust of wind sets every branch and twig to swaying, and in an instant turns the whole white tree into a spouting and spraying explosion of flashing gems of every conceivable color; and there it stands and sways this way and that, flash! flash! a dancing and glancing world of rubies, emeralds, diamonds, sapphires, the most radiant spectacle, the most blinding spectacle, the divinest, the most exquisite, the most intoxicating vision of fire and color and intolerable and unimaginable splendor that ever any eye has rested upon in this world, or will ever rest upon outside of the gates of heaven."

Winter about to swing its brutish cubes

—Howard Moss, *"Cardinal"*

Besides freezing rain, several kinds of ice fall or materialize out of the sky. First, there is *sleet,* which is frozen, not freezing, rain, the raindrops having on their way to Earth passed through a deep band of cold air which caused them to turn to ice before touching down. Meteorologists call sleet "ice pellets." The pellets are transparent or translucent, usually smaller than a third of an inch in diameter, and dense. Falling fast and hitting hard, they often make a clicking sound on impact and bounce several inches high.

Then there's *rime,* or "frozen breath of the earth upon its beard" (Thoreau). Rime forms when moving air pushes supercooled cloud or fog droplets against exposed objects on the ground, particularly narrow objects—twigs, towers, wires, lampposts, and poles—and the droplets freeze on contact. Fine-grained and opaque, rime can form as a compact mass, or as feathers (some buildups look like Indian war bonnets), or as spikes. In his *Field Guide to Snow Crystals,* Edward R. LaChapelle includes a photo of a ski pole he left outdoors one foggy winter night and which had accumulated so much rime by morning that it looked like an oversized, overused toothbrush. The rime usually builds on one side of the object, which gives an onlooker the impression that it was blown back by the wind; it actually grows *into* the wind.

Looking at the rime, you might be able to tell not only the direction from which the wind blew during a storm, by the way its spikes or feathers are pointing, but the speed at which the wind blew, by how thick the rime layer is. On Mount Washington, where winds are strong and fog is common in winter, the rime on research shacks, instruments, and vehicles gets to be several feet thick. "I've had rime ice form on *me* up there," says Charles Ryerson, atmospheric icing specialist at CRREL. "Every little thread, every little hair will show up as white. My beard, my eyebrows, a feather on my down parka are coated with rime."

There's also *graupel.* It's the official name for this kind of ice, but when was the last time you heard anyone say, "Hey, look at the graupel coming down"? Graupel forms when rime freezes onto a snow crystal in the air instead of onto an object on the ground (the crystal's fall provides the "wind"). Small particles of rime envelop the crystal, hiding its hexagonal shape, until it becomes a tufted, rounded, or cone-shaped lump: graupel. Graupel tends to be small, under a fifth of an inch, and to have many air bubbles in it. Squeeze a piece of graupel and it will give under the pressure like a pea.

Then there's *hail.*

> *Look, children,*
> *hail-stones!*
> *Let's rush out!*
>
> —Basho

One reason to rush out when you see hailstones is that if you pick up large ones before they melt and break or saw them apart and look closely at them, you may be able to learn something about their life histories. "A hailstone carries on its surface and in the differences of its internal structure the story of its formation and fall," wrote Charles Knight and his wife Nancy, also at the National Center for Atmospheric Research in Boulder, Colorado, both of them former "hail chasers." Over several decades, they picked up, or received in the mail from people who themselves picked up and saved in their freezers to show to neighbors or insurance companies, many hundreds of hailstones. The lives that hailstones lead are short and brutish. They grow in thunderstorms, the largest ones in the largest and most severe thunderstorms, where winds can reach vertical speeds of 100 miles an hour. A hailstone may start as a frozen raindrop, or as a particle of graupel, or as a fragment of a larger hailstone that splits apart. Around this ice "embryo," the hailstone grows by colliding with supercooled droplets in a cloud (usually cumulonimbus) which freeze and stick to it—in other words, by riming. A big hailstone is not, as many people think, a collection of small hailstones glommed together into one big hailstone, like a snowball, although some hailstones in their general lumpiness look that way. A hailstone grows by accruing liquid water, not ice.

To grow large, what a hailstone needs first of all is many water droplets to collide with. It also needs strong vertical winds, which can carry it to high altitudes, 30,000 or 35,000 feet, where temperatures are very cold even in summer. (You don't often see hail in winter, the season of other icy precipitation, or in polar regions either, since air over cold ground tends to be stable, and what produces strong vertical winds is unstable air.) The hailstone also needs time to take advantage of the cold and the droplet supply to make ice. For it to get that time, there should be a rough match between its falling speed, determined by its weight, and the speed of the updraft it is caught in. In this case, Edward Lozowski, professor of earth and atmospheric sciences at the University of Alberta at Edmonton, explains, the hailstone will "hover like a helicopter," adding layer upon layer of ice until it becomes so heavy that it falls faster than the wind is carrying it upward and sinks to Earth. When falling speed and updraft speed balance for a long time—in a hailstone's life a long time is 10 or 20 minutes—the hailstone can grow into a giant.

One summer day, August 27, 1973, Mrs. Skorupa of Cedoux, Saskatchewan, picked up a hailstone that landed on her farm during a storm in which hail killed several of her geese. The hailstone was the size of a McIntosh apple and shaped like a bow tie. Soon afterward, Lozowski showed up in Cedoux, asking around for hailstones that people had saved, and she gave him hers. Back in his lab, he cut a thin slice from the stone and examined it under a microscope. It had a round core about the size of a dime with a dark, clear crescent shape inside it; the crescent was the embryo, either a large, deformed raindrop, Lozowski figured, or a splinter from another hailstone that had broken apart while aloft. Around the core was a dark region of ice—dark because the ice was clear, with few bubbles to scatter light. The paucity of bubbles indicated that the ice had frozen slowly, at a temperature near the freezing point, which gave the dissolved air in the droplets time to escape.

Encircling the dark layer was a "very white" one, Lozowski noted—white because it was full of small bubbles, which indicated that the droplets in that region had frozen at colder temperatures than those in the dark one, too fast for air to escape. Beyond that very white layer were several other layers, some milky in appearance and spongy in texture. These layers were formed, Lozowski deduced, when droplets collided with the hailstone in such profusion that some didn't freeze completely and dissipate the latent heat of fusion before others hit the same spots, and those spots, still warmed by the latent heat, stayed right at 32°F, and the surface stayed wet, with minuscule fingers of ice reaching into the liquid. If in that state the hailstone had hit the ground, it would have splatted. Most hailstones bounce, though; some can bounce higher than your head. Those stones probably form by "dry growth," with each droplet freezing completely and getting rid of latent heat before others hit the same spot, and the ice becomes solid instead of spongy.

People often compare hailstones to onions, with each layer representing a trip upward in a cloud and many layers representing many trips. Hailstones do have layers but most have fewer layers than most onions, according to Lozowski, and the layers form not because the hailstones are riding up and down in a cloud but because there are changes in temperature or droplet supply in the part of the cloud where they happen to be. "Most of the layers don't require excursions of altitudes," Charles Knight states, "just subtle changes in conditions." Lozowski agrees. "My intuition," he says, "is that most hailstones make only one trip. I'm 99.9 percent sure they don't go up and down like elevators."

In cross-section, Mrs. Skorupa's hailstone looked like a petaled flower. Lozowski counted 18 major "lobes" of ice in the very white layer, extending radially outward from it, with each lobe separated from the next by lines of bubbles, or bands of clear ice, or air gaps. Most large hailstones develop such lobes ("columns"), some from water on the surface that gets thrown up into points as the hailstone rotates, like water off a spinning tire.

Mrs. Skorupa's hailstone got its bow-tie shape from the way it fell. Some hailstones fall simply, holding the same orientation all the way down, and thus grow on only one side, the bottom side, which is the one that hits the most droplets. They are usually shaped like cones and are small, 4/5 of an inch or less across. (What distinguishes a cone-shaped graupel particle from a cone-shaped hailstone is size; hail is more than a fifth of an inch across and graupel is less.) Some other hailstones spiral through the air, some tumble, and some oscillate, swinging to and fro like falling leaves.

Once hailstones become sizable, however, they are often unstable and start rotating around different axes, so that ice builds more uniformly on them, producing a spherical shape. The sphere often develops a dimple on either side of its axis of rotation and becomes a flattened sphere, or a "hamburger," Lozowski says. By then, its axis of rotation is likely to be horizontal rather than vertical, with the hamburger lying "on its side so the mustard will roll out." The dimples that formed on Mrs. Skorupa's hailstone as it spun were so pronounced that the hamburger turned into a bow tie.

Although hailstones are not just like onions, scientists who want to convey an idea of the size and shape of the ones they are discussing often resort to comparisons with other vegetables as well as with fruit, grain, and sports balls. During a storm in Canada on July 6, 1973, hail that hit Mr. G. Wilson's farm near Wetaskiwin, Alberta, was the size of "peas and grapes," Lozowski and L. Wojtiw of the Alberta Research Council reported, "with the shape of raspberries and peas." Some of the hailstones were like marbles and walnuts. The Canadian record for size is held by a hailstone that fell during the same storm that killed Mrs. Skorupa's geese. It landed on Mr. H. Gawel's wheat farm in Cedoux. When he picked it up, Mr. Gawel stated, it had had spiky lobes on it, but as he carried it by hand to the freezer it melted half an inch. With intact spikes, Lozowski estimates, it could have weighed as much as a pound (450 grams).

The largest hailstone ever recorded—and undoubtedly larger ones have landed on Earth but nobody was around to weigh them—burrowed into the dirt of a yard in Coffeyville, Kansas, on September 3, 1970.* It was 17 1/2 inches around, the size of a grapefruit, and weighed 1 2/3 pounds (766 grams). The person in whose yard it landed didn't rush out and pick it up but waited until the storm was over. A hailstone that heavy would have been moving at 100 miles an hour, Charles Knight estimates, and if it had hit somebody on the head, it would have killed him. "Hailstones go clear through roofs sometimes," Knight points out. "In Wyoming, hailstones kill antelope fairly often, more often than people know since there aren't many people around to witness it."

Another form of atmospheric ice is *snow* ("Frost knots on an airy gauze"—Robert Frost; "Feathered rain"—William Strode; "Actualized air"—John Updike). Snow forms in a very different way from hail: water evaporates off a supercooled droplet in some cloud, then the vapor condenses directly onto an ice crystal in the same cloud, skipping the liquid phase.

In 1951 the International Association of Hydrology's Commission on Snow and Ice listed seven major shapes, or "habits," that snow crystals take: plates, stellar crystals, columns, needles, spatial dendrites (ones with arms and branches that stick up off the flat plane of the crystal), capped columns, and irregular (which is anything that isn't something else). In 1966, C. Magono and C. W. Lee collected snow crystals from a mountainside where, as you might suspect, air temperature and humidity

*A hailstone the size of a honeydew melon landed in Aurora, Nebraska, on June 22, 2003, and was quickly claimed as the largest, but the winning weight and size couldn't be verified because part of the stone was lost when it hit a roof gutter and part melted in a faulty cooler on its way to the lab. The people of Aurora were disappointed, reported a spokesman for the National Climatic Data Center.

differed at different elevations, and after studying them closely they increased the number of snow shapes to over 80, including many hybrids. Stellar and dendritic crystals, the classic lacy snow crystals of Christmas cards and childhood memories, are only a few among many possible snow shapes, and not even the most common.

SNOW CRYSTALS.

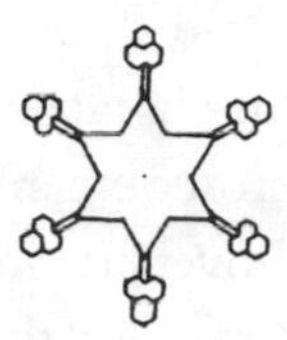

Snow crystals drawn by whaling captain William Scoresby Jr., 1820

A snowflake, by the way, is several snow crystals that are stuck together, which tends to happen when the air is warm and the crystals wet. If there's only one particle of snow, it's a snow crystal. Most snow crystals that touch the ground have already turned into flakes. The fact that a snow crystal is six-sided was recognized at least as long ago as the 2nd century B.C., when a Chinese scribe observed that while flowers of plants generally have five points, "flowers" of snow are always "six-pointed." In the early 17th century, astronomer Johannes Kepler noted in a pamphlet, *A New Year's Gift, or On the Six-Cornered Snowflake,* that snowflakes are six-sided and symmetrical but admitted he didn't know the reason. "Why six-sided?" he asked. "If it happens by chance, why do they not fall just as well with five corners or with seven?" In the mid-17th century, philosopher René Descartes made detailed sketches of snowflakes, emphasizing their symmetry but coming to the wrong conclusion about their six-sidedness. Later, other people took to drawing and photographing as well as describing snowflakes, including the Japanese feudal lord Toshitsura Doi, who printed what he saw through a "Dutch glass" in his 1832 book *Illustration of Snow Blossoms.*

One of the greatest observers and recorders of snow-crystal shapes was Wilson A. Bentley, a Vermont farmer who took photographs of snow crystals under a microscope starting in 1885 and for almost 50 years thereafter, mostly for aesthetic purposes, to capture the beauty of their forms. During a snowfall, he would catch a crystal on a small, smooth black board and examine it with a magnifying glass to see if it was "promising," W. J. Humphreys recounted in his text for *Snow Crystals,* a book containing over 2,000 of Bentley's snow photos. If he judged it promising, he would transfer it by means of a wooden splint onto a glass plate and with "the gentle stroke of a small wing feather" induce the crystal to stick to the glass.

The crystals in Bentley's book fulfilled the promise he saw in them, being endlessly varied and lovely and, one is tempted to say, inventive. But how, John Hallett, professor of atmospheric physics at the University of Nevada at Reno, asks rhetorically, "can there be such a powerful ordering process on the molecular level as is characteristic of ice in all its forms and yet such diversity in the macroscopic results?" In the late 1930s, Ukichiro Nakaya at Hokkaido University in Japan showed by experiment that the main thing that determines the shape of a snow crystal is the temperature of the air. He suspended a rabbit hair in a stream of moist air, exposed the hair to different temperatures, and took note of what ice crystals grew on it.

At air temperatures between 32°F and 27°F, thin hexagonal plates grew on the hair. Between 27°F and 23°F, needles did; between 23°F and 18°F, hollow columns; between 18°F and 10°F, hexagonal plates; and between 10°F and 3°F, fernlike stars. At air temperatures between 3°F and −13°F, hexagonal plates grew (again), and between −13°F and −58°F hollow columns did (again).

The categories were specific. Transitions were mostly sharp, within a degree or two Fahrenheit. Whenever Nakaya moved his rabbit hair up or down into a slightly different temperature zone, any new growth that took place went into creating a different shape. How fast the growth took place was determined mostly by how much water vapor was available; the more vapor, the faster the growth. Also, the more vapor, the more elaborations and complications and secondary features the snow crystal developed. Although low humidity encouraged growth across the flat faces, high humidity encouraged growth at the tips, corners, and edges. Nakaya began to refer to snow crystals, expressing as they do to such a fine degree the environments from which they come, as "hieroglyphs of heaven."

Most snow crystals that touch Earth (which by LaChapelle's definition they must: snow crystals are "ice crystals in the atmosphere which grow large enough to fall and reach the ground") are not really symmetrical. Those in Bentley's photographs appear to be so—symmetry is part of their beauty—but they are not. As John Hallett notes, "any given crystal will have grown over a long fall trajectory in the atmosphere" ("long" meaning up to six miles) and will have encountered a variety of conditions on the way, including shifting temperatures (usually rising).

In a recent paper, Hallett and Charles Knight explained that even when a snow crystal's arms—arms being "the first six growths from the corners of a hexagonal plate"—are symmetrical, the *branches*—which are the "side growths from the arms"—are not likely to be. The explanation they proposed for this phenomenon is that if temperature fluctuations during the crystal's fall are large enough, the arms will grow equally, but if the fluctuations are short-range—as short-range as, say, "a very small eddy or the approach of a cloud droplet"—one arm could be induced to grow and not another. In the absence of such fluctuations, Hallett and Knight concluded, "the law of the jungle prevails for the side arms, with smaller ones excluded by competition for heat loss/mass gain compared to any branch which has a momentary advantage." Next time you're out contemplating a gentle snowfall, think of the law of the jungle operating within it!

Other things may happen during a snow crystal's long earthward journey. Water droplets may stick to it and form bumps of rime. Winds may break off one branch, or three, or five. Other crystals may hit it and become attached. It may descend in a spiral path, with one edge angled more into the airstream than the others, and since a leading edge grows faster than a trailing edge (having more water vapor available to it), a crystal that holds this orientation could end up developing only one arm. "When at last the snow crystal or its remnants come to rest under an observer's microscope," LaChapelle pointed out, "it can bear little resemblance to the original particle formed high in the atmosphere."

Knowing that so much can go into the creation of a single snow crystal would seem to give scientific underpinning to the popular idea that no two snow crystals are alike. Even if two adjoining snow crystals in the same cloud started falling at the very same instant, wouldn't they have minutely different environmental experiences on their way to Earth? Might not their very proximity to each other produce a tiny fluctuation that would cause a branch on one crystal to lengthen? Knight notes that there are 10,000,000,000,000,000,000 molecules in a typical snow crystal, more or less, and the number of ways they could arrange themselves into shapes are incalculable. Yet who can say for sure that a crystal has never been duplicated? To use more zeroes, David Phillips, senior climatologist at Environment Canada, has estimated that over the lifetime of Earth, there have fallen on its surface 100,000,000,000,000,000,000, 000,000,000,000,000 snowflakes, more or less.

I wield the flail of the slashing hail,
And whiten the green plains under

—Percy Bysshe Shelley,
"The Cloud"

Any of the types of ice that are born in the sky can cause trouble on the ground—if they ever reach it. Some ice evaporates while aloft, and some melts and falls as rain. Much of the rain in temperate regions, in fact, and some of the rain in the tropics, probably started out as ice in the atmosphere. When ice does touch the ground (splatting, bouncing, coasting to a feather finish), or forms on the ground after rain hits it, people have to deal with the consequences: locked moving parts, sealed openings, added weight on things, slippery highways, and aerial bombardment by little, hard, fast-moving ball-shaped missiles.

In August 1992, the barley on Clair Manshreck's 800-acre farm in Deloraine, Manitoba, was waist-high and almost ready for harvest. Then a hailstorm passed through, and within 20 minutes, Manshreck says, "you couldn't tell there'd ever been a crop." His neighbor, George Weidenhamer, had 120 acres planted in sunflowers. After the hailstorm, Weidenhamer says, there were "no leaves left, no flowers, just sticks about four inches high sticking up all over."

The damage hail does seems almost biblical in its selectivity. According to a U.S. National Science Foundation report, "hailstreaks can be disasters that wipe out crops at ten farms leaving a thousand farms lying unharmed between them." A hailstreak is an area hit by a single volume of hail produced by a storm, the average one being only about two-thirds of a mile wide and six miles long; a storm can have one or many of them.

"Hail damages most crops grown in this country," the NSF report continues, with fruits the most easily damaged. A mere 90 seconds was all it took for hail to wipe out Ed Epp's entire peach crop one summer at Niagara-on-the-Lake, Ontario, by tearing holes through the leaves and bruising every peach. Epp told a reporter, "That crop was worth $2 million. Now, it's worth nothing." The second most easily damaged is tobacco, but in terms of amounts ruined, the big losers are wheat, cotton, and corn. Unfortunately for farmers, the same warm, moist conditions that are great for growing these crops are also great for growing hailstones.

Also at risk from falling hailstones are barn roofs, house siding, tree foliage, and automobiles. In August 1996, in Landmark, Manitoba, car dealer Sid Reimer had on display in his outdoor lot 60 new and used cars and trucks; within 15 minutes hailstones had put dents in every single one of them. "The dents were a half inch deep and one to two inches in diameter," he reported. "On the hood you might have 20 dents."

> *Slowly, slowly, nobody wants a mess.*
> *I float over the black roads, pure ice.*
>
> —Margaret Atwood,
> *"Small Poems for the Winter Solstice"*

One day C. Allen Wortley of the University of Wisconsin was driving along a highway that ran through Indiana and Ohio when he saw within the space of about 100 miles at least 100 semi-trailers—"that's right, I counted them"—lying in ditches or on the median, upside down or on their sides. Although the highway surface was dry, Wortley knew that rain had fallen earlier, and he figured the rain must have frozen to the road so fast that the drivers might have started slowing down but hadn't pulled over. The wind picked up and "caught all the lightly loaded or empty semi's and sent them sliding out of control. True icing conditions," Wortley concludes, "make highway travel impossible. Get off and sit it out."*

Incidentally, the reason that BRIDGE FREEZES BEFORE ROAD, as the signs say, is because cold air under the bridge adds to the chill of the driving surface while the road leading up to and away from the bridge has plenty of dirt under it, which probably still holds some of the warmth of the summer and autumn past.

> *Twenty miles an hour, over fallen heaven.*
>
> —Ted Hughes, *"Robbing Myself"*

*Each winter North Americans put 10 to 15 million tons of salt on their roads to keep ice and snow off them. That tonnage is 10 percent of all the salt used in the entire world every year for all purposes combined.

What falls on roadways also falls on railways. I asked CRREL engineer George Blaisdell what happens to trains running on tracks that have a thick layer of ice on top: could the trains derail or slide? He reminded me of what happens to pennies that kids put on tops of rails before trains come along. The pennies become thin as gum wrappers. Imagine what a train would do to ice! The contact area between a rigid wheel and a rigid rail is only about two square inches, and with a locomotive weighing 200,000 pounds and having 12 wheels, that would mean more than four tons per square inch bearing down on the ice. "The train cleans ice right off the tracks," Blaisdell concludes. "Ice can't exist there."

Ice can exist, however, on rail switches and couplers. "As little as a few hundredths of an inch can keep a switch from closing and derail a train," CRREL climatologist Charles Ryerson points out. To deice switches, a Metro-North Railroad spokesman says, "we try everything, no holds barred." Everything includes leaf blowers, roofing torches, car-wash wands, and hair dryers. In winter, Metro-North keeps its own supply of hair dryers.

He casteth forth his ice like
morsels

—Psalm 147:16

What falls on roadways and railways falls on runways and taxiways, and on the airplanes that use them. During a light snowfall in March 1994, a USAir Fokker F28 twin-engine jet rolled down a runway at New York's LaGuardia Airport, swerved, hit an embankment, and plunged into a bay, killing 27 people and injuring 24. Its wings had been deiced twice while it was waiting at the gate, the second time a half hour before the plane pulled onto the runway, but investigators concluded that it was mainly ice on the wings that caused the crash. The wait had been too long and the deicer fluid too weak to keep ice from forming and disrupting air flow over the wings, reducing lift and keeping the F28 from reaching flying speed.

Sometimes airplanes pick up ice in the air rather than on the ground. In his book *Fate Is the Hunter,* Ernie Gann tells the story of a night flight he made as copilot in a DC-2 carrying eight passengers from Nashville to New York City. Shortly after takeoff, the DC-2 enters a cloud. Outside air temperature is 30 degrees, ideal for icing. Ice begins coating the windshield. "Forward, we are as blind as if the oblique sections of glass were marble slabs," Gann wrote. The wings take on ice; boots on their leading edges inflate and knock it off. "So this is true ice," Gann thinks. "It looks more like pie crust."

Next, the propellers ice up. "Chunks the size of baseballs are being hurled against the resounding aluminum." Then the rudder pedals won't move; the rudder at the rear of the plane is frozen. The other pilot has "a very rough time with the controls. Now the sweat is dripping from his cheekbones and he is breathing heavily. . . . He is afraid." So is Gann. Within minutes, airspeed drops from 170 mph to 120 mph. If it drops to 100 mph, the plane will no longer stay in the air. Already it is "foundering like a ship." The wing's leading edge is "one long, unbroken bar of ice"; ice has built up too fast for the boots to knock it off.

The plane bucks "viciously," on the edge of a stall, and falls another 2,000 feet. "The Blue Ridge Mountains are buried in the night below. We are already below the level of the highest peaks." Ice collects on the air-intake openings. "Without air the engine dies as surely as a drowning human being." The engines are "grizzle-bearded in ice." "This is the way you die."

They do not die but barely make it into Cincinnati on empty tanks. Once on the ground, "it took the mechanics two hours of hard labor to knock the ice from our wings, engine cowlings, and empennage. In most places it had reached a thickness of four inches." In his log entry, next to the notation for duration of flight, Gann wrote a single word: "Ice."

There are two kinds of ice that form on airplanes in flight, rime and glaze. Rime forms when the temperature of the cloud the plane is passing through is cold, 14°F or lower, and the surfaces of the plane are therefore cold as well, so that when the plane smashes into supercooled cloud droplets at hundreds of miles an hour, the droplets freeze quickly to the cold surfaces. The fast-frozen rime conforms pretty well to the shape of the wings' leading edges. "The ice stretches the airfoil a little bit forward," John Hansman, director of the aeronautical systems laboratory at the Massachusetts Institute of Technology, admits, "which is bad, but there's still a pretty decent airfoil."

Glaze, however, *reshapes* the airfoil and is therefore more dangerous than rime. It forms at warmer temperatures, closer to freezing, and the droplets don't freeze immediately upon contact but spread over the surface. "Imagine stream lines coming at the airfoil," Hansman instructs. "One stream line goes over the wing, one goes under the wing, and one goes in between—and stops. That's the stagnation point, where it stops. Not much cooling goes on there since most cooling comes from air flowing past; therefore there's not much ice. Most of the ice growth is above and below the stagnation point, and you get these two horns sticking out of the airfoil, creating an immense amount of drag. Since the horns are made of ice and are cold, once they start to grow they become efficient collectors of ice themselves, droplet catchers."

As it happens, where there's ice in clouds, you don't get much icing. "Ice crystals will just bounce off the wings," Hansman explains. "They aren't a particular threat." Above 15,000 feet clouds tend to be "glaciated" so icing usually occurs below that. Ernie Gann's encounter with ice took place in the era when transport planes had propellers and flew at lower altitudes than most do now. Jets can usually climb quickly through freezing levels to levels where freezing has already taken place.

Yet the problem of in-flight icing hasn't gone away, even for commercial planes. On Halloween day in 1994, an American Eagle turboprop was holding at 9,400 feet over the Indiana countryside, awaiting clearance to land in Chicago, when its right wing dropped abruptly. It rolled 70 degrees, rolled upright, rolled upside down, dived, and crashed in a field of soybeans. All 68 people on board died. The Eagle's wings were particularly vulnerable to icing because they were thin and narrow. "Big, fat wings don't pick up as much ice as skinny ones," Hansman notes. "The smaller the object, the more efficiently it collects ice."

Researchers are working on strategies for combating ice aloft. Victor Petrenko at Dartmouth devised a system in which wings would be coated with a film containing a fine network of imbedded foil electrodes, and whenever the wings iced up, a current passing through the electrodes would split the water molecules into their constituent atoms of hydrogen and oxygen. "You'd only need to eliminate a few molecules to break the ice grip," he claims. The same system would work, he suggests, for car windshields, house roofs, road signs, and refrigerators.

this ice-ribbed ship
Bearded and blown, an albatross of frost

—Sylvia Plath, *"A Winter Ship"*

What falls on runways, etc., falls on waterways and on what's floating in them. Every year along the coasts of the North Atlantic, North Pacific, and Arctic oceans, seven to ten ships are lost because of icing. Some of the icing comes from water that falls out of the air naturally, including snow, fog, and rain, but most of it, 90 percent, comes from sea spray. Spray consists either of splashings from waves the ship plunges into or gets slapped by or of clouds of droplets the wind has whipped off wave crests ("spindrift") and carried toward the ship through air that chills them. The smallest vessels, bouncing around the most in rough seas and getting splashed the most, are at greatest risk for icing. The ice burden raises the vessel's center of gravity and inclines it to wallow and roll, and the extra weight causes the bow to dig even deeper into waves and churn up yet more spray. "It's a positive feedback," CRREL's Charles Ryerson says. The ship's surfaces become treacherously slick for people to walk on, and chunks of ice break off from overhead structures and land on their heads. "Even the lifeboats become encased in ice," Ryerson says. "You're sitting on an ice tomb."

> *And when they go down; they go down fast you know . . .*
> *Funny how fast the ice comes off; comes off far, far, and deep below.*
>
> —Ryan Blackmore, *"Night of Freezing Spray"*

Nobody has yet figured out how to stop ice from forming on ships, or how to get rid of it quickly when it does. The salt in saltwater can't be counted on to keep the ice from sticking the way that salt sprinkled on highways can; the amounts involved are too small. Researchers have looked hard for "icephobic" materials that can be painted on ship structures to keep ice from adhering to them or, once a thin layer of ice has formed, to get the ice to "self-shed," but without success. "There's no surface so slippery that ice won't stick to it," Hansman states flatly. "Unless the surface is perfectly smooth at the molecular level, it will still catch a few molecules of water, which will turn into ice and become the starting point for more growth."

People have also tried installing electrical heat systems and inflatable boots on ships, but such items are useful only in limited places, such as around masts. Better weather forecasting would probably help (the most congenial temperature for ship icing, according to Ryerson, is 18°F) and so would cleaner boat designs. But usually what it comes down to is whacking the ice with something: sledgehammers, axes, shovels, ice picks, baseball bats. Particularly baseball bats. Nowadays the chief tool against the deadly hazards of shipboard icing is the Louisville Slugger.

CHAPTER THIRTY-ONE

ATMOSPHERE II

Icicles filled the long window
with barbaric glass.

—Wallace Stevens,
"Thirteen Ways of Looking at a Blackbird"

THERE'S A CHILLING DRAWING in a *Nature* magazine of 1881 of the *Phoenix,* a Danish steamer caught in a winter gale while carrying cargo and mail from Copenhagen to Iceland, in addition to thick ice on masts and rigging and a huge mound of ice on the forecastle, the etching shows icicles hanging from all the yards. The longest icicles are hanging on the side the ship is listing toward. The icicles probably contributed to greater ship icing by capturing water runoff that might otherwise be blown away and by keeping it around long enough for it to freeze. (Eventually, to save his men, the captain ran the *Phoenix,* "buried as it was under an enormous mass of ice," aground on a sunken rock, and all hands took to boats, risking frostbite, starvation, and drowning. They saved nothing but their lives, according to *Nature,* plus "the English mail, and a bundle of blankets which . . . was found to be useless—frozen into a solid lump.")

Icicles, like rime, grow *into* the air, yet unlike rime they don't get their water *out* of the air. Snow melting off a rooftop, water seeping down a cliff face, sea spray running along a ship's yard will do. An icicle, by definition, forms when cold water drips off an overhang into air that is colder than 32°F and freezes. It sounds simple, but it isn't so simple.

Say it's a cold day yet the sun is shining. Sunlight melts snow on a rooftop, and drops of the melt trickle down to the edge of the roof, roll over it, and hang in the frigid air. The stage is set. First, the sides of the drop freeze, into a shell of ice only a few platelets—a few thousandths of an inch—thick. Then more drops of snowmelt roll off the roof and flow over the surface of the tiny ice shell—the incipient icicle—and part of that water freezes, thickening the shell, and part of that water joins the drop at the end, the sides of which freeze in turn, lengthening the shell. As this keeps on happening, there develops inside the shell a tube, or inverted cup, that holds water and is probably kept in place by air pressure. NCAR's Charles Knight suggests that if you take a long toothpick or unfolded paper clip and poke upward at the end of a small, growing icicle, chances are your pick or clip will go into the tube as far as a couple of inches. Like a soda straw, the tube stays the same width its entire length since its width is determined by the diameter of the drop. The tube doesn't usually extend all the way to the top of the icicle since water in the upper part starts freezing, slowly, from the outside in.

From time to time, the drop at the end falls off, exposing the shell briefly to the freezing air, to be replaced by another drop. No matter what the conditions, however, the end of an icicle can't get wider than the drop, which is why icicles are pointy at the bottom. Just as long as meltwater keeps coming off the roof and replacing the drop at the tip and cold air keeps turning the sides of the drop into ice, the icicle grows longer. And as long as water keeps flowing down the walls of the icicle on its way to the tip and covers the walls with a film of water that freezes in the cold air, the icicle grows wider. Although lengthening and widening go on simultaneously, they are two very different processes.

Lengthening happens much faster than widening, eight to 32 times faster, Norikazu Maeno and Tsuneya Takahashi from The Institute of Low Temperature Science at Japan's Hokkaido University determined, which accounts for those carroty shapes that icicles take. The lopsided growth rates are probably due to the fact that more water collects at the tip than along the sides, where the film is no more than a few thousandths of an inch thick.

Oddly, an icicle lengthens faster when there's not much water flowing down it than when there's a lot. "It is an interesting paradox," Edward Lozowski and K. K. Chung of the University of Alberta point out, "that the maximum growth rate of the pure icicle occurs just before the drip rate goes to zero, and then it suddenly stops growing." When the drip rate is fast, they deduced, drops fall more frequently from the tip, and the replacement drops are warmer at first than the previously exposed drops were and thus take longer to freeze.

You can tell by touching an icicle if it is still growing; it will be wet and have soft walls. The walls of an icicle typically form not as solid ice but as a network of ice "fingers" (on a schematic diagram, they look more like branching bushes) which extend into the outer film of water, the way they do in spongy hailstones.

A brilliant beard of ice
Hangs from the edge of the roof

—Robert Pinsky, *"Icicles"*

When snowmelt stops rolling off the roof, the partially frozen icicle stops growing and becomes completely frozen, on the outside as well as inside the tube and on the tip. (Interestingly, the walls of the icicle get rid of latent heat through a *liquid* layer; the drop at the tip gets rid of the heat directly to the *air;* and the water inside the tube gets rid of heat through the *ice,* by conduction upward all the way to the root of the icicle and out onto the roof edge. Icicle-making is a complicated business.) Once the icicle is completely frozen, it can still change, the walls smoothing out as the outermost layers sublimate away. Icicle walls usually develop "wrinkles," plus an occasional spike. A spike can be as long as 3/4 of an inch and probably forms when the outer layer of the film of water freezes (due to a shortage of down-flowing meltwater) before the inner layer does, then when the inner layer freezes, the force of its expansion into ice cracks the frozen outer layer, which sends water squeezing out through the crack, to freeze in midsquirt.

Scientists have figured out what causes the spikes, but they aren't sure about the wrinkles. Wrinkles are horizontal ribs or bulges that often go all the way around the body of an icicle. Usually they appear in series, with each wrinkle about 2/5 of an inch from the next with hollows in between, for a wavy silhouette. Wrinkles tend to be solid ice while hollows are spongy; Knight says he has "easily" stuck knife blades a third or half an inch deep into hollows. The wrinkles are full of bubbles—it's thought that they formed in places where the water froze fast—and the hollows tend to be clear—it's thought that they formed where water froze slowly. Although it's easy to appreciate that once a protrusion—a wrinkle—develops on the wall of an icicle it will grow bigger, since any flow of water over the bump will be thinner than the rest of the walls and thus more exposed to cold air, it is not easy to understand why the protrusions developed in the first place, and at such regular intervals.

Something else icicles often get is bubbles running up the center, tiny ones trapped when water inside the tube froze; these give a transparent icicle a long, whitish core, like a bone under X-ray. As for saltwater icicles, they are shorter and broader than "their freshwater cousins," Lozowski and Chung report. They're also more milky-looking, because of brine pockets, and have bigger wrinkles.

When icicles hang by the wall.

—William Shakespeare, *"Love's Labour's Lost"*

Icicles can show up almost anywhere on land, sea, or in the air. In March of 1999, when Bertrand Piccard and Brian Jones, the first people to fly around the world in a balloon, were cruising along high over the "red vastness of the Sahara," Piccard wrote, they discovered icicles ten feet long hanging from both sides of their capsule. Piccard climbed out the hatch and chopped the icicles off with a fire ax, "watching them tumble and turn as they fell toward the impossibly empty sands of Mali below."

Long, narrow, and pointed as they are, icicles look like stilettos and occasionally function like them. A friend reports that while he was visiting a steel mill in Pittsburgh a worker's head was lethally speared by a falling icicle. In winter, some rail companies run boxcars with metal bars on the roof through tunnels ahead of trains, just to scrape away the icicles on the tunnel ceilings that could smash the windshield of the engineers' cab.

Some icicles even grow upward. Several years ago readers of the British magazine *Weather* wrote a series of letters in which they reported finding strange objects, usually in the early morning, usually in their gardens. Colin Clark of Bruton, Somerset, told of discovering a vertical "ice tower" atop a bowl of frozen water left outside on a cold night in January and insisted that there was "no overhanging vegetation or wires from which the growth could have originated."

An explanation for such an "unusual ice formation" was offered by Andrew Thain of East Sheen, London. When water froze against the sides of the container in question, he suggested, the ice cover had cracked, and water under it "which was forced up through the hole quickly froze forming the tip of the 'icicle.' More water from beneath was pushed up and similarly froze." A photo that Clark enclosed with *his* letter to *Weather* lent weight to Thain's theory. In it, readers could see the "central duct through which the water would have to pass" on its way to the pinnacled tip, as well as smaller ducts radiating outward along the length of the tower, which may have contributed to its breadth. ("Anti-icicles," Lozowski calls the towers.)

"This phenomenon," Thain stated firmly, "occurs very rarely."

Night comes and the stiff dew.

—W. D. Snodgrass, *"Heart's Needle"*

Yet another kind of ice that forms in the atmosphere is *frost;* it's the ice version of dew. ("The frost makes a flower, / The dew makes a star"—Sylvia Plath, "Death & Co.") When water vapor condenses as liquid droplets on the ground, or on objects close to the ground—grass, trees, rocks, dog bones—it's dew. When vapor condenses on the ground or the objects as ice, it's frost.* Frost usually appears after cold, clear, still nights when the ground has given up large amounts of heat to the sky through radiational cooling, and when there's plenty of moisture in the air. "A fisherman says they were much finer in the morning," Thoreau wrote of a cluster of leaflike frost crystals he was admiring one day. Shortly before dawn is usually the time when the air is at its coldest and most saturated, therefore when frost is most likely to form. (You can distinguish frost from rime because, with its facets, frost glitters while rime, with its bumps, does not.)

Frost and snow have much in common; John Hallett calls frost "snow growing close to the ground." Both of them form directly from water vapor, and many of their crystals are similar. But snow crystals tend to be smaller, since they're always falling and don't have as much time to grow. They're also more symmetrical, since they grow in the more uniform environment of a cloud. On the ground, small localized winds come and go, puff and die, and follow irregular paths over uneven ground. Even on apparently calm nights, surface winds may move at a rate of several inches a second, which is fast enough to cause a change in the way a crystal grows, by changing the temperature and humidity of the air around it. According to Hallett, vapor deposited slowly in very cold temperatures produces flat, faceted crystals, like leaves or plates, while vapor deposited more rapidly at warmer temperatures produces "spikey-looking things," crystals with sprouts at the corners. As vapor keeps condensing on and around the spikey things, they get denser and bushier, with interlocking branches, producing by "mutual interfer-

*Sometimes the dew itself freezes ("[A]nd the dew / that stiffens quietly to quartz, / Upon her amber shoe"—Emily Dickinson). And sometimes frost condenses on top of the frozen dew!

ence . . . an intricate, porous matrix," even more intricate than snow crystals.

Frost often grows around highly localized sources of moisture: animal burrows ("small doormats, as it were, of fibrous white, / like summer cobwebs"—John Updike, "Frost"), road drains, and open streams. Austin Hogan, atmospheric scientist at CRREL, once found a "fur-lined privy seat." You can figure it out: high humidity below, cold air above, a rich growth of frost spikes in between. In Barrow, Alaska, I came across piles of plowed snow at the end of several streets, still unmelted in mid-May and grimy with gravel and soot, yet over their dark surfaces had grown dozens and dozens of fresh frost crystals, as white as goose feathers, as shiny as glass, some so delicate they trembled in the slightest breeze, setting off sparks as the light caught them, all of them brightening and prettifying the dingy snow piles; it was enough to make a person believe for a time in innocence and renewal.

In that case, the moisture source was snow. A smooth snow surface, Edward LaChapelle observed, is often covered with a "fuzzy blanket" of frost, which produces an "extremely fast skiing surface." As the skis glide across, the frost gives off "a characteristic rustling sound."

> The frozen breath of the river at a myriad breathing holes.
>
> —Henry David Thoreau, Journal (*Winter*)

Thoreau of course saw plenty of frost on his winter rambles and of course described it. Often it took the form of frost flowers on top of lake or river ice. "The new ice over the channel is . . . covered with handsome rosettes two or three inches in diameter," he wrote, "where the vapor which rose through froze and crystallized." Some of the frost features he spotted reminded him of bits of asbestos, others of spilled oats ("some boreal grain"), and still others of "very large leaves. . . . They are, on a close examination," he reported, "surprisingly perfect leaves, like ferns."

In protected places of prolonged cold, such as glacial caves and crevasses, frost crystals can grow to be extremely large and complex and months or even years old. Arthur DeVries, the biologist who discovered antifreeze proteins in Antarctic fish, and Charles Knight, the atmospheric scientist who chased hail, visited several glacier caves in Antarctica one austral summer in order to study the crystals in the places where they grew; they considered the crystals scientifically intriguing as well as aesthetically dazzling. Until then, there had been few detailed descriptions of frost crystals. "Frost is so common that this . . . was surprising," they pointed out in a journal article, "though perhaps its very commonness is the explanation."

One of the glacial caves they examined was inside the Erebus glacier tongue (like the larger ice shelves, tongues float upon the sea). The lower part of this cave, with its floor of sea ice in close proximity to water, was warmer than the upper part, which had 15 or 20 feet of cold glacial ice above it. Therefore, moisture that evaporated off the floor and the walls in the lower part would condense onto the ceiling and walls in the upper part, as frost crystals. The wide temperature spread and the rich moisture source helped some frost crystals to grow . . . and *grow.* One was 20 inches across. There was a variety of spirals and scrolls, projecting spikes, hanging columns, capped columns, and hexagonal cups. Many of "the best" crystals, in the two scientists' opinion, grew from icicles.

During another field season, I got a chance to sketch some cave crystals that DeVries had picked off the walls of ice caves (very very carefully) and placed in a freezer at McMurdo. My favorite was a partial cup with a hexagonal scroll at either end; it reminded me of curlicues on a Mayan frieze. One scroll was a bit smaller than the other. "Sometimes only one spiral develops well," DeVries and Knight had explained, "because the other gets 'shadowed' by other growth and loses its vapor supply." Barclay Kamb and Hermann Engelhardt, too, preserved frost crystals that mountain guides had plucked from crevasses near their camp on WAIS. A particularly spectacular one was composed of more than 100 small, stepped plates, stacked high in the middle, lower toward the edges, the whole as glittery and exquisitely crafted as a diamond brooch.

Of ferns and blossoms and fine spray of pines,
Oak-leaf and acorn and fantastic vines.

—T. B. Aldrich, *"Frost-Work"*

Even more visually evocative than frost growing in snow caves and rimming holes in riverbanks is frost on windowpanes. When outdoors the temperature is well below freezing and indoors there's plenty of moisture, patterns form on the inner sides of glass windows that "rival the most delicate plumes and outdo the finest lace of hand or loom," W. J. Humphreys declared. Thoreau referred to "frost sheaves, surpassing any cut or ground glass." John Rowlands, author of *Cache Lake Country,* reported seeing designs that resembled, among other things, "the spreading tail of a peacock," "a spider web with what looked like a spider near the center," and a design that put him in mind of "the geometry problems I had in school many years ago."

Frost can form inside a house on objects other than windowpanes if the cold is great enough and the woodpile small enough. On February 7, 1855, Thoreau wrote in his journal: "The coldest night for a long, long time. Sheets froze stiff about the face. . . . People dreaded to go to bed. . . . The latches are white with frost, and every nail-head in entries, etc., has a white head." It wasn't just frost: "My pail of water was frozen in the morning so that I could not break it. . . . Bread, meat, milk, cheese, etc., all frozen." In January a year later, Thoreau had this to say: "You lie with your feet or legs curled up, waiting for the morning, the sheets shining with frost about your mouth. Water left by the stove is frozen thick, and what you sprinkle in bathing falls on the floor, ice. The house plants are all frozen, and soon droop and turn black." The winter after that: "Ink froze; had to break the ice in my pail with a hammer."

CHAPTER THIRTY-TWO

ATMOSPHERE III

Suddenly I saw the cold and rook-delighting heaven
That seemed as though ice burned and was but the more ice

—William Butler Yeats, *"The Cold Heaven"*

ICE IN THE ATMOSPHERE can affect humans without ever touching Earth. (Yeats's poem *"The Cold Heaven"* was, he told Maud Gonne when she asked, an effort to convey the feelings that the "cold and detachedly beautiful winter sky" provoked in him, feelings of loneliness, awareness of past mistakes, and memories "of love crossed long ago.") For example, there's lightning. Ice is thought to make lightning possible, by separating electrical charges in storm clouds. The rough scenario goes:

When large particles of ice—soft hail, graupel—fall in storm clouds, they collide with lighter ice crystals that are being blown upward by vertical winds. The large hail particles strip electrons off the surface of the smaller ice crystals, and the top of the cloud where the rising ice crystals go becomes mostly positively charged and the middle and bottom of the cloud into which the hail particles fall become mostly negatively charged. At some point, the charge separation is breached; an avalanche of electrons zips toward a positive area; lightning strikes; thunder rolls. The particulars are extremely complex and not understood—"two hundred years after Franklin and we still don't know the physics of cloud electrification," Martha B. Baker, geophysicist at the University of Washington, says—but it's widely believed that ice generates the difference in electric potential necessary for the creation of lightning. One meteorologist says he's never seen a thunderstorm-producing cloud that didn't have rapidly forming ice in it.

Ice crystals also play a small but important role in the eco-drama of ozone holes. In 1985, British scientists in Antarctica reported that the earth's ozone layer, which protects organisms from the sun's harmful ultraviolet rays, had thinned over their research station by as much as 40 percent. It thinned during the months of September and October, Antarctic springtime, and recovered in November. In the years that followed, the hole grew steadily larger, eventually as large as the ground area of the United States, and took longer to re-form. In other parts of the world, including the Arctic, the ozone layer has been found to be thinning, too, although the effect is still greatest over the Antarctic. Chlorofluorocarbons from human use of such things as refrigerants and aerosol sprays are the main culprit. As they drift into the upper atmosphere and are broken down by sunlight, they release atoms of chlorine, which attack and destroy molecules of ozone; a single chlorine atom can destroy thousands of ozone molecules.

Where ice crystals come in is that they provide surfaces where chemical reactions can take place that release the ozone-destroying chlorine atoms from the harmless compounds they were incorporated into after wreaking their initial damage. These ice-crystal surfaces are highly active even at very cold stratospheric temperatures, as Steven George at the University of Colorado has pointed out. "The ice surface does not just sit there," he explains. "You've got water molecules raining down and water molecules leaving at incredibly high rates." Chemicals in the atmosphere are very rapidly adsorbed onto this active surface and very rapidly buried in the ice. "Molecules on the surface easily exchange with crystals in the underlying bulk."

Another way ice crystals in the air affect people on the ground is by their influence on climate. Two-thirds of the Earth is covered by clouds at any one time; and a quarter of the clouds are cirrus, and cirrus are made up of ice crystals. (They are those wispy, streaky, gauzy clouds that are 20,000 to 35,000 feet above the Earth's surface and *look* it.) In addition, many other clouds are "mixed": they contain ice crystals as well as water droplets. Clouds both cool the Earth, by reflecting incoming radiation from the sun, and warm the Earth, by absorbing radiation from Earth and re-emitting it. Which of these effects dominates in the overall climate picture is hard to determine—"it's not at all obvious which way it works," says Andrew Heymsfield, research scientist at the National Center for Atmospheric Research, who has been studying cirrus for over 30 years—in part because ice crystals are so variable and sensitive to their environments. All cloud droplets arise in pretty much the same way and have the same (globular) shape, but, as Baker points out, "ice particles form and grow by a number of different mechanisms, have a wide range of shapes, and play a range of climatic roles."

Cirrus crystals tend to be smaller than snow crystals, which grow at lower, warmer altitudes, but they can nevertheless have complicated shapes including—Heymsfield names a few—bullet rosettes, columns branching from a point, and columns that are hollow at both ends "like hourglasses" only not pinched in the middle. The cirrus formation known as "mares' tails" gets its look of wavy hair strands because some of the ice crystals being blown sideways by strong winds in the main cloud fall to a lower level where winds are weaker, and the threads of cloud trail gracefully downward. Heymsfield assumes the crystals in the mares' tails are rosettes. "Rosettes start growing first," he says, "and they grow the fastest and largest. Any time I fly through a mares' tail, that's what I see, rosettes."

The highest of all earthly clouds aren't cirrus, though, but noctilucent, or "night-glowing." They are so high, five times higher than any other clouds, 50 miles above the Earth, that sunlight keeps reflecting off them after everything in the sky below is dark. The temperature at that altitude is around −190°F, and not surprisingly the clouds are made up entirely of ice crystals, minuscule ones about the size of the particles in cigarette smoke. Noctilucent clouds were probably first seen in the summer of 1885 in northern Europe, and probably seen then only because water vapor injected into the air by the eruption of the Krakatoa volcano two years earlier had brightened them over the visibility threshold. Ordinarily, very little water vapor makes it to that altitude, and the clouds had probably been so insubstantial before the eruption that they blended in with the twilight background.

After those first sightings, the clouds appeared for only a few years, "then got dimmer and dimmer and dimmer," Gary E. Thomas, professor in the department of astrophysical, planetary and atmospheric sciences at the University of Colorado at Boulder, says. "By the turn of the century they were almost gone." After that, they were reported sporadically, then in the 1960s they seemed to become more numerous. Now they are seen maybe 20 to 40 times a year, but only during the summer months and only in northern countries—Scandinavia, Canada, Russia, Scotland—generally between latitudes 50°N and 65°N. They resemble cirrus clouds in that they are thin, with a filigree structure and sometimes a corrugated pattern, but their color is silvery white with a bluish cast, the blue produced by the steep angle of the sunlight. "I think they are the most beautiful clouds," Thomas says.

How can a person get to see these rare and beautiful clouds? Thomas was asked. Go to about 54° North, he answered, which is the ideal latitude for viewing them (or 54° South, in which case you'd most likely be on a ship at sea), during the months of July or August and look northward around midnight. At that time and in those places the sun will be not very far below the horizon and the ordinary blue of the sky will be gone. Above a thin red band of twilight running along the rim of the earth you will see a deep band of black sky and above that, if you're lucky, the shining clouds. "They are very dramatic when they are visible," Thomas says. "Sometimes you can read a book by them."

Or you could take a seat on the left side of an airplane that's going from North America to Europe and in the middle of the night peer out your window to the north. On Thomas's last trip across the Atlantic he did that and saw the eerie bluish glow and wanted to tell the pilots about it but wasn't allowed to. In his experience, pilots don't show much interest in noctilucent clouds anyway, perhaps mistaking them for common cirrus, despite the fact that the clouds are, as he points out, "*above* the airplane!"

On its way to Europe, the airplane from which you are scanning the sky for noctilucent clouds will probably be making its own ice clouds. At the altitudes where jets fly, it doesn't take much vapor to saturate the very cold air, and when the engines spew hot, humid exhaust, about a quarter of which is water vapor, into that air, it becomes fully saturated. Soot and other particles in the exhaust, acting as ice nuclei, quickly convert the vapor into ice crystals, which trail behind the plane, in a narrow whitish line: a contrail (the "con" is for condensation). If the ice crystals evaporate quickly—that is, if the contrail is short—it means the air at that level is dry, but if the line is long and stays around for a while and spreads out, it suggests the air up there is humid and a front is moving in.

Test pilot Jackie Cochran deliberately drew a contrail on the sky just before she became the first woman to fly at Mach 1, in 1953. Before putting her Sabrejet F-86 into a vertical, full-power dive, she flew over a ranch house near the Mexican border where she knew her husband was waiting and watching. "I . . . leave a contrail of ice crystals as a signal to Floyd," she wrote in her autobiography. She was then at 45,000 feet, "the top of the world," and as she wrote in the sky with ice crystals, she was certain "Floyd sees me." Then she headed down and broke the sound barrier, twice.

Contrails often turn into cirrus clouds. As Andrew Heymsfield explains it, latent heat released by the newly forming ice crystals produces air motion which causes the crystals to fan out, and with the extra packet of moisture delivered by the airplane's exhaust, the tiny crystals grow. "It may not be moist enough up there to generate ice crystals in the first place," Heymsfield says, "but it's moist enough for them to grow if formed." Satellite images taken on September 12, 2001, when commercial airplanes were grounded in North America, showed dramatically how the contrails made by nine military airplanes crisscrossing several eastern states spread over a period of five hours into an enormous veil of cirrus, covering 24,000 square miles. NASA is studying these priceless images as part of its effort to learn what effect aircraft have on climate; on an ordinary day, there would have been—crisscrossing the same eastern states and making their own separate contrails—700 to 800 jets instead of nine.

Automobiles make ice clouds, too. In parts of the world where winter temperatures are usually very cold and there are cars around to pump out hot and humid exhaust—the Alaskan interior, Siberia, northern Canada—"ice fogs" are common. Every pound of gasoline burned produces a pound of water vapor, Carl Benson, professor emeritus of geology and geophysics at the University of Alaska at Fairbanks, reports. (Benson is known not only for being an expert on ice fogs but as an enthusiastic performer of the ice scientists' anthem, "Ice Is Nice," sung at academic Christmas parties to the tune of "Onward Christian Soldiers," with trumpet and trombone accompaniment and an eleven-word lyric: "Ice is nice and good for you, snow makes glaciers grow." Repeat. Repeat.) Crystals in ice fogs form extremely fast, Benson says; cooling rates can be hundreds of degrees a second, or a million times faster than in air without exhaust. An ice fog looks like a regular fog; its crystals are so tiny—diamond dust is ten times bigger, snow crystals 100 times bigger—that they float. The ice fog can be quite dense, though, with "visibility measured in arms' lengths."

Airplanes too can create ground-dwelling ice fogs—contrails without the trails. There are times, Guy Murchie wrote in his book *Song of the Sky,* when a "single airplane warming up at one end of an arctic runway is enough to transform clear air into dangerous ice fog over a whole airport in ten minutes." Even smoke from fireplaces or stoves can trigger ice fogs in cold air close to the ground. "Every time you burn anything in Alaska," CRREL atmospheric scientist Austin Hogan declares, "it begins to blow ice smoke rings."

And ice underfoot is mica

—William Meredith,
"Talking Back (to W. H. Auden)"

Away from the ground, the commonest ice nuclei in the atmosphere come from the ground: soil particles, mostly clay and mica. In his poem William Meredith was referring to ice's mica-like glitter, but mica crystals also have an icelike molecular structure, which water molecules can freeze around. At the South Pole, scientists have trapped snow crystals and detected at the center of each a speck of clay, thousands of miles from any sources of clay, in Africa or South America. Dust storms in Asia have been called "ice-nucleus storms." Bits of vegetation can serve as ice nuclei too; one of the best, most icelike, is poplar mulch.

Still, the atmosphere has a poor supply of ice nuclei per unit area compared to what's in a lake or even in an ice-cube tray. In one maritime cloud, scientists found the concentration of ice crystals to be 100,000 times greater than the concentration of potential ice nucleators, which puzzled them. Eventually they concluded that ice nuclei were being generated by ice itself, splinters or fragments produced when, for instance, ice crystals collide with one another and break apart.

After World War II, some people started adding ice-nucleating particles to clouds intentionally. The thinking was that the particles would change a large number of droplets in those clouds to ice crystals, which would grow into graupel particles, melt on their way to the ground (it was summertime), and drench the crops that needed it. In 1947, according to physicist Peter Hobbs, Bernard Vonnegut of General Electric "searched through X-ray crystallographic data for materials which had cell dimensions and crystal symmetries as close as possible to that of ice" and came up with silver iodide. It had only a "small lattice misfit." One day Vonnegut put crystals of silver iodide inside pieces of burning charcoal, then dropped the charcoal from an airplane into a cloud whose temperature was 14°F. The cloud was a large one, almost a thousand feet thick and covering 6 1/2 square miles. The result was impressive to Hobbs. "The cloud was converted into ice crystals," he declared, "by less than 30 g [about an ounce] of silver iodide!"

The reason for changing droplets into crystals is that water droplets are often too small and lightweight to fall out as rain. Ice crystals start smaller but grow faster. In fact, they grow at the expense of the droplets. As they enlarge, they sink, and on the way downward collide with water droplets, which freeze to them and make them still larger. They keep falling and converting water droplets and vapor to ice until they reach a layer of air that's warm enough to melt them and convert them into raindrops.

Nobody knows for sure whether this "cloud seeding" helps or not.

Who's to say how much rain would have fallen anyway? "There's great natural variability," Roelof Bruintjes, a scientist at the National Center for Atmospheric Research in Boulder, points out. "No cloud is the same as another cloud, and no cloud ever will be the same as another cloud." Yet rain is such an emotional issue that people will try anything, he notes, even if they think it may work only half the time. "Or even 10 percent of the time."

and the hoary frost of heaven, who hath gendered it?

—Job 39:29

Another way that ice crystals in the atmosphere can affect human beings on the ground is by creating displays of light they can marvel at: circles, spots, streaks, arcs, and crosses, all formed by sunlight shining through or off ice crystals floating in the air. The ice crystals involved are mostly simple ones, Robert Greenler, emeritus professor of physics at the University of Wisconsin–Milwaukee, noted in his book *Rainbows, Halos, and Glories,* solid hexagonal plates and columns (or "pencils," as he refers to the columns, each resembling "a common wooden pencil before sharpening"). The most common sky pattern that ice crystals produce when light hits them, and the one most commonly seen, is the 22-degree halo.

"After a few days of fine bright spring weather, the barometer falls and a south wind begins to blow," Dutch professor M. Minnaert wrote in his 1954 book, *The Nature of Light and Colour in the Open Air,* describing the conditions in which the 22-degree halo is likely to occur. "High clouds, fragile and feathery, rise out of the west, the sky gradually becomes milky white, made opalescent by veils of cirrostratus. The sun seems to shine through ground glass, its outline no longer sharp but merging into its surroundings. There is a peculiar, uncertain light over the landscape; I 'feel' that there must be a halo round the sun!

"And as a rule I am right."

When Minnaert was right, what he saw was a large, bright ring around the sun with its inner edge 22 degrees away from the sun, as he would have perceived it from the ground. If you see such a ring, you can verify the 22-degree angle by extending one of your arms in front of you, spreading the fingers of that hand wide, placing the top of your thumb over the sun, and noting that the tip of your little finger touches the halo's inner edge. The moon, too, can have a 22-degree halo around it, although its halo will be paler than the sun's.* Minnaert was aware of only one person who was lucky enough to see at the very same moment a halo around the setting sun and a halo around the rising full moon.

The inside edge of a 22-degree halo is sharper than the outside edge, and reddish, with a band of yellow next to it, then maybe green, then white. The fact that the light is broken up into colors is evidence that it passed through prisms of airborne ice crystals and didn't just reflect off them, in which case the halo would be all white. The types of crystals that produce a 22-degree halo are pencils and plates, Greenler deduced, small ones tumbling randomly through the air. "In every direction we look," Minnaert wrote, "innumerable minute prisms of this kind float about in every possible orientation." Since they are prisms, they redirect the light that hits them. The sun's rays as they enter a hexagonal crystal through one face and pass out through another face at a 60° angle to it are bent, or refracted, just as rays are when they pass into and out of water in your swimming pool, giving the impression that the leg you've dangled over the edge is broken and angled toward you.

For a halo to be complete, the clouds (usually cirrus or cirrostratus) should contain crystals falling in "every possible orientation" so that some with the necessary orientation will be in position to redirect light toward what Greenler calls "the waiting eye." If the crystals aren't falling randomly, the eye sees only part of a halo.

All this is inferred; nobody actually sees individual crystals and light rays in action. In recent years, Greenler and his colleagues have used computer programs to figure where in the sky rays should appear given certain shapes and orientations of crystals and elevations of the sun. Once the computer has come up with likely patterns in the sky, the scientists can test them against what actually appears.

Along with a portion of a 22-degree halo, a person might see "sun dogs" (or "mock suns," "false suns," "parhelia"). Sun dogs are bright spots that appear on the ring of the halo at the same altitude as the sun, on either side of the sun, sometimes looking as large as the sun. They may display even more colors than the halo. You might see sun dogs without the halo, or see only one sun dog, or see sun dogs with white tails. The ones with tails are thought to form when sunlight refracts through plate crystals falling not randomly but with their flat bases

*A halo is not the same thing as a corona, which is a much smaller disk of light around the moon or sun, only a few moon diameters across, the precise number of diameters variable because the corona is formed by light diffracted by water droplets or ice crystals in an intervening cloud, and the size of the droplets varies.

nearly horizontal. "If this seems strange," Greenler wrote, "think of the way a leaf or a spread-eagled sky diver drifts to earth." When the sun is low and the sky full of plate crystals falling in this manner, only the crystals at the sides of the halo, at the height of the sun, are oriented so they direct light to the waiting eye.

Upper parts of halos can take on intensified glows, too, arcs that rise from the halos but curve away from them. These are called "tangent" arcs since they don't lose touch with the halos. Depending on how high the sun is, they take on different shapes: V's, flattened V's with drooping ends, arcs curved upward in a way that makes them look like small bowls sitting on large hoops. Tangent arcs are produced by sunlight refracting off *pencil* crystals falling in still air with their *long axes horizontal*. "If that surprises you," Greenler wrote, "throw a blade of grass in the air and see how it falls." Both ice pencils and grass blades orient themselves to maximize, rather than minimize, air resistance as they fall.

Arcs can also appear at the bottom of halos, in which case they may be below the horizon and you won't see them unless you're in an airplane looking down through an "ice cloud," Greenler points out. Sometimes both lower and upper tangent arcs curve so far around that they meet and form a halo around the halo but one still tied top and bottom to it, a "circumscribing" halo.

Such visual treats are fairly common, yet "many people live their entire lives without noticing them," Greenler says with obvious disappointment. He estimates that 70 to 80 days a year he sees at least one sky effect—and he works indoors. "If you look," he told a city dweller who also works indoors (me), "I guarantee that you will see a sun dog within two weeks." Four weeks later I saw from a cafe overlooking a river an awesomely large 22-degree halo, but—Greenler was right—nobody else was looking at it. Minnaert claimed that a practiced observer in the part of the world where he lived, northern Europe, could see on average one halo every four days, if he's "on the watch all day long," and "the most observant see halos on 200 days a year!"

Ice crystals sometimes *reflect* rather than refract the sun's rays, as if they were tiny mirrors. Reflection is responsible for sun pillars, which are vertical columns of light—"feathers," Minnaert called them—that stick up above or below, or both above and below, the sun when the sun is close to the horizon in the process of rising or setting. The pillars usually form when sunlight bounces off plate crystals whose flat bases are aligned nearly horizontally. To direct light to the waiting eye, the crystals must be tipped slightly. The visual effect is similar to the "glitter paths" that light rays from a setting sun make on the rippled surface of a lake as they, too, bounce off a series of near-horizontal surfaces.

If the sun is fairly high and the sun pillar is below it, the pillar will be so short it's more like a stretched-out spot, a "subsun." You may catch sight of a subsun while you're on a mountain or in an airplane that's passing over clouds containing ice. "The subsun spot moves along with the airplane through its climbs, dives, or turns," Greenler noted, "but disappears when the ice clouds vanish." Some people think what they're seeing is a UFO.

There are even "streetlight pillars." On cold nights when water vapor freezes near the ground, artificial lights can reflect off the flat plate crystals just as sunlight would. Because a street lamp is much closer to the viewer than the sun is, its pillars can extend much higher—in principle to any height, since the plates needn't tilt to be seen.

Reflection is also responsible for the parhelic circle, a line of white light that runs all the way around the horizon at the level of the sun; occasionally it meets a sun pillar to form a cross. "To remark that the superstitious have made the most of this is superfluous!" Minnaert wrote and told this story. "On July 14, 1865, the alpinist [Edward] Whymper and his companions were the first to reach the top of the Matterhorn, but on the way back four of the men slipped and fell headlong down a precipice. [They died.] Towards the evening Whymper saw an awe-inspiring circle of light with three crosses in the sky: 'the ghostly apparitions of light hung motionless, it was a strange and awesome sight, unique to me [Whymper] and indescribably imposing at such a moment.' "

As common as 22-degree halos are, scientists have wondered why they aren't more common. Why shouldn't cirrus clouds, which are themselves common ("ubiquitous," Baker says), produce more 22-degree halos than they do? Recently, Heymsfield and other cloud scientists flew through some cirrus-cloud systems in the Midwest that were known to produce halos and, using oil-coated plates, collected samples of the ice crystals there. On examining their catch, they were surprised to find only a small number of the simple solid columns thought to be ideal for creating 22-degree halos but a great many complex, intricate crystals, including bullet rosettes and hollow-ended columns. Heymsfield concluded that the hollow-ended crystals could generate 22-degree halos but only with blurred inner edges, and that rosettes could produce only "unexceptional" ones. Neither type would create a 46-degree halo, which is even bigger (and fainter) than a 22-degree one, two handspans from the sun rather than one.

Still other, rarer sky phenomena are produced when light and ice interact in the atmosphere: mock suns of the mock suns. Halos of 8 degrees, 18 degrees, 32 degrees, and 90 degrees. Anthelions, which are hazy concentrations of light in the region of the sky away from the sun, two of which when crossed look like a big overhead X. Subhelic arcs, a pair of arcs meeting at the subsun, produced perhaps by light rays reflecting off two inside faces of a pencil crystal before exiting, as in a kaleidoscope. When you consider all the paths that sunlight can take after hitting ice crystals of different shapes and orientations, according to how high the sun is at the time and where on the crystal surface the light enters (side face or end face) and at what angle it enters or whether it enters at all or is just reflected or whether after it enters it is bent *and* reflected off internal faces . . . Well, the sky's the limit.

Once in a long while, light patterns combine with spectacular effect. In April of 1857 Captain Leopold M'Clintock and the crew of the ship *Fox* were in Baffin Bay looking for the missing Franklin when they witnessed a complex sky show around the moon. The effects probably heightened the emotions the men were feeling as they conducted a funeral for one of their own who had fallen down a hatchway. As the man's body was "committed to the deep" through a hole in the ice and a ship's bell tolled mournfully under a murky, overcast sky, there appeared a sight "seldom seen even here," M'Clintock wrote, "a complete halo encircling the moon, through which passed a horizontal band of pale light that encompassed the heavens. Above the moon appeared the segments of two other halos, and there were also six mock moons. The misty atmosphere lent a very ghastly hue to this singular display, . . . which lasted for rather more than an hour."

M'Clintock wrote about what he saw, but other people in the past have also drawn what they saw, some of the displays so elaborate they became famous. In 1790, Tobias Lowitz of Saint Petersburg, Russia, observed and sketched over the course of five hours not only a 22-degree halo and a 46-degree halo, a circumscribed halo, infralateral arcs, anthelic arcs, a parhelic circle, and two parhelia but a pair of arcs that had never been described before and were afterward named for him, Lowitz's arcs, a pair of which were below the parhelic circle. In 1819–20, while searching for a Northwest Passage, British explorer

W. E. Parry saw and sketched many of the same features plus, above the sun, a novel arc of his own, now called Parry's arc.

Indulging in some "sky archeology" with the computer, Greenler and his colleagues analyzed some famous displays "piece by piece, to get insight into the whole." The Lowitz sky show could be explained by sunlight passing through plate crystals that were spinning around their long axes as they fell. The Parry arcs might be accounted for by sunlight passing through the hexagonal arms of a snowflake-type crystal. Over England on April 14, 1974, there appeared what became known as the "Easter Sunday" display, during which skygazers reported seeing as many as six concentric halos, which may have involved rare pyramid ice crystals.

The suggestion has even been made that Ezekiel's "wheel in the middle of a wheel," "rings so high that they were dreadful," "full of eyes round about them four," was an early description of a complex sky show, the interaction of sunlight and ice high in the heavens.

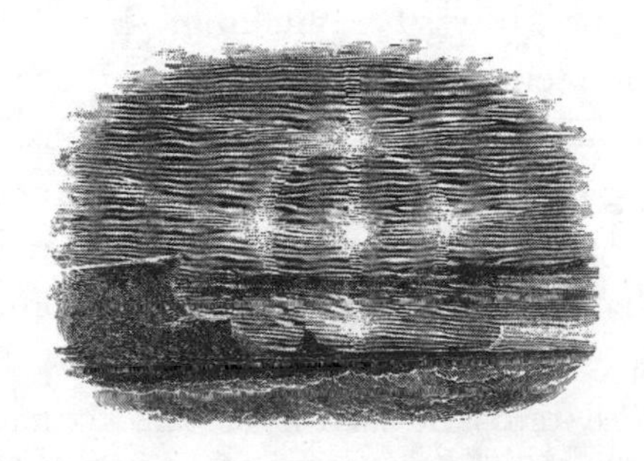

CHAPTER THIRTY-THREE

SPACE I

We'll kiss, grow old, walk around.
Light months will fly over us
Like snowy stars.

—Anna Akhmatova,
"The road is black . . ."

IN 1971 WHEN ASTRONAUT Alan B. Shepard stood on the surface of the moon looking back at the world he had blasted off from 33 1/2 hours earlier, he wept. "It was breathtaking," he recalled. "Looking up in the black sky and seeing the brilliant planet . . . the ice caps over the poles, the white clouds, the blue water . . . gorgeous!" He wept not just because Earth was gorgeous, with its display of water in two phases, but because it looked defenseless, floating alone in the immense dark vacuum of space. What had seemed so limitless at home was, seen "from here, from the moon, . . . in fact, very finite, very fragile." The only body in our solar system known for certain to have liquid water, a requirement for life as we know it, Earth could "be blown away so easily!" Shepard fretted. "A meteor, a cataclysmic volcano, man's own uncaring outpourings of poison."

Although Earth is the only body in the solar system known for certain to have *liquid* water, it is definitely not the only one to have water. It's just that the water found elsewhere up to now is in the form of ice or vapor. All of the giant planets—Jupiter, Saturn, Neptune, and Uranus—have water ice in them as well as in or on almost all of their major moons. Saturn's rings consist almost entirely of water ice. Mars has water ice at its north and south poles and almost certainly in its soil in enormous amounts. Comets are made up in large part of ice. Even Mercury, the second-hottest of all the planets, may have water ice sequestered on it, as may the dry, dust-covered moon where Shepard shed his own watery tears.

Water ice is, in fact, the most abundant solid compound in the solar system. Jonathan I. Lunine, professor of planetary sciences at the University of Arizona, calls it "ubiquitous . . . an important, permanent part of most of the other objects in the solar system." (By "other" he means other than Earth, where ice is "ephemeral": it hasn't always been here.) Water ice is abundant because, for one thing, two of the three most common elements in space are hydrogen and oxygen (the third is helium), which combine easily to form water molecules. For another thing, once the water molecules have formed, they are "very stable, very durable, even in the rigors of space," Dale P. Cruikshank, research scientist at the NASA Ames Research Center, points out, "partly because space is so cold they stay there forever."

What's important for the way our solar system turned out is that temperatures in the primordial disk of gas and dust out of which the planets evolved cooled enough at a certain point that water vapor beyond the region where Mars and the asteroid belt are now began condensing into ice. That circumstance was probably what allowed the outer planets to grow into giants, and it probably happened this way:

In that cooling primordial disk, water molecules hit a rocky grain and froze to it, the way water molecules freeze to soot particles or leaf bits in clouds on Earth, then the ice-coated rocky grain collided with other icy grains and stuck to them in clumps, and those icy clumps hit others and either broke apart or stuck to them. "We don't even know the sticking process," Tobias Owen, professor of astronomy at the University of Hawaii, admits, "but somehow the grains built into iceballs. They really were iceballs." When the iceballs reached a size about ten times the present mass of Earth, they had enough gravity to attract gases (hydrogen, helium) from the surrounding cloud, and with that added mass they were able to pull in more gases and grow still larger. Jupiter became the largest planet in the solar system because it formed at a distance from the emerging sun that was (1) far enough away that water vapor could freeze and (2) close enough that it could complete an orbit faster than planets farther out and gobble up more material in its feeding zone.

Meanwhile, in the inner solar system, temperatures were so high that water vapor tended to stay in the vapor state and eventually drifted away from the planets. As a result, Mercury, Venus, Earth, and Mars assembled themselves mostly out of rock and metal. Lacking access to a supply of ice grains, that ready and abundant source of solids available to the outer planets, they did not grow into giants.

It all comes back to the hydrogen bond. Because of its strength, water molecules form ice crystals at warmer temperatures than most other gases in space do, so that while, say, methane, carbon monoxide, and nitrogen freeze only at great distances from the sun, in extremely cold regions, water vapor freezes in more moderately cold regions, closer to the sun. Without the hydrogen bond, Lunine points out, "the solar system from Jupiter outward might well have been much emptier than it actually is, with only rocky grains available to form solid bodies. Even the giant planets might have been absent." In which case, our own fragile, gorgeous, blue-and-white world might have turned out to be the largest planet in the solar system.

and moon a frozen bird's eye

—Michael Ondaatje,
The Collected Works of Billy the Kid

All of the giant, outer planets have "icy moons." Jupiter has three major ones, Saturn seven, Uranus five, and Neptune one. (The planets also have many small icy moons—ice-rich "shards," as one astronomer calls them—but not much is known about them.) Water ice makes up anywhere from 30 to 70 percent of the mass of these moons, judging by their low densities. The ice is usually on the outside; being lighter than rock, it rose in relation to the rock and concentrated in the upper layers. "Ice therefore replaces the more familiar silicate rocks as the bedrock-forming substance on these satellites," Paul M. Schenk, planetary geologist at the Lunar and Planetary Institute in Houston, points out. Being iced over doesn't mean being inert, however. "The outer solar system may be locked in a permanent ice age," Schenk declares, "but the bodies populating this region of space are surprisingly complex and active."

And diverse. The ways that ice presents itself on these moons are remarkably various. For instance:

• ***Titan,*** Saturn's largest moon, has a mantle of water ice so thick that the mass of the ice alone nearly equals the mass of our entire moon. Since the upper ice layers would weigh heavily on deeper ones, some of those high-density forms of ice not found on Earth, except in labs, probably make up much of Titan's icy mantle, with the particular form of ice in each layer determined by the temperature and pressure at that depth.

• Saturn's moon ***Iapetus*** has one bright half and one very dark half, a division that's unique in the solar system and mystifying to scientists. A dark look to a moon usually means the matter on top is dirt, a bright look that it's ice. A *very* bright look means the ice is pure. One idea for how the moon got to be two-toned is that dust blown in from a neighboring moon (Phoebe) became implanted on the side of Iapetus that faces in the direction of its orbit, blackening the ice on that side but leaving ice on the trailing side bright.

• Saturn's ***Mimas,*** no bigger across than the state of Utah, has a shell of almost pure water ice that's nearly as rigid as granite and as little given to wear. One of its craters, 80 miles wide and looking to Earth viewers like a big eyeball, has remained intact for four billion years. "Cold water ice is a reasonably strong material," William B. McKinnon, professor at the Washington University in St. Louis, explains in his contribution to the book *The New Solar System.* "Although never as strong as igneous rock, water ice is able to absorb the punishment of a hypervelocity impact, form a crater, and retain that crater shape for geologic lengths of time."

• Saturn's ***Enceladus,*** is the brightest moon in the solar system, 100 percent on the albedo scale, which suggests that it has a fresh coating of ice on top, frost particles spread across its surface by an "ice volcano." Like volcanism as we know it on Earth, ice volcanism is an eruption onto a surface of material from underneath, except that molten or melted ice takes the place of molten rock. The icy lava spews or flows over old landscapes, smoothing them, burying them, brightening their surfaces. On small, low-gravity Enceladus, McKinnon notes, "eruption from a single vent could easily coat the entire satellite."

But what could make the ice on Enceladus mobile enough to erupt? For a start, there's tidal heating. As Enceladus passes in its eccentric orbit around Saturn, the planet's gravity tugs at the moon unevenly, and this repeated kneading of Enceladus's substance causes it to heat up, like a paper clip being flexed. In addition, there's antifreeze, or, as McKinnon calls it, "superantifreeze." Ammonia trapped inside ice crystals would lower the ice's melting point drastically, by *180°F!* Although the kneading would probably not warm the ice enough for it to melt on its own (temperatures are very low out by Saturn), the ice wouldn't need to be all that warm if it contained antifreeze; it could melt while still very cold.

• The coldest moon in the solar system is Neptune's huge moon ***Triton.*** At 2.5 billion miles from the sun, it has a surface temperature of −391°F as well as a surface texture that has been compared to a cantaloupe's skin and a water-ice crust hundreds of miles thick. In that extreme of cold, "water ice will not be the familiar stuff of glaciers, ice sheets and snowballs," John Stansberry, astronomer at the University of Arizona, points out. "Instead it forms the rock-hard bones of the landscape." Also in that extreme of cold, gases such as nitrogen, carbon dioxide, and carbon monoxide will freeze into ices. During each 165-year seasonal cycle on Triton, thin skins of these other ices migrate for hundreds of miles across the hard bones of water ice.

Its surface underfoot is ancient ice,
thus frozen firm four billion years ago

—John Updike, *"The Moons of Jupiter"*

• ***Callisto,*** the subject of the above two lines of John Updike's and the farthest out of Jupiter's moons, is a fairly uniform mix of ice and rock, with a shallow layer of rock on top. "It just sat while the ice sublimed away and it got this dark, cruddy-looking surface," Robert T. Pappalardo, planetary scientist at the University of Colorado at Boulder, says almost in complaint. "It looks deader than dead," Paul Schenk agrees. Scientists assume that nothing much has been going on inside it for billions of years. "A dull history," says Pappalardo.

• But another of Jupiter's moons, ***Ganymede,*** only a little larger than Callisto (both are about the size of the planet Mercury), gives the impression that "something dynamic is going on inside there," Torrence V. Johnson, senior research scientist at the Jet Propulsion Laboratory and chief project scientist for the Galileo spacecraft mission, reports. (During its eight years in orbit around Jupiter, *Galileo* made many flybys of each of the large icy moons.) As seen by *Galileo,* Ganymede's surface is a patchwork of bright (icy) and dark (clayey) terrain that "might crunch under some future astronaut's boot like frozen mud," according to Pappalardo. Some of the bright terrain is covered with long, parallel ridges and valleys that were apparently produced by *stretching.* The outer layer of ice was stretched and broken along fault lines. "Ganymede has just been torn apart by faulting," Pappalardo observes. Many of the faults are extremely fine, a half mile long or so, but some are so large they resemble rift zones on Earth. All signal "a fierce and violent past, during which severe icequakes shuddered through the satellite."

What could cause a moon's surface to stretch, though? "Ganymede must have expanded somehow," Pappalardo answers, "causing an extension all over the surface." But what would cause a moon to expand? Pappalardo speculates that before Ganymede heated up, high-density forms of water ice deep in its interior, including Ice V and Ice VI, could have been displaced by rock sinking toward the center, and as the dense ices moved outward toward regions of lower pressure, they changed into ordinary, lower-density Ice I, and the expansion of that change enlarged the whole moon.

• Yet it is ***Europa,*** the smallest of Jupiter's major moons, that is generating the most excitement these days. Scientists have excellent reasons to suspect that beneath the moon's icy crust, and not very far beneath it either, lies an ocean of *liquid water,* or at least a slurry of water and ice, or slush. Through gravity measurements, they have determined that Europa has an outer shell of H_2O about 60 miles thick above a rocky mantle, but they don't know what physical state all that H_2O is in. At the surface, it is certainly ice, since the temperature at the moon's equator is −260°F. But on high-resolution images taken by Galileo—resolution so high that an object the size of a truck shows up on them—are features that indicate there's something under the very hard, very cold ice at the surface besides more hard, cold ice. These features include:

A crater whose floor is almost as high as the surrounding terrain, suggesting that some soft ice, or an ice-water slush, rose from a shallow depth to fill it.

Iceberg-like chunks that seemed to have broken off of (calved from?) an expanse of icy plains, as evidenced by the fact that both chunks and plains have the same pattern of crisscrossing, overlapping ridges (like "a string-wrapped baseball without its cover," as Ronald Greeley, professor of geology at Arizona State University, describes it). It appears that many chunks would fit to the edges of other chunks, like pieces of a jigsaw puzzle, if they were moved and rotated; therefore it seems likely that to reach their present positions the chunks have already moved and rotated, which they could have done only if they were once suspended in a runny or slushy matrix.

"Pull-apart features" (what even verbally adept scientists call these things), dark bands separating continent-size plates of ice whose edges would also fit like jigsaw-puzzle pieces if they were brought together. Did ice on either side of a crack slide away from it, as tectonic plates do on Earth, after which a more ductile, dark ice filled the crack and hardened?

"Splash areas" around impact craters, like ones you'd get if you threw stones through the frozen skim of a pond and water sloshed up out of the holes and refroze around it. *What sloshed?* Was it partly melted ice, its melting point lowered by salts or ammonia? Or, as in an earthly pond, water?

"Freckles," or a variety of spots, pits, and domes, the domes formed perhaps by blobs of warm ice rising, like wax blobs in a lava lamp, through colder ice and pushing the surface ice up into a mound.

All these surface features suggest that at some time in Europa's history there was a warm and "accommodating" material underneath. To find out if there still is, some time in the next decade and a half a spacecraft will orbit Europa and measure, among other things, the moon's shape. Shape matters because if Europa has a small tidal bulge, say three feet high, it would mean that all of the 60-mile-deep shell of H_2O is frozen, since a thick layer of solid ice can't do much flexing. But if the tidal bulge is large, say 100 feet high, it would mean that the ice part of the shell is thin and below it might be a global sea (with possibly more water in it than exists on Earth).

If there is such a sea (with a slushy layer sandwiched between ice and water), as seems "reasonably probable," Pappalardo says, it will be the first time that liquid water has definitively been found anywhere in the solar system besides Earth. That raises the possibility that Europa could also be the first place in the solar system besides Earth to be found to harbor life. In recent years, biologists have discovered primitive lifeforms in highly unlikely places in our world, between rocks ten miles below ground; in the dark, volcanically heated depths of the sea; in the ice of Siberia and the perennially frozen lakes of Antarctica. There's a new appreciation for what things are necessary for life to get started and survive, and sunlight and oxygen seem not to be among them.

But water still is. In the future, probably at a time when there's a new generation of astroscientists, a lander will be setting down on the surface of Europa, digging up some ice, and analyzing it for biochemicals. Or a hydrobot might be melting its way down through ice and slush to a watery underworld, where it will go hunting for microbes.

In the view of some scientists, the evidence for liquid water on Saturn's moon Enceladus is nearly as strong as it is for water on Europa. "In the last couple of years, we've found reason to suspect that there are liquid oceans in *many* icy moons," Pappalardo says. "Apparently the moons don't have to be too warm, just have enough salt or ammonia to keep them from freezing." Titan and Ganymede could have, at the least, briny, wet layers. Even Callisto, sullen, undifferentiated (some have said "boring") Callisto, was revealed by recent magnetic-field measurements to have what could be salty water inside.

Jonathan Lunine reflects on the difference a phase makes. "There is . . . much magic in the subtle difference between water as liquid and water as ice," he writes; "without the former there might be no intelligence to contemplate the latter's importance in the solar system."

the whole range flew
gliding on interstellar ice

—A. R. Ammons, *"Mountain Liar"*

One of Saturn's icy moons could have produced Saturn's rings. In that creation story, an icy comet hit an icy moon and broke it to smithereens, and the icy pieces of the moon circled the planet in rings. Or, possibly, the icy comet itself broke apart and produced the icy pieces. "It would have had to be a pretty big comet," Jeffrey N. Cuzzi, space scientist at the NASA Ames Research Center, points out. "Almost as big as our moon." Otherwise, he explains, it wouldn't have had the internal heat to separate out its own ingredients and produce a "nice, white, icy outer shell." (A moon that broke up would have had to be pretty big, too, for the same reason.) In another, long popular creation story, Saturn's rings are leftovers from the formation of the solar system, which because of Saturn's powerful gravitational influence were not able to pull themselves together into a ball. In other words, the rings are a moon manqué.

Whatever their beginnings, Saturn's rings are made up now almost entirely—90 to 99 percent—of water ice. The ice is in the form of small particles, a fraction of an inch to 40 feet across, "gazillions" of them, Cuzzi says, circling Saturn's equator in a disk. The disk is 170,000 miles across but only about 100 feet thick, which gives it a width-to-thickness ratio like that of "a sheet of paper the size of San Francisco," Cuzzi points out. That still adds up to a lot of ice.

As might be expected of something consisting of almost pure ice, Saturn's rings are extremely bright. They reflect so much sunlight that if they were as close to Earth as our moon is and receiving as much sunshine as it does, they would light up our night sky almost like daylight. To a person on Earth looking at them through a telescope 400 million miles away, the rings appear white. On images taken by *Voyager 2* during its 1981 flyby and later color-enhanced, the rings are garish shades of red, blue, green, and purple. What these shades told scientists was that the rings aren't pure ice but have small and varying amounts of non-ice material in them. Curious about what their "real" color would be if the rings were seen by the human eye under the same degree of illumination as the Earth gets, Cuzzi and Paul R. Estrada, a postdoctoral associate at

Ames, applied color theory to *Voyager* data and determined that the color is not white nor even the yellowish-tan that it appears to be on non-color-enhanced *Voyager* images but an extremely pale reddish-tan, more red than tan, "salmon," Cuzzi says. He has since revised his description to "pale salmon."

One reason Cuzzi and Estrada cared about color was that it might reveal what the parts of the rings that don't consist of ice do consist of. "We asked ourselves, what can you add only one percent of to ice and get this red color?" No other moon of Saturn is that color. Silicates "just won't do," Cuzzi says. Neither will iron; it's present in amounts too small to impart redness. Where that red can be found is on Triton and on a few other icy moons as well as in some comets and asteroids and in patches on icy Pluto. The color comes from carbon-based compounds that when concentrated form a gunk that NASA's Dale Cruikshank has termed "the dark red stuff." The compounds were probably brought to the rings by the comet that provided the original ice (in that creation scenario), with more being added by later comet hits. "The stuff is a pollutant," Cuzzi states, "just like the grime and soot that turn the ice in New York City darker and grayer."

So why isn't the ice darker? If comets kept on arriving at the rate that scientists believe they did after the initial bombardment and unloading pollutants, why is the ice in the rings so pure? By now the rings should be as dark as the rings of Uranus, which are almost as dark as charcoal. Cuzzi and others have come to the conclusion that Saturn's rings are pale because they are young, far younger than the solar system, only a couple of hundred million years old instead of 4.6 billion years. The "age of the rings" on Saturn corresponds to "the age of the fish" on Earth. As the rings grow older, they should darken.

Like other young things, the rings of Saturn are active and mutable. When *Voyager 1* sent images back to Earth in 1980, astronomers were "dazzled and stunned," according to Clark R. Chapman, planetary scientist at the Southwest Research Institute in Boulder, Colorado. Where they expected to see a simple disk sticking out in a flat plane around Saturn's equator, like the hat brim in kids' drawings, they found a disk that was not at all simple and not flat either but elaborate and with weird contours. "The intricate structure of innumerable ringlets was totally unanticipated," Chapman wrote in his book *Planets of Rock and Ice: From Mercury to the Moons of Saturn,* conveying his excitement by ellipses. "The bewildering complexity of rings within rings within rings . . . intertwined rings shepherded by tiny satellites . . . shadowy spokes on approach that became iridescent on departure . . . gaps and divisions, some seemingly randomly spaced, others in an orderly geometric progression . . . rings of snow boulders, others of fine ice dust . . . eccentric rings, tilted rings, and clumpy rings."

Like a great ring of pure and endless light

—Henry Vaughan, *"The World"*

As their structure is now understood, the rings have three major regions, A, B, and C, the outermost being A and the innermost C, plus four minor rings, D through G. Each region is made up of tens of thousands of smaller rings, or ringlets. In orbit among the rings, and affecting their behavior, are a couple dozen "ringmoons," which are also composed of ice, probably piles of ice rubble, judging by the fact that their density is half that of solid ice. The alphabetic regions have different characteristics from one another. The B ring, for instance, has a complex filigree structure, "like record grooves," Cuzzi says. "We don't know what causes that." The C ring has a long-wavelength undulation in the middle, symmetrical, flat-topped "plateaus," and several empty gaps. We don't understand them, either." The E ring is faint, gossamer, fuzzy, diffuse, made up almost entirely of extremely tiny ice particles, which may have been rocketed off the surface of nearby Enceladus by a meteorite hit or spewed by an ice volcano, with only one particle size—the extremely tiny—joining the ring's orbit. Without this continual resupply of "ice dust," the tenuous E ring would probably disappear.

The youngest of the rings is probably the F ring. Its age can be measured in human terms; it's about 100 years old. "Probably another collisional fragment," Cuzzi says of it. The ice grains in F form strands that move back and forth in relation to each other, as if "dancing in unison," excited perhaps by the nearby orbiting presence of the moon Prometheus. Where the strands of ice grains overlap, they produce what in *Voyager 1* images looked like bright clumps and knots. In the mere year and a half between the visits of *Voyager 1* and *2*, the clumps and knots shifted positions quite noticeably.

Embedded in the A ring is a "ring-clearing moonlet." Just by being there, the moonlet repels ice grains and pushes them to either side of its orbital path, creating a gap in the ring. A few scientists have speculated that ringmoons can act in tandem, as a pair of "shepherds," one on each side of a ring, keeping the ice grains in it from spreading viscously outward.

The B ring has "spokes" running across it, dazzlingly. In *Voyager 2* images, the spokes went "reeling around Saturn as if on a gigantic merry-go-round," Chapman wrote. They raced around the planet as in a "dizzying movie," he also wrote. The spokes are probably produced when Saturn's rotating magnetic field charges the rings with static electricity, then, like a comb picking up paper scraps, lifts ice dust off them and blows the dust radially across the face of the B ring, casting dark, wheel-like shadows upon it.

"A lot is going on," Cuzzi concludes. Chunks of ice—some as big as a barn, some as little as a pearl—keep coming together, breaking apart, and moving, moving, moving in response to the gravity of passing moons and meteorite hits and Saturn's magnetic field. Cuzzi, who has been studying Saturn's rings for 30 years, considers them "lots of fun." In June of 2004, spacecraft *Cassini* reached the region of the rings and began taking pictures with cameras a thousand times more sensitive than those on *Voyager 2.* Once scientists have studied the images, "we won't have to be guessing any more how old the rings are," Cuzzi predicts, "or what they're made of, besides ice."

Other giant planets have rings, too, though not as large and spectacular as Saturn's nor, presumably, as much fun. Neptune's and Uranus's rings are dark and narrow, formed perhaps when small moons got knocked apart and ground up by impacts. Jupiter's rings are of rock dust and so insubstantial that astronomers didn't discover them until 1979. Even the Earth might have had a ring of sorts around it once, about the time that the moon formed. If, as some scientists propose, a Mars-size asteroid smashed into a coalescing mass of rocky material and debris was flung out of it that became our moon, the hit could have left in its wake a ball of molten rock—Earth—with a ring of loose rubble orbiting it.

"A ring 'round the Earth!" Immanuel Kant exclaimed in his *Theory of the Heavens.* "What a beautiful sight for those who were created to inhabit the Earth as a paradise!" The ring 'round the earth Kant envisioned was not of rocky rubble but, like Saturn's, of water-based particles, maybe frozen water particles, and as such useful to a God who concerned himself with human affairs. Conveniently (for God), the earth's ring had "this property of being able to be broken up on occasion, if need were, to punish the world which had made itself unworthy of such beauty, with a Deluge."

CHAPTER THIRTY-FOUR

SPACE II

The polestar (Polaris) during the Dark Ages
was called the "Nail of the Heavens"—
an iron nail hammered into the frozen sky.

—Nicholas Christopher,
"24" (from *5° and Other Poems*)

As for Mars, it has a wimpy, next-to-nothing ring, from dust leaking off the surface. The whole planet is dusty, barren, desertified. "Poor Mars!" University of Hawaii astronomer Tobias Owen once exclaimed in print. "Its fate was sealed by its small size." Only one-eleventh as large as Earth, it didn't have the gravity to hang on to more than one percent of the atmosphere it started out with. It's also very cold, on average −76°F at the equator, −193°F at the poles in winter, so the meager atmosphere can retain only the meagerest amount of water. Michael H. Carr, geologist for the U.S. Geological Survey and author of the book *Water on Mars,* estimates that if all the water vapor in Mars's atmosphere were to be precipitated out onto the surface at once, it would form a layer a 33-millionth of a foot thick. A flea's leg is thicker than that.

Yet there are marks on Mars's surface—what look like dried-up river courses, deltas, lakebeds, ocean basins—which indicate that in the past there were vast amounts of water on Mars. Floods greater than any floods that ever occurred on Earth, torrents that carved great channels in the ground, one of them so long that if it were on Earth it would extend from New York to Los Angeles. By measuring (with the aid of spacecraft photos) the amount of rock that would have been eroded from such channels, scientists have calculated how much water was involved in the floods. In another cover-the-planet way of illustrating volume, Carr estimates that if all that water were spread evenly over the surface of Mars, it would stand a third of a mile deep.

Where did so much water go? It couldn't all have drifted off the planet; the mechanisms for losing it to space are inefficient, Carr points out. Where is it then? The only way to account for most of the missing water is to conclude that it is buried in the ground in the form of ice. "There's lots of ice on Mars," Carr declares. "It's just not on the surface." In some regions, ice could make up half the volume of the Martian soil. There could be giant ice lenses or entire frozen lakes underground, as on Earth. Permafrosted layers could be a mile and a half thick. The ice probably formed early in Mars's history when its climate was cooling and the water that had been flowing across its surface slowed, pooled or soaked into the ground, then froze and gradually became covered with rock and dust. Carr suspects that ice in the ground at the middle latitudes would slowly sublimate away since the ground temperature there is above the planetary frost point of −100°F, but at higher, very cold latitudes, beyond 30° or 40° North and South, the ice probably "just stays and stays. If you put a chunk of ice in the ground there and covered it with dust, it would last forever."

Clues to ice present aren't as graphically displayed on the Martian surface as clues to water past, but they are there. In spacecraft images, geologists have identified features characteristic of permafrosted regions on Earth: A rounding and softening of some of the topography, where the surface material seems to have flowed. Rocks at the bottom of slopes that, instead of piling up where they fell, appear to have moved outward after they fell, the ice-rich ground creeping and carrying the rocks with it. Polygons of patterned ground.

Mars's poles are now known to have caps of water ice on them. In winter, both caps are covered by thinner ones of dry (carbon dioxide) ice, but in summer the north pole's layer of dry ice evaporates and exposes the water ice. Seen in spacecraft images, there's a swirl of dark lines in the white cap—probably exposed valleys, according to Carr. What probably happened was that sunlight hit one slope of a crack and caused the ice there to vaporize, which left a thin layer of darker rocky material behind, then the vapor condensed on the other, shadow side. The valleys thus moved, albeit slowly, a fraction of an inch a year. The movement could explain the smooth, crater-free look of the north polar region; it acts as an eraser.

Ice at the north pole is what controls the water content of Mars's atmosphere, such as it is. In summer, sunlight strikes the pole directly, heating the water ice to temperatures above the frost point and causing it to vaporize. Water enters the Martian atmosphere in "a big rush," Carr says, laughing at the word "rush" since the amounts involved are minuscule. "Still, it's very dramatic," he says. Water content over the polar region goes up tenfold, from 0.0004 of an inch deep to 0.004 of an inch. Mars does have clouds, thin ones like our cirrus, made up of small crystals of water ice, bluish white in contrast to the butterscotch color of the planet's dust clouds. Mars also has frost. It deposits on cooled ground at night and evaporates on warmed ground the next day. In photos taken in 1977 by *Viking Lander 2,* there are white patches at the base of boulders that any earthling would recognize.

Mars is the heavenly body on which people have pinned their greatest hopes for finding life outside Earth. Around the turn of the last century, the American amateur astronomer Percival Lowell construed the long, straight, crisscrossing lines he saw through his telescope as canals built by intelligent and cooperative beings so they could transport water from the poles to the equatorial regions, where they were trying to live and grow vegetables. In 1938 when Orson Welles broadcast a radio dramatization of H. G. Wells's 1898 novel *The War of the Worlds* in which Martians—tentacled beings with eyes like luminous disks and drooling V-shape mouths and oily, fungoid skin—invade Earth, it was so alive with probability that not a few listeners fled their homes in terror. In Ray Bradbury's 1950 *The Martian Chronicles,* the natives lived in houses made of crystal pillars with fruit growing out of the walls. Planetary scientists now studying Mars grew up under the influence of these accounts, Laurence Bergreen points out in his recent book, *Voyage to Mars: NASA's Search for Life Beyond Earth.* It's not that they believed the accounts but that they were inspired by them.

A search for life on Mars begins with a search for water. Although water is considered essential for life to exist, it may be in peculiar places and states. "Ice can be a great refugium for bacteria," John Priscu of Montana State University, the biologist who discovered bacteria in permanently frozen lakes of Antarctica's Dry Valleys, observes. "They can stay frozen for a long time and survive. There could be a whole biosphere going on under the surface."

Recently, *Odyssey,* one of three spacecraft orbiting Mars, detected in the ground surrounding both polar caps (much as Carr predicted) what scientists consider to be water ice. The ground ice covers a large area, "the size of North America," according to Jeffrey Plaut, planetary geologist at the Jet Propulsion Laboratory and project scientist for *Odyssey.* "It seems to be everywhere from 60 degrees North and South to the poles." The ice was in the top four feet or so of soil (*Odyssey*'s instruments couldn't penetrate further), which suggests that it is millions, not billions, of years old; that is, it doesn't date from the time when there was lots of water on Mars soaking in. In those four feet, there is more ice than dirt, which Plaut finds puzzling. "It could be remnant snow." Even more recently, the European Space Agency's *Mars Express,* with a high-resolution camera on board, found on the surface much closer to the equator signs not of a body of water that came and went eons ago but of one that might still be around, in frozen form. Plates, ridges, and rubble typical of pack ice on Earth cover an area the size of the North Sea.

Carr thinks it unlikely that there is any life on Mars, or ever has been, but insists that "we'll do our best to find it." The best way to look for signs of ancient Martian life, he says, would be to dig up old sediments, bring them back to Earth, and examine them for fossils of microbes and complex organic molecules. The best places to do the digging would be where water had been brought to the surface from far below, perhaps by volcanic activity. "If life ever did arise on Mars," Carr explains, "it would have gone into deep, underground aquifers, miles below the surface, where geothermal heat may have melted the ice and produced liquid water." Researchers are now devising "cryobots" to do the digging; Caltech's Hermann Engelhardt, for one, is creating a small version of the hot-water ice drill he used on WAIS in Antarctica.

It may not seem all that outlandish that there should be water ice on frigid Mars, but on torrid Mercury? Mercury is the closest of all the planets to the sun, only 58 million miles from it, and its surface temperature can climb above 800°F, hot enough to melt lead. Even smaller than Mars, Mercury has even less atmosphere. There's no indication that it ever had running water; its craters look unaltered since their formation. It has been called a corpse of a planet. Although it has a slow spin and thus a long night, and temperatures on the night side can get down to −450°F, the night isn't endless. The dark side eventually turns into the bright side and heats up again. Therefore, when in 1991 radar probes aimed from Earth at both poles got return signals similar to those from known icy sites, such as Mars's north pole and Jupiter's icy moons (high reflectivity, polarization signatures), astronomers were "surprised and stunned," one of them admitted.

"People kind of laughed [when we told them]," says Duane Muhleman, professor emeritus of planetary science at the California Institute of Technology and a member of one of the teams that did the radar probing. Although it seemed highly unlikely that water ice was responsible for the signals, alternative sources, including lava flows and extremely cold rocks, seemed even more unlikely. If water ice *was* producing the signals, how did it get on Mercury, in the first place, and stay on Mercury, in the second?

Probably like this: At some time in Mercury's history, water would have made its way to the surface either from inside the planet, through volcanic activity, or—more likely—from outside it, as comets struck it and released ice. Some of the water molecules in the ice would have been broken apart by ultraviolet rays and some would have escaped to space, but some, maybe 5 to 10 percent, would have made their way to the poles, by "hopping." Hopping consists of individual molecules lifting off the surface as sunlight vaporizes them then condensing back onto the surface as they are cooled in the overlying vacuum, then rising again and falling, bouncing up and down, like Ping-Pong balls. The water molecules would have kept on hopping until they reached the poles, where they achieved a stable state by freezing. They would freeze because, with the planet's axis perpendicular to its orbit, sunlight strikes the poles at a very low angle and temperatures run as low as −100°F, and although −100° isn't cold enough to stop the hopping, there happen to be craters at the poles so deep that the angled sunlight never reaches inside them, and they have been in permanent shadow for billions of years. Temperatures there could get down to −400°F, or close to absolute zero. Any water molecules that bounced in would quit bouncing. They would stick around "forever," says Muhleman. "You know that 70s song, 'Hotel California'?" asks Jeff Kargel, a geologist at the U.S. Geological Survey. Upon being told no, he sings, " 'You can check out any time you like, but you can never leave.' "

It was night. It was bitter. But stars
like birches splintered the black forest sky.

. . . .

There is a lake of ice on the moon.

Pamela Sutton,
"There Is a Lake of Ice on the Moon"

The dark regions on the lit face of our moon were long thought to be seas, but when people got a good look at them through telescopes they turned out to be dried-out plains. And when astronauts actually set foot on the moon, they found nothing that suggested water was there, or had ever been there, in any of its phases. Then in the late 1990s, spacecraft circling the moon got readings from both poles for hydrogen, and hydrogen, to scientists, usually means water ice. It seemed to be another case of "Hotel California." The moon, too, has craters near its poles as well as slanting sunlight and perpetual darkness in sections of the craters. Temperatures in those sections could run as low as −300°F, and any water molecules that bounced their way across the lunar surface and into the craters would stay in them, in cold storage, under an insulating coat of dirt, for hundreds of thousands of years. One estimate of the amount of almost pure water ice at the two poles is 10 billion tons.

Space-travel enthusiasts would like to mine the ice. They point out that ice can be broken down into hydrogen and oxygen and used for rocket fuel, which astronauts on their way out to explore the solar system (starting with Mars) could pick up on lunar pit stops. But recent radar readings from Earth indicate that ice on the moon is in thin layers, mixed with dirt, making up only 1 to 2 percent of the soil in shadowed regions. By contrast, ice deposits on Mercury are believed to be at least three feet deep covering miles and miles of crater floor. "We don't have that smoking gun that Mercury provides with its bright crater floors," Bruce Campbell, planetary scientist at the Smithsonian's National Air and Space Museum, states. In low concentrations, even billions of tons of moon ice would be tough to mine.

A better source of water than lunar polar ice for refueling space explorers, Kargel suggests, would be asteroids, some of which have ice inside them and orbits that approach Earth. Still, ice on the moon is "a bird in the hand."

I saw a peacock with a fiery tail
I saw a blazing comet drop down hail

—Anonymous,
"I Saw a Peacock with a Fiery Tail"

Comets are related to asteroids—both are very small bodies that didn't end up in planets—but comets consist mostly of ice and asteroids mostly of rock. The material in comets is thought to be left over from the formation of the solar system 4.6 billion years ago and little changed since; most asteroids have differentiated somewhat, like planet wannabes. While asteroids number in the hundreds of thousands and inhabit a zone between Mars and Jupiter, comets number in the *trillions* and reside at the far, dark, extremely cold (−364°F to −437°F) edges of the solar system, 50,000 to 100,000 times farther from the sun than Earth is, in a vast cloud called Oort ("Oort" for Jan H. Oort, a Dutch astronomer who proposed the cloud's existence in 1950). Also in a recently discovered belt beyond the orbit of Neptune called Kuiper. ("Kuiper" for Gerard P. Kuiper, a Dutch American astronomer who proposed the belt's existence in 1951.) In the Kuiper belt are numerous large icy objects—"ice dwarfs," "icy planetesimals"—which would if they were brought nearer to the sun turn into comets. Pluto, long counted as the ninth and outermost planet in the solar system and the favorite of American schoolchildren, should in some astronomers' opinion be demoted to the status of a Kuiper belt object. It is much smaller than all the other planets; it has an orbit that is tilted from theirs; and it is icy.*

The Oortian skies are not crowded. The cloud is so vast that each comet is millions of miles from any neighbor. Traveling only about as fast as a middle-of-the-pack marathon runner, and being at such a great remove from the sun, a typical Oort comet takes millions of years to orbit the sun a single time. Now and then a passing star gives the cloud a gravitational shaking, which knocks some comets out of their orbits ("a garden gently raked by stellar perturbations," Oort said of his cloud). Once nudged from their old orbital paths, some comets head deeper into space, out of the solar system entirely, and some head deeper into the solar system, toward the sun. On rare occasions, people have seen these intruders as whitish fuzz or streaks against the dark sky and generally not rejoiced at the sight. "Vile stars," comets have been called, responsible for locust plagues, wars, and the high price of rice.

"There is an overwhelming sadness to the literature on comets," Carl Sagan and Ann Druyan pointed out in their book *Comet*; "any comet at any time viewed from anywhere on Earth is assured of some tragedy for which it can be held accountable." Comets probably came by their iniquitous reputation because, as Harvard astronomer Fred L. Whipple noted in his book *The Mystery of Comets,* they materialize in any part of the sky, without warning, seemingly erratically, unpredictably, "easily interpreted as an indication of wilful intent, undoubtedly malicious."

Dispassionate observers spoke up. In the 16th century Danish astronomer Tycho Brahe determined that comets lie well outside the atmosphere of Earth, and in the early 18th century, Edmond Halley worked out the rough orbits around the sun of a couple dozen comets, noting that one of them kept coming around regularly, at 76-year inter-

*Recently, astronomers found a Kuiper belt object larger than Pluto and didn't hesitate to call it a planet; they expect to find others.

vals. However, it was a very long time before anybody figured out what comets are made of, that what seems to blaze consists mostly of ice. In his 1755 *Theory of the Heavens,* philosopher-metaphysician Immanuel Kant argued that since comets form in the far reaches of space and have "vapor heads and tails," they probably consist of particles "of the lightest kind."

"But now a wrong turn," Sagan and Druyan wrote, adding compassionately, "You want to reach across the centuries and give him a word of encouragement." The great philosopher reasoned that comets must not "consist of the *very* lightest stuff" because the vapor heads and tails appear when the comets are in the region of Earth's orbit, from where they would receive only moderate heat from the sun's rays. Kant failed to apprehend *what* solid stuff could evaporate when only moderately heated. Picture him, "quill pen in hand, wondering about what this extraordinary 'lightest stuff' might be, trying to grasp the quintessential, when it was all around him," Sagan and Druyan wrote. "He had only to look out his window, perhaps at the very moment he was writing these lines, to see a cloud of vapor rising off the frozen River Pregel, and he would have had his answer: Ice. Ordinary water ice. Not an exotic substance, at least on Earth. No special celestial stuff. Just ice."

Two centuries would pass before Whipple came up with his model of comets as "icy conglomerates," or water ice mixed with rocky grains and the ices of other gases. It was the same year that Oort proposed the existence of his cloud, and the prevailing model of a comet then was the "flying sand bank," or swarm of separate, small rocks traveling in orbits parallel to each other, accompanied by some gaseous material. Whipple argued that a comet is not many objects but a single one composed mostly of ice. He called it a "dirty snowball." (Astronomers still do likewise.)

Whipple didn't actually detect the ice in comets; he based his ideas about their makeup on their behavior, including the way they sometimes slowed down or speeded up in their orbits in manners not explainable by the laws of gravity ("fishes darting through the cosmic ocean," Johannes Kepler called comets). Whipple figured that "jetting" was responsible for the darting. As he envisioned it, the jetting occurs when a dirty ball of ice rotates in orbit and the side toward the sun warms (as happens on Earth of a summer afternoon) and a patch of heated ice evaporates from that side, carrying off with it dust and gas, thus propelling the comet in the opposite direction, according to Newton's principle of action and reaction. "A snowball in sunlight is really a small jet engine," Whipple wrote. One strategy for changing the path of a comet on a collision path with Earth would be to heat a patch of its ice with nuclear power, converting the ice to steam and nudging the comet off course.

Two decades more would pass before Whipple's iceball model was confirmed by observation. During the swing of Halley's comet past Earth in 1986—its fourth since the days when Kant was puzzling about the light stuff—spectrometers on Earth picked up the wavelength of water in it. They picked up the wavelength not in the comet's core but in its "coma," the globular cloud of gas, dust, and ice flakes that a comet gives off as it warms on approach to the sun and ultraviolet light causes it to fluoresce. The particles making up the coma are no bigger than bacteria, but there are billions of them, and with little gravity to restrain them, they spread out over millions of square miles of space, presenting a large target for sunlight to illuminate.

Besides a core and a coma, a comet has a tail (or two or three; white, yellow, or blue), which is the part of the coma that's blown away from it by the solar wind. Although the tail looks to earthbound observers like evidence of speedy travel, like a rocket trail, it always points away from the sun even when the comet itself is also heading away from the sun. In 1996, Comet Hyakutake, which had a core less than two miles wide, had a tail 350 million miles long, probably the longest tail ever. When we are looking at a comet, it's the coma and tail we're seeing, not the core.

Still another decade would pass before ice was detected in a core. "As the feeble light from the comet [Hale-Bopp] . . . registered on our instrument," Dale Cruikshank wrote, "we gradually discerned the spectral signature of frozen water! This was the first time that frozen water had ever been seen directly on a comet." He was thrilled.

Despite their reputation for disaster-mongering, comets are not very prepossessing objects. They are small, the smallest objects in the solar system, a half mile to six miles across. They have almost zero gravity. If you were standing on a comet, Sagan and Druyan calculated, you'd weigh about what a lima bean does on Earth. (What would H. G. Wells make of that? A lima bean on a comet!) If you were standing on a comet and tossed your spoon straight up, you could never catch it because it wouldn't come down.

Although water ice is the most plentiful ice in comets, it's not the only ice. Others include carbon dioxide ice, carbon monoxide ice, ammonia ice, methane ice, and nitrogen ice. ("Complex ice," Whipple called comet ice.) The other ices freeze at lower temperatures than water does, carbon dioxide at −71°F, for example, nitrogen at −210°F, and they vaporize at lower temperatures, too, when the comets are a long way out. In 1973 when Comet Kohoutek was first spotted, it was so bright it was expected to be unusually brilliant by the time it got closer to Earth (and therefore the sun), and people booked cruises so they could see it in its full glory away from city lights. But Kohoutek was a fizzle. By the time people on board ships got a look at it, its exotic ices had long ago left it.

Comet cores are dark, a surprise to those who have seen comets as white marks on the night sky. "They are some of the darkest things in the solar system," Karen Meech, associate astronomer at the University of Hawaii, says. "As dark as charcoal." Darker, actually: while comets reflect 3 to 4 percent of the light that hits them, charcoal reflects 10 percent. "As dark as black velvet," astrophysicist John C. Brandt suggests. Meech illustrates the point to her students by having them make their own comets. "Take a big bucketful of finely shaved ice," she instructs, "finer than in a Sno Cone. Add a few tablespoons of ordinary dirt to it. Wait an hour. By then the ice will be melting and darkening, like a snowbank in winter. I usually have a student mash it into a comet-like shape."

In real comets, the water ice doesn't melt into a liquid since an atmosphere is required for that to happen and comets exist in a vacuum. But some of it evaporates, beginning when the comet is near the asteroid belt, coming up on Mars. Compared to the more volatile ices, which shoot off the surface like fountains, water ice tends to leave the comet "like steam coming off a pond at dawn," Dale Cruikshank says. "I think of it as a benign, low-energy thing." A typical comet in the inner solar system loses ten feet of ice with every swing past the sun. At that rate, Halley's comet will no longer be swinging by in 76,000 years.

When water ice leaves a comet, it takes some dirt with it, but often it leaves a lot more dirt behind, in a layer that insulates the remaining ice. Sometimes the dirt builds into such a thick outer layer that the ice has no access to the surface, in which case the comet can't form a coma and tail. It is dead, a rocky asteroid with some water ice locked inside. A third to a half of all asteroids passing near Earth's orbit are thought to be dead comets.

Some comets end their lives more violently. In July 2000, Comet Linear was making its closest approach to the sun when it disintegrated, "spewing its mass into space like a puffball in the wind," as one newspaper report described the event, "leaving nothing but dust behind." Comets sometimes crash onto moons or planets, like the one believed to have struck Earth 65 million years ago, expelling so much of its dirt load that skies darkened around the globe, temperatures dropped, vegetation quit growing, and coldblooded dinosaurs, unable to find food, died off, leaving the field open for the more adaptable mammals to take over. "We owe our very existence," Sagan and Druyan wrote, "—every one of us bigger than a mouse anyway—to the extinction of the dinosaurs" and therefore to the agent of that extinction, an icy comet or asteroid.

The iceballs that reached the Oort cloud (by being flung there through gravitational forces, "like an automatic pitching machine throwing baseballs into the bleachers once a minute for a hundred years," in Sagan and Druyan's image) carried with them not only rocky particles but organic ones. In recent decades, the spectrographic signatures of several dozen organic chemicals have been discovered in comets, prompting some scientists to suggest that the ratio of ingredients be revised to one-third ice, one-third rock dust, and one-third carbon. (A few scientists even took to calling dirty snowballs "icy dirt.") The discoveries led to the adoption of a "new paradigm," as Cruikshank calls it, in which the ingredients for life would not have come from inside the earth (gases, water vapor, and carbon compounds leaked onto the surface by volcanoes) but from outside it, delivered by comets that dropped onto its young surface. The earth was so hot during its first half billion years or so that it lost most of the water and lighter materials it started out with and became a barren, rocky ball—but comets kept bringing in new supplies and, once the earth cooled, it kept them.

"Turn on the water faucet in your kitchen and notice what comes out," Cruikshank instructs readers of his essay "Ice from the Beginning of Time." "Next, take a deep breath and exhale. Finally, look at yourself in a mirror. In all three cases, you are experiencing material that formerly resided in the icy interiors of one or more comets that was brought to Earth at some time early in the history of our planet."

First, the water. Tobias Owen estimates that half the earth's current supply of water came from comet ice and the other half from volcanoes, which released water vapor trapped between rocks and in the earth's interior. Then, the breath. "Why is there air?" Owen asks, like a cheerleader working a crowd. He answers his own question exuberantly: "Because there are comets!" Comets were probably the major source for the gases that make up our current atmosphere—they contain nitrogen and carbon in the right proportion—with meteorites and the earth's rocks making contributions. (Green plants provide the oxygen.)

Finally, the mirror. The person looking back at you is made up of material that was brought to Earth by comets. "Everything alive," Sagan and Druyan elaborated this view, "would derive ultimately from comets—all the plants and animals and microbes and men and women."

Part of the new paradigm is that the material in comets is similar to material outside the solar system, in deep space, and may even have formed there first. In recent decades, telescopes trained on the dark, cold wastelands between stars have identified the signatures of more than 100 different molecules (starting with water ice, which Wendy M. Calvin of the University of Nevada at Reno calls "the universal ice"), and they turn out to be almost exactly the same ones that have been identified in comets. Cruikshank likes to say that "through cometary messengers from far beyond the planetary region of our solar system, life on Earth is connected to stardust." He likes even better to quote singer-songwriter Joni Mitchell who, in her 1960s anthem "Woodstock," sang, "We are stardust."

Within the last decade, a series of spacecraft have been launched toward comets to give scientists a chance to study their topography and take samples of the light stuff. "Nobody knows quite how a comet is put together," Peter Schloerb, professor of physics and astronomy at the University of Massachusetts at Amherst, points out. Because of the comets' low gravity and the uneven way ices evaporate off them, "we may see some really fantastic landscapes," he says, "like fairy castles." Fred Whipple envisaged areas of comets "raised above the general terrain, like geological dikes, or perhaps the mesas of the American Southwest," as well as "dust sculptures . . . taller than our highest skyscrapers." And right he was. When NASA's *Stardust* space probe flew by Comet Wild (pronounced "Vilt") in 2004, the images it sent back showed not only mesas and pinnacles over 300 feet high but depressions with steep walls shaped like giant footprints, overhangs and buttes, pointy-topped spires, and sinkholes, ridges, flaps, and pits. The feature-rich surface caused some scientists to reconsider the theory advanced after Comet Shoemaker-Levy went to pieces on close encounter with Jupiter: that comets are loose agglomerations of very small dust and ice grains—"rubble piles"—barely held together in weak gravity. Wild's rugged landscape suggested that at least some of them have considerable strength and cohesion.

On the other hand, when NASA's *Deep Impact* slammed into Comet Tempel 1 at 23,000 miles an hour on July 4, 2005 (as intended, so scientists can find out what lies inside), what spewed out was a plume of talcum-like dust and minuscule ice grains, the whole comet largely empty space, with the strength of "lemon meringue." Sending probes to many different comets—the European Space Agency's *Rosetta* is even now speeding toward Comet Churyumov-Gerasimenko where it will set a three-legged lander on the surface—is justified because, as Owen explains, "the comets seem to have unique personalities."

To Peter Schloerb, the water-ice component of a comet is "a bit boring." The other ices, he says, should tell scientists more than water ice can about how comets and the solar system formed. Yet "it's water ice that controls all the activity," Karen Meech points out. She contends that the water ice in comets formed at extremely low temperatures as amorphous, not crystalline, ice. She has seen comets flare, or light up, and display the signature for water when they are at distances from the sun too great for regular, hexagonal ice to vaporize; what's happening, she suggests, is that the amorphous ice is converting to regular hexagonal ice and in the process giving off energy and releasing gases trapped in its nooks and crannies ("amorphous ice is very irregular"). As the ice warms, "the spaces between the molecules of water close up," she says, "and that squishes out the guest molecules." Cruikshank comments: "This shows the dynamic behavior of ordinary old water ice, that at such a low temperature it's doing things."

Ordinary old water ice is the vehicle, then—the interplanetary cargo ship—for carrying more exotic materials around the solar system. Abundant and stable, "it provides the structure," Owen points out. Without water ice, the ingredients needed to get life started on Earth might never have reached the planet at the critical moment.

"It is possible," Owen concludes, "no ice, no us."

CHAPTER THIRTY-FIVE

ICE AGES

Some say the world will end in fire,
Some say in ice.

—Robert Frost, *"Fire and Ice"*

ONCE UPON A TIME, the very place where I sit in front of my superannuated computer writing this book on ice was filled with ice. Solid ice extended from the ground to a thousand feet overhead. Horizontally it reached into Brooklyn, Queens, and Staten Island. Its weight depressed the earth under it by several hundred feet. Winds whipped over the crevassed, dirty top of it and carried its chill into the barren countryside around it. Embedded in it were rocks as small as talcum grains, as big as houses, which it had scraped off the ground as it moved south out of Canada. Streams of meltwater flowed from it in summer and pooled in lakes at its edges. Some of the ice in it was 10,000 years old. It would stick around for a few thousand years, then slowly melt away, leaving all the rocks behind.

The time was the peak of the last ice age, 20,000 years ago. The ice in New York City was at the southern edge of the Laurentide ice sheet, whose center was in eastern Canada but which in the American Midwest reached as far south as the Ohio River valley and farther west joined the Cordilleran ice sheet of the Rocky Mountains and to the north merged with the Innuitian ice sheet on the Canadian High Arctic islands which, in its turn, reached out and touched the Greenland ice sheet. Across the Atlantic, most of Scandinavia was covered with ice, and so were Scotland and Ireland and the north of England and parts of Germany and Poland and western Russia. The Alps, Pyrenees, Urals, Himalayas, and Andes all had enormous amounts of ice on them. So did western Patagonia. Iceland was all iced over. Antarctica had 10 percent more ice on it than it does now. There was ice on some tropical mountains where there is none today, Mauna Loa in Hawaii, Mount Elgon in Uganda.

All in all, at the peak of the last ice age, one-third of the world's land area lay beneath a layer of ice that was on average one mile thick. That made for a total volume of *17 million cubic miles* of glacial ice on Earth. Because of all the water that was sucked out of the oceans to produce the additional ice, sea levels around the world dropped 400 feet. On the surface of the sea there was yet more ice, which covered a large portion of the world's oceans to a thickness of several feet.

Today the only remnants of the ice sheets are on Greenland (25 percent smaller now) and Antarctica, plus two tiny bulldogs of glaciers on islands in the Canadian High Arctic, the Barnes and the Penny. There are, of course, thousands of other glaciers scattered around the world in various states of health, but they formed anew after the old ones melted. For the last 10,000 years or so we have been in a period of comparative warmth and ice downsize known as an "interglaciation"; inherent in the word is the assumption that another glaciation will follow. It's an assumption based on the observation that for the last million years or so, glaciations (also known as ice ages) and interglaciations have alternated with each other in an almost rhythmical fashion, with glaciations lasting roughly 90,000 years and interglaciations roughly 10,000. Since our interglaciation has lasted 10,000 years so far, can we consider ourselves due for another one? What accounts for that ice age rhythm? And why do ice sheets grow and shrink anyhow?

"Anyone who could figure that out would win the Nobel Prize," says George H. Denton, professor of geology at the University of Maine. "We've been trying since 1837"—the year that Louis Agassiz informed shocked Swiss colleagues that ice had inundated their beloved Alps during an *"Eiszeit"* (ice age). The cause of ice ages couldn't be just cold; during the last one, for example, two regions of the far north that would seem naturals to have had plenty of ice on them, inland Alaska and most of Siberia, did not. Apparently, they were too dry. The East Antarctic Ice Sheet, for another example, had more snow on it during a particular interglaciation than during several glaciations, according to a recent study, presumably because warmer air can retain more moisture and therefore can—as long as the temperature is still below freezing—produce more snow.

Still, it's *mostly* a case of cold, but then the question arises, what causes the cold?

"The welter of proposals regarding causes of ice-age conditions resembles a bucket of worms," geologist Robert P. Sharp observed in his book *Living Ice: Understanding Glaciers and Glaciation*. Scientists have considered sunspot activity, volcanic eruptions, meteorite impacts, seafloor spreading, the drift of continents over the poles, the rise of mountains, changes in ocean currents and precipitation patterns and carbon dioxide levels in the atmosphere, even the effects of the glaciers themselves. But when they tried to link the occurrence of these phenomena to the actual timing of ice ages, Windsor Chorlton noted in his book *Ice Ages,* some "are hardly more satisfactory than when cricket scores are used."

One factor acknowledged to be somewhat more satisfactory than cricket scores is extraterrestrial in origin. It is the amount of sunlight that reaches the northern hemisphere in summer according to variations in the tilt of Earth's spin axis, the wobble (or precession) of its axis, and the eccentricity of its orbit (how close it gets to the sun). Not the southern hemisphere: it's 90 percent ocean, and ice sheets need land. Not the winter; ice sheets grow not so much because more snow falls in winter but because less snow melts in summer. In the 1920s, a Serbian mathematician, Milutin Milankovitch, worked out with great precision the intervals at which each of the three orbital conditions could be expected to deliver the least amount of sunlight to various latitudes of the northern hemisphere in summer, and therefore when ice sheets would be most likely to form there. He came up with every 22,000, 41,000, and 100,000 years.

However, nobody knew whether ice sheets had actually formed according to these "Milankovitch cycles" until the 1960s when corroboration came from some tiny hard-shelled sea creatures called forams. By studying foram fossils dug from the seabeds at various depths, marine scientists found that they could tell whether or not there had been ice sheets around at the time the forams died. What's more they could tell how big the ice sheets were.* After dating the fossils, they concluded that ice sheets in the northern hemisphere had built up at intervals of . . . 22,000, 41,000, and 100,000 years.

Orbital positioning, then, seemed to be setting the rhythm for ice ages—but it couldn't explain them entirely. One problem is that although the 100,000-year cycle, which is governed by the eccentricity of the

*To find out about the ice sheets, scientists compared the ratio in the forams' shells of heavy oxygen (^{18}O) to ordinary oxygen (^{16}O). Since ^{16}O is lighter than ^{18}O, the reasoning went, more of it than of ^{18}O would have evaporated from the ocean surface and become incorporated in snow and fallen onto land and turned to ice, leaving behind more ^{18}O in the water for the forams to use when making their shells. The bigger the ice sheets at the time, then, the more ^{18}O would stay behind in the water and end up in the tiny shells.

Earth's orbit, has the strongest correspondence with the actual timing of ice ages over the last million years, eccentricity has the *least* influence of the three orbital factors on the seasonal flux of sunlight. "Just hiccups," George Denton says. This is known as the "100K problem." Another sticking point, Denton points out, is that judging by treeline evidence the southern hemisphere has cooled and warmed in the past at roughly the same times as the northern hemisphere, which is not what you'd expect if orbital effects were the whole story. In addition, orbital cycles should produce a smooth curve, with changes in summer sunlight taking place gradually, incrementally, yet the "ice curve" is sawtoothed, as Denton and Wallace S. Broecker, professor of geochemistry at Columbia University's Lamont-Doherty Earth Observatory, pointed out. "The ice grows episodically for nearly 100,000 years," they wrote, "and then crashes in a few thousand."

Scientists are therefore still rooting around in the worm bucket, looking for other elements to add to the orbital ones to explain the comings and goings of ice ages. One important element seems to be ice's high albedo, or reflectivity. Once glaciers start growing, the increase in white surface means less sunlight is absorbed in an area, and that leads to a drop in temperatures, an increase in snow compared to rain, and less melting of the fallen snow. Growth of glaciers ensures more growth: cooling leads to greater cooling. This doesn't explain what triggers the development of glaciers, but the underlying assumption that ice sheets can grow through this positive feedback (as well as shrink through negative feedback, when the surface turns dirty or dark land is exposed around it) is widely accepted.

As to whether or not ice sheets per se can nudge the Earth over into an ice age, the consensus seems to be that it depends on their size. In some scientists' view, the Little Ice Age—which was not really an ice age, even a little one, but a cold snap within our present warm period (average temperatures dropped only a Fahrenheit degree or two)—came very close to setting off a *real* ice age, through a substantial buildup of glacial ice in northern Canada. (In Robert Sharp's definition, a real ice age requires that ice sheets approaching 400,000 square miles in area and thousands of feet in thickness appear on *nonpolar* continents; our only ice sheets now are on polar continents.) "An abortive glaciation," one scientist called the LIA. As temperatures dropped during it, glaciers on islands of the Canadian archipelago and snowfields on the Labrador plateaus started to grow and spread toward each other. Arthur Dyke of the Geological Survey of Canada ventured, "Maybe if the glaciers had been twice as big, or maybe only 20 percent bigger, a positive feedback loop would have got going and sustained an ice age."

Ice sheets can also by virtue of their size increase the size of climate swings. (Ice ages aren't unrelievedly cold nor interglaciations constantly warm, as ice cores have shown us; there are warm spells in cold periods and cold spells—like the Little Ice Age—during warm ones.) When Jerry F. McManus, associate scientist at the Woods Hole Oceanographic Institution, and colleagues analyzed the North Atlantic seabed record over the last half million years, they found that temperature swings during glaciations were far more extreme than during interglaciations, and what made the difference, it seemed, was the presence of large ice sheets. To their surprise, there appeared to be a threshold, an "important and persistent physical threshold," beyond which only a small change in ice volume set off a large change in climate. What could this threshold represent? they wondered. *Total amount of global ice?* Perhaps: a change in the proportion of water locked up in solid as compared to liquid form would affect the hydrological cycle. *Ice growth across some key strait, like the one between northern Greenland and Canada?* Maybe: ice barriers can block ocean currents. *Height and shape of the ice sheets themselves?* William Ruddiman, climate scientist at the University of Virginia, suggests that by the time the last glaciation reached its peak, the ice sheet in North America had developed plateau-like domes of ice "comparable in size to the Tibetan or Colorado plateau," and these split the jet stream in two, with one band passing north of the ice and one south of it. Strong winds from the northern band carried cold air over the North Atlantic and "even froze the ocean surface as far south as 50 degrees north, the latitude of France."

Still another way that ice sheets are thought to influence climate is through their waste products, so to speak—their icebergs and meltwater. It would work this way:

The circulation of the North Atlantic during our present warm period resembles a conveyor belt. (During previous warm periods, it did, too.) Water from the tropics flows north and warms western Europe beyond what would be expected by its latitude; Paris has about the same mean annual temperature as Montreal although it's almost ten degrees farther north. "Climatically, most of Europe is moved artificially closer to the equator," Jerry McManus notes, "which partly explains the accident of European civilization." Wallace Broecker was amazed to discover when he did the calculations that the current delivers to the North Atlantic almost a third as much heat as the sun does.

When the current reaches the vicinity of Iceland, the water becomes cooler and saltier (through evaporation), therefore denser and heavier. It sinks toward the ocean floor and starts flowing back the way it came, rising diffusely farther south. As long as water keeps sinking in high latitudes, water keeps on flowing from low latitudes to replace it. Where icebergs and meltwater from the ice sheets come in is that when they are added to the sea, they dilute the saltwater with fresh water, making it lighter and less likely to sink. If water doesn't sink in the north, water from the south has no place to flow into. The conveyor belt slows, or stops. Europe shifts climatically away from the equator, and closer to the pole.

There's evidence that during the last ice age, fresh water from ice-sheet breakdown did "kick the conveyor belt," in Denton's phrase. The amount of fresh water involved would not have been paltry. Denton and Broecker estimate that while the Laurentide ice sheet was melting, a process that took around 7,000 years, it released meltwater "at about the same rate as today's Amazon River." It also released fleets of icebergs, which sailed hundreds of miles into the Atlantic, melting along the way. There's a question whether cooling of the North Atlantic would have resulted in cooling around the globe (and the onset of an ice age) or only, as James W. C. White of the University of Colorado at Boulder suggested, a "halo of effects."

For most of the earth's history, there has been no ice here at all. The "normal" state of our planet could thus be considered to be iceless. During the first half of its 4.6-billion-year lifetime, the earth apparently contained no ice even at the poles, despite the fact that the sun was far weaker then than it is now. (Geological proof of a total lack of ice is hard to come by since the earth's surface has been so massively reworked.) Scientists puzzling over why the planet didn't just freeze over completely in those "dim-sun" times (a condition from which some computer models suggest it would never have recovered) have surmised that there was more CO_2 in the atmosphere then, which trapped outgoing radiation, compensating for the weak incoming radiation.

During the second half of the earth's lifetime, there have been several long cold periods, lasting millions of years each, during which there were substantial ice buildups, but these didn't last nearly as long as the warm, ice-free periods between them. The most recent long cold period started about 55 million years ago. (It is officially considered to be an ice age, with glaciations and interglaciations coming and going within it, but the glaciations themselves have come to be known informally as ice ages, too.) Before about 55 million years ago, temperatures around the world were about 10°F warmer than they are now, with temperatures at the poles 15°F or 20°F warmer (temperature effects are exaggerated at the poles). Palmetto-like trees were growing above the Arctic Circle and crocodile-like creatures swam along the north slope of Alaska. Then the earth started cooling. By about 35 million years ago Antarctica had its first ice sheet. The "least-astonishment" explanation for the cooling, according to Ruddiman, was a drop in CO_2 in the atmosphere, and one possible explanation for that drop was a rise in rock. As continental plates slid around, one with India on it slammed into Asia between 55 and 40 million years ago and forced the earth upward, producing the Tibetan plateau and the Himalayan mountains. The uplift slowly pulled CO_2 out of the atmosphere, according to a theory put forth by Maureen E. Raymo, research professor at Boston University:

Those higher, steeper slopes, more exposed to the atmosphere and in the path of passing storms, weathered faster, and dissolved CO_2 from the rock was carried by rainwater into the sea, where it was taken up by tiny creatures like plankton and coral. When the creatures died, they took the CO_2 with them to the seafloor, and there it stayed. Thus was CO_2 taken out of the atmosphere, where it helped warm the earth, and sequestered in the seabed, where it did not.

"It seems every time there were ice ages there was a tectonic event," Denton remarks. The raising of the Andes, the opening and closing of sea gateways like the isthmus of Panama could also have changed global temperatures, through their effect on atmospheric CO_2 and ocean currents. It all happened slowly. After Antarctica got its first ice sheet, more than 30 million years passed before the northern hemisphere got its first large ice sheet. That was two and a half million years ago. Orbital cycles had been operating all along, of course, but it wasn't until then that the northern hemisphere had cooled to the point where it could respond to the orbital cycles by producing large amounts of ice.

During the million and a half years after that, ice ages appeared in response to orbital forcing "like a ticking grandfather clock," Denton says, every 40,000 years. But starting about a million years ago, "it became more complicated," according to White. The 40,000-year cycle didn't disappear but it was overwhelmed by the 100,000-year cycle.

"Something fundamentally different happened at the million-year mark," he notes. "The magnitude of the ice sheets shifted, and we don't know why."

Some scientists suspect that what happened a million years ago was that the long-term decline in CO_2 levels allowed the earth to get cold enough for ice sheets to grow large enough to survive the shorter-cycle warmings; and once the ice sheets got large, they could, by their influence on the ocean and atmosphere, amplify the 100,000-year eccentricity cycle and make it dominant, despite its weak, "hiccup" status. The key to the "100K problem," then, may lie with the ice sheets themselves. It could have taken 100,000 years for the ice sheets to grow large enough to create a shift in the ocean-atmosphere system and cause the climate to "flip," according to Denton and Broecker. The reason it could have taken the ice sheets that long to grow that large, Jerry McManus explains, is simply that "ice sheets can grow only as fast as it can snow each winter."

So we are living in abnormal times as far as the earth's experience of ice is concerned. There is ice on the continent over the South Pole as well as on the sea over the North Pole (usually). Ten percent of the world's supply of fresh water is still in frozen form. Furthermore, there are signs on a large portion of the land in the northern hemisphere where no ice exists now—except for seasonal lake covers and icicles—that in the not very distant past, geologically speaking, ice was a powerful, overweening presence there.

"Have you no memories, O Darkly Bright?"

—Hart Crane, *"North Labrador"*

The biggest ice sheet of the last ice age was the Laurentide ice sheet, the one that buried New York City. It was the biggest ice sheet ever, so far as we know. It was as big or bigger than the combined Antarctic ice sheets are now, almost 200 feet "sea-level equivalent," which is a way of saying that at its maximum its total volume would, when melted into the ocean, have raised the ocean's surface around the globe by 200 feet. *"A very big old honking piece of ice,"* White calls it (he's from Tennessee).

Scientists know that the Laurentide ice sheet was big and would have resembled those on Antarctica today, but there are plenty of other things they don't know about it and would like to. A small group of them—glaciologists, physicists, geologists, and paleoclimatologists—are trying to figure out such things as where the ice sheet got started, how thick it became in various places, what its margins were during the years of its ascendance, where it changed speed and direction, and how it finally came undone and transferred its mass to the lakes and oceans—all without there being any actual ice for them to measure, or drill cores from, or make radar soundings of, or take aerial photographs of. Essentially, they are trying to reconstruct an ice sheet in absentia.

The task is a huge, complex, and messy one with long time lines, like the ice sheet itself. Shawn Marshall, associate professor of geography at the University of Calgary, expects to spend at least two decades making round-the-clock computer calculations to test and refine ideas, 1,700 billion calculations a day 365 days a year for an annual total of over 600 trillion calculations, in order to come up with a mathematical model of the ice sheet over its lifetime. *Why do it?* Why take the time and trouble to reimagine a vanished slab of ice, no matter how big it was? "In the debate about climate change we need to understand natural variability," Marshall says, "and for that we need to know what the ice sheet was like and what it was doing before humans got here." After all, he points out, the last ice age was "the most dramatic recent climate change on Earth."

Although long gone, the ice sheet did leave plenty of clues about itself behind. Having a thick layer of ice sitting on top of a landscape for thousands of years was no small matter. For one thing, the ice was heavy. Richard R. Peltier, university professor of physics at the University of Toronto, who is also engaged in ice-sheet reconstruction, figured that the ice depressed the Earth's crust to roughly a third of its own thickness. (That means the ground under my apartment building would have been pushed downward 300 feet.) All the land covered by ice sheets during the last ice age is still rebounding from the load. Peltier uses the amount of rebound to estimate how thick the ice sheet was in a particular place. In Hudson Bay, where the crust is rising at 1/3 to 2/3 of an inch a year, he reckons that the ice got to be 11,500 feet thick, the thickest of any place on the ice sheet. Thickest but not highest: Hudson Bay was an "ice trough" or "ice valley" into which ice flowed and built up from that low point. The ice stood only 6,500 to 9,000 feet above sea level there. The highest place on the ice sheet was probably southeast of Hudson Bay, where the surface was about 10,500 feet—almost two miles—above sea level.

The Laurentide wasn't just a heavy, cold blanket spread across the landscape, however. Like all glacial ice, it moved. In most places, it moved more than once over the same terrain, advancing and retreating and readvancing in response to changes in the intensity of summer sunlight as well as in snowfall and winds and conditions at the bed. As this Godzilla of an ice sheet marched across the North American continent, it crushed vegetation (the first time through), scraped away topsoil, pulverized rock, pried boulders loose from valley walls and carried them off, molded rocky debris and dust into ridges many miles long, and scratched bedrock with stones it held in its basal grip. When it retreated (it didn't actually move in reverse but melted back), it released rocks it had scooped up or shoved ahead of it and dropped boulders it had borne up to 500 miles from home.

Some of the imprints it made on the Earth during its brutal passage tell scientists which direction it was headed (for instance, the tapered end of little hills that the ice sculpted into teardrop shapes—drumlins—point the way the ice sheet was going). Some marks, like "trains" of boulders and moraines, reveal where the margins of the ice sheet were. But when the ice passed through an area or stopped is tricky to figure out. According to Arthur Dyke, scientists look for fossils of marine shells embedded in moraines along the coast, "then if we're lucky, we follow flow features like drumlins back to the interior until we encounter what looks like a younger margin, where we look for more fossils." They also look for bits of trees and moss scooped up by the moving ice and date those. Recently, some scientists began running isotope tests on boulders, to get an idea how long they were exposed to the sun and therefore when the ice sheet might have plucked them from valley walls. Some marks, though, scientists can only date relatively; where one set of scratches crosses another set on bedrock, for instance, "all we can say," admits Dyke, "is 'this before that,' 'this after that.' "

One thing is certain: margins and directions of flow changed often. "The ice sheet wasn't like a big mushroom slowly growing," Garry Clarke, professor of physics at the University of British Columbia, notes. "There were lots of actions, comings and goings, warm periods in the middle." If a time-lapse movie could have been taken from overhead, the Laurentide would have looked more like an amoeba than a mushroom, with wavy, undulating edges, lobes pushing out and pulling back, a giant pulsating blob. When Geoffrey Boulton, Regius Professor of Geology at the University of Edinburgh, examined satellite images of areas once covered by the ice sheet, he found "palimpsests" of former lines of movement under more recent ones and was struck by how often the lines diverged. From one advance to the next, he found, the ice switched headings by as much as 90 degrees.

Palimpsests notwithstanding, later ice advances usually erased marks made by earlier ones. It's not surprising, then, that the best reconstructions made of the ice sheet so far are for the most recent period, from the time of its greatest extent 20,000 years ago to its extinction 7,000 years ago. During this period of deglaciation, Dyke says, "we can pin down the ice margin to within a century." He and Victor Prest, also at the Geological Survey of Canada, have drawn up an authoritative series of maps that portray the ice sheet at several-thousand-year intervals during this period. On them are such topographical features as ice domes, ice divides, and saddles. "Domes" were points from which ice flowed out radially, "divides" lines from which ice flowed out laterally, and "saddles" places where divides intersected. The survey's map for 20,000 years ago, for instance, displays three domes (over Quebec-Labrador, Keewatin, and Foxe Basin), one "superdivide," and a network of smaller divides. The domes wandered around during deglaciation as snowfall patterns changed and the melting ice was funneled out through ice streams; a couple of domes migrated a distance of over 500 miles.*

Also on that map are fingers of ice reaching deep into Illinois, Indiana, Iowa, the Dakotas, and Nebraska. "We suspect that it [the American heartland] got invaded very late in the day," Clarke says. That far south, the underside of the ice sheet and the bed beneath it were most likely warm, wet, and slippery. "As the ice moves down into Michigan, you can see it is making a Napoleonic, overextended rush," he explains. Over the hard rock of the Canadian Shield in central and eastern Canada, the ice probably stayed frozen to the bed during most of the ice age and moved only by creep; "the ice just sat there and got thick." But closer to the margins of the ice sheet, where the bed consisted of soft soils (silts, sands, clays), the ice slid as well as crept. And once it started sliding, it slid faster. The heat of the friction melted the bed underneath, and the ice's rate of flow increased as much as 100-fold. "It was like a big marshmallow you bring to the melting point," Clarke points out. "It could switch fast." As the ice spread out and thinned, or else ran out of ice to drain from farther north, it refroze to the bed. Some beds probably alternated between being warm and cold.

All in all, Shawn Marshall concludes, "the southern margins were a really, really energetic system, with lots of melting and rain in summer and lots of snow in winter." He has bicycled along the edge of the Vat-

*Bodies of ice around the Laurentide ice sheet and sometimes in touch with it but not part of it, including the Appalachian ice complex over Maine and New Brunswick, the Newfoundland and Nova Scotia ice caps, and the Innuitian and Cordilleran ice sheets, had their own domes and divides.

najökull ice cap in Iceland and suggests that the ice over New York City at the peak of the ice age might have looked the way that ice cap does now: "dirty, cracked up, riddled by crevasses, with exorbitant quantities of summer meltwater draining into the crevasses, then bursting out of the margins in a large network of braided streams, which shift and pulse."

During the first 5,000 years after the Laurentide reached its peak size, it shrank slowly, mostly along its southern and western margins—the southern margin because of the warmer temperatures, the western margin because of the way the winds blew (west to east). As Dyke explains it, the western edge of the ice sheet began melting, then meltwater formed large lakes on top of the ice sheet, then much of that water evaporated. Winds blew the moisture-laden air eastward over the ice sheet, which cooled it, and the moisture turned to snow, which fell on the ice sheet. Thus, Dyke concludes, "the western part of the ice sheet cannibalized itself to feed the eastern."

After 5,000 years, the melting speeded up. On the Geological Survey map for 12,000 years ago, you can see large "proglacial" (meaning next to the ice sheet, fed by the ice sheet) lakes forming. One was Lake Agassiz, the largest lake there ever has been in the history of the world, so far as we know. At one time, it covered 135,000 square miles, an area the size of Italy, and had a drainage system so vast that fish from the Mississippi valley could swim to Quebec through it. "It seems likely that the majority of fishes living anywhere in the interior of Canada and the northern United States today have ancestors that, at one time or another, lived in Lake Agassiz," E. C. Pielou wrote in her book *After the Ice Age: The Return of Life to Glaciated North America.* Where Agassiz was walled in by the ice sheet, it was probably deep, 3,000 feet or so. Tongues of ice probably reached out onto its surface, like ice shelves, and icebergs probably floated across it—in summer, that is. A recent study suggests that Agassiz was frozen nine to ten months of the year.

The ice sheet produced a colossal amount of meltwater in a short amount of time. According to Wallace Broecker, the outflow of meltwater from all of the ice sheets of the northern hemisphere during their period of decline was nearly half as much as is discharged today from all of the hemisphere's rivers combined. At first, most of the Laurentide's meltwater drained out through the Mississippi River into the Gulf of Mexico. But once the southern edge of the ice sheet had receded past the Great Lakes, the meltwater started pouring out through the Saint Lawrence River valley into the North Atlantic. There, Broecker contends, it stopped the conveyor belt, which in the lengthening summer sunshine had started up again and rewarmed Europe and brought on the growth of forests there. (One definition of an ice age is botanical; an interglaciation, wrote geologist John Imbrie, co-author of *Ice Ages: Solving the Mystery,* is "an interval of time during which oak and other deciduous trees are widespread in Europe. It is the demise of these oak forests that signals the beginning of an ice age.") Once the drainage route shifted to the North Atlantic and fresh meltwater diluted saltwater there, surface currents no longer flowed north out of the tropics. Europe cooled. The forests died. The tufted little tundra flowers called dryas bloomed. Over a thousand years later when a lobe of ice blocked the Saint Lawrence's North Atlantic exit and the flow of meltwater from the Laurentide petered out, the conveyor belt switched on again. Europe warmed. The forests grew. The dryas flowers vanished. Broecker considers this cold snap after deglaciation had begun—the Younger Dryas—a "smoking gun" for the proposition that changes in the North Atlantic current, brought on by infusions of fresh water from melting ice sheets, affected climate in a big way.

Even in decline, the ice sheet kept on acting like an amoeba. Its outlines "wobbl[ed] unsteadily as they contracted to nothing," Pielou wrote. "The ice continued to advance in some regions for thousands of years after it had begun to shrink elsewhere." Sometimes the retreating ice sheet left "dead" ice behind. Blocks of ice broke from the ice front, then debris covered the blocks, and trees and plants grew on top of the debris. The buried ice stuck around for thousands of years. In a negative feedback process, "melting ice slowed its own melting," Pielou explained. As ice at the surface melted, stones that had been embedded in it collected on the surface. Thus, "the thin spots in the original insulation automatically repaired themselves." When the buried ice finally wasted away, it left behind hollows, which when filled with water became the prairie ponds so popular today with migrating ducks.

By 7,000 years ago, as seen on a Geological Survey map, only four pieces of the Laurentide ice sheet were left: a shmoo-shaped slab in Labrador less than 600 miles across, two pickle-shaped stretches on Baffin Island and the Melville Peninsula, and a tiny scrap of ice on nearby Southampton Island. By 5,000 years ago, all but the two on Baffin—the Barnes and Penny—were gone. Scientists have pondered why the peak of the ice age came so near its end—or, to put it another way, why the ice sheet collapsed so fast. "There are enough positive feedbacks—the cold, the albedo—you'd think it would persist," Marshall says. Dyke agrees: "Everything being equal, ice sheets want to stay ice." So does McManus:

"Why is it that in one Milankovitch cycle only a fraction of the ice goes away," he asks, "and in the next cycle as well, then *whhisshh!* the entire Laurentide disappears?"

Shawn Marshall suggests that it might have something to do with ice "overextending" itself. "The ice sheet came so far down it became unstable," he says. "It was too large to be self-sustaining." When ice moves in and occupies an area, he explains, it starts off by freezing to the permafrosted bed. Over time, as the thickening ice cover reduces the dissipation of geothermal heat, the bed warms. "So throughout its life," Marshall concludes, "an ever-increasing percentage of the Laurentide's base is at the melting point," 40 or 50 percent of it 40,000 years ago, 80 percent 14,000 years ago. It was thus late in the day when the ice sheet became both large and warm-based enough to spread south and west onto the soft soils of the continental heartland, at which point, sliding fast and thinning drastically, it would have been vulnerable to deglaciation at the next climate warming. Meanwhile, the core of the ice sheet, the central domes still frozen to the Canadian Shield, might themselves have become vulnerable when the ice to the south tapped into their frozen stores and drew ice out "at a greedy rate."

The model of the Laurentide ice sheet that Marshall and Clarke are creating will cover not just the final period, following its peak, but its entire lifetime, birth to death, plus the periods just preceding and following it, or a total of 120,000 years. Their unique approach is to divide the area once overlain by the ice sheet into 12,000 grids, each roughly 30 miles square. For each grid they enter into the computer the particulars of the topography that's there today, not just averages as some others have done but actual high and low points, slope angles, etc. Then they factor in two equations, one describing the way ice flows under its own weight and the other giving climate information, based on general models and the Greenland and Antarctic ice-core findings.

From these starting points, they can deduce how much snow fell within a grid, which way the snow flowed off a high point, and how much snow melted in deep valleys. Eventually, they will be able to produce "snapshots" of the ice sheet as it was forming and moving around and melting back during the whole span of the ice age, one snapshot for every grid for every five- to ten-year period for the entire 120,000 years. That will add up to millions of snapshots, which someday they hope to put together into a movie. "We could speed up the whole 120,000 years into something you could see in five minutes," Marshall says. "It would be like watching a river. It could be terrifying."

According to what they've learned so far, the Laurentide ice sheet had four birthplaces: Baffin Island, the Torngat Mountains and Ungava Plateau of Labrador ("they get a lot of snow," says Marshall), an area west of Hudson Bay in the Keewatin district of the Northwest Territories ("it's just cold"), and the Quebec highlands ("with the Great Lakes and ocean around, there's lots of snow"). All four places began to "light up," or grow a lot of ice, at the same time. Marshall is also intrigued to see popping up on his computer screen "a bit of an island" of ice just north of Lake Superior. Is it a "model result"? he wonders. Or a fifth birthplace? "It's pretty wet there," he notes. "Maybe it started to get some snow."

Why would four or five places begin lighting up at once almost 120,000 years ago? It *was* a time of vulnerability in the Milankovitch cycles, but of itself that probably wouldn't have been enough to get an ice sheet started. Were there lower levels of CO_2 in the atmosphere? CO_2 levels *were* lower during the ice age than before it started, but not when it started. "I have a bias toward the ice," Marshall says. "It looks as if the ice was doing things first"—things like covering up vegetation that produces CO_2. But what put the ice there in the first place? Large amounts of moisture around the North Atlantic probably helped. Ocean temperatures were still warm from the interglacial, and there probably wasn't much sea ice around. Moisture from evaporation of the open water would have turned to snow in the far north, and the snow fell on land—the beginning of an ice sheet. "Additional moisture was more important than additional cold," Marshall concludes. "It was plenty cold up there already."

From the four (or five) slabs of progenitor ice, the ice sheet grew quickly, for 5,000 or 10,000 years. By 110,000 years ago, according to the Marshall-Clarke model, there was a large buildup of ice on northern Canada, but over the next 10,000 years it shrank to almost nothing. By 90,000 years ago it had grown back to about the size it had been at 110,000, but by 80,000 years ago it had shrunk again. It used to be thought that ice-sheet growth consisted of a mostly steady buildup with some sawtoothed fluctuations along the way, but here were these huge "wiggles," as Marshall calls them. He says the wiggles reflect new information from the Greenland ice cores about how variable the ice age climate was. The large fluctuations were a dramatic response of the ice sheet to changes in summer sunlight, produced by the shorter 21,000-year orbital cycles superimposed on the longer cycles. "Ice sheets are not big, lugubrious features," Jerry McManus declares. "They are quite a bit more dynamic than we expected."

Starting 60,000 years ago, the Laurentide didn't behave as "fitfully" as it had before, according to Marshall. The wiggles became less pronounced, but they didn't disappear. The ice sheet didn't, either. After that point, "it was pretty much there all the time."

Among reconstructors of the ice sheet, there are disagreements about such things as how many domes it had, how thick its ice was, how fast it moved, and whether its bed was warm or not. (Peltier, for one, argues that it wasn't the bed that became soft and deformable but the ice.) Someday when these differences are worked out, and more fieldwork is done, and the trillions of computer calculations are made and the millions of snapshots produced, it should make a great movie. "The Long, Active Life of the Largest Ice Sheet in the History of the World." There's the drama of multiple births, the growth spurts, the Godzillan marches, the amoeba-like shimmying at the edges, the Napoleonic rushes southward, the wandering domes, the self-cannibalization of the western flank, the Amazonian gushes of meltwater, the flooding and pooling and damming, the release of fleets of icebergs, the profligate scattering of rocks, the rapid forfeiture of substance, the wasting away of all but two of the last patches of living ice.

And there is sure to be a sequel. Nobody doubts that the ice will return. When it will return, though, is anybody's guess. The guesses range from 500 years in the future to 50,000 years. "Tough call," Ruddiman admits. "There's just enough uncertainty about how long interglaciations have lasted." Recently, McManus and George Kukla, senior scientist at the Lamont-Doherty Earth Observatory, found evidence in seabed cores that the last interglaciation lasted almost twice as long as had been assumed, 20,000 years instead of 10,000. An earlier interglaciation may be an even better analogue for our own than the last one, McManus points out, since orbital alignments in that period were very close to today's. Then as now, the Earth's orbit was nearly circular, and the cool northern summers essential to getting an ice age started were thus not very pronounced, and the Milankovich cycle could have "skipped a beat." That interglaciation took place 400,000 years ago and lasted *30,000* years, or three times as long as ours has so far.

"If you go by the Milankovitch cycles," McManus concludes, "we are about due for an ice age. It might be tomorrow or a thousand years from now, but we are due." However, if you use previous warm periods as a "blueprint" for our own, we are not due. "We may be headed for an extended period of warmth in spite of the Milankovitch cycles. It could be 20,000 years before ice age cold sets in."

How about global warming due to a buildup in the atmosphere of greenhouse gases from the burning of fossil fuels—could it change the timing? One scientist suggested the warming could lead us into a "superinterglacial." Estimates are, however, that at the present rate of use our fossil fuels will be gone in several hundred years. "Maybe that's the best reason to conserve hydrocarbons," Dyke muses. "Maybe thousands of years from now when we need them to counter the orbital effects, people will say, 'Why did those troglodytes use them all up?' "

When the Sun flares out, and the Earth turns to ice

—Nicholas Christopher, "*33*" (from *5° and Other Poems*)

Now and then I try to imagine another ice sheet the size of the Laurentide invading New York City, ice oozing down the avenues, knocking over streetlights, bulldozing apartment buildings, plucking bridges off their stanchions, cleaving the Empire State Building off at the 84th floor, just below the observation deck. But it took tens of thousands of years for the Laurentide to reach the size and state of mobility where it could reach this far south, and by the time that many years have passed after a new ice sheet is born on some wet, cold, middling-high spots in eastern Canada, the Empire State Building, not to mention my apartment building and book and bones, will be long gone.

In the Norse creation story, a very cold, empty abyss (Ginnungagap) lies between a region of mist (Niflheim) and a region of fire (Muspellheim). As Barbara Leonie Picard related in *Tales of the Norse Gods and Heroes:*

"When it touched the cold that rose from the chasm, the damp mist from Niflheim turned to blocks of ice which fell with a terrible sound into Ginnungagap; and as the fiery sparks from Muspellheim fell upon these ice blocks they sent up steam which turned to hoar-frost as it rose into the cold air. And thus, with blocks of ice and with rime, the chasm was slowly filled."

From this frosty, rimy place emerged gods and men as well as their enemies, the ice giants. After eons of meddling in the gods' affairs, the ice giants joined forces with fire giants and storm giants and a host of monsters to do bloody battle against men and gods on the day of doom known as Ragnarok, in which all would perish:

"First would there be three winters more terrible than any that had ever gone before, with snow and ice and biting winds and no power in the sun . . . one long winter-time with never a respite. And at the end of that winter, Skoll, the wolf who had ever pursued the sun, would leap upon it and devour it . . . there would be darkness in the world. . . .

"And in the mighty conflict which would follow, all the earth . . . would shake with the clang and cry of war. . . . Though Odin would fight long and bravely with Fenris-Wolf, in the end that mighty monster would be too strong for him, and the wolf with his gaping jaws would devour the father of the gods . . .

"Then fire from Muspellheim would sweep over all, and thus would everything be destroyed; and it would indeed be the end of all things."

The world ended in fire. Ice was the prelude.

CHAPTER THIRTY-SIX

LAKE OF THE WOODS

Canada is only a dozen miles north, and the ice sheet that left its tracks all over the region has not gone for good, but only withdrawn. Something in the air, even in August, says it will be back.

—Wallace Stegner, *Crossing to Safety*

KENORA IS A TOWN on the north shore of Lake of the Woods, a lake 70 miles long with its southern part in the United States—Minnesota—and its northern part in Canada—Ontario, plus a slice of Manitoba. Kenora is just far enough north that a person might be able to spot a noctilucent cloud from there if he or she knew what to look for, could in fact tell a lenticular glow from the lucent sway of an aurora. The town started out as a Hudson Bay trading post, on the fur-pickup route of the voyageurs, and now has 15,000 people living in it, with 10,000 Indians, Ojibway and Cree, on nearby reserves. An Indian account has it that Wendigo, an evil spirit with a heart of ice, created the lake, then turned itself into a rock and sank to the bottom to rest, rising now and then to sabotage some boat. The paleoclimatologists' account is that the basin for the lake was refashioned from an old one when the Laurentide ice sheet passed through, melted, and formed Lake Agassiz, then Agassiz drained away.

Every summer, I spend a month on an island near Kenora, usually the month of August. I spend it at my friend Jamie's camp, where one day I discovered the tin icebox he used to fill as a child, buying a 50-pound block from an icehouse and pulling it to the dock in his little red wagon. In August, the weather in Kenora can get good and hot, and townspeople make widespread and exuberant use of the lake, swimming and boating and waterskiing in it. Yet as the month wears on, there *is* something in the air, as Stegner said, the evening chill arriving a bit earlier in the day perhaps, or a few yellow birch leaves appearing among the green. Even before the month's end, lake-water temperatures have started dropping, on what will presumably be their long, slow slide toward the state of 39°F poised equilibrium when all it will take is a calm, clear, cold night or two to transform the top into a film of ice. In *Cache Lake Country,* John Rowlands noted that the Ojib-Cree call the time when the thin ice that forms first in coves spreads out on quiet nights into open lakes "Kuskatinayoui," or "Ice Moon."

I'm not around to see that happen, but sometimes I feel I might as well be. Even in the heart of summer, Kenorans talk about what has been done, or will be done, in wintertime. I hear plans, reminiscences, cautionary tales. From them I get a little idea of what it must be like to live in a community where ice commands serious attention, where it is a strong, pervasive presence.

diamant streets . . .

—Richard Lassels, *The Voyage of Italy*

"Rattle rattle rattle rattle," Dale Lentinen says of the sound a boat makes as it pushes its way through a half inch of fresh crust. "Sounds like a tin can in a tin bucket." She lives in a house on the same island where Jamie has his camp but stays there year-round, so the way she gets about depends on how the ice has evolved. During the awkward period when there's ice on the lake but it's too thin to take a load, she and her husband, John, keep on using their boat, but as soon as the ice is three or four inches thick they switch to a snowmobile, and, when it's nine to ten inches thick, to a truck. In spring, it goes in reverse, truck to snowmobile back to boat, except that between snowmobile and boat, when the ice is still thick but candled and weak, they use a windjammer. John built the windjammer himself, tin-plating the bottoms and sides because, as Dale says, "the ice would cut a wooden boat to pieces."

. . . or a crystal pavement

—Lassels, *Voyage of Italy*

When it comes time for the truck, John makes his own ice road, two miles long, from the island to town. He makes the road almost as broad as 42nd Street in New York City, about 80 feet across, even though most of the time his truck will be the only vehicle on it. The main reason for the breadth of ice roads is so vehicles will stay well away from the piles of plowed snow on either side. A few years ago, a man from Kenora on his way to do some ice fishing with his eight-year-old son was driving on an ice road when his truck slid into a snowbank and flipped end over end and the man flew out and the truck landed on him and peeled off his face, as one witness related, and the blood got frozen into the ice, and when cars passed over that spot during spring melt "a quarter mile of road was running all red."

Even as John is out making his ice road, other people around the lake are making theirs, Indians wanting to haul firewood off islands, contractors planning to fix up lakeside cottages. It's evidently cheaper to truck equipment and materials over ice in winter—or, as one city official tells it, to "wait for winter to build its own, extended wharf out to the work"—than it is to barge them through open water in summer. Contractor Perry Heatherington has made ice roads up to 60 miles long to reach cottages he needs to work on.

> *Come on, Baby, we'll go cruise up and down*
> *on the Ice Road.*
>
> —Michael Faubion, *"Ice Road"* (song)

An ice road is a public road; one person may do all the work of making it, but everyone else feels free to use it. On weekends people pile into their cars and trucks and go roving on the frozen lake. "You relax, listen to the radio, take a picnic, go fishing," information officer Leo Heyens says. "In winter, there's more traffic on ice roads than on the streets of Kenora." The lake has 14,000 islands in it, rocky outcroppings that Wendigo supposedly scattered around to serve as windbreaks for Indian canoes, some with long score lines made when stones embedded in the underside of the Laurentide ice sheet scraped across them, some with discolorations left by the beaches of old Lake Agassiz, high above today's waterline, like rings around a bathtub.

"We choose our island and go exploring," musician Rob Lindstrom says. "It's an adventure, like going off the edge of the mountain. *There be dragons.*" Some people's idea of a good time is going nowhere at all, just taking advantage of the slick, cleared surface to spin their cars. "It's called 'doing doughnuts,' " fisheries technician Gavin Olson explains. "Crank the steering wheel, hit the brakes, away you go, just hold on. Seven is my record, in a Ford Escort."

In the Moon
Of Frost in the Tepees,
There were two stars
That got free.

—Paul Muldoon,
"The Year of the Sloes, for Ishi"

In February 2001, two sweethearts, she 18, he 23, were riding across the lake in a snowmobile, on their way back from a friend's house. It was past midnight, and they decided to take a shortcut, over snow-covered ice, away from the thick ice roads. At a place where a current must have thinned the ice further, the snowmobile hit slush, then open water, and sank. "She screamed hysterically," went an account in the *Toronto Globe and Mail* a couple of days later. He "kept reassuring her, trying to calm her down. Both of them tried, countless times, to get hold of a solid piece of ice," something they could drag themselves up onto. "But everywhere they touched, the ice snapped off in their hands. . . . 'The ice was breaking away forever,' " she remembered. "She eventually crawled up onto her stomach and rolled away. . . . With her help, he made it too. Limp, they lay on the ice. Severe frostbite was beginning." She tried to rise but couldn't "because her long blonde hair was stuck—glued, really, to the ice. It snapped—a sound like a particularly stubborn ice-cube tray when it's cracked open—as she ripped her hair off the surface. . . . Her legs, frozen solid, sounded like bowling pins clanking together."

They struggled through snow to an uninhabited shoreside cabin, broke in, and tried to keep warm by holding each other, but the next morning when two men in a windjammer picked them up, she had severe frostbite in her fingers and toes and he was dead of cardiac arrest, from hypothermia. The shortcut he took, disastrously, happened to be one he had taken, safely, twice before in the previous 24 hours.

silent icicles,
Quietly shining to the quiet Moon.

—Samuel Taylor Coleridge,
"Frost at Midnight"

Kenora is known as the Ice Candle Capital of Canada (at least in Kenora). On Christmas Eve, thousands of people go to the Lake of the Woods Cemetery to put "ice candles" on the graves of their loved ones, making what for many of them is a sad time of year less sad. In all, they set out about 5,000 ice candles. Starting a month before Christmas, volunteers make 100 of them a night by pouring water into five-gallon oil pails, letting the water freeze along the top and sides, overturning the pails, and dumping the unfrozen water out of the center. Some people buy these, but most people make their own. The homemade ones can be square or round, tall or squat, clear or frosted (depending on how fast the water froze), the ice sometimes colored by Kool-Aid (red or green) or embedded with things, minnows (for a fisherman), pine boughs, hairbows. On Christmas Eve, as a trumpeter and bagpiper play, townspeople light the long-lasting candles they've put inside their ice holders that burn all the way through Boxing Day.

"It is an absolutely stunning sight," says Barb Manson, cemetery manager. "It makes me cry every year." Ice candles, she explains, are an old Scandinavian tradition; her grandmother was a Finnish Lapp and made ice candles. Because of them, a visit to the graveyard on Christmas Eve is now so natural and popular in Kenora "you can't get into our cemetery at 1 a.m."

> Then the whole cow magically emerges out of the ice and is dragged on its side, its four legs straight and hard in the air
>
> —Michael Ondaatje, *In the Skin of a Lion*

Until recently, Dale and John Lentinen kept cows on their island and in winter would water them on the ice. They'd chop a hole about 50 feet from shore—any closer and the ice might be frozen to the bed—and the cows would stand around the hole lapping up cold water. If there wasn't much snow on the ice, John would rough up the surface with a bar so the cows could keep their footing. He'd also dig a trough in the ice leading away from the hole so the calves could drink without risk of falling in. "One time a calf slid into the hole and started to go under the ice," John relates. "I grabbed him by the tail just as he was heading down."

Contractor Tim Thorburn reports that wolves and wild dogs often chase deer out onto the ice—intentionally. In autumn, the ice is often smooth and bare and a deer, with its "hard feet," can't stand up on it. "It takes a few steps and its legs spread out and it lands on its belly." Even without wolves closing in, a fall on ice can be deadly for a deer; the contortion can break its pelvis. "One year it didn't snow before Christmas, and fifteen deer got killed on the ice," Thorburn says. "But once the ice has a little snow on it, a deer gets traction and it can outrun a wolf."

Dishevelled breast feathers worse
Than ice inside my closed hand.

—W. S. Merwin,
"Some Winter Sparrows"

One day Tim's father, Denver, was out hunting on the frozen lake when he "read" a story on top. He'd been following mink tracks across the ice for a couple of hours when he came across imprints made by wings . . . and no more mink tracks. He kept on walking and after a while saw a big pile of feathers ahead—and more mink tracks. Filling in the blanks represented by the long stretch of unmarked snow and ice, he could say that a bald eagle picked up the mink in its claws and flew away with it; the mink fought back while dangling in the air; the eagle landed to regroup; the mink kept on fighting; the eagle flew away minus some feathers and minus the mink; the mink carried on. So did Denver; "mink were worth a lot of money in those days," he says.

"Few animals *won't* travel on ice," biologist Scott Lockhart points out. "The bears come out at the end of March and look along the shoreline for things to eat—newly sprouted seedlings, grasses, emerging clover. They walk on ice the same as if it was land. Foxes are always out there looking for food scraps; they feed on gut piles the fishermen leave behind." Contractor Perry Heatherington once saw a couple of otters slipping in and out of a hole in the ice only two or three inches wide, sliding down and popping up, "the action wearing away the ice."

One animal that won't travel on ice if it can help it is the beaver. "A beaver sticks out on ice like a sore thumb," Lil Anderson, a wildlife rehabilitator in Kenora, points out. "Even a bald eagle will take a small one. With all that high-energy fat, it's like a giant sunflower seed." Also, with all that fat, it's still chewable when dead. "It doesn't freeze rock-hard, like a deer or moose."

an ice-vise, you could
say, vice-ice—a crown of ice

—A. R. Ammons,
"The King of Ice"

Alice Whiting, wife of a contractor, tells about the time she was driving along the lake shore and saw two deer out on the ice "with their horns locked." They were frozen half in, half out of the ice. She figured they had been fighting on the high bank, got their antlers caught, tumbled onto the lake, broke through the ice, and died together, their antlers still entwined.

> *I have frozen enough with terror,*
> *Better summon up a Bach chaconne*
>
> —Anna Akhmatova,
> *"Poem Without a Hero"*

Gavin Olson, the man who does doughnuts, likes to take his yellow Labrador retriever out onto the ice, for the sake of the retriever. "He associates ice with fun," Olson explains. "To him a field means work. On ice he knows he doesn't have to hunt, he doesn't have to scare anything up. He can chase his tail." Back when Kenorans kept horses instead of snowmobiles, they'd often race horses on the ice. The workhorses would clear a track for the racehorses, pulling squared timbers behind them. The track was a mile-long straightaway, and spectators would stand around on the ice and bet with cigars.

> The perspiration on their coats began to turn to frost. . . . White horses frozen on a field of ice and snow.
>
> —Alistair MacLeod, *No Great Mischief*

Until environmentalists complained, timber companies would store logs *in* the ice as well as on it, what Denver Thorburn called "making ice cubes with toothpicks in them." Trucks would haul logs from the bush and dump them on frozen bays; the weight of the logs would sink the ice; water would rise and freeze around the logs; trucks would dump more logs on top of those. When the ice became too thick to sink under the logs' weight, men would pump water up from below, "all day and all night for days and days," flooding the surface, Denver recounted. "On maybe six feet of ice there could be six to eight feet of wood, thousands of cords." In spring, when warm air and sunshine melted the ice and set the "toothpicks" free, boats would tow them by boom to the mill.

> *Winter . . .*
> *built himself a magic house of glass*
>
> —Charles G. D. Roberts, *"Ice"*

Denver showed me a home movie of an ice harvest in progress (he was one of the harvesters). I could just make out little horses, saws, ramps, pulleys. Ice harvesting was common on the lake until the mid-1950s, he explained, with icehouses lining the shores, including one that belonged to the Canadian Pacific Railway. That icehouse was so "humongous," retiree John Edwards remembers, "we'd be hauling ice eight hours a day in semi-trailers and *still* it would take us darn near a month to fill it!" All trains heading east or west stopped in Kenora to take on ice for cooling the passenger cars. On hot summer days, when the ice gang opened the doors of iceboxes under the cars to fill them, sometimes all they'd find of the ice loaded at the last icing station a mere 150 miles back was meltwater.

this train is speeding
along tracks of ice—not iron—
that are melting in its wake.

—Nicholas Christopher,
"23" (from 5° *and Other Poems*)

During World War II, Lake of the Woods had six German prisoner-of-war camps around it, and the guards, mostly Canadian World War I vets ("easygoing guys, very friendly," former POW Hans Kaiser recalls), would march the Germans, many of whom had been captured in the desert, across the ice into the woods, where they would cut logs for pay. A guard once marched several of them into Kenora to the dentist and while waiting for them all to have their teeth fixed got so drunk that he conked out and the prisoners had to pull him back over the ice on a toboggan. "One of the fellows was carrying the guard's rifle so that it wouldn't get slush on it," a former prisoner recalls. "But when they got close to camp, he realised that he'd better lay the rifle" down. The only known escape attempt was from Yellow Girl Camp, both men frozen before they got very far, their bodies discovered among the trees in spring.

Spring shall bloom where now the ice is,
. . . .
And my garden teem with spices.

—Christina Rossetti, *"Amen"*

Some people tell me that if you fall through the ice in spring, you're in worse trouble than if you fall through in winter. "You don't leave a hole in springtime," is how Rick McFarlane, captain of the tour boat *Kenora,* explains it. "The ice is like millions of icicles, and when you step on some of them you punch through and the rest move in over the top of you." "You can't swim in it," contractor Ernie Whiting agrees. "Dogs or animals that get in seldom get out." Once candling starts, the ice can go fast. "In two weeks you could go from driving a gravel truck to where you couldn't drive a half-ton," Tim Thorburn says. The ice goes in splotchy fashion, too—here and not there. In some parts of the lake people can be riding around in cars while in another part they're riding around in boats.

[they] live most upon fish . . .
. . . and eat grass and ice with delight.

—Nicholas Christopher,
"18" (from *5° and Other Poems*)

One family on Lake of the Woods still harvests its own ice. They're from Texas, with oil, gas, and cattle money and their own island. They come north in summer to fish and escape the even hotter summers in Texas. The ice is strictly for drinks, their cook confides. "We ice down the beers. We ice down the sodas. We make cocktails with ice. We put ice in the chest so the family can have ice water when they're fishing." The ice is usually cut before Christmas when it's about a foot thick, according to foreman Leonard Boucha. "We used to wait until spring, but the ice got so thick by then it took four guys to move one cake." From a chest on a fishing boat, he lifted a chunk of ice and held it up to the light. "You'll never get *that* from a machine!" he exclaimed. "See how *clear* it is?"

When I showed interest, Boucha took me to see the icehouse. It sits on a small rise, in the middle of the island, among trees, a wooden shack with a corrugated tin roof and—the mod touch—fiberglass insulation in the walls. A couple of the Texans came with us; they were curious and not a little proud. Boucha opened the icehouse door and we peered into the darkened interior, at a tall, brownish mound. There were seven tiers of ice there, he told us, 18 blocks to a tier, with a foot of sawdust on top. He started shoveling sawdust and throwing it aside, as if he were mucking out a horse stall. After a few minutes, he pulled a block of ice out of the top tier. It still had sawdust clinging to it.

With his hand he brushed away what he could of the dry, brown, opaque, insulating sawdust, slowly exposing the blue-gray, translucent, sharp-edged, glistening object underneath. The contrast was breathtaking. Even the Texans were in awe. In the half light of a shack on a small island in a lake remaining from the last ice age, a descendant of Persian domes and English haystacks, where the cold of winter is preserved for the needs and pleasures of summers, the block glowed richly against its ragged brown coating, like a gem loosed from its burlap wrapping:

Solidified azure.
Crystal jewelry.
Barbaric glass.
Dancing emerald.
Splendid pearl.
Pillar of light.
Opal and beryl.
The divinest, the most exquisite, the most intoxicating vision of fire and color and unimaginable splendor.
Diamond of the desert.

The largest diamond in the world.

Acknowledgments

For this book, I interviewed many dozens of scientists and other experts on ice of one sort or another who generously shared their observations and experiences with me. I am extremely grateful to them: their contributions form the core of the book. I cannot possibly list all of them here, but the following are ones who did me the additional favor of reviewing chapters, or parts of chapters, for accuracy: George Ashton, Sam Colbeck, Glenn Cota, Jeffrey Cuzzi, Jack Duman, Andrew Gage, Craig George, Yves Gilbert, Lorne Gold, Tony Gow, Robert Greenler, Charles Guy, Gordon Hamilton, Armand Karow, Brendan Kelly, Dennis Klug, Charles Knight, Gerald Kooyman, Edward Lozowski, J. Ross Mackay, Shawn Marshall, William J. Mills Jr., Donald Murphy, Bernard Nagengast, Thomas Osterkamp, Robert Pappalardo, Don Perovich, Austin Post, Davis Reid, Charles Ryerson, Bud Somerville, Ian Stirling, Gary Thomas, Lonnie Thompson, and Peter Zendt. A few of them—they know who they are—bore an extra burden by having to answer more than their share of my most vexatious questions. I would especially like to thank those scientists who furnished me with their photographs from the field.

In addition, I am indebted to Edna and Terry Dancy, who supported me in a wise and gracious manner during my stay at Pleasant Lake, New Hampshire. Also to Guy Guthridge of the Artists and Writers Program of the National Science Foundation, who, despite the fact that my application was late, gave me the chance to spend six weeks in Antarctica; to David Rosenthal, the artist in the program the year that I was the writer, who later put me up in his house in Cordova, Alaska; to Patricia Vornberger, my dorm roommate at the U.S. base at McMurdo, who knew her way around The Ice and tried to clue me in; and to Nirmal Sinha of the National Research Council of Canada, who first welcomed me to the ice island and got me interested in ice.

Many writer friends have given me comradely tips about the publishing enterprise plus the occasional nod of commiseration: Katrine Ames, Anita Been, Gary and Carol Carey, Claudia Carlin, Jean Carper, Kate Coleman, Naomi Cutner, Myron and Sabine Farber, Sue Haven, Mimi Hirsh and David Montenegro, Harriet Huber, Suzanne Little, John Lyon, John Mitchell, Cynthia Moss, Lucia Nevai, Elsa Rassbach, Phil Ross, Diane Weathers, and Delta Willis. One of them came up with the excellent slogan "Have an Ice Day," and another donated a bottle of *vin de glacière* to the cause.

Still other friends were advice-givers and cheerleaders from the start of the project: Davis Bernstein, Bea Brown, Sandy Erickson, Marcia and Saul Gordon, Joyce Gosnell, Mark Haven, Joan Headly, Shirley Kelley, Jacques Lecours and Claire Brassard, Bob and Pam Leitman, Vernon Lewis, Peter Lyman and Judith Moses, Alan Oberlin, Sandra Olansky, Barbara Ross, Sandy Straus, Noelle Thompson, Judith Turner, and Liz Wainstock. One of these friends, a Montrealer, offered an important corrective to my obsession: "Some people hate ice, you know."

I was amazed at the amount of work besides my own that went into making this a book; the publishing team at Knopf was magnificent. My editor, Ann Close, cut the manuscript so judiciously that I barely felt the pain, and on other matters, large and small, she was clearly in my corner. Her assistant, Millicent Bennett, was organized, knowledgeable, and endlessly reassuring. Robert Olsson made wonderful choices about design (he also drew the delightful ice-dripping ICE on the title page). Chip Kidd created a beautiful and original cover. Victoria Pearson was exacting but admirably flexible in her handling of the book through the production process. Rebecca Heisman took on the task of getting a large number of permissions with a minimum of *sturm* or *drang.* Wendy Weil, my agent, was always encouraging and usually calming.

Special thanks to Jamie Fenwick, who introduced me to Canadiens' hockey, to Lake of the Woods, and to the Internet, and who was honest in his judgments of my text and imaginative in keeping me well-fed.

Warm thanks, too, to my family, particularly my sister and brother, Molly Rudy and Bill Gosnell, who sent encouraging messages my way over the years. My cousin Art Gosnell, who kept trying to speed up the project, went so far one time as to write this poem: "My cousin, Mariana, is nice / She has written for pay, once or twice / She also has flown / cross country alone / But still has no book about ice!"

Now I have, and I thank all who helped me produce it.

Notes

Unless otherwise noted, all the quotes in this book by the following explorers were taken from a major—in some cases only—book each wrote:

Richard Byrd, *Alone*; Frederick Cook, *My Attainment of the Pole*; Douglas Mawson, *Home of the Blizzard*; Fridtjof Nansen, *Farthest North*; Robert Peary, *The North Pole*; Ernest Shackleton, *South*; Vilhjalmar Stefansson, *The Friendly Arctic*; and Will Steger, *North to the Pole*.

INTRODUCTION

5 The origin of the word "crystal" in a misunderstanding about ice is mentioned in *The Snowflake: Winter's Secret Beauty* by Kenneth Libbrecht (Stillwater, Minn.: Voyageur Press, 2003); "I blame Pliny," Libbrecht writes (p. 38).

5 Galileo's demonstration that ice floats is described in James Reston Jr.'s *Galileo: A Life* (New York: HarperCollins, 1994), pp. 122–25.

6 The paleoclimatologist quoted on water's triple point is Jerry F. McManus, associate scientist at the Woods Hole Oceanographic Institution.

6 The oceanographer quoted on our climate's temperature range was J. W. Kanwisher of the Woods Hole Oceanographic Institution, writing in a chapter on freezing in intertidal animals in *Cryobiology*, p. 487.

6 The idea of an open polar sea, which was "ancient, deep-rooted, widespread, and tenaciously preserved," according to Farley Mowat, is discussed in Mowat's book *The Polar Passion*, in which Isaac Hayes, an explorer who tried to reach that sea, is also quoted (p. 117).

CHAPTER ONE: LAKES

10 Estimates of temperatures at which water would freeze and boil if it were a "normal" substance are provided in *The Global Water Cycle: Geochemistry and Environment* by Elizabeth Kay Berner and Robert A. Berner, p. 8.

13 Roentgen's and Rowland's proposals about water's structure are mentioned, briefly, in N. E. Dorsey's book *Properties of Ordinary Water-Substance*, p. 164.

13 James Trefil, *Meditations at 10,000 Feet: A Scientist in the Mountains*. (New York: Charles Scribner's Sons, 1986), p. 128.

17 "Knowing Ice" appeared in *Audubon*, January 1980.

18 "The Pond at the Center of the Universe" appeared in *Ohio Magazine*, June 1991.

20 All W. A. Bentley quotes came from *The Ice of Lake Erie Around South Bass Island, 1936–1964*, by Thomas and Marina Langlois.

21 The early researcher was H. T. Barnes, writing in 1929, as quoted in the Langloises' *Ice of Lake Erie*, p. 20.

23 Hurley's comment about "ice flowers," as he referred to them, can be found in *South with Endurance: Shackleton's Antarctic Expedition, 1914–1917: The Photographs of Frank Hurley* (New York: Simon & Schuster, 2001), p. 275.

25 The use of ice patterns to forecast rice harvests in Japan is touched on in C. A. Knight's "Slush on Lakes," p. 462.

25 The ice-fishing crows are described on pp. 70–71 of Alexander F. Skutch's *The Minds of Birds* (College Station: Texas A&M Univ. Press, 1996).

27 The table giving rates of thickening of ice covers is in Dorsey's *Properties of Ordinary Water-Substance,* p. 409.

CHAPTER TWO: RIVERS

34 It was V. J. Schaefer, writing in "The Formation of Frazil and Anchor Ice in Cold Water," *Transactions of the American Geophysical Union* (1950), who marveled at frazil's capacity to shut down a 30,000-kilowatt hydroplant in less than an hour.

43 The CRREL report outlining problems of winter shipping on American waterways is *RIM: River Ice Management,* September 1981.

43 J. C. Tatinclaux, former chief of CRREL's ice engineering branch, provided the rough estimate of how much money ice's presence adds to the annual cost of shipping on U.S. rivers.

43 A news story about the Lisbon ship winning the golden cane, written by Alexander Norris, appeared in the *Montreal Gazette* of January 2, 1998.

44 Bob Splendorio of the New York State Power Authority supplied background on the Lake Erie–Niagara River ice boom.

CHAPTER THREE: GREAT LAKES

All statements by Thomas and Marina Langlois are taken from their publication, *The Ice of Lake Erie.*

47 The record high seiche was mentioned in Lawrence Pringle's *Rivers and Lakes* (Alexandria, Va.: Time-Life Books, 1985), p. 149.

51 The fluctuating extent of ice coverage of the Great Lakes is documented in Assel et al., *Great Lakes Ice Atlas.*

51 Dwight Boyer, *Great Stories of the Great Lakes* (Cleveland: Freshwater Press, 1966), pp. 211–13.

52 The article on Isle Royale wolves appeared in the December/January 2001 issue of *National Wildlife.*

53 The Rasputin quote about sacred sea is on p. 188 of his *Siberia on Fire* (translated from the Russian by G. Mikkelson and M. Winchell. DeKalb: Northern Illinois Univ. Press, 1989). Rasputin's quote about walking on the frozen lake is on pp. 176–77 of his *Siberia, Siberia* (translated from the Russian by M. Winchell. Evanston, Ill.: Northwestern Univ. Press, 1997).

54 Thubron's reference to the kingfisher blue of Baikal is on p. 156 of his *In Siberia* (New York: HarperCollins, 1999) and to bactrian camels on p. 159.

CHAPTER FOUR: LOADING

61 Doug Struthers of the Manitoba highways and transportation ministry provided details on that province's winter road network.

62 Information on rail service across the frozen Saint Lawrence River came primarily from two articles, "A Railway on Ice" in the May 1944 *Canadian Pacific Railways Staff Bulletin* and "The Ice Railway" in the July 1960 Canadian Railroad Historical Association's *News Report,* written by Robert R. Brown.

65 A description, with sketches, of the evolution of pie shapes (or "point-loaded wedges") appears on p. 184 of *River and Lake Ice Engineering* (G. D. Ashton, ed.), where the quote about foolhardiness also appears.

71 The incident of the boy who fell through ice in Central Park was recounted by Jim Dwyer in *New York Newsday,* March 22, 1993

71 Hugh Gray's 1809 *Letters from Canada* about saving horses is excerpted in Cavell and Reid's *When Winter Was King,* pp. 16–17.

CHAPTER FIVE: BREAKUP

Bernard Michel's statements come from his *Winter Regime of Rivers and Lakes.*

73 The *Times Argus* coverage of the Montpelier flood began on March 11, 1992.

80 There's something else besides candling that sunlight does to ice: melts "negative crystals" within it. "Take a slab of lake ice and place it in the path of a concentrated sunbeam," John Tyndall advised in *The Forms of Water.* "Part of the beam is stopped," he explained, "part of it goes through; the former produces internal liquefaction, the latter has no effect whatever. . . . From separate spots of the ice, little shining points are seen to sparkle forth. Every one of those points is surrounded by a beautiful liquid flower with six petals." Since the ice's melt takes up less room than the ice had, each point—the center of each flower—is a vacuum, and it is the surface of this empty space, according to Tyndall, that "shines in the sun with the lustre of burnished silver." The internal flowers are formed naturally whenever "the sun shines upon the ice of every lake, sometimes in myriads, and so small as to require a magnifying glass to see them." Today they are called "Tyndall figures."

CHAPTER SIX: ALPS

All of John Tyndall's statements were taken from his *Forms of Water.*

All of Twain's remarks in this chapter (including quotes from d'Arve) came from his *Tramp Abroad* (1880. Reprint, New York: Penguin, 1997).

86 Parfit's quote about blue ice is from "Ice Station," *Flying* (December 1985).

88 The text of Agassiz's lecture, on "The Ancient Glaciers of the Tropics," appeared (with applause bursts noted) in the *New-York Weekly Tribune* of February 20, 1867.

88 The progression of believers in glacier movement from local alpine hunter to internationally known biologist is discussed in *Ice Ages* by Windsor Chorlton, pp. 85–86.

CHAPTER SEVEN: SURGING GLACIERS

98 The elongation of a weighted beam of ice is described in "Tensile Creep-Rupture of Polycrystalline Ice," a paper submitted by John L. Burdick of the University of Alaska at Fairbanks to the Proceedings of the Third International Conference on Port and Ocean Engineering under Arctic Conditions in 1975.

99 The contributions of basal sliding to the movement of the Athabasca and Saskatchewan glaciers are compared in *Glaciers* by Hambrey and Alean, p. 57.

99 Tyndall describes the passage of a weighted wire through ice as "an experiment of singular beauty and interest," performed by a Mr. Bottomley, and reports that although the line where the wire passed was visible, "the two severed pieces of ice are so firmly frozen together that they will break elsewhere as soon as along the surface of regelation."

100 For a discussion of features characteristic of surging glaciers, including loops, see Austin Post's "Exceptional Advances of the Muldrow, Black Rapids, and Susitna Glaciers"; for folds, see his "Periodic Surge Origin of Folded Medial Moraines on Bering Piedmont Glacier, Alaska."

101 The large-scale Kutiah and Bråsvellbreen surges are mentioned in Sharp's *Living Ice,* p. 71.

103 Vitus Bering's failure to notice the glacier that would one day bear his name is remarked on in Muller and Fleisher's "Surging History and Potential for Renewed Retreat."

CHAPTER EIGHT: WEST ANTARCTIC ICE SHEET

109 Mercer's dire warnings are in his 1978 "West Antarctic Ice Sheet" paper.

117 The Alley-Whillans statement about WAIS changing *now* appears in their 1991 article, "Changes in the West Antarctic Ice Sheet."

118 Since Ice Stream B was renamed to memorialize Ian Whillans, other ice streams and ice ridges have been renamed to honor other glaciologists who worked on them: A is now Mercer Ice Stream, for instance; C is Kamb Ice Stream; the ridge between B and C is Engelhardt Ice Ridge; D is Bindschadler Ice Stream and F is Echelmeyer Ice Stream.

CHAPTER NINE: CORING

All but one of Richard Alley's comments are from his book *The Two-Mile Time Machine.* The book is a good primer on the science of ice coring in general, as is the brochure "Ice Core Contributions to Global Change Research" by the NSF's Ice Core Working Group. The Alley quote about Kilimanjaro at the end of the chapter was taken from a *New York Times* article, "Warming's Likely Victims" by Andrew Revkin, February 19, 2001.

123 A snapshot description of Camp Century appears in *Greenland Then and Now* by Erik Erngaard (Copenhagen: Lademann, 1972), p. 226.

125 The statement that Mayewski (who is now at the University of Maine at Orono) made about strained levels in ice sheets came from a lecture he gave at the American Museum of Natural History on November 4, 1998.

127 For a report on GRIP findings by GRIP participants, see "Climate Instability During the Last Interglacial Period Recorded in the GRIP Ice Core," *Nature,* vol. 364 (July 15, 1993).

127 Peel's comment was reported in W. Sullivan's "Study of Greenland Ice Finds Rapid Change in Past Climate," *New York Times,* July 15, 1993.

132 Ice age quotes from *Orlando: A Biography (1928)* are on pp. 198–99 of the 1993 reprint (London: Penguin).

132 In 1983, Wilfried Haeberli set up, also in Zurich, the World Glacier Monitoring Service, another group meant to keep track of the state of glaciers, not this time out of a concern about climate cooling but climate warming.

133 For impartial testimony that Thompson has not been wasting his time, see "Cooling the Tropics," by W. S. Broecker, *Nature,* July 20, 1995.

CHAPTER TEN: ON GLACIERS

136 The first pole at the South Pole was erected during the first winter-over there; as Paul Siple reported in *90° South* (p. 184), it was a bamboo pole with orange and black ascending stripes and a 16-inch mirrored glass ball from New Zealand which had been parachuted in without breaking.

140 Scott was "exceedingly unlucky" with the weather at the end of his trek, according to Susan Solomon, senior scientist at the National Oceanic and Atmospheric Administration and an expert on ozone holes. In her book *The Coldest March: Scott's Fatal Antarctic Expedition* (New Haven: Yale Univ. Press, 2001), she points out that temperatures were not only low but exceptionally low. "From the end of February until the last temperatures were recorded on March 19," she writes, "Scott and his men struggled through three weeks when almost every daily minimum temperature was a bitter and debilitating 10–20°F colder than what can now be shown to be typical . . . in that region"—and with such cold "went Scott's chances for survival."

140 In Roland Huntford's introduction to *The South Pole Expedition: The Amundsen Photographs* (New York: Atlantic Monthly, 1987), he wrote that "in the course of a few days, Amundsen had demythologized Antarctic travel. It *was,* after all," he explained, "only ski racing writ large" (p. 43).

143 The Commonwealth Trans-Antarctic Expedition's troubles with crevasses are described in Fuchs's article, "The Crossing of Antarctica," *National Geographic,* January 1959, as well as in Hillary's book *Nothing Venture, Nothing Win* (New York: Coward, McCann & Geoghegan, 1975).

145 Since my visit and talk with Hothem, several modern buildings with windows and stilts have been erected at the South Pole, and the dome's facilities are gradually being transferred to them. By 2017, a person at the southernmost center of the world may have to lie on snow where the infirmary used to be.

CHAPTER ELEVEN: ICEBERGS I

For information on ice's role in the sinking of the *Titanic* and the sinking's aftermath—rescue, inquiry, early schemes for raising, etc.—I relied largely on Lord's *Night to Remember* and *Night Lives On.* He interviewed dozens of survivors for his book, and I was able to interview him, in his memento-filled apartment.

The International Iceberg Patrol's chief scientist, Donald Murphy, provided most of the background on IIP activities as well much of it on North Atlantic icebergs. Also useful in both regards was the U.S. Coast Guard booklet *International Ice Patrol.*

149 The photo taken by a steward on the *Prinz Adalbert* is printed in *"Unsinkable": The Full Story of RMS Titanic* by Daniel Allen Butler, who considers the striped ice "almost certainly the iceberg struck by the *Titanic*" (Mechanicsburg, Pa.: Stackpole Books, 1998), facing p. 149.

151 The rivet revelations are in William J. Broad's *New York Times* article, "Faulty Rivets Emerge as Clues to *Titanic* Disaster," January 27, 1998.

155 The *Hanshedtoft* account is in the *International Ice Patrol* booklet, p. 4.

155 There's a reference to "ice manufactory" in *Arctic Whalers, Icy Seas* by W. Gillis Ross (Toronto: Irwin Publishing, 1985), p. 60.

155 The U.S. Navy's 1954 manual *Ice Observations* shows the "relative sizes of ice masses of land origin."

157 The Castle Berg photo, with caption, appears in *1910–1916 Antarctic Photographs, Herbert Ponting and Frank Hurley* (New York: St. Martin's, 1980), p. 58.

159 W. Bradford, *The Arctic Regions* (London: Chiswick Press, 1873).

159 The *Anchorage Daily News* account was written by Pamela Doto, June 25, 1993.

161 Osborn's comparison of ice floes and icebergs appears in Mowat's *Ordeal by Ice* on p. 257.

161 The commentator on Davis's iceberg photos is Alexandra Anderson-Spivy; her essay appears in a guide to a Johnson County Community College Gallery of Art exhibit in summer 1994.

162 Cowper wrote of borrowed beams (including "the ruby's fiery glow") in his poem "On the Ice Islands Seen Floating in the German Ocean."

162 Porter's photo of a jade-green iceberg is on p. 128 of his book *Antarctica* (New York: E. P. Dutton, 1978).

162 The article "Bottle-Green Iceberg near the South Shetland Islands," by K. N. Moulton and R. L. Cameron, appeared in the *Antarctic Journal,* no. 11 (1976).

163 Scientists on the Australian ship—S. G. Warren, C. S. Roesler, V. I. Morgan, R. E. Brandt, I. D. Goodwin, and I. Allison—reported their findings in the paper "Green Icebergs Formed by Freezing of Organic-Rich Seawater to the Base of Antarctic Ice Shelves," *Journal of Geophysical Research,* vol. 98, no. C4 (April 15, 1993).

163 The IIP's Murphy suggests that "bioluminescence" would be a better term than phosphorescence for the lights Cook saw.

CHAPTER TWELVE: ICEBERGS II

164 Estimates of standing room and drinking water that B-9 could provide are given in EOSAT Landsat Data User Notes, December 1987.

167 Easy watering from an iceberg is described by James Cook in Beaglehole's *Life of . . . Cook,* p. 317.

167 Isaacs's proposal for berg towing is referred to in "Iceberg Water for California" by J. C. Burt, *Science Digest,* vol. 39 (1956).

168 The estimate of total mass of all Antarctic icebergs dumped in the ocean annually comes from the NSF's Antarctic Program.

169 See Brian Hill's Website, *www.icedata.ca,* for a list of iceberg-ship collisions in the North Atlantic from 1686 to present (perpetrator and damage usually given). "Database of Ship Collisions with Icebergs," Institute for Ocean Technology, National Research Council of Canada.

169 The first recorded encounter with an iceberg may have been the one experienced by Saint Brendan of Ireland, who claimed to have crossed the Atlantic in the 6th century and come upon, as J. F. Webb wrote in his translation of *The Age of Bede . . . with the Voyage of St Brendan* (London: Penguin Books, 1965): "A column rising out of the sea . . . pure crystal . . . higher than the sky" with a canopy "the colour of silver and . . . harder than marble."

173 Gladys's fate is described in Lever's "Hazard Potential of Glacial Ice Masses," *C-CORE News* (March 1988).

175 Crocker's account of a berg explosion appears in "Size Distributions of Bergy Bits and Growlers Calved from Deteriorating Icebergs," *Cold Regions Science and Technology,* vol. 22 (1993).

175 James Cook's quote about the danger of small pieces of ice is in Beaglehole's *Life of . . . Cook,* p. 321.

175–176 Darwin's 1855 iceberg essay appears in *The Collected Papers of Charles Darwin,* Paul H. Barrett, ed. (Chicago: University of Chicago Press, 1977).

176 The spokesman was Stephen E. Bruneau, manager of Newfoundland operations at North Atlantic Pipeline Partners.

176–177 The relation between sea-ice extent and iceberg numbers in the Grand Banks is explained in "Iceberg Severity off Eastern North America: Its Relationship to Sea Ice Variability and Climate Change" by J. R. Marko, D. B. Fissel, P. Wadhams, P. M. Kelly, and R. D. Brown, *Journal of Climate,* vol. 7 (September 1994).

178 In *Travels in Alaska,* Muir reports on the calving rate of his namesake glacier on p. 214, on the jail of ice on p. 110, on the fleet of icebergs in a fjord on p. 115, and on the ousel overhead on p. 140.

CHAPTER THIRTEEN: SEA ICE I

181 Statements by and about Frederick Cook near 70°*S* are on pp. 317–20 of Beaglehole's *Life of . . . Cook* and at 70°*N* on pp. 622–23.

183 In his *Arctic Manual,* Stefansson compares fresh and sea ice on p. 355, describes the grinding down of ice floes on p. 359, contrasts first- and multiyear ice on pp. 360 and 240 as well as ridges of both on p. 360. (The quote on p. 5 about sea ice splashing like ice cream, however, is from his *Friendly Arctic,* p. 297.)

185 The navy's 1954 *Ice Observations* manual that compared berg sizes to familiar items does the same for floes.

187 Nelson's list of Eskimo words for sea ice is on pp. 398–403 of *Hunters of Northern Ice.*

187 Pinker refers to the Eskimo language "hoax" on pp. 64–65 of his book *The Language Instinct* (New York: William Morrow, 1994).

189 The account of Barrow's ice override is on p. 302 of Brower's *Fifty Years.*

190 The reference to Inupiat oral history is in *Arctic Anthropology,* vol. 21, no. 1 (1984), p. 11. Kane's "white-marble" quote appears on p. 177 of Berton's *Arctic Grail.*

190 Ridge-building is described on pp. 4–5 of Lansing's *Endurance.*

193 Nansen describes his first sighting of sea ice in his book *The First Crossing of Greenland* (London: Longmans, Green, 1893), pp. 80–81.

193 Lansing's description of sea-ice noises are on p. 5 of *Endurance.*

194 Fiennes's comments on ice-block shapes appear in "What It Feels Like to Get Frostbite," *Esquire,* June 2001.

196 Ross's account of the *Victory*'s imprisonment in ice is on pp. 191–229 of Mowat's *Ordeal.*

197 Berton wrote of the ice flow that trapped Franklin's ships on page 329 of *The Arctic Grail.*

198 The four Northeast Passages are illustrated in *Northern Sea Route and Icebreaking Technology,* p. 63.

198 The quotes from Payer about boring through ice and being mere insects are from Jeannette Minsky's *To the Arctic!* The rest of his quotes are from *Realm of the Ice King.*

199 The role of Mercator in perpetuating the notion of an open polar sea is noted in Mowat's *Polar Passion* on p. 39.

200 DeLong's quotes can be found on pp. 197–98 of Mirsky's *To The Arctic!*

CHAPTER FOURTEEN: SEA ICE II

203 The *Sibiryakov* and *Chelyuskin* voyages are mentioned in *Northern Sea Route and Icebreaking Technology,* p. 11, as well as in Mirsky's *To the Arctic!* p. 280.

205 "Hap" Arnold's quote is from Lowell Thomas Jr.'s article, "Scientists Ride Ice Islands on Arctic Odysseys," *National Geographic,* November 1965.

205 The breakup of sea ice floes with U.S. research stations on them are described in Weeks and Maher's *Ice Island: Polar Science,* on pp. 20 and 60.

205 Joseph O. Fletcher, "Three Months on an Arctic Ice Island," *National Geographic,* April 1953.

206 The explorer who compared the ice rollers of Ellesmere to water rollers on a beach was Pelham Aldrich, on the ship *Alert* in the winter of 1875–76, as quoted in "Arctic Ice Islands" by L. S. Koenig et al., *Arctic,* vol. 5, no. 2 (July 1952).

207 In April 2003, a dozen years after the *Soviets* abandoned their last drifting research station, North Pole XXXI, the *Russians* set up a new one, North Pole XXXII, on a floe in the Arctic Ocean. Within a year XXXII had to be abandoned because strong winds shoved the ice into tall ridges which crashed down onto station buildings. Details are given in "Rescue On for Russian Crew After Arctic Camp Collapses" by Andrew C. Revkin, *New York Times,* March 5, 2004.

211 James F. Calvert, "Up Through the Ice of the North Pole," *National Geographic,* July 1959.

213 SHEBA project reports include "Year on Ice Gives Climate Insights" by D. K. Perovich et al., *Eos, Transactions, American Geophysical Union,* vol. 80, no. 41 (October 12, 1999), and "The Seasonal Evolution of Arctic Sea Ice Albedo" by D. K. Perovich, T. C. Grenfell, B. Light, and P. V. Hobbs, *Journal of Geophysical Research,* vol. 10 (2002).

215 In late 2004, a four-year study by 300 experts on the Arctic culminated in the release of the Arctic Climate Impact Assessment (ACIA), in which drastic changes in the region were documented and even more drastic changes predicted. Among other things, ACIA reported that the Arctic is warming almost twice as fast as the rest of the Earth and by the end of the century should be at least 7°F warmer than it is now. The part that has warmed the most is Alaska and the Canadian Yukon. The Greenland ice sheet will melt substantially and raise sea levels. Permafrost will thaw and damage buildings, train tracks, airport runways, and oil lines in Siberia. Ice roads will have shorter seasons and be less safe. Areas of tundra will shrink severely. Many of the effects of warming in the Arctic, including widespread melting of summer sea ice, will themselves intensify global warming.

216–217 The massive timbering of ships against the ice is described for the *Fram* on p. 152 of Huntford's *Nansen* and for Shackleton's *Endurance* on p. 19 of Lansing's book.

218 The observer is John D. Harbron, writing in "Modern Icebreakers," *Scientific American,* vol. 249, no. 6 (December 1983).

218 The *New York Times* story, "Ages-Old Icecap," was written by J. N. Wilford, August 18, 2000.

218 Amundsen's description of broken-up ice at the North Pole is on p. 142 of his *First Crossing of the Polar Sea;* the photo of the broken ice is facing p. 304.

221 Canadian Coast Guard, *Ice Navigation in Canadian Waters* (Canadian Government Publishing Centre, 1987).

CHAPTER FIFTEEN: GROUND ICE I

225 C. A. Knight, letter, *Weatherwise,* vol. 33, no. 4, (August 1980).

226 H. Ozawa and S. Kinosita, "Segregated Ice Growth on a Microporous Filter," *Journal of Colloid and Interface Science,* vol. 132, no. 1 (October 1, 1989).

228 For a discussion of upfreezing, read pp. 79–91 of Washburn's book *Geocryology.* For a *demonstration* of upfreezing, watch a video of time-lapse photos, taken by CRREL researchers, of a rock painted red and buried in New Hampshire silt inside a plexiglass tank, then exposed to cold air and water piped in from the bottom. When the photos are speeded up, you can see the freezing level work its way downward from the surface of the soil, with ice lenses appearing here and there as horizontal lines—narrow, dark, wavy, and wormlike. The rock rises slowly through the ground, like a lazy balloon, a quarter inch a day, three inches in 12 days.

229–230 Note CRREL paper, "Defining the Cold Regions of the Northern Hemisphere," by R. E. Bates and Michael A. Bilello (CRREL Technical Report 178, June 1966).

231 S. W. Muller named and defined permafrost in 1947 and a dozen years later admitted to Washburn that the term came from a slip of his own tongue "that he was encouraged to formalize," *Geocryology,* p. 21.

231 The figure of less than one percent and the Washburn quote about the significance of that one percent are on p. 25 of *Geocryology.*

231 Jack Ives, "Permafrost," chapter in *Arctic and Alpine Environments,* p. 177.

231 The Isham observations are in Fred Bruemmer's "Life upon the Permafrost," *Natural History,* April 1987.

232 Pierre Berton, *The Klondike Fever: The Life and Death of the Last Great Gold Rush* (New York: Alfred A. Knopf, 1958), p. 20.

233 The finding of a baby mammoth in Siberian permafrost is described in "Open Season on the Wooly Mammoth" by Randall Hyman, *International Wildlife,* May/June 1995.

CHAPTER SIXTEEN: GROUND ICE II

234 Porsild is quoted by Fred Breummer in "Life Upon the Permafrost," *Natural History,* April 1987.

234 Speck, *Samuel Hearne,* p. 155.

236 Richardson's sketch of a pingo, and a photo of it, and a discussion of names for it, can be found in Mackay's "Pingos of the Tuktoyaktuk Peninsula Area" on p. 4. Richardson's own description of the pingo, along with his sketch, are in his book *Arctic Searching Expedition: Journal of a Boat-Voyage Through Rupert's Land and the Arctic Sea* (New York: Harper & Bros., 1852).

236 DeSainville's quote is on p. 6 of Mackay's "Pingos of the Tuktoyaktuk."

238 The requirement that massive ice have 250 percent more ice by weight than soil was set by the Permafrost Subcommittee of the Associate Committee on Geotechnical Research in 1988, as Mackay and Dallimore noted in "Massive Ice of the Tuktoyaktuk Area, Western Arctic Coast, Canada," p. 1235.

239 Farley Mowat, *The Siberians* (New York: Little, Brown in association with Atlantic Monthly Press, 1970), p. 152.

239 Williams, *Pipelines and Primafrost,* p. 28.

239 Ives, "Permafrost," *Arctic and Alpine Environments,* p. 189.

240 Mowat, *Siberians,* p. 153.

244 A. Delano, *Across the Plains and Among the Diggings* (1850; reprint, New York: Wilson-Erickson, 1936).

244 K. L. Holmes, ed., *Covered Wagon Women: Diaries and Letters from the Western Trails 1840–1890,* vols. 2 and 3 (Glendale, Calif.: Arthur H. Clark, 1983).

CHAPTER SEVENTEEN: PLANTS

246 Robert Frost, "Birches."

247 The ability of Dahurian larches—which compose the most northerly forests in the world—to withstand extreme low temperatures is mentioned in Marchand's *Life in the Cold,* pp. 42–43.

248 J. Levitt, "Winter Hardiness in Plants," chapter 11 of the book *Cryobiology,* p. 501.

256 J. Bennett, *The Harrowsmith Northern Gardener* (Camden East, Ontario: Camden House, 1982).

256 Donald Hall, *Here at Eagle Pond* (Boston: Houghton Mifflin, 1990).

CHAPTER EIGHTEEN: ANIMALS I

258 The contributor to *The Bowhead Whale* mentioned here was Victor B. Scheffer, retired biologist with the U.S. Fish and Wildlife Service, who wrote the introduction.

259 *Uiñiq (The Open Lead).* (North Slope Borough, Barrow, Alaska, July 1989).

262 The engineer eavesdropping on whale banter was Christopher Clark, who wrote up his findings and impressions in "Singing in the Ice" (*The Living Bird Quarterly,* Autumn 1988); "Listening for the Bowhead" (*Whalewatcher,* vol. 21, no. 1, Spring 1987); and "Moving with the Heard" (*Natural History,* March 1991).

263 Brower, on walking across sea ice for two weeks: *Fifty Years Below Zero,* pp. 191–98.

265 Almost all the information about polar bears not otherwise attributed here comes from Ian Stirling, either in interviews or in his book *Polar Bears.*

268 The study that determined what types of ice polar bears like (and don't) is reported on in "Habitat Preferences of Polar Bears in the Western Canadian Arctic in Late Winter and Spring" by I. Stirling, D. Andriashek, and W. Calvert, *Polar Record,* vol. 29, no. 168 (1993).

270 Nelson on ice-throwing by bears: *Hunters of the Northern Ice,* p. 191.

271 Brower on polar bears crossing thin ice: *Fifty Years,* p. 247.

271 For more on the fix that Hudson Bay bears find themselves in, see "Possible Impacts of Climatic Warming on Polar Bears" by I. Stirling and A. E. Derocher (*Arctic,* vol. 46, no. 3, Sept. 1993), as well as "Running Out of Ice?" by I. Stirling (*Natural History,* March 2000).

CHAPTER NINETEEN: ANIMALS II

273–274 Feazel, *White Bear: Encounters with the Master of the Arctic Ice.* (New York: Henry Holt, 1990).

274 The long-time researcher mentioned is Brendan Kelly, seal expert at the University of Alaska Southeast.

276 Nelson on looking into seal holes: *Hunters of the Northern Ice,* p. 232.

276 For more details on how seals locate breathing holes, see "Under-Ice Movements and the Sensory Basis of Hole Finding by Ringed and Weddell Seals" by D. Wartzok, R. Elsner, H. Stone, B. P. Kelly, and R. W. Davis, *Canadian Journal of Zoology,* vol. 70, no. 9 (1992).

277 For a good summary of which seals prefer which ice types, see J. J. Burns's "Remarks on the Distribution and Natural History of Pagophilic Pinnipeds in the Bering and Chukchi Seas." For advantages that ice in general offers seals in general, see F. H. Fay's "Role of Ice in the Ecology of Marine Mammals of the Bering Sea."

278 Stefansson describes the crawling method of hunting seals on pp. 452–54 of his *Arctic Manual.*

281 Sketches of a Weddell seal with teeth sunk into the underside of ice covering its breathing hole can be found in Kooyman's "Weddell Seal" article; other comments about the seal's ice-reaming teeth are in his *Weddell Seal* book (pp. 89, 90, and 117).

CHAPTER TWENTY: ANIMALS III

289 Wilson's journal entries are in the appendix to Scott's *Last Expedition,* with the quotes about emperor penguins on p. 483.

295 Kooyman's first impressions of the Cape Washington colony appear in his "Emperors of Antarctica."

296 Kooyman's study of key elements in a successful colony is explained in his "Breeding Habitats."

297 M. W. Browne describes Kooyman's night flight over an emperor colony in "Eggs on Feet and Far from Shelter, Male Penguins Do a Shuffle," *New York Times,* September 27, 1994.

298 The paleontologist commenting on the Adélies' appearance is G. G. Simpson, in his book *Penguins: Past and Present, Here and There* (New Haven, Conn.: Yale University Press, 1976), p. 35.

299 Peterson, *Penguins* (Boston: Houghton Mifflin, 1979), p. 149.

304 The reference to deep water in Labrador is in DeVries's paper "Biological Antifreeze Agents."

305 The way protein antifreezes in fish work, DeVries explains, is by throwing a "net" of molecules over a nascent ice crystal and binding to it, thereby keeping additional water molecules from themselves binding to the crystal and enlarging it. "The proteins make the crystal face very bumpy," he says. "It's analogous to dividing the surface into many microcrystals. The highly curved water molecules don't want to stay there; they'd rather stay in the water. To make them stay, you'd have to reduce temperatures so that the hydrogen bonds would be strong enough to hold them."

CHAPTER TWENTY-ONE: ANIMALS IV

307 Facts on freezing in insects came largely from Duman et al., "Adaptations of Insects," and K. and J. Storey's "Frozen and Alive," as well as from interviews with John Duman and Janet Storey.

310 Richard Conniff, *Spineless Wonders: Strange Tales from the Invertebrate World* (New York: Henry Holt, 1996), p. 153.

310 The step-by-step explanation of how frogs freeze (and thaw) is in the Storeys' "Frozen and Alive."

311 Speck, on frozen frogs: *Samuel Hearne and the Northwest Passage,* p. 158.

312 The painted turtle tale is told in "Out Cold" by J. and K. Storey, *Natural History,* January 1992.

312 J. W. Kanwisher, in the book *Cryobiology,* pp. 488–89.

313 David Rains Wallace, *Idle Weeds: The Life of an Ohio Sandstone Ridge* (Columbus: Ohio State University Press, 1980), p. 23.

314 For (much) more on Divoky and his Cooper Island guillemots, see "George Divoky's Planet" by Darcy Frey, *New York Times Magazine,* January 6, 2002.

314 Alfred Marshall Bailey, *Birds of Arctic Alaska* (Colorado Museum of Natural History, Popular Series, no. 8, April 1, 1948).

314–315 Balogh, "Secret Life of the Spectacled Eider" (*National Wildlife,* April/May 1997) and "Spectacle on Ice" (*Bird Watcher's Digest,* January/February 1997).

315 For Brower's fox quotes, see *Fifty Years,* p. 246.

316 Nelson's remarks about how well equipped polar bears are for eating frozen food are on p. 190 of *Hunters of the Northern Ice.*

316 K. and G. Harrison, *Birds of Winter* (New York: Random House, 1990), p. 39.

CHAPTER TWENTY-TWO: HUMAN I

319 Jack London, "To Build a Fire" (1910). Reprinted in *To Build a Fire and Other Stories* (New York: Bantam, 1986).

320 Stuck on "strong cold": *Ten Thousand Miles,* pp. 59, 67–68.

321 All Lapp references are from Turi, *Turi's Book.*

322 Harold Meryman, "Mechanics of Freezing in Living Cells," *Science,* vol. 124, no. 322 (September 21, 1956).

323 The Scott search party report can be found in chapter 21 of *Last Expedition,* "The Finding of the Dead," by surgeon E. L. Atkinson, pp. 466–69.

324 Various names for cold-injured feet are given in *Cold Injuries with Special References to German Experiences During WWII,* by Hans Killian (1952), p. 42.

324 Among the opportunities for getting frostbite in Alaska, as listed in the "Cold Injury" issue of *Alaska Medicine,* were hunting accident, fall through ice in river, assault ("fled in bare feet"), changing truck tire, trapping in snowstorm, and runaway dog team.

326 Peter Marchand on humans as tropical beings: *Life in the Cold,* pp. 186–87, 200.

326 Stuck's sword and deadly weapon quotes are on pp. 59–60 of his *Ten Thousand Miles,* and his information about noses is on p. 87.

327 In his *Arctic Manual,* Stefansson offered advice on using a warm hand to thaw a cold face on p. 292, what underwear to wear on p. 248, and why not to keep a beard on p. 293.

328 Munro's 500th jump is described in *Whiskey Whiskey Papa: Chronicling the Exciting Life and Times of a Pilot's Pilot,* by Norman Avery (self-published), p. 115.

CHAPTER TWENTY-THREE: HUMAN II

332 The expert was Dr. Charles J. Devine, whose statement about ice's prolongation of tissue life appeared in the *New York Times* "Doctor's World" column, by Lawrence K. Altman, July 13, 1993.

337 Ralph C. Merkle, "The Molecular Repair of the Brain," *Cryonics,* vol. 15, nos. 1 and 2 (January and April 1994).

342 The Iceman's shoes got particularly high marks from investigators. As Burkhard Bilger tells the story in "Sole Survivor" (*New Yorker,* February 14 and 21, 2005), mountain climbers and shoe experts hiked over rock and ice for three days wearing modern duplicates of the Iceman's old shoes—deerskin uppers, calfskin bindings, bearskin soles, braided-linden-bark netting stuffed with hay and moss—and concluded, one reported, "There is no mountain in Europe that couldn't be conquered in these shoes."

342 The director of the Bolzano museum was Alex Susanna (Associated Press, July 25, 2001).

343 Information on the hunter in British Columbia came from "Out of the Ice" by H. Pringle (*Canadian Geographic,* July/August 2002); on the girl in Peru from "Move Over, Iceman, for New Star from the Andes," by J. N. Wilford (*New York Times,* October 25, 1995); and on the woman in Siberia from "A Mummy Unearthed from the Pastures of Heaven," by N. Polosmak (*National Geographic,* October 1994).

CHAPTER TWENTY-FOUR: GAMES I

344 Studs Terkel, *Working* (New York: Pantheon, 1974), pp. 505–506.

344 William Shakespeare, *King John,* Act iv, sc. ii.

345 In the chapter on surface properties in his book *Ice Physics* (pp. 392–460), Hobbs mentions the work of both Faraday (on a liquid-like layer) and Bowden and Hughes (on friction) as ways of explaining ice's slipperiness.

349 On skating in Central Park: "Parks in History: New York on Ice," by M. Siegel, *Leaflet* (Winter 1985).

349 Cavell and Reid on the first covered rink: *When Winter Was King,* p. 19.

349 John Cheever, *Oh What a Paradise It Seems* (New York: Alfred A. Knopf, 1982), pp. 7–8.

352 It was Matthew Murphy, author of "Dan Clapp's Skeeters," who wrote of a "harmonic convergence of the elements."

354 In Cavell and Reid's *When Winter Was King,* Montmorency Falls is described, and illustrated several times, on pp. 10, 16, 27, 64, 74.

356 The sportswriter mentioned was Bruce Bennett, sports columnist for the *Duluth News-Tribune,* whom James P. Sterba quoted in the *Wall Street Journal* of February 17, 1992.

357 *A Hog on Ice and Other Curious Expressions* (Scranton, Pa.: HarperCollins, 1985).

358 Stephen King's musings on slippery ice are from "What Are You Afraid Of?" (*New Yorker,* September 7, 1998) by Mark Singer, who had sheepishly asked King about ghosts.

CHAPTER TWENTY-FIVE: GAMES II

359 The *Fox's* rounders game is mentioned on p. 291 of Mowat's *Ordeal by Ice,* the *Fram's* rope-dancing on p. 634 of Sverdrup's report in the appendix of Nansen's *Farthest North,* and the *Pioneer's* walkabouts on p. 256 of Mowat's *Ordeal.*

360 Cavell and Reid, *When Winter Was King,* pp. 15, 33.

361 Chouinard's quote about ice being an unnatural medium for man is on p. 180 of his *Climbing Ice;* the one about ice being plastic is on p. 136; and the one about ice being renewable is on p. 186.

363–364 Vince Rhodes, "Arctic Adventure," *Sport Diver* (December 1994).

364 Jim Mastro, "Oasis Under the Ice," *International Wildlife* (November/December 1997).

365 Yves Gilbert's parachutist quote is from M. Beaudry's "Trial by Ice," in *Equinox,* vol. 9 (January/February 1990).

CHAPTER TWENTY-SIX: USES I

371 Photos of pretend ices are in "The Mediterranean: Sea of Man's Fate," *National Geographic,* December 1982.

371 Molly O'Neill, "The Ice Capades," *New York Times,* August 22, 1993.

371 M. Tekulsky, *Make Your Own Ice Pops with Juices, Puddings, Yogurts, Ice Cream and More!* (New York: Berkley Books, 1994).

375 For more on Persian icehouses as well as Persian methods of making ice and habits of eating their distinctive ices (tamarind, pomegranate), see E. Beazley and M. Harverson's *Living with the Desert: Working Buildings of the Iranian Plateau* (Warminster, UK: Aris and Phillips, 1982).

375 Eric Sloane, *An Age of Barns* (New York: Funk & Wagnalls, 1967).

376 Beethoven's comment about the scarcity of ice comes from his letter to musician-friend Nikolaus Simrock, quoted in Maynard Solomon's *Beethoven* (New York: Schirmer, 1977), p. 86.

376 Emily Eden's plaintive letter is quoted in David's *Harvest of the Cold Months,* p. 272.

377 Bernard Nagengast, ventilation engineer, historian, and coauthor of *Heat and Cold,* suggests that Gorrie may have been inspired in his approach to ice-making by an incident that took place in a Hungarian coal mine in 1755. The miners pumped water out of a shaft, using the weight of a column of water as a piston, then removed the weight, which according to a contemporary account caused water and air to shoot upward as "ice pellets" with such force that they could "pierce a hat, if held against them, like pistol bullets."

378 The bit of history, with photos, of muffin tin as first ice-cube tray appears on pp. 213 and 214 of Donaldson and Nagengast, *Heat and Cold.* The ad about the speed of ice-cube production determining one's social status is on p. 242 of *Heat and Cold.*

378 David, *Harvest,* pp. 376–77.

379 J. J. Harris, "Cooling an Auditorium by the Use of Ice," *Transactions of The American Society of Heating and Ventilating Engineers,* vol. 9 (1903).

379–380 "The Ice Gang" by "One of Them," *Canadian Pacific Staff Bulletin* (October 1946).

380 Rocking chair that cools is shown on p. 269 of Donaldson and Nagengast, *Heat and Cold.*

CHAPTER TWENTY-SEVEN: USES II

384 The quote from Melvina Dean, a *Titanic* survivor who as of this writing is still living, appeared in the *London Times* on August 20, 1997, according to *The Oxford Dictionary of Twentieth Century Quotations,* E. Knowles, ed. (Oxford: Oxford University Press, 1998).

384 First rubber ice-cube tray, caption for figure 20, *Heat & Cold.*

385 Anne Sexton, "The Money Swing."

385 Sebastian Junger, *The Perfect Storm: A True Story of Men Against the Sea* (New York: W. W. Norton, 1997).

387 Early quick-freezing systems are discussed in the *Air Conditioning Refrigerating Data Book,* pp. 1–02 and 1–03.

388 "Freezers: The Rewards of Using Them Wisely," *New York Times,* May 18, 1992.

389–390 Márquez, *One Hundred Years of Solitude,* trans. from the Spanish by G. Rabassa (New York: Harper and Row, 1970).

CHAPTER TWENTY-EIGHT: USES III

391 Concerning 19th-century ice carving, see p. 8 of Amendola's book *Ice Carving Made Easy,* and concerning ice-carving lesson books, see pp. 9, 11, and 12 of his book.

393 Douglas Yates, *Alaska Ice Postcard Collection: New Ice Carving from Fairbanks, Alaska* (Ester, Alaska: 64th Parallel Press, 1990).

394 Mina Curtiss, *A Forgotten Empress: Anna Ivanovna and Her Era, 1730–1740* (New York: Frederick Ungar, 1974).

395 Most of the information on palaces at their peak was drawn from Anderes and Agranoff's *Ice Palaces.*

398 Bob Spitz, "Cold Comfort Charm," *Condé Nast Traveler,* August 1996.

398 James Houston, *Confessions of an Igloo Dweller* (Toronto: McClelland & Stewart, 1955), p. 52.

398 Stefansson on igloos: *Arctic Manual,* pp. 161–63.

399 On making an ice window for an igloo, See Houston's *Confessions,* p. 54, and on the moon looking through that window, pp. 169–70.

400 F. I. Mackenzie's quote is from McEvoy's "Operation Habbakuk: 'Professor' Pyke's Secret Weapon," *The Beaver,* April/May 1994.

400 Habakkuk (the code name's misspelling a typist's error, never erased) was a minor prophet of the 7th century B.C. in whose eponymous book of the Bible can be found words befitting a project involving bergships: "wonder marvelously: for I will work a work in your days, which you will not believe, though it be told you."

405–406 *X-men* Series (Marvel Enterprises, New York, N.Y.).

CHAPTER TWENTY-NINE: OTHER FORMS OF ICE

For good overviews on other forms of ice, see Whalley's "Hydrogen Bond in Ice" as well as chapter 3 of N. H. Fletcher's *Chemical Physics of Ice.*

408 K. Vonnegut, *Cat's Cradle* (New York: Delacorte, 1963).

410 For more on clathrates whose guest molecule is methane, see "The Crystal Fuel," *Natural History,* May 1997. Air can be a guest molecule, too (see Coring chapter, p. 126) and so can carbon dioxide, perhaps, on the poles of Mars.

411 Svishchev's quote about Jupiter is from Szpir's "Bits of Ice XI and Ice XII."

412 Concerning cubic ice on Earth, see "Scheiner's Halo: Evidence for Ice Ic in the Atmosphere," by E. Whalley (*Science,* vol. 211, Jan. 23, 1981), as well as—an alternate explanation—"Scheiner's Halo: Cubic Ice or Polycrystalline Hexagonal Ice?" by A. J. Weinheimer and C. A. Knight (*Journal of the Atmospheric Sciences,* vol. 44, no. 21, November 1987).

CHAPTER THIRTY: ATMOSPHERE I

414 "Ice Man Cometh," Ice Storm Report, Entergy, New Orleans, 1994.

414 *Power Workers' News,* Toronto, vol. 5, issue 4 (February 1998).

414 "The Ice Storm of '98," *Montreal Gazette,* January 24, 1998.

415 Bernd Heinrich, *The Trees in My Forest* (New York: HarperCollins, 1998).

415 Editorial, *Toronto Globe and Mail,* January 13, 1998.

415 Mark Twain, *Following the Equator: A Journey Around the World* (1897; reprint, New York: Harper and Bros., n.d).

421 Johannes Kepler, *On the Six-Cornered Snowflake* (1611; reprint, Oxford, U.K.: Clarendon Press, 1966).

421 For early observers of snow-crystal shapes, see B. J. Mason's commentary on Kepler's *On the Six-Cornered Snowflake* in the 1966 Clarendon reprint (pp. 47–50), as well as Hobbs's account in *Ice Physics,* pp. 524–26.

422 For one of many discussions of Nakaya's work, see Hallett's "How Snow Crystals Grow."

424 The National Science Foundation report referred to is "The Hail Problem," *Hail Suppression: Impact and Issues* (1977).

426 Ernie Gann, *Fate Is the Hunter* (New York: Simon and Schuster, 1961), pp. 79–107.

CHAPTER THIRTY-ONE: ATMOSPHERE II

430 J. A. Allen, "A Singular Case of Shipwreck," *Nature,* June 2, 1881.

431 N. Maeno, L. Makkonen, K. Nishimura, K. Kosugi, and T. Takahashi, "Growth Rate of Icicles," *Journal of Glaciology,* vol. 40, no. 135 (1994).

433 K. K. Chung and E. P. Lozowski, "On the Growth of Marine Icicles," *Atmosphere-Ocean,* vol. 28, no. 4 (1990).

433 Bertrand Piccard, "Around at Last!" *National Geographic,* September 1999.

434 Some of the letters on "anti-icicles" appear in the November 1985, August 1986, November 1987, and January 1991 issues of *Weather.*

434 Thoreau on frost forms: *Winter,* p. 8.

434 For Hallett's comments on frost, see his "How Snow Crystals Grow."

435 Edward LaChapelle on frost over snow: *Field Guide,* p. 96.

436 Humphreys's description of frost on windowpanes appears on p. 20 of *Snow Crystals,* Thoreau's on p. 151 of *Winter,* and Rowlands's on p. 249 of *Cache Lake Country.*

CHAPTER THIRTY-TWO: ATMOSPHERE III

438 Yeats's feelings conveyed to Maude Gonne were in turn conveyed to interested parties by Yeats's biographer, A. N. Jeffares, according to the *Norton Anthology of English Literature* (vol. 2, 3rd ed., M. H. Abrams et al. eds., New York: W. W. Norton, 1974), p. 1921.

441 *Jackie Cochran: An Autobiography* by Jacqueline Cochran and Maryann Bucknum Brinley (Toronto: Bantam, 1987), p. 279.

442 For particulars on how nine military airplanes left narrow contrails that grew into widespread cirrus, see "Briefly Empty Skies Offer Climate Clues," by Andrew C. Revkin, *New York Times,* October 30, 2001.

442 C. S. Benson, "Ice Fog," *Weather,* vol. 25, no. 1 (January 1970). In Fairbanks, Benson once used a "supermarket parking lot full of idling automobiles" as a collection site for ice crystals.

442 Guy Murchie, *Song of the Sky,* 2nd rev. ed. (New York: Ziff-Davis, 1979), pp. 238–39.

443 Peter Hobbs on cloud-seeding: *Ice Physics,* pp. 501 and 650.

CHAPTER THIRTY-THREE: SPACE I

450 A. Shepard and D. Slayton with J. Barbree and H. Benedict, *Moon Shot: The Inside Story of America's Race to the Moon* (Atlanta: Turner Publishing, 1994).

451 All Jonathan Lunine's quotes are from his introduction in a special issue of *Ad Astra,* "Ice in the Solar System."

452 Paul M. Schenk, "Ice Among the Gas Giants," in "Ice in the Solar System."

453 All William B. McKinnon's quotes are from chapter 22, "Midsize Icy Satellites," in Beatty et al., *The New Solar System.*

453 John Stansberry, "Triton and Pluto: On the Frozen Fringe," in "Ice in the Solar System."

454 Pappalardo's quote about severe icequakes comes from his "Ganymede and Callisto," chapter 19 in Beatty et al., *The New Solar System.*

455 Greeley's baseball simile is from his "Europa," chapter 18 in Beatty et al., *The New Solar System.*

458 In *Planets of Rock and Ice,* C. R. Chapman describes the complexity of Saturn's rings on p. 158 and their dazzling spokes on pp. 160 and 162.

CHAPTER THIRTY-FOUR: SPACE II

461 Owen mentions Mars's minimal atmosphere in "How the Earth Got Its Atmosphere," in special issue of *Ad Astra* ("Ice in the Solar System").

467 Whipple mentions the apparent willfulness of comets on pp. 1–2 of *Mystery of Comets,* the jetting about they do on p. 148–49, and the fantastic landscapes they may contain on pp. 169–70.

470 Brandt's "black velvet" comparison is from his "Comets," chapter 24 of Beatty et al., *The New Solar System.*

471 Cruikshank's "Ice from the Beginning of Time" is in "Ice in the Solar System" in *Ad Astra.*

472 Calvin's comment on water ice as universal ice appears in " 'Ices' throughout the Solar System: A Tour of Condensable Species," her article "Ices in the Solar System," *The Planetary Report.*

472 Cruikshank's quote about life on Earth being connected to stardust appears in his "Stardust Memories."

472 For report on *Stardust* images of Comet Wild 2, see Weaver, "Not a Rubble Pile?"

CHAPTER THIRTY-FIVE: ICE AGES

477 Quote on episodic growth of ice sheets is from Broecker and Denton, "What Drives Glacial Cycles?"

478 The "persistent threshold" is mentioned in McManus et al., "0.5-Million-Year Record."

479 For more on the enormous fleets of icebergs launched periodically by the North American ice sheet, see G. Bond, H. Heinrich, et al., "Evidence for Massive Discharges of Icebergs into the North Atlantic Ocean during the Last Glacial Period," *Nature,* vol. 360 (November 19, 1992).

482 Using a different method from Peltier's, Marshall and Clarke estimate that the ice in Hudson Bay got to be 12,800 feet thick, or more than 1,000 feet thicker than Peltier estimated.

485 In *After the Ice Age,* Pielou's comment about Lake Agassiz being the migrating hub for fish is on p. 195, about the wobbling outlines of the ice sheet on p. 107, and about the slow melt of dead ice on p. 180.

CHAPTER THIRTY-SIX: LAKE OF THE WOODS

492 Despite being regular features of polar and subpolar regions, auroras have nothing whatever to do with ice in the atmosphere. Unless you choose to put credence in Scandinavian folk explanations such as the one that says they are "sunlight reflecting off . . . swans trapped in the polar ice flapping desperately to free themselves" (L. Jago, *The Northern Lights,* New York: Alfred A. Knopf, 2001).

492 There's a terrific description of the Wendigo phenomenon in Margaret Atwood's book, *Strange Things: The Malevolent North in Canadian Literature* (Oxford, U.K.: Clarendon Press, 1995). According to the traditional beliefs of eastern woodland Indians,

mostly Ojibway and Cree, she explains, the Wendigo has not only a heart of ice but sometimes a body of ice as well. It also has "prodigious strength," claws, eyes that roll in blood, and a "ravenous hunger for human flesh." You can kill a Wendigo "in the usual ways—knifing, strangulation, and so forth," but you must "remove and burn the heart of ice in order to melt it."

495 G. MacDonald, " 'We Aren't Going to Make It,' " *Toronto Globe and Mail,* February 24, 2001.

495 J. Gibson, "Ice Candle Display Marks Christmas Eve," *Kenora Miner and News,* January 7, 2004.

499 J. MacDonald, "Prisoners on the Lake," *Lake of the Woods District Area News,* vol. 26, no. 3 (May 1996).

501 *Solidified azure.* (Henry David Thoreau)
Crystal jewelry. (Elisha Kent Kane)
Barbaric glass. (Wallace Stevens)
Dancing emerald. (Mark Twain)
Splendid pearl. (William Cowper)
Pillar of light. (Robert Greenler)
Opal and beryl. (Henry David Thoreau)
The divinest, the most exquisite . . . (Mark Twain)
Diamond of the desert. (Alonzo Delano)
The largest diamond in the world. (Gabriel García Márquez)

Select Bibliography

The sources listed below represent only a portion of those I consulted and found helpful in preparing this book. To list them all would have required a second, if smaller, book.

Adams, W. P., and N. T. Roulet. "Illustration of the Roles of Snow in the Evolution of the Winter Cover of a Lake." *Arctic,* vol. 33, no. 1 (1980).

Alley, R. B. "Icing the North Atlantic." *Nature,* vol. 392 (March 26, 1998).

Alley, R. B., and I. M. Whillans. "Changes in the West Antarctic Ice Sheet." *Science,* vol. 254 (November 15, 1991).

Alley, R. B., D. D. Blankenship, C. R. Bentley, and S. T. Rooney. "Deformation of Till Beneath Ice Stream B, West Antarctica." *Nature,* vol. 322 (July 3, 1986).

Alley, R. B., et al. "Abrupt Increase in Greenland Snow Accumulation at the End of the Younger Dryas Event." *Nature,* vol. 362 (April 8, 1993).

Amendola, J. *Ice Carving Made Easy.* 2nd ed. New York: Van Nostrand Reinhold, 1994.

American Society of Refrigerating Engineers. *Air Conditioning Refrigerating Data Book.* Applications volume, 6th ed., 1956.

Amundsen, R. *My Life as an Explorer.* New York: Doubleday, Page, 1927.

Amundsen, R. *The South Pole: An Account of the Norwegian Antarctic Expedition in the "Fram," 1910–1912,* 2 vols. London: John Murray, 1913.

Amundsen, R., and L. Ellsworth. *First Crossing of the Polar Sea.* Garden City, N. Y.: Doubleday, Doran, 1928.

Anderes, F., and A. Agranoff. *Ice Palaces.* New York: Abbeville Press, 1983.

Anderson, W. R., with Clay Blair Jr. *Nautilus 90 North.* New York: Signet Books, 1959.

Anisimov, O. A., and F. E. Nelson. "Permafrost Distribution in the Northern Hemisphere Under Scenarios of Climatic Change." *Global and Planetary Change,* vol. 14, (1996).

Arctic Climate Impact Assessment (ACIA). Overview Report. *Impacts of a Warming Arctic.* Cambridge, U.K.: Cambridge Univ. Press, 2004.

Ashton, G. D. "Freshwater Ice Growth, Motion, and Decay." In *Dynamics of Snow and Ice Masses.* San Diego: Academic Press, 1980.

Ashton, G. D. "River Ice." *American Scientist,* vol. 67 (January-February 1979).

Ashton, G. D., ed. *River and Lake Ice Engineering.* Highlands Ranch, Colo.: Water Resources Publications, 1986.

Ashton, G. D., and J. F. Kennedy. "Ripples on Underside of River Ice Covers." *Journal of the Hydraulics Division, Proceedings of the American Society of Civil Engineers,* 1972.

Assel, R. A., F. H. Quinn, G. A. Leshkevich, and S. J. Bolsenga. *Great Lakes Ice Atlas.* Ann Arbor, Mich.: Great Lakes Environmental Research Laboratory, NOAA, 1983.

Bailey, R. H. *Glacier.* Alexandria, Va.: Time-Life Books, 1982.

Baker, M. B. "Cloud Microphysics and Climate." *Science,* vol. 276 (May 16, 1997).

Baker, M. B., and J. G. Dash. "Charge Transfer in Thunderstorms and the Surface Melting of Ice." *Journal of Crystal Growth,* vol. 97 (1989).

Balch, E. S. *Glacières or Freezing Caverns.* New York: Johnson Reprint Corp., 1970.

Ball, P., H., S. Gaskill, and R. J. Lopez. "Iceberg Motion: An Analysis of Two Data Sets Collected at Drill Sites in the Labrador Sea." C-CORE Technical Report 81–2, 1981.

Barnes, P. W., S. E. Rawlinson, and E. Reimnitz. "Coastal Geomorphology of Arctic Alaska." *Arctic Coastal Processes and Slope Protection Design.* American Society of Civil Engineers, May 1988.

Baust, J., A. A. Gage, H. Ma, and C-M. Zhang. "Minimally Invasive Cryosurgery—Technological Advances." *Cryobiology* magazine, vol. 34 (1997).

Beaglehole, J. C. *The Life of Captain James Cook.* Stanford, Calif.: Stanford Univ. Press, 1974.

Beatty, J. K., C. C. Petersen, and A. Chaikin, eds. *The New Solar System,* 4th ed. Cambridge: Cambridge Univ. Press, 1999.

Benoit, M., M. Bernasconi, P. Focher, and M. Parrinello. "New High-Pressure Phase of Ice." *Physical Review Letters,* vol. 76, no. 16 (April 15, 1996).

Benson, C. S. "Ice Fog." *Weather,* vol. 25, no. 1 (January 1970).

Bentley, C. R. "Ice on the Fast Track." *Nature,* vol. 394 (July 2, 1998).

Bentley, C. R. "Rapid Sea-Level Rise Soon from West Antarctic Ice Sheet Collapse?" *Science,* vol. 275 (February 21, 1997).

Bentley, W. A., and W. J. Humphreys. *Snow Crystals.* New York: McGraw-Hill, 1931. Reprint, New York: Dover, 1962.

Bergreen, L. *Voyage to Mars: NASA's Search for Life Beyond Earth.* New York: Riverhead Books, 2000.

Berner, E. K., and R. A. Berner. *The Global Water Cycle: Geochemistry and Environment.* Englewood Cliffs, N.J.: Prentice-Hall, 1987.

Berton, P. *The Arctic Grail: The Quest for the North West Passage and the North Pole, 1818–1909.* Toronto: McClelland and Stewart, 1988.

Berton, P. *Niagara: A History of the Falls.* Toronto: McClelland and Stewart, 1992.

Bindschadler, R. "Future of the West Antarctic Ice Sheet." *Science,* vol. 282 (October 16, 1998).

Bindschadler, R. A., and C. R. Bentley. "On Thin Ice?" *Scientific American,* December 2002.

Blaisdell, G. L., R. M. Lang, G. Crist, K. Kurtti, R. J. Harbin, and D. Flora. *Construction, Maintenance, and Operation of a Glacial Runway, McMurdo Station, Antarctica.* CRREL Monograph 98–1, March 1998.

Boulton, G. S. "Two Cores Are Better than One." *Nature,* December 9, 1993.

Boulton, G. S., and C. D. Clark. "The Laurentide Ice Sheet Through the Last Glacial Cycle: The Topology of Drift Lineations as a Key to the Dynamic Behaviour of Former Ice Sheets." *Transactions of the Royal Society of Edinburgh: Earth Sciences,* vol. 81 (1990).

Boynton, W. V., et al. "Distribution of Hydrogen in the Near-Surface of Mars: Evidence for Subsurface Ice Deposits." *Sciencexpress,* May 30, 2002.

Branson, J., D. M. Lawler, and J. W. Glen. "The Laboratory Simulation of Needle Ice." In *Physics and Chemistry of Ice,* N. Maeno and T. Hondoh, eds. Sapporo: Hokkaido Univ. Press, 1992.

Broecker, W. S. "The Biggest Chill." *Natural History,* October 1987.

Broecker, W. S. "Cooling the Tropics." *Nature,* vol. 376 (July 20, 1995).

Broecker, W. S., and G. H. Denton. "What Drives Glacial Cycles?" *Scientific American,* vol. 262, no. 1 (January 1990).

Brower, C. D. *Fifty Years Below Zero: A Lifetime of Adventure in the Far North.* New York: Dodd, Mead, 1942.

Brown, C., with H. Horwood. *Death on the Ice: The Great Newfoundland Sealing Disaster of 1914.* Toronto: Doubleday Canada, 1974.

Brown, R. *Voyage of the Iceberg: The Story of the Iceberg That Sank the Titanic.* New York: Beaufort Books, 1983.

Burns, J. J. "Remarks on the Distribution and Natural History of Pagophilic Pinnipeds in the Bering and Chukchi Seas." *Journal of Mammalogy,* vol. 51, no. 3 (August 28, 1970).

Burns, J. J., J. J. Montague, and C. J. Cowles, eds. *The Bowhead Whale.* Special Publication No. 2. Lawrence, Kansas: Society for Marine Mammalogy, 1993.

Byrd, R. E. *Alone.* New York: G. P. Putnam's Sons, 1938.

Campbell, B. A., D. B. Campbell, J. F. Chandler, A. A. Hine, M. C. Nolan, and P. J. Perillat. "Radar Imaging of the Lunar Poles." *Nature,* vol. 426 (November 13, 2003).

Capossela, J. *Ice Fishing: A Complete Guide, Basic to Advanced.* Woodstock, Vt.: Countryman Press, 1992.

Carr, M. H. *Water on Mars.* New York: Oxford Univ. Press, 1966.

Carr, M. H., et al. "Evidence for a Subsurface Ocean on Europa." *Nature,* vol. 391 (January 22, 1998).

Cavalieri, D. J., P. Gloersen, C. L. Parkinson, J. C. Comiso, and H. J. Zwally. "Observed Hemispheric Asymmetry in Global Sea Ice Changes." *Science,* vol. 278 (November 7, 1997).

Cavell, E., and D. Reid. *When Winter Was King: The Image of Winter in Nineteenth Century Canada.* Banff, Alberta: Altitude Publishing in association with Whyte Museum of the Canadian Rockies, 1988.

Chantraine, P. *The Living Ice: The Story of the Seals and the Men Who Hunt Them in the Gulf of St. Lawrence.* Trans. from the French by David Lobdell. Toronto: McClelland and Stewart, 1980.

Chapman, C. R. *Planets of Rock and Ice: From Mercury to the Moons of Saturn.* New York: Scribner, 1982.

Cherry-Garrard, A. *The Worst Journey in the World.* New York: Carroll & Graf, 1989.

Chou, I-M., J. G. Blank, A. F. Goncharov, H-K. Mao, and R. J. Hemley. "In Situ Observations of a High-Pressure Phase of H_2O Ice." *Science,* vol. 281 (August 7, 1998).

Chouinard, Y. *Climbing Ice.* San Francisco: Sierra Club, 1978.

Clark, J. I., M. El-Tahan, and R. Khan. "Icebergs—Where They Are and Where They Are Going." XVI Fairmont Workshop, Canadian Petroleum Association, 1990.

Clarke, Garry K. C. "A Short History of Scientific Investigations on Glaciers." *Journal of Glaciology,* special issue, 1987.

Clarke, T. S., C. Liu, N. E. Lord, and C. R. Bentley. "Evidence for a Recently Abandoned Shear Margin Adjacent to Ice Stream B2, Antarctica, from Ice-penetrating Radar Measurements." *Journal of Geophysical Research,* vol. 105, no. B6 (June 10, 2000).

Cook, F. A. *My Attainment of the Pole.* New York: Mitchell Kennerley, 1913.

Cota, G. F., L. Legendre, M. Gosselin, and R. G. Ingram. "Ecology of Bottom Ice Algae: I. Environmental Controls and Variability; II. Dynamics, Distributions and Productivity; III. Comparative Physiology." *Journal of Marine Systems,* vol. 2 (1991).

Crocker, G. B., and A. B. Cammaert. "Measurements of Bergy Bit and Growler Populations off Canada's East Coast." *IAHR Ice Symposium 1994,* Trondheim, Norway.

Crowley, T. J. "The Geologic Record of Climate Change." *Reviews of Geophysics and Space Physics,* vol. 21, no. 4 (May 1983).

Cruikshank, D. P. "Stardust Memories." *Science,* vol. 275 (March 28, 1997).

Cullen, B. "Testimony from the Iceman." *Smithsonian,* February 2003.

Cuzzi, J. N. "Evolution of Planetary Ringmoon Systems." *Earth, Moon, and Planets,* vol. 67 (1995).

Cuzzi, J. N., et al. "Saturn's Rings: Pre-Cassini Status and Mission Goals." *Space Science Reviews,* vol. 118 (2002).

Daly, S. F. *Frazil Ice Dynamics.* CRREL Monograph 84–1, 1984.

Dansgaard, W., J. W. C. White, and S. J. Johnsen. "The Abrupt Termination of the Younger Dryas Climate Event." *Nature,* vol. 339 (June 15, 1989).

Dansgaard, W., S. J. Johnsen, J. Möller, and C. C. Langway Jr. "One Thousand Centuries of Climatic Record from Camp Century on the Greenland Ice Sheet." *Science,* vol. 166 (October 17, 1969).

Dasch, P., ed. *Icy Worlds of the Solar System.* Cambridge, U.K.: Cambridge Univ. Press, 2004.

Dash, J. G. "Frost Heave and the Surface Melting of Ice." In *Phase Transitions in Surface Films 2,* H. Taub, ed. New York: Plenum Press, 1991.

David, E. *Harvest of the Cold Months: The Social History of Ice and Ices.* New York: Viking, 1995.

Day, B. *Glacier Pilot.* Sausalito, Calif.: Comstock, 1957.

Denton, G. H., and T. J. Hughes, eds. *The Last Great Ice Sheets.* New York: John Wiley & Sons, 1981.

DeVries, A. L. "Biological Antifreeze Agents in Coldwater Fishes." *Comparative Biochemistry and Physiology,* vol. 73A, no. 4 (1982).

Dickson, B. "All Change in the Arctic." *Nature,* vol. 397 (February 4, 1999).

Dodge, M. M. *Hans Brinker or The Silver Skates.* Boston: Mutual Book, n.d.

Donaldson, B., and B. Nagengast. *Heat and Cold: Mastering the Great Indoors, a Selective History of Heating, Ventilation, Refrigeration and Air-Conditioning.* Atlanta: American Society of Heating, Refrigerating and Air-Conditioning Engineers, 1994.

Dong, Y. Y., and John Hallett. "Droplet Accretion During Rime Growth and the Formation of Secondary Ice Crystals." *O. J. R. Meteorol. Soc.,* vol. 115 (1989).

Dorsey, N. E. *Properties of Ordinary Water-Substance in All Its Phases: Water-Vapor, Water, and All the Ices.* New York: Reinhold, 1940.

Duman, J. G., and T. M. Olsen. "Thermal Hysteresis Protein Activity in Bacteria, Fungi, and Phylogenetically Diverse Plants." *Cryobiology,* vol. 30 (1993).

Duman, J. G., D. W. Wu, L. Xu, D. Tursman, and T. M. Olsen. "Adaptations of Insects to Subzero Temperatures." *Quarterly Review of Biology,* vol. 66, no. 4 (December 1991).

Dyke, A. S., and V. K. Prest. "Late Wisconsinan and Holocene History of the Laurentide Ice Sheet." *Géographie Physique et Quaternaire,* vol. 41, no. 2 (1987).

Dyson, J. L. *The World of Ice.* New York: Alfred A. Knopf, 1962.

Echelmeyer, K. A., W. D. Harrison, C. Larsen, and J. E. Mitchell. "The Role of the Margins in the Dynamics of an Active Ice Stream." *Journal of Glaciology,* vol. 40, no. 136 (1994).

Eisenberg, D., and W. Kauzmann. *The Structure and Properties of Water.* New York: Oxford Univ. Press, 1969.

Ellis, M. *Ice and Icehouses Through the Ages.* Southampton, U.K.: Southampton Univ. Industrial Archaeology Group, 1982.

El-Tahan, M., and H. W. El-Tahan. "Forecast of Iceberg Ensemble Drift." 15th Annual Offshore Technology Conference, Houston, Texas, 1983.

Embick, A. *Blue Ice and Black Gold: A Climbers Guide to the Frozen Waterfalls of Valdez, Alaska.* Valdez: Valdez Alpine Books, 1989.

Engelhardt, H., N. Humphrey, B. Kamb, M. Fahnestock. "Physical Conditions at the Base of a Fast Moving Antarctic Ice Stream." *Science,* vol. 248 (April 6, 1990).

Engelhardt, H., and B. Kamb. "Structure of Ice IV, a Metastable High-Pressure Phase." *Journal of Chemical Physics,* vol. 75, no. 12 (December 15, 1981).

England, G. A. *The Greatest Hunt in the World.* Montreal: Tundra Books, Collins Publishers, 1969.

EPICA Community Members. "Eight Glacial Cycles from an Antarctic Ice Core." *Nature,* vol. 429 (June 10, 2004).

Estrada, P. R., and J. N. Cuzzi. "Voyager Observations of the Color of Saturn's Rings." *Icarus,* vol. 122 (1996).

Fahnestock, R. K., D. J. Crowley, M. Wilson, and H. Schneider. "Ice Volcanoes of the Lake Erie Shore near Dunkirk, New York, U.S.A." *Journal of Glaciology,* vol. 12, no. 64 (1973).

Fahy, G. M. "Vitrification: A New Approach to Organ Cryopreservation." In *Transplantation: Approaches to Graft Rejection.* New York: Alan R. Liss, 1986.

Fay, F. H. "The Role of Ice in the Ecology of Marine Mammals of the Bering Sea." In *Oceanography of the Bering Sea,* D. W. Hood and E. J. Kelley, eds. Fairbanks: Institute of Marine Science, Univ. of Alaska, Fairbanks, 1974.

Feldman, W. C., et al. "Fluxes of Fast and Epithermal Neutrons from Lunar Prospector: Evidence for Water Ice at the Lunar Poles." *Science,* vol. 281 (September 4, 1998).

Ferguson, S. A. *Glaciers of North America: A Field Guide.* Golden, Colo.: Fulcrum Publishing, 1992.

Fisher, D. A., N. Reeh, and K. Langley. "Objective Reconstructions of the Late Wisconsinan Laurentide Ice Sheet and the Significance of Deformable Beds." *Géographie Physique et Quaternaire,* vol. 39, no. 3 (1985).

Fletcher, N. H. *The Chemical Physics of Ice.* Cambridge: Cambridge Univ. Press, 1970.

Fowler, B. *Iceman: Uncovering the Life and Times of a Prehistoric Man Found in an Alpine Glacier.* New York: Random House, 2000.

Fraser, W. R., W. Z. Trivelpiece, D. G. Ainley, and S. G. Trivelpiece. "Increases in Antarctic Penguin Populations: Reduced Competition with Whales or a Loss of Sea Ice Due to Environmental Warming?" *Polar Biology,* vol. 11 (1992).

Gage, A. A. "Cryosurgery." In *Encyclopedia of Medical Devices and Instrumentation,* vol. 2., J. G. Webster, ed. New York: John Wiley, 1988.

Garrison, D. L., C. W. Sullivan, and S. F. Ackley. "Sea Ice Microbial Communities in Antarctica." *BioScience,* vol. 36, no. 4 (April 1986).

Geer, I. W. "The Not-So-Ordinary Icicle." *Weatherwise,* vol. 34 (December 1981).

George, J. C., C. Clark, G. M. Carroll, and W. T. Ellison. "Observations on the Ice-Breaking and Ice Navigation Behavior of Migrating Bowhead Whales (*Balaena mysticetus*) near Point Barrow, Alaska, Spring 1985." *Arctic,* vol. 42, no. 1 (March 1989).

George, S. M., and F. E. Livingston. "Dynamic Ice Surface in the Polar Stratosphere." *Surface Review and Letters,* vol. 4, no. 4 (1997).

Gold, L. W. "Building Ships from Ice: Habbakuk and After." *Interdisciplinary Science Reviews,* vol. 29, no. 4 (2004).

Gold, L. W. *The Canadian Habbakuk Project.* Cambridge: International Glaciological Society, 1993.

Gold, L. W. "Field Study on the Load Bearing Capacity of Ice Covers." Tech. Paper 98, Div. of Building Research, National Research Council of Canada, 1960.

Gold, L. W. "Fifty Years of Progress in Ice Engineering," *Journal of Glaciology,* special issue, 1987.

Gow, A. J. "Orientation Textures in Ice Sheets of Quietly Frozen Lakes." *Journal of Crystal Growth,* vol. 74 (1986).

Gow, A. J., and J. W. Govoni. *Ice Growth on Post Pond, 1973–1982.* CRREL Report 83-4, 1983.

Gow, A. J., and D. Langston. *Growth History of Lake Ice in Relation to Its Stratigraphic, Crystalline and Mechanical Structure.* CRREL Report 77-1, 1977.

Greenler, R. *Rainbows, Halos, and Glories.* Cambridge: Cambridge Univ. Press, 1980.

Hallet, B. "Circles of Stone." In *1969 Yearbook of Science and the Future, Encyclopedia Brittanica.*

Hallet, B. "Self-Organization in Freezing Soils: From Microscopic Ice Lenses to Patterned Ground." *Canadian Journal of Physics,* vol. 68 (1990).

Hallett, J. "How Snow Crystals Grow." *American Scientist,* vol. 72 (November–December 1984).

Hallett, J., and C. A. Knight. "On the Symmetry of Snow Dendrites." *Atmospheric Research,* vol. 32 (1994).

Hambrey, M., and J. Alean. *Glaciers.* Cambridge: Cambridge Univ. Press, 1992.

Heymsfield, A. J. "Cloud Physics." In *Encyclopedia of Physical Science and Technology,* vol. 3. San Diego: Academic Press, 1992.

Hill, D. D., and E. R. Hughes. *Ice Harvesting in Early America.* New Hartford, N.Y.: New Hartford Historical Society, 1977.

Hobbs, P. V. *Ice Physics.* Oxford, U.K.: Clarendon Press, 1974.

Huges, T. "Is the West Antarctic Ice Sheet Disintegrating?" *Journal of Geophysical Research,* vol. 78, no. 33 (1973).

Huntford, R. *Nansen: The Explorer as Hero.* London: Gerald Duckworth, 1997.

"Ice and Water" Committee. *Ice and Water: The Flood of 1992—Montpelier, Vermont.* 1992.

Ice Core Working Group, National Science Foundation. "Ice Core Contributions to Global Change Research: Past Successes and Future Directions." Durham, N.H.: National Ice Core Laboratory at Univ. of New Hampshire, 1998.

"Ice in the Solar System." Special issue, *Ad Astra,* vol. 7, no. 6 (November-December 1995).

"Ices in the Solar System." Special issue, *The Planetary Report,* vol. 19, no. 2 (March-April 1999).

Iglauer, E. *Denison's Ice Road.* New York: E. P. Dutton, 1974.

Imbrie, J., and K. P. Imbrie. *Ice Ages: Solving the Mystery.* Short Hills, N. J.: Enslow, 1979.

Imbrie, J., et al. "On the Structure and Origin of Major Glaciation Cycles: The 100,000-Year Cycle." *Paleoceanography,* vol. 8, no. 6 (December 1993).

Intergovernmental Panel on Climate Change (IPCC). Second Assessment Report. *Climate Change 1995: The Science of Climate Change.* Cambridge, U.K.: Cambridge Univ. Press, 1995.

Intergovernmental Panel on Climate Change (IPCC). Third Assessment Report. *Climate Change 2001: The Scientific Basis.* Cambridge, U.K.: Cambridge Univ. Press, 2001.

Ives, J. D., and R. G. Barry, eds. *Arctic and Alpine Environments.* London: Methuen, 1974.

Jacobs, S. "The Voyage of Iceberg B-9." *American Scientist,* vol. 80 (January–February 1992).

Jeffries, M. O. "Arctic Ice Shelves and Ice Islands: Origin, Growth and Disintegration, Physical Characteristics, Structural-Stratigraphic Variability, and Dynamics." *Reviews of Geophysics,* vol. 30, no. 3 (August 1992).

Jeffries, M. O., and M. A. Shaw. "The Drift of Ice Islands from the Arctic Ocean into the Channels of the Canadian Arctic Archipelago: The History of Hobson's Choice Ice Island." *Polar Record,* vol. 29, no. 171 (1993).

Kamb, B., C. F. Raymond, W. D. Harrison, H. Engelhardt, K. A. Echelmeyer, N. Humphrey, M. M. Brugman, and T. Pfeffer. "Glacier Surge Mechanism: 1982–1983 Surge of Variegated Glacier, Alaska." *Science,* vol. 227, no. 4686 (February 1, 1985).

Kane, E. K. *Arctic Explorations: The Second Grinnell Expedition in Search of Sir John Franklin, in the Years 1853, '54, '55,* 2 vols. Philadelphia: Childs and Peterson, 1856 and 1857.

Karow, A. M. "Chemical Cryoprotection of Metazoan Cells." *BioScience,* vol. 41, no. 3 (March 1991).

Kelly, B. P. "Climate Change and Ice Breeding Pinnipeds." In *"Fingerprints" of Climate Change,* ed. Walther et al. New York: Kluwer Academic/Plenum, 2001.

Kelly, B. P., and D. Wartzok. "Ringed Seal Diving Behavior in the Breeding Season." *Canadian Journal of Zoology,* vol. 74, no. 8 (1996).

Kerr, R. A. "How Ice Age Climate Got the Shakes." *Science,* vol. 260 (May 14, 1993).

Kimball, W. "The Iceman." Unpublished mss. From "Florida Collection," Florida State Library, 1920s.

Klug, D. D., Y. P. Handa, J. S. Tse, and E. Whalley. "Formation of Slush on Floating Ice." *Cold Regions Science and Technology,* vol. 15 (1988).

Klug, D. D., Y. P. Handa, J. S. Tse, and E. Whalley. "Transformation of Ice VIII to Amorphous Ice by 'Melting' at Low Temperature." *Journal of Chemical Physics,* vol. 90, no. 4 (February 15, 1989).

Knight, C. A. *The Freezing of Supercooled Liquids.* Princeton, N.J.: D. Van Nostrand, 1967.

Knight, C. A. "Icicles as Crystallization Phenomena." *Journal of Crystal Growth,* vol. 49 (1980).

Knight, C. A. "Slush on Lakes." In *Structure and Dynamics of Partially Solidified Systems,* ed. D. E. Loper. New York: Plenum Press, 1987.

Knight, C. A., and A. L. DeVries. "Growth Forms of Large Frost Crystals in the Antarctic." *Journal of Glaciology,* vol. 31, no. 108 (1985).

Knight, C., and N. Knight. "Hailstones." *Scientific American,* vol. 224, no. 4 (April 1971).

Knight, C., and N. Knight. "Breeding Habitats of Emperor Penguins in the Western Ross Sea." *Antarctic Science,* vol. 5, no. 2 (1993).

Kooyman, G. L. "Breeding Habitats of Emperor Penguins in the Western Ross Sea." *Antarctic Science,* vol. 5, no. 2 (1993).

Kooyman, G. L. "Emperors of Antarctica." *Equinox,* no. 54 (1990).

Kooyman, G. L. "The Weddell Seal." *Scientific American,* vol. 221, no. 2 (August 1969).

Kooyman, G. L. *Weddell Seal: Consummate Diver.* Cambridge: Cambridge Univ. Press, 1981.

Kooyman, G. L., D. Croll, S. Stone, and S. Smith. "Emperor Penguin Colony at Cape Washington, Antarctica." *Polar Record,* vol. 26, no. 157 (1990).

Kovacs, K. M. "Harp and Hooded Seals—A Case Study in the Determinants of Mating Systems in Pinnipeds." In *Whales, Seals, Fish and Man,* A. S. Blix, L. Wallöe, and Ö. Ulltang, eds. Amsterdam: Elsevier, 1995.

Krakauer, J. "Straight Up Ice." *National Geographic,* December 1966.

Krantz, W. B. "Self-Organization Manifest as Patterned Ground in Recurrently Frozen Soils." *Earth Science Reviews,* vol. 29 (1990).

Kukla, G., J. F. McManus, D-D. Rousseau, and I. Chuine. "How Long and How Stable Was the Last Interglacial?" *Quaternary Science Reviews,* vol. 16 (1997).

LaChapelle, E. R. *Field Guide to Snow Crystals.* Seattle: Univ. of Washington Press, 1969.

Lampe, D. *Pyke: The Unknown Genius.* London: Evans Bros., 1959.

Langlois, T. H., and M. H. Langlois. *The Ice of Lake Erie Around South Bass Island, 1936–1964.* Columbus: Ohio State Univ. in cooperation with Center for Lake Erie Research and Ohio Sea Grant Program, Tech. Report 165, 1985.

Lansing, A. *Endurance: Shackleton's Incredible Voyage.* New York: McGraw-Hill, 1959.

Lavigne, D. M., and K. M. Kovacs. *Harps and Hoods: Ice-breeding Seals of the Northwest Atlantic.* Waterloo, Ontario: Univ. of Waterloo Press, 1988.

Laws, R. M. "The Ecology of the Southern Ocean." *American Scientist,* vol. 73 (January-February 1985).

Lehman, S. "Sudden End of an Interglacial." *Nature,* vol. 390 (November 1997).

Lentfer, J. W., ed. *Selected Marine Mammals of Alaska: Species Accounts with Research and Management Recommendations.* With chapters on, among other mammals, the bowhead whale, the Pacific walrus, the polar bear, and the ringed, bearded, ribbon, and harbor seals. Washington, D.C.: Marine Mammal Commission, 1988.

Leonie, B. *Tales of the Norse Gods and Heroes.* London: Oxford Univ. Press, 1953.

Lethcoe, N. R. *An Observer's Guide to the Glaciers of Prince William Sound, Alaska.* Valdez: Prince William Sound Books, 1987.

Lewis, M., and W. Clark. *The Journals of Lewis and Clark.* B. DeVoto, ed. Boston: Houghton Mifflin, 1953.

Lingle, C. S., A. Post, U. C. Herzfeld, B. F. Molnia, R. M. Krimmel, and J. J. Roush. "Bering Glacier Surge and Iceberg-calving Mechanism at Vitus Lake, Alaska, U.S.A." Letter in *Journal of Glaciology,* vol. 39, no. 133 (1993).

Lobban, C., J. L. Finney, and W. F. Kuhs. "The Structure of a New Phase of Ice." *Nature,* vol. 391 (January 15, 1998).

Lock, G. S. H. *The Growth and Decay of Ice.* Cambridge: Cambridge Univ. Press, 1990.

Lopez, B. *Arctic Dreams: Imagination and Desire in a Northern Landscape.* New York: Scribner's, 1986.

Lord, W. *The Night Lives On.* New York: William Morrow, 1986.

Lord, W. *A Night to Remember.* New York: Henry Holt, 1955.

Lowe, J. *Ice World: Techniques and Experiences of Modern Ice Climbing.* Seattle: Mountaineers, 1996.

Mackay, J. R. "Pingos of the Tuktoyaktuk Peninsula Area, Northwest Territories." *Géographie physique et Quaternaire,* vol. 33, no. 1 (1979).

Mackay, J. R. "Pingos and Pingo Ice of the Western Arctic Coast, Canada." *Terra,* vol. 106, no. 1 (1994).

Mackay, J. R., and S. R. Dallimore. "Massive Ice of the Tuktoyaktuk Area, Western Arctic Coast, Canada." *Canadian Journal of Earth Sciences,* vol. 29, no. 6 (1992).

Makkonen, L. "A Model of Icicle Growth." *Journal of Glaciology,* vol. 34, no. 116 (1988).

Marchand, P. J. *Life in the Cold: An Introduction to Winter Ecology.* Hanover, N.H.: Univ. Press of New England, 1987.

Masterson, D. M., and R. M. W. Frederking. "Local Contact Pressures in Ship/Ice and Structure/Ice Interactions." *Cold Regions Science and Technology,* vol. 21 (1993).

Mawson, D. *The Home of the Blizzard: Being the Story of the Australasian Antarctic Expedition, 1911–1914.* 2 vols. 1915.

Maykut, G. A. *An Introduction to Ice in the Polar Oceans.* Seattle: Univ. of Washington, 1985.

Maykut, G. A., T. C. Grenfell, and W. F. Weeks. "On Estimating Spatial and Temporal Variations in the Properties of Ice in the Polar Oceans." *Journal of Marine Systems,* vol. 3 (1992).

McManus, J. F. "A Great-Grand-Daddy of Ice Cores." *Nature,* vol. 429 (June 10, 2004).

McManus, J. F., D. W. Oppo, and J. L. Cullen. "Influence of Thermohaline Circulation on the Duration and Demise of Peak Interglacial Climate in the Circum-North Atlantic Region." *EOS, Transactions, American Geophysics Union,* vol. 79, no. 17 (1998).

McManus, J. F., D. W. Oppo, and J. L. Cullen. "A 0.5-Million-Year Record of Millennial-Scale Climate Variability in the North Atlantic." *Science,* vol. 283 (February 12, 1999).

Mercer, J. H. "West Antarctic Ice Sheet and CO_2 Greenhouse Effect: A Threat of Disaster." *Nature,* vol. 271 (January 26, 1978).

Meryman, H. T., ed. *Cryobiology.* San Diego, Calif.: Academic Press, 1966.
Michel, B. *Winter Regime of Rivers and Lakes.* CRREL Monograph III-B1a, 1971.
Mills, W. J., and R. S. Pozos. "Low Temperature Effects on Humans." *Encyclopedia of Human Biology,* vol. 4. San Diego: Academic Press, 1991.
Mills, W. J., et al. "Cold Injury: A Collection of Papers." *Alaska Medicine,* vol. 35, no. 1 (January-February-March 1993).
Minnaert, M. *The Nature of Light and Colour in the Open Air.* Trans. from the Dutch by H. M. Kremer-Priest, revision K. E. B. Jay. New York: Dover, 1954.
Mirsky, J. *To the Arctic! The Story of Northern Exploration from Earliest Times to the Present.* Chicago: University of Chicago Press, 1970.
Mowat, F., ed. *Ordeal by Ice: The Search for the Northwest Passage.* Toronto: McClelland and Stewart, 1973.
Mowat, F., ed. *The Polar Passion: The Quest for the North Pole.* Toronto: McClelland and Stewart, 1967.
Muir, J. *Travels in Alaska.* New York: Modern Library, 2002.
Mulherin, N., D. Sodhi, and E. Smallidge. *Northern Sea Route and Icebreaking Technology: An Overview of Current Conditions.* CRREL, June 1994.
Muller, E. H., and P. J. Fleisher. "Surging History and Potential for Renewed Retreat: Bering Glacier, Alaska, U.S.A." *Arctic and Alpine Research,* vol. 27, no. 1 (1995).
Müller-Schwarze, D. *The Behavior of Penguins: Adapted to Ice and Tropics.* Albany, N.Y.: State Univ. of New York Press, 1984.
Murphy, M. "Dan Clapp's Skeeters: An Iceboating Family Comes of Age." *WoodenBoat,* no. 110 (January-February 1993).
Murray, J. B., et al. "Evidence from the Mars Express High Resolution Stereo Camera for a Frozen Sea Close to Mars' Equator." *Nature,* vol. 434 (March 2005).
Nagengast, B. "John Gorrie: Pioneer of Cooling and Ice Making." *ASHRAE Journal,* vol. 3 (January 1991).
Nakaya, U. *Snow Crystals: Natural and Artificial.* Cambridge: Harvard Univ. Press, 1954.
Nansen, F. *Farthest North,* 2 vols. 1898.
Nelson, R. K. *Hunters of the Northern Ice.* Chicago: Univ. of Chicago Press, 1969.
"New Light on the Solar System." Special issue, *Scientific American,* vol. 13, no. 3 (2003).
Nicol, S., and I. Allison, "The Frozen Skin of the Southern Ocean." *American Scientist,* vol. 85 (September–October 1997).
North Greenland Ice Core Project members. "High-Resolution Record of Northern Hemisphere Climate Extending into the Last Interglacial Period." *Nature,* vol. 431 (September 2004).
North Slope Subsistence Study, Barrow, 1987, 1988, 1989. Minerals Management Service, U. S. Dept. of the Interior, Alaska Outer Continental Shelf Region, 1991.
Oppenheimer, M. "Global Warming and the Stability of the West Antarctic Ice Sheet." *Nature,* vol. 393 (May 28, 1998).
Osterkamp, T. E. "A Thermal History of Permafrost in Alaska." *Proceedings of the 8th International Permafrost Conference in Zurich, Switzerland* (July 2003).
Osterkamp, T. E., L. Viereck, Y. Shur, M. T. Jorgenson, C. Racine, A. Doyle, and R. D. Boone. "Observations of Thermokarst and Its Impact on Boreal Forests in Alaska." *Arctic, Antarctic, and Alpine Research,* vol. 32, no. 3 (2000).
Pappalardo, R. T., et al. "Does Europa Have a Subsurface Ocean? Evaluation of the Geological Evidence." *Journal of Geophysical Research—Planets,* vol. 104 (1999).
Pappalardo, R. T., et al. "Geological Evidence for Solid-State Convection in Europa's Ice Shell." *Nature,* vol. 391 (January 22, 1998).

Parkinson, C. L. "Trends in the Length of the Southern Ocean Sea-Ice Season, 1979–99." *Annals of Glaciology,* vol. 34 (2002).

Parkinson, C. L., and D. J. Cavalieri. "A 21 Year Record of Arctic Sea-Ice Extents and Their Regional, Seasonal and Monthly Variability and Trends." *Annals of Glaciology,* vol. 34 (2002).

Parsons, L. R., T. A. Wheaton, and D. P. H. Tucker. "Florida Freezes and the Role of Water in Citrus Cold Protection." *HortScience,* vol. 21, no. 1 (February 1986).

Paterson, W. S. B. *The Physics of Glaciers.* Oxford, U.K.: Pergamon Press, 1981.

Peary, R. E. *The North Pole: Its Discovery in 1909 Under the Auspices of the Peary Arctic Club.* New York: Dover, 1986.

Peltier, W. R. "Postglacial Variations in the Level of the Sea: Implications for Climate Dynamics and Solid-Earth Geophysics." *Reviews of Geophysics,* vol. 36, no. 4 (November 1998).

Pendleton, Y. J., and D. P. Cruikshank. "Life from the Stars?" *Sky and Telescope,* March 1994.

Perovich, D. K., and J. A. Richter-Menge. "Surface Characteristics of Lead Ice." *Journal of Geophysical Research,* vol. 99, no. C8 (August 15, 1994).

Perutz, M. F. "A Description of the Iceberg Aircraft Carrier and the Bearing of the Mechanical Properties of Frozen Wood Pulp upon Some Problems of Glacier Flow." *Journal of Glaciology,* vol. 1. no. 3 (March 1948).

Petit, J. R., et al. "Climate and Atmospheric History of the Past 420,000 Years from the Vostok Ice Core, Antarctica." *Nature,* vol. 399 (June 3, 1999).

Pielou, E. C. *After the Ice Age: The Return of Life to Glaciated North America.* Chicago: Univ. of Chicago Press, 1991.

Post, A. "Distribution of Surging Glaciers in Western North America." *Journal of Glaciology,* vol. 8, no. 53 (1969).

Post, A. "The Exceptional Advances of the Muldrow, Black Rapids, and Susitna Glaciers." *Journal of Geophysical Research,* vol. 65, no. 11 (September 1960).

Post, A. "Periodic Surge Origin of Folded Moraines on Bering Piedmont Glacier, Alaska." *Journal of Glaciology,* vol. 11, no. 62 (1972).

Post, A., and E. R. LaChapelle. *Glacier Ice.* Seattle: Univ. of Washington Press, in association with the International Glaciological Society, Cambridge, England, 2000.

Prévost, J. *Écologie du Manchot empereur.* Paris: Hermann, 1961.

Pyne, S. J. *The Ice: A Journey to Antarctica.* Iowa City: Univ. of Iowa Press, 1986.

Quetin, L. B., and R. M. Ross. "Behavioral and Physiological Characteristics of the Antarctic Krill, *Euphausia superba.*" *American Zoologist,* vol. 31 (1991).

Raymo, M. E., D. W. Oppo, and W. Curry. "The Mid-Pleistocene Climate Transition: A Deep Sea Carbon Isotopic Perspective." *Paleoceanography,* vol. 12, no. 4 (August 1997).

Raymond, C. F. "How Do Glaciers Surge? A Review." *Journal of Geophysical Research,* vol. 92, no. B9 (August 10, 1987).

Raymond, J. A., P. Wilson, and A. L. DeVries. "Inhibition of Growth of Nonbasal Planes in Ice by Fish Antifreezes." *Proc. Nat. Acad. Sci. USA,* vol. 86 (February 1989).

Reid, D. S. "Optimizing the Quality of Frozen Foods." *Food Technology,* July 1990.

Reimnitz, E., E. W. Kempema, and P. W. Barnes. "Anchor Ice, Seabed Freezing, and Sediment Dynamics in Shallow Arctic Seas." *Journal of Geophysical Research,* vol. 92, no. C13 (December 15, 1987).

Rignot, E. J. "Fast Recession of a West Antarctic Glacier." *Science,* vol. 281 (July 24, 1998).

Rignot, E., and S. S. Jacobs. "Rapid Bottom Melting Widespread near Antarctic Ice Sheet Grounding Lines." *Science,* vol. 296 (June 14, 2002).

Rignot, E., and R. H. Thomas. "Mass Balance of Polar Ice Sheets." *Science,* vol. 297 (August 30, 2002).

Robin, G. de Q., and C. Swithinbank. "Fifty Years of Progress in Understanding Ice Sheets." *Journal of Glaciology,* special issue, 1987.

Rothrock, D. A., Y. Yu, and G. A. Maykut. "Thinning of the Sea-Ice Cover." *Geophysical Research Letters,* vol. 26, no. 23 (December 1, 1999).

Rowlands, J. J. *Cache Lake Country: Life in the North Woods.* New York: W. W. Norton, 1947.

Ruddiman, W. F., and J. E. Kutzbach. "Plateau Uplift and Climatic Change." *Scientific American,* vol. 264, no. 3 (March 1991).

Ruge, R. A. "The History and Development of the Iceboat." *WoodenBoat.* part 1, 1600–1930, no. 38 (January-February 1981), part 2, 1930–1980, no. 39 (March-April 1981).

Ryden, H. *Lily Pond: Four Years with a Family of Beavers.* New York: William Morrow, 1989.

Sagan, C., and A. Druyan. *Comet.* New York: Random House, 1985.

Scambos, T. A., J. A. Bohlander, C. A. Shuman, and P. Skvarca. "Glacier Acceleration and Thinning after Ice Shelf Collapse in the Larsen B Embayment, Antarctica." *Geophysical Research Letters,* vol. 31 (2004).

Sceats, M. G., and S. A. Rice. "Amorphous Solid Water and Its Relationship to Liquid Water: A Random Network Model for Water." In *Water: A Comprehensive Treatise,* vol. 7, ed. F. Franks. New York: Plenum Press, 1982.

Scott, R. F. *Scott's Last Expedition: The Personal Journals of Captain R. F. Scott, R.N., C.V.O., on His Journey to the South Pole.* New York: Dodd, Mead, 1923.

Shackleton, E. *South: The Story of Shackleton's Last Expedition, 1914–1917.* London: Century, 1983.

Sharp, R. P. *Living Ice: Understanding Glaciers and Glaciation.* Cambridge: Cambridge Univ. Press, 1988.

Sigfusson, S. *Sigfusson's Roads.* Winnipeg: Watson & Dwyer, 1992.

Siple, P. *90° South: The Story of the American South Pole Conquest.* New York: G. P. Putnam's Sons, 1959.

Sodhi, D. S., and M. El-Tahan. "Prediction of an Iceberg Drift Trajectory During a Storm." *Annals of Glaciology,* vol. 1 (1980).

Speck, G. *Samuel Hearne and the Northwest Passage.* Caxton, Idaho: Caldwell, 1963.

Spindler, K. *Arctic Manual.* New York: Macmillan, 1944.

Spindler, K. *The Man in the Ice: The Amazing Inside Story of the 5,000-Year-Old Body Found Trapped in a Glacier in the Alps.* Trans. from the German by Ewald Osers. London: Weidenfeld & Nicolson, 1994.

Stefansson, V. *The Friendly Arctic: The Story of Five Years in Polar Regions.* New York: Macmillan, 1921.

Steger, W. with P. Schurke. *North to the Pole.* New York: Crown, 1987.

Steponkus, P. L. "Cold Acclimation and Freezing Injury from a Perspective of the Plasma Membrane." In *Environmental Injury to Plants.* San Diego: Academic Press, 1990.

Steponkus, P. L., M. Uemura, and M. S. Webb. "Redesigning Crops for Increased Tolerance to Freezing Stress." In *Interacting Stresses on Plants in a Changing Climate,* M. B. Jackson and C. R. Black, eds. NATO ASI Series, vol. 1, no. 16. Heidelberg: Springer-Verlag, 1993.

Stirling, I. *Polar Bears.* Ann Arbor: Univ. of Michigan Press, 1988.

Storey, K. B., and J. M. Storey. "Frozen and Alive." *Scientific American.* December 1990.

Stuck, H. *Ten Thousand Miles with a Dog Sled: A Narrative of Winter Travel in Interior Alaska.* Lincoln: Univ. of Nebraska Press, 1988.

Swinzow, G. K. "On Winter Warfare." CRREL, March 1982.

Szpir, M. "Bits of Ice XI and Ice XII." *American Scientist,* September–October 1996.

Taubes, G. "Neutrino Watchers Go to Extremes." *Science,* vol. 263 (January 7, 1994).

Taylor, K. C., G. W. Lamorey, G. A. Doyle, R. B. Alley, P. M. Grootes, P. A. Mayewski, J. W. C. White, and L. K. Barlow. "The 'Flickering Switch' of Late Pleistocene Climate Change." *Nature,* vol. 361 (February 4, 1993).

Testa, J. W. "Over-Winter Movements and Diving Behavior of Female Weddell Seals (*Leptonychotes weddellii*) in the Southwestern Ross Sea, Antarctica." *Canadian Journal of Zoology,* vol. 72 (1994).

Thomas, G. W., J. J. Olivero, E. J. Jensen, W. Schroeder, and O. B. Toon. "Relation Between Increasing Methane and the Presence of Ice Clouds at the Mesopause." *Nature,* vol. 338 (April 6, 1989).

Thomas, R., et al. "Accelerated Sea-Level Rise from West Antarctica." *Science,* vol. 306 (October 8, 2004).

Thomashow, M. F. "*Arabidopsis thaliana* As Model for Studying Mechanisms of Plant Cold Tolerance." In *Arabidopsis.* Cold Spring Harbor, N.Y.: Cold Spring Harbor Laboratory Press, 1994.

Thompson, L. G., E. Mosley-Thompson, X. Wu, and Z Xie. "Wisconsin/Würm Glacial Stage Ice in the Subtropical Dunde Ice Cap, China." *GeoJournal* vol. 17, no. 4 (April 17, 1988).

Thompson, L. G., E. Mosley-Thompson, M. E. Davis, J. F. Bolzan, J. Dai, T. Yao, N. Gundestrup, X. Wu, L. Klein, and Z. Xie. "Holocene–Late Pleistocene Climatic Ice Core Records from Zinghai-Tibetan Plateau." *Science,* vol. 246 (October 27, 1989).

Thompson, L. G., et al. "A 25,000-Year Tropical Climate History from Bolivian Ice Cores." *Science,* vol. 282 (December 4, 1998).

Thoreau, H. D. *Winter: From the Journal of Henry David Thoreau.* H. B. O. Blake, ed. Williamstown, Mass.: Corner House Publishers, 1973.

Turi, J. *Turi's Book of Lappland.* Trans. from the Danish by E. G. Nash. London: Harper and Bros., 1931.

Tutton, A. E. H. *The High Alps: A Natural History of Ice and Snow.* London: Kegan Paul, Trench, Trubner, 1927.

Tyndall, J. *The Forms of Water in Clouds and Rivers, Ice and Glaciers.* Akron, Ohio: Werner, n.d.

Upper, C. D., and G. Vali. "The Discovery of Bacterial Ice Nucleation and Its Role in the Injury of Plants by Frost." In *Biological Ice Nucleation and Its Applications,* R. E. Lee Jr., G. J. Warren, and Lawrence V. Gusta, eds. St. Paul: APS Press, 1995.

U.S. Army Corps of Engineers. *Breaking Ice Jams.* CRREL, 1985.

Venkatesh, S., and M. El-Tahan, "Iceberg Life Expectancies in the Grand Banks and Labrador Sea." *Cold Regions Science and Technology,* vol. 15 (1988).

Walker, J. "Icicles Ensheathe a Number of Puzzles: Just How Does the Water Freeze?" *Scientific American,* vol. 258, no. 5 (May 1988).

Washburn, A. L. *Geocryology: A Survey of Periglacial Processes and Environments.* New York: John Wiley, 1973.

Water: A Comprehensive Treatise, vol. 7, ed. F. Franks. New York: Plenum Press, 1982.

Weaver, H. A. "Not a Rubble Pile?" *Science,* vol. 304 (June 18, 2004).

Weightman, G. *The Frozen Water Trade: A True Story.* New York: Hyperion, 2003.

Weiser, C. J. "Cold Resistance and Injury in Woody Plants." *Science,* vol. 169, no. 3952 (September 25, 1970).

Whalley, E. "The Hydrogen Bond in Ice." In *The Hydrogen Bond/III: Dynamics, Thermody-*

namics and Special Systems, P. Schuster, G. Zundel, and C. Sandorfy; eds. Amsterdam: North Holland Publishing, 1976.

Whillans, I. M., and C. J. van der Veen. "The Role of Lateral Drag in the Dynamics of Ice Stream B, Antarctica." *Journal of Glaciology,* vol. 43, no. 144 (1997).

White, J. W. C. "Don't Touch That Dial." *Nature,* vol. 364 (July 15, 1993).

Williams, P. J. *Pipelines and Permafrost: Science in a Cold Climate.* Ottawa: Carleton Univ. Press, 1986.

Wojtiw, L., and E. P. Lozowski. "Record Canadian Hailstones." *Bulletin of the American Meteorological Society,* vol. 56, no. 12 (December 1975).

Index

ILLUSTRATION CREDITS

Grateful acknowledgment is made to the following for permission to reprint illustrations:

INSERT

Three closeups of ice courtesy of David Hirmes.

Slush holes courtesy of Charles Knight.

Ice shove in Oil City courtesy of Larry Berlin; Frozen Hudson River courtesy of N. J. Fenwick.

Surging glacier photo by Austin Post, U. S. Geological Survey; Icehenge courtesy of Bruce Koci; Wedding cake ice courtesy of Lonnie G. Thompson.

Two-sailed iceberg photo courtesy of the Canadian Ice Service; Horseshoe iceberg photo courtesy of the U. S. Coast Guard International Ice Patrol; Ship dented by iceberg reprinted by permission of *The Telegram*.

Iceberg in sea ice courtesy of the U. S. Coast Guard International Ice Patrol; Submarine at North Pole courtesy of the U. S. Navy.

Pingo courtesy of J. Ross Mackay; Circles of stone courtesy of Bernard Hallet, photo first appeared in 1989 Yearbook of Science and the Future, Encyclopedia Brittanica, 86–97.

Ice ribbon courtesy of Brian Swanson; Hair ice courtesy of Kathleen Jansen; Otter courtesy of Michael Quinton; Polar bear photo by Michio Hoshino/Minden Pictures.

Eskimo whaling boat courtesy of Bill Hess; Penguins courtesy of Gerald L. Kooyman.

Ice sitting contest © 2005 Ripley Entertainment, Inc.; Ice climber courtesy of George Hurley; Child on fur courtesy of the American Museum of Natural History Library.

"Glace de la Concorde" courtesy of Gamma Press; "Missie Hattie Atwater as 'Fanciful Fans' posed for a composite," Montreal, QC, 1870 [accession number I-43635.1], Notman Photographic Archives, McCord Museum of Canadian History, Montreal; Hockey player courtesy of Eric Nesterenko.

Ice plane courtesy of Joseph O'Donoghue.

Snowflake and frost photos by William A. Bentley, courtesy of Dover Publications; Hailstone courtesy of Charles Knight.

Cave crystals courtesy of Charles Knight; Marine icing courtesy of the Bedford Institute of Oceanography; Rime ice photo copyright of the Royal Meteorological Society.

Three photos of ice in space courtesy of NASA.

IN-TEXT ILLUSTRATIONS

Ice molecules: After Ashton 1986 (after Hobbs 1974)

Meandering iceberg: From C-CORE report "Iceberg Motion" by P. Ball, H. S. Gaskill, and R. J. Lopez

Ice IV diagram: From "Structure of Ice IV, a Metastable High-Pressure Phase" by H. Engelhardt and B. Kamb (*Journal of Chemical Physics,* December 15, 1981)

Most etchings at the end of chapters are taken from Elisha Kent Kane's *Arctic Explorations: The Second Grinnell Expedition.* The black guillemot in the ANIMALS IV chapter (page 318) comes from Frederick Cook's *My Attainment of the Pole.* René Descartes's sketches of snow crystals in the ATMOSPHERE I chapter (page 429) appear in U. Nakaya's book *Snow Crystals.* The ice-harvesting tools in the USES I chapter (page 380) come from J. C. Jones Jr.'s *America's Icemen: an Illustrated History of the United States Natural Ice Industry 1665–1925.* And the ice sheet in the ICE AGES chapter (page 491) appears in Fridtjof Nansen's *The First Crossing of Greenland.*

Permissions Acknowledgments

Grateful acknowledgment is made to the following for permission to reprint previously published material:

Alfred A. Knopf: Excerpt from "Two Grotesques" from *The Mail from Anywhere* by Brad Leithauser. Copyright © 1990 by Brad Leithauser. Reprinted by permission of Alfred A. Knopf, a division of Random House Inc.

Alfred A. Knopf and Faber and Faber Limited: Excerpt from "The Reader" from *The Collected Poems of Wallace Stevens* by Wallace Stevens. Reprinted by permission of Alfred A. Knopf, a division of Random House Inc. and Faber and Faber Limited.

Nicholas Christopher: Excerpts from the poems "31," "23" and "18" from *5 Degrees and Other Poems* by Nicholas Christopher. Copyright © 1995 by Nicholas Christopher. Reprinted by permission of the author.

Elsa Franklin Literary Agent: Excerpts from *Niagara* by Pierre Berton. Reprinted by permission of Elsa Franklin Literary Agent.

Faber and Faber Limited: Excerpt from "The Heron" and "The Mosquito" from *Under the North Star* by Ted Hughes. Copyright © 1981 by Ted Hughes. Reprinted by permission of Faber and Faber Limited.

Farrar, Straus and Giroux, LLC.: Excerpt from "The Imaginary Iceberg" from *The Complete Poems 1927–1979* by Elizabeth Bishop. Copyright © 1979, 1983 by Alice Helen Methfessel. Reprinted by permission of Farrar, Straus and Giroux, LLC.

Farrar, Straus and Giroux, LLC. and Carcanet Press Limited: Excerpt from "Eclogue IV: Winter" from *Collected Poems in English* by Joseph Brodsky. Copyright © 2000 by the Estate of Joseph Brodsky. Reprinted by permission of Farrar, Straus and Giroux, LLC. and Carcanet Press Limited.

Farrar, Straus and Giroux, LLC. and Faber and Faber Limited: Excerpt from "October Dawn" from *Collected Poems* by Ted Hughes. Copyright © 2003 by The Estate of Ted Hughes. Reprinted by permission of Farrar, Straus and Giroux, LLC. and Faber and Faber Limited.

Farrar, Straus and Giroux, LLC. and Paul Muldoon: Excerpt from "The Year of the Soles, for Ishi" from *Poems 1968–1998* by Paul Muldoon. Copyright © 2001 by Paul Muldoon. Reprinted by permission of Farrar, Straus and Giroux, LLC. and the author.

Michael Faubion: Excerpt from the song "Ice Road" by Michael Faubion. Reprinted by permission of the author.

Georges Borchardt, Inc. and Carcanet Press Limited: Excerpt from "Into the Dusk-Charged Air" from *Rivers and Mountains* by John Ashbery. UK edition was titled *Selected Poems.* Copyright © 1962, 1966 by John Ashbery. Reprinted by permission of Georges Borchardt, Inc., on behalf of the author and Carcanet Press Limited.

Harcourt, Inc.: Excerpt from "Year's End" from *Ceremony and Other Poems* by Richard Wilbur. Copyright © 1949 and renewed 1977 by Richard Wilbur. Excerpt from "Wyeth's Milk Cans" from *New and Collected Poems* by Richard Wilbur. Copyright © 1988 by Richard Wilbur. Reprinted by permission of Harcourt, Inc.

Harcourt, Inc. and Faber and Faber Limited: Excerpt from "Little Gidding" from *Four Quartets* by T. S. Eliot. UK edition was titled *Collected Poems 1909–1962.* Copyright © 1942 by T. S. Eliot and renewed 1970 by Esme Valerie Eliot. Excerpt from "April 5, 1974" from *The Mind Reader* by Richard Wilbur. UK edition was titled *New and Collected Poems.* Copyright © 1975 by Richard Wilbur. Reprinted by permission of Harcourt, Inc. and Faber and Faber Limited.

Harvard University Press: Excerpt from "337: I know a place where summer strives" and "640: I cannot live without you" from *The Poems of Emily Dickinson* by Emily Dickinson, Thomas H. Johnson, ed. (Cambridge Mass.: The Belknap Press of Harvard University Press). Copyright © 1951, 1955, 1979, 1983 by the President and Fellows of Harvard College. Reprinted by permission of the publishers and Trustees of Amherst College and Harvard University Press.

J. Gordon Shillingford Publishing: Excerpts from *Sigfussion's Roads* by Svein Sigfusson. Reprinted by permission of J. Gordon Shillingford Publishing.

Oxford University Press: Excerpt from "The Beginning of All Things" from *Tales of the Norse Gods* retold by Barbara Leonie Picard (OUP, 2001). Copyright © 1953 by Barbara Leonie Picard. Reprinted by permission of Oxford University Press.

The Random House Group Ltd.: Excerpt from "The Road is Black" from *Selected Poems* by Anna Akhmatova, published by Secker & Warburg. Reprinted by permission of The Random House Group Ltd.

Ian Stirling: Excerpt from *Polar Bears* by Ian Stirling. Reprinted by permission of the author.

University of Toronto Press: Excerpt from "My Breath" by Orpingalik from *Northern Voices: Inuit Writing in English* edited by Penny Petrone. Reprinted by permission of University of Toronto Press.

Wesleyan University Press: Excerpt from "Seal" from *Hinge & Sign: Poems 1968–1993* by Heather McHugh (Wesleyan University Press, 1994). Copyright © 1994 by Heather McHugh. Reprinted by permission of Wesleyan University Press.

ABOUT THE AUTHOR

Mariana Gosnell was born and grew up in Columbus, Ohio, and graduated from Ohio Wesleyan University with a major in fine art. She worked for many years at *Newsweek,* where she reported on medicine and science. She lives in New York City.

A NOTE ON THE TYPE

This book was set in a version of the well-known Monotype face Bembo. This letter was cut for the celebrated Venetian printer Aldus Manutius by Francesco Griffo, and first used in Pietro Cardinal Bembo's De Aetna of 1495.

The companion italic is an adaptation of the chancery script type designed by the calligrapher and printer Lodovico degli Arrighi.

COMPOSED BY NORTH MARKET STREET GRAPHICS, LANCASTER, PENNSYLVANIA

PRINTED AND BOUND BY BERRYVILLE GRAPHICS, BERRYVILLE, VIRGINIA

DESIGNED BY ROBERT C. OLSSON